T0326397

Hearing Loss

Hearing Loss

Causes, Prevention, and Treatment

Jos J. Eggermont
University of Calgary, Calgary, AB, Canada

Academic Press is an imprint of Elsevier
125 London Wall, London EC2Y 5AS, United Kingdom
525 B Street, Suite 1800, San Diego, CA 92101-4495, United States
50 Hampshire Street, 5th Floor, Cambridge, MA 02139, United States
The Boulevard, Langford Lane, Kidlington, Oxford OX5 1GB, United Kingdom

Notices

Knowledge and best practice in this field are constantly changing. As new research and experience broaden our understanding, changes in research methods, professional practices, or medical treatment may become necessary.

Practitioners and researchers must always rely on their own experience and knowledge in evaluating and using any information, methods, compounds, or experiments described herein. In using such information or methods they should be mindful of their own safety and the safety of others, including parties for whom they have a professional responsibility.

To the fullest extent of the law, neither the Publisher nor the authors, contributors, or editors, assume any liability for any injury and/or damage to persons or property as a matter of products liability, negligence or otherwise, or from any use or operation of any methods, products, instructions, or ideas contained in the material herein.

British Library Cataloguing-in-Publication Data

A catalogue record for this book is available from the British Library

Library of Congress Cataloging-in-Publication Data

A catalog record for this book is available from the Library of Congress

ISBN: 978-0-12-805398-0

For Information on all Academic Press publications
visit our website at https://www.elsevier.com/books-and-journals

Publisher: Mara Conner
Acquisition Editor: Melanie Tucker
Editorial Project Manager: Kathy Padilla
Production Project Manager: Sue Jakeman
Designer: Mark Rogers

Typeset by MPS Limited, Chennai, India

Contents

2. Brain Plasticity and Perceptual Learning

3. Multisensory Processing

Part II
The Problem

4. Hearing Problems

5. Types of Hearing Loss

8. Early Diagnosis and Prevention of Hearing Loss

Part IV
The Treatments

9. Hearing Aids

Preface

Hearing is often taken for granted, and so its functions become noted only when they start to deteriorate. Prior to that there may already be warning signs in the form of tinnitus. Suddenly, one faces problems understanding speech, especially in noisy environments. This is annoying because those environments often accommodate social encounters, cocktail parties, sports venues, and bars. Hearing loss is foremost a communicative disorder, but it also diminishes the warning function that it normally has.

Hearing allows us to localize threatening sounds such as a car speeding in our direction or a barking dog close by. Sound localization also allows us to orient to interesting sounds, for instance to distinguish a musical instrument in an orchestra, or a voice in a cocktail party, and focus our attention to it.

The function of hearing is clearly present in all vertebrates. Even if we do not know exactly what they hear, we can infer it from the sounds they make. Bird song can be annoying (magpies) or pleasant (canaries) but we can assume that their conspecifics can hear it. Animals that normally do not make much noise such as cats have sensitive, broad frequency, hearing and use it to localize the scurrying mice and vocal songbirds. One can conversely assume that the potential prey uses hearing to avoid the predators.

Everyone who has heard a chorus of frogs or cicades understands the communication and socialization aspect of sound, particularly if the underlying intention is to call potential mates. Thus, hearing serves to localize and identify sounds emitted by conspecifics that may lead to mating, in humans potentially after socializing and chatting. Hearing warns about danger, from predators to approaching vehicles, and most of these functions are severely diminished in case of hearing loss.

Why do we have hearing loss? If we would be able to ask this to a frog or a finch, the answer would potentially be "what is hearing loss?" Just as sharks have a batch of teeth in reserve to replace the worn out current set, non-mammalian vertebrates can replace the hair cells in the inner ear when they are damaged by loud sounds or aging. Mammals have lost this gift, because evolution dictated that the ability to hear sounds with higher frequencies had more advantage than loosing some hearing sensitivity. Of course noise trauma was rare in the evolutionary period. The super-numerous replacement cells were converted to form a high-frequency region in the inner ear allowing hearing for frequencies above say 5–10 kHz, the upper region in birds. Currently, a large effort is put into finding out how to tweak genetic mechanisms to regain the possibility to replace damaged cochlear hair cells and so cure hearing loss and deafness.

My approach to the topic of this book is that the substrate of hearing sensitivity loss is in the ear whereas that for hearing problems is in the brain. The combination of the two aspects forms the communication disorder problem. Any amelioration has to deal with restoring the hearing sensitivity via hearing aids or cochlear implants, but also has to deal with the cognitive problems that result from the hearing loss or may be separate there from as in aging. We also will look at the feedback from the brain to the ear, which may result from attention and stress, through the corticofugal activity to midbrain, brainstem, and even the cochlea. Such central action may include protective effects for noise-induced hearing loss and presbycusis.

The book is comprehensive in the sense that it may be used as a standalone text in last year undergraduate and graduate courses in audiology. It starts with three basic science chapters, refreshing and updating knowledge about the auditory system, brain plasticity, and multisensory interaction. These chapters form the basis for training approaches in cochlear-, brainstem-, and midbrain-implant use and may also help understanding acclimatization effects on hearing aid use.

The next two chapters elucidate the problems associated with various types of hearing loss. These are problems that are manifest in addition to or may even occur without audiometric hearing loss. Chapters 6–8, Causes of Acquired Hearing Loss, Epidemiology and Genetics of Hearing Loss and Tinnitus, and Early Diagnosis and Prevention of Hearing Loss detail the causes of hearing loss, the genetics, the early diagnosis and ways to prevent hearing loss. Chapters 9–11, Hearing Aids, Implantable Hearing Aids, and Cochlear Implants describe the currently used treatments—surgical ones excluded—based on hearing aids, implantable hearing aids, and cochlear implants.

The final two chapters cover new developments in brainstem and midbrain implants that restore hearing in cases where the auditory nerves are destroyed or missing, and the ongoing promising efforts to design gene therapy for arresting hereditary hearing loss and to recreate cochlear hair cells. Extensive references, in total about 1300, more than normally encountered in a textbook, will make the book also extremely useful as a reference about the silent epidemic of hearing loss, its consequences, prevention, and treatment.

Also I have used excerpts and elaborations thereupon from three of my previous books: The Neuroscience of Tinnitus (2012), Noise and the Brain (2014), and Auditory Temporal Processing and its Disorders (2015). In many cases I simply could not say it any better, and the material is very relevant in the context of hearing loss and accompanying problems, including these here made this book self-contained. The excerpts also serve to draw attention to the more extensive descriptions in the original texts.

Writing the treatment chapters, required considerable study of the more technical literature and gave considerable insight into the benefits and problems associated with hearing aids and cochlear implants. It is clear that hearing aids can still be improved, and that cochlear implants, despite being the best neural prosthesis, will continue to be improved. Besides ameliorating audiometric hearing loss, which already results in greatly improving the quality of life, treatments particularly need to address the challenges of the elderly. Treating these problems must go beyond amplification and should include increasing the signal-to-noise ratio, and more awareness of multisensory integration in the rehabilitation process.

I benefitted immensely from the input of colleagues with great expertise in the various topics I describe in this book and were willing to critically read a chapter or sometimes two. These are in alphabetic order: Bob Burkard (University of Buffalo), Marschall Chasin, Barbara Cone (University of Arizona), Johan Frijns (Leiden University), Karen Gordon (University of Toronto), Bob Harrison (University of Toronto), Andrej Kral (Medical University Hannover), Hubert Lim (University of Minnesota), Steve Lomber (Western University, London Ont.), Ray Munguia (Purdue University), Frank Musiek (University of Arizona), Richard Smith (University of Iowa), Ad Snik (Radboud University, Nijmegen), and Kelly Tremblay (University of Washington). Without their supportive criticism, corrections, and suggestions the book would have been much less balanced and accurate. Mary, my wife for more than 50 years, a writer, translator, and editor, read the entire book and corrected numerous typos, grammatical inconsistencies, and often suggested changes in style and emphasis. I thank all for helping with this book.

Jos J. Eggermont
Calgary, July 2016

List of Abbreviations

AAF	anterior auditory field
A1	primary auditory cortex
A/D	analog-to-digital
ABI	auditory brainstem implant
ABR	auditory brainstem response
AC	air conduction
ADL	activities of daily living
AMEI	active middle ear implant
AMI	auditory midbrain implant
AMPA	α-amino-3-hydroxy-5-methyl-4-isoxazolepropionic acid
ANF	auditory nerve fiber
ANP	auditory neuropathy
ANSD	auditory neuropathy spectrum disorder
AP	auditory processing
APD	auditory processing disorder
ARHI	age-related hearing impairment
ASSR	auditory steady-state response
ATP	adenosine triphosphate
AV	audiovisual; adenoviral
AVCN	anteroventral cochlear nucleus
BAHA	bone-anchored hearing aid
BC	bone conduction
BCD	bone-conduction device
BM	basilar membrane
BOLD	blood oxygen level dependent
BTE	behind the ear
CA	conceptional age
CAEP	cortical auditory evoked potential
CAP	compound action potential
cCMV	congenital cytomegalovirus
cDNA	complementary DNA
CF	characteristic frequency
CHAMP	cochlear hydrops analysis masking procedure
CHL	conductive hearing loss
CI	confidence interval, cochlear implant

CIC complete in the canal aid
CIS continuous interleaved sampling
CL caudolateral belt area
CM cochlear microphonic; caudomedial belt area
CMV cytomegalovirus
CN cochlear nucleus
CNS central nervous system
COCB crossed olivocochlear bundle
CP cisplatin
CROS contralateral routing of signals
CSF cerebrospinal fluid
CSOM chronic suppurative otitis media
dB decibel
dB(A) A-weighted sound level
DCoN dorsal column nuclei
DCN dorsal cochlear nucleus
DFNA autosomal dominant modes of transmission deafness
DFNB autosomal recessive modes of transmission deafness
DFNX X chromosome-linked modes of transmission deafness
DNA deoxyribonucleic acid
DNLL dorsal nucleus of the lateral lemniscus
DPOAE distortion product otoacoustic emission
DPT desynchronizing pulse train
DZ dorsal zone
eABR electrically evoked ABR
eACC electrically evoked auditory change complex
EAE enhanced acoustic environment
eCAP electrically evoked CAP
ECochG electrocochleography
EEG electroencephalography
eMLR electrically evoked MLR
ERP event-related potentials
FAES auditory field of the anterior ectosylvian sulcus
FDA Food and Drug Administration
FFR frequency-following response
FM frequency modulation
fMRI functional MRI
FMT floating mass transducer
FTC frequency-tuning curve
GBC globular bushy cell
GM gray matter
GWAS genome-wide association study
HA hearing aid

HG Heschl's gyrus
HINT hearing in noise test
HIV *human immunodeficiency virus*
HL hearing level
HR hazard ratio
HSV Herpes simplex virus
HT hearing threshold
Hz Hertz
IC inferior colliculus
ICC central nucleus of the inferior colliculus
ICX cortical division of the inferior colliculus
IDT intensity-discrimination task
IHA implantable hearing aid
IHC inner hair cell
IIC invisible in the ear canal aid (also CIC)
ILD interaural level difference
IPI inter pulse interval
ITE in the ear aid
ITD interaural time difference
kHz kilohertz
KI knockin
KO knockout
LDL loudness discomfort level
LFP local field potential
LL lateral lemniscus
LOFT linear octave frequency transposition
MEG magneto-encephalography
MEMR middle ear muscle reflex
MET mechano-electrical transmission; also middle ear transducer
MGB medial geniculate body
MGBd dorsal MGB
MGBm magnocellular region of the MGB
MGBv ventral division of the medial geniculate body
MLR middle latency response
MMN mismatch negativity
MNTB medial nucleus of the trapezoid body
MRI magnetic resonance imaging
MSO medial superior olive
MUA multiunit activity
μM micromolar
μV microvolt
NAC *N*-acetylcysteine
NAP narrow-band CAP

NF-2	neurofibromatosis 2
NH	normal hearing
NHANES	National Health and Nutritional Examination Surveys
NICU	neonatal intensive care unit
NIHL	noise-induced hearing loss
NLFC	nonlinear frequency compression
OAE	otoacoustic emission
OCB	olivocochlear bundle
OHC	outer hair cell
OR	odds ratio
OSHA	Occupational Safety and Health Administration
PABI	penetrating ABI
PAF	posterior auditory field
PET	positron emission tomography
PLD	personal listening devices
PLLS	posterolateral lateral suprasylvian area
PP	planum polare
PSP	postsynaptic potential
pSTS	posterior superior temporal sulcus
PT	planum temporale
PTA	pure-tone average
PTS	permanent threshold shifts
PVCN	posteroventral cochlear nucleus
rAOM	recurrent acute otitis media
RAS	reticular activating system
RFC	risk factor screening
Ri	retroinsular cortical area
RLF	rate-level functions
ROI	region of interest
ROS	reactive oxygen species
QoL	quality of life
SAM	sinusoidally amplitude modulation
SBC	spherical bushy cell
SC	superior colliculus
SD	standard deviation
SFAP	single fiber action potential
SFR	spontaneous firing rate
SGN	spiral ganglion neuron
SII	speech intelligibility index
SL	sensation level
SNHL	sensorineural hearing loss
SNP	single nucleotide polymorphism
SNR	signal-to-noise ratio

SOAE spontaneous otoacoustic emission
SOC superior olivary complex
SP summating potential
SPL sound pressure level
SP5 spinal trigeminal nucleus
SR spontaneous release
SRM spatial release from masking
SRT speech reception threshold
SSD single-sided deafness
STG superior temporal gyrus
STS supra temporal sulcus
TEOAE transient evoked otoacoustic emission
TGFB transforming growth factor beta
TI temporal integration
TLE temporal lobe epilepsy
TM tympanic membrane (ear drum)
TMTF temporal modulation transfer function
TRT tinnitus retraining therapy
TTS temporary threshold shifts
UNHS Universal Newborn Hearing Screening
USH Usher syndrome
V1 primary visual area
V2 secondary visual area
VBM voxel-based morphometry
VCN ventral cochlear nucleus
VNLL ventral nucleus of the lateral lemniscus
VNTB ventral nucleus of the trapezoid body
VOT voice-onset time
VS vestibular schwannoma
VSB Vibrant Soundbridge
VZV varicella zoster virus
WAI wideband acoustic immittance
WDRC wide dynamic-range compression
WHO World Health Organization
WT wild type

Part I

The Basics

Chapter 1

Hearing Basics

Hearing loss comprises reduced sensitivity for pure tones (the audiogram) and problems in the understanding of speech. The loss of sensitivity results from deficits in the transmission of sound via the middle ear and/or loss of transduction of mechanical vibrations into electrical nerve activity in the inner ear. Problems of speech understanding mainly result from deficits in the synchronization of auditory nerve fibers' (ANFs) and central nervous system activity. This can be the result of problems in the auditory periphery but may also occur in the presence of nearly normal audiometric hearing. In order to appreciate the interaction of the audibility and "understanding" aspects of hearing loss, I will, besides presenting a condensed review of the auditory system, pay detailed attention to new findings pertaining to the important role of the ribbon synapses in the inner hair cells (IHCs), parallel processing in the ascending auditory system, and finally the importance of the efferent system.

1.1 HEARING SENSITIVITY IN THE ANIMAL KINGDOM

Every animal that grunts, croaks, whistles, sings, barks, meows, or speaks can hear. Most of the hearing animal species that we are familiar with are vertebrates, however insects also have keen hearing. For cicades and crickets that come as no surprise as these form choruses to enliven our nights. It may also turn out that the humble fruit fly, *Drosophila*, whose song is barely audible (Shorey, 1962), and the favorite of geneticists, may become important to elucidate the genetics of hearing loss (Christie et al., 2013).

A common way to quantify the hearing sensitivity (or loss) in humans is by way of the audiogram—a plot of the threshold level of hearing at a fixed series of (typically octave-spaced) frequencies between 125 Hz and 8 kHz when used in a clinical setting. In research settings, a wider and more finely spaced range of frequencies is employed (Fig. 1.1). In research settings, the just audible sound pressure level (dB SPL) is plotted, whereas in clinical settings the loss of sensitivity relative to a normalized value is represented (dB HL). To avoid confusion we call the research representation the "hearing field."

Hearing Loss. DOI: http://dx.doi.org/10.1016/B978-0-12-805398-0.00001-3

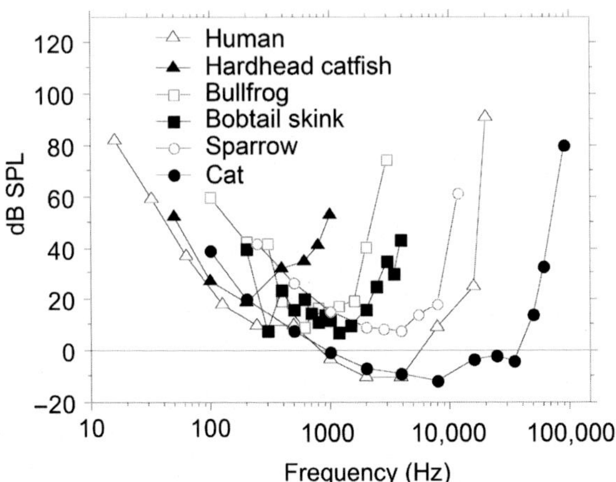

FIGURE 1.1 Representative hearing fields from the five vertebrate classes: hardhead catfish, bullfrog, sparrow, cat, bobtail skink, and human. *Data of hardhead catfish from Popper, A.N., Fay, R.R., 2011. Rethinking sound detection by fishes. Hear. Res. 273, 25–36; bullfrog, sparrow, and cat data from Fay, R.R., 1988. Hearing in Vertebrates. A Psychophysics Databook. Hill-Fay Associates, Winnetka, IL; bobtail skink data from Manley, G.A., 2002. Evolution of structure and function of the hearing organ of lizards. J. Neurobiol. 53, 202–211; human data from Heffner, H.E., Heffner, R.S., 2007. Hearing ranges of laboratory animals. J. Am. Assoc. Lab. Anim. Sci. 46, 20–22 (Manley, 2002; Fay, 1988).*

As Fig. 1.1 shows, hearing sensitivity differs considerably between vertebrates, even between mammals. Small mammals often have better high-frequency hearing than humans, with the 60 dB SPL upper limits of the hearing field ranging from 34.5 kHz for the Japanese macaque, and about 60 kHz for the cat to more than 120 kHz for the horseshoe bat (Heffner and Heffner, 2007). One reason for this variation may be that small mammals need to hear higher frequencies than larger mammals do in order to make use of sound localization cues provided by the frequency-dependent attenuating effect of the head and pinnae on sound. As a result, mammals with small heads generally have better high-frequency hearing than mammals with large heads, such as the elephant. Almost all mammals have poorer low-frequency hearing than humans, with the 60 dB lower limits ranging from 28 Hz for the Japanese macaque to 2.3 kHz for the domestic mouse (Heffner and Heffner, 2007; not shown). Only the Indian elephant, with a 60-dB low-frequency limit of 17 Hz, is known to have significantly better low-frequency hearing than humans, reaching into the infrasound range (Garstang, 2004).

Birds are among the most vocal vertebrates and have excellent hearing sensitivity. However, a striking feature of bird hearing is that the high-frequency limit, which falls between 6 and 12 kHz—even for small birds—is well below those of most mammals, including humans. Fig. 1.1 shows a typical bird audiogram represented by the sparrow. Among reptiles, lizards such as the bobtail skink (Fig. 1.1) are the best hearing species and are up to 30 dB more sensitive than alligators and crocodiles.

Anurans (frogs and toads) are very vocal amphibians: In specific parts of the year, depending on the species, their behavior is dominated by sound. As I wrote earlier (Eggermont, 1988): "Sound guides toads and frogs to breeding sites, sound is used to advertise the presence of males to other males by territorial calls, and sound makes the male position known to females through mating or advertisement calls. To have the desired effect these calls must be identified as well as localized. Frogs and toads are remarkably good in localizing conspecific males, especially when we take into account their small head size and the fact that hardly any of the territorial or mating calls has sufficient energy in the frequency region above 5 kHz to be audible at some distance." Especially, the bullfrog's threshold is relatively low at 10 dB SPL around 600 Hz (Fig. 1.1).

Teleost fishes, the largest group of living vertebrates, include both vocal and nonvocal species. This is especially evident for some by their intense sound production during the breeding season (Bass and McKibben, 2003). Except for the hardhead catfish (Fig. 1.1) that hears sounds at approximately 20 dB SPL for 200 Hz, most fishes have thresholds around 40 dB SPL, and with few exceptions do not hear sounds above 2 kHz (Popper and Fay, 2011).

Nearly all insects have high-frequency hearing (Fonseca et al., 2000). For instance, the hearing ranges for crickets are 0.1–60 kHz, for grasshoppers 0.2–50 kHz, for flies 1–40 kHz, and for cicades 0.1–25 kHz. Tiger moths are typically most sensitive to ultrasound frequencies between 30 and 50 kHz. The frequency sensitivity of the ears of moth species is often matched to the sonar emitted by the bats preying upon them (Conner and Corcoran, 2012).

1.2 THE MAMMALIAN MIDDLE EAR

"The auditory periphery of mammals is one of the most remarkable examples of a biomechanical system. It is highly evolved, with tremendous mechanical complexity" (Puria and Steele, 2008).

Transmission of sound energy from air to fluid typically results in considerable loss as a result of reflection from the fluid surface and estimated at about 99.7% of the incoming energy. This is compensated by the pressure gain provided by the ratio of the areas of the tympanic membrane (TM) (typical 0.55 cm^2) and the stapes footplate (typical 0.032 cm^2) for human, which is approximately 17, and the lever action of the middle ear bones which contributes a factor approximately 1.3 (Dallos, 1973). This would theoretically result in a combined gain of a factor 22 (about 27 dB). In practice, the gain is considerably less and maximal between 20 and 25 dB in the 800–1500 Hz range (Dallos, 1973).

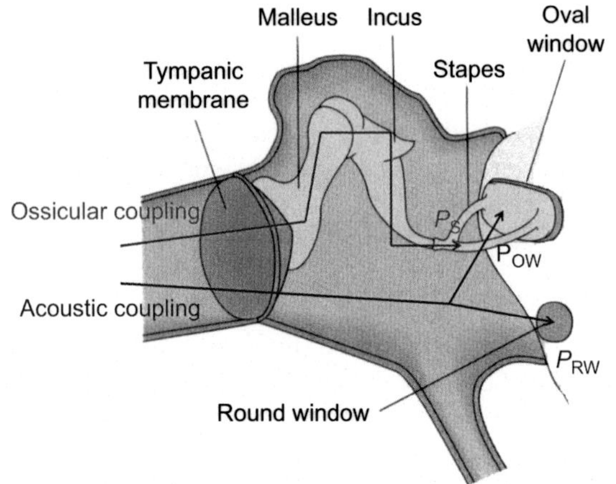

FIGURE 1.2 Illustration of ossicular and acoustic coupling from ambient sound to the cochlea. In the normal middle ear, sound pressure in the ear canal results in TM motion transmitted through the malleus and incus to produce a force at the stapes head (*red* path). This force applied over the area of the footplate produces a pressure P_S. P_S represents the ossicular coupling of sound to the cochlea. TM motion also compresses and dilates the air in the middle ear cavity, creating sound pressure in the middle ear (*blue* paths), which just outside the OW equals (P_{OW}) and at the RW (P_{RW}). The difference between these two acoustic pressures, $\Delta P = P_{OW} - P_{RW}$ represents the acoustic coupling of sound to the cochlea. In the normal ear, this acoustic coupling is negligibly small.

Merchant et al. (1997) extensively described the middle ear action as the result of two mechanisms: ossicular and acoustic coupling. Ossicular coupling incorporates the gain in sound pressure that occurs through the TM and ossicular chain. In the normal middle ear, sound pressure in the ear canal results in TM motion transmitted through the malleus and incus to produce a force at the stapes head (Fig. 1.2). This force applied over the area of the footplate produces a pressure P_S. P_S represents the ossicular coupling of sound to the cochlea. Acoustic coupling refers to the difference in sound pressures acting directly on the oval window (OW), P_{OW}, and round window (RW), P_{RW}. In normal ears, acoustic coupling, $\Delta P = (P_{OW} - P_{RW})$, is negligibly small, but it can play a significant role in some diseased middle ears (Peake et al., 1992).

1.3 THE MAMMALIAN INNER EAR

Until 1971, the basilar membrane (BM) was considered to be a linear device with broad mechanical tuning, as originally found already in the 1940s by von Békésy (1960). The bridge to the narrow ANF tuning was even long thereafter considered the result of a "second filter" (Evans and Klinke, 1982). Thus, it took a while before the results from Rhode (1971) indicating that the BM was a sharply tuned nonlinear filtering device were accepted. Appreciating these dramatic changes in viewing the working of the cochlea, Davis (1983) wrote: "We are in the midst of a major breakthrough in auditory physiology. Recent experiments force us, I believe, to accept a revolutionary new hypothesis concerning the action of the cochlea namely, that an active process increases the vibration of the basilar membrane by energy provided somehow in the organ of Corti." Then, another crucial discovery was that the outer hair cells (OHCs), in response to depolarization, were capable of producing a mechanical force on the BM (Brownell, 1984) later called the "cochlear amplifier," but this is in essence the "second filter."

1.3.1 Basilar Membrane Mechanics

The BM presents the first level of frequency analysis in the cochlea because of its changing stiffness and mass from base to apex. High-frequency sound produces maximal BM movement at the "base" of the cochlea (near the stapes) whereas low-frequency sound also activates the apical parts of the BM. Thus each site on the BM has a characteristic frequency (CF), to which it responds maximally in a strict tonotopic order (Robles and Ruggero, 2001). BM movements produce motion of hair cell stereocilia, which open and close transduction channels therein. This results in the generation of hair cell receptor potentials and the excitation of ANFs.

In a normal ear the movement of the BM is nonlinear, i.e., the amplitude of its movement is not proportional to the SPL of the sound, but increases proportionally less for increments in higher SPLs. In a deaf ear, the BM movement is called passive (Békésy-labeled wave and envelope in Fig. 1.3) as it just reacts to the SPL, without cochlear amplification. Davis (1983) proposed a model for the activation of IHCs that combined a passive BM movement and an active one resulting from the action of a cochlear amplifier. The passive BM movement only activates the IHCs at levels of approximately 40 dB above normal hearing threshold (von Békésy, 1960). At lower sound levels and up to about 60 dB above threshold, the cochlear amplifier provides a mechanical amplification of BM movement in a narrow segment of the BM near the apical end of the passive traveling wave envelope (Fig. 1.3). The OHC motor action provides this amplification. Davis (1983) noted that both the classical high-intensity system and the active low-level cochlear amplifier system compress the large dynamic range of hearing into a much narrower range of mechanical movement of BM and consequently the cilia of the IHCs. Robles and Ruggero (2001) found that the high sensitivity and sharp-frequency tuning, as well as compression and other nonlinearities (two-tone suppression and intermodulation distortion), are highly labile, suggesting the action of a vulnerable cochlear amplifier. Davis (1983) underestimated the effect of the cochlear amplifier considerably (Fig. 1.3) as the next section shows.

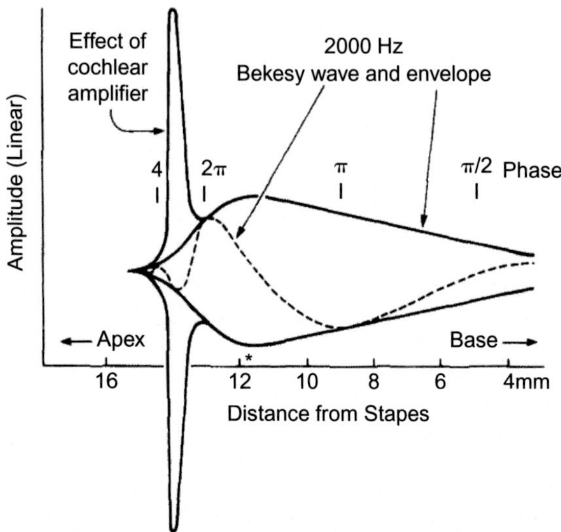

FIGURE 1.3 A traveling wave, as described by von Békésy (1960), is shown by the dashed line and its envelope by the heavy full line. The wave travels from right (base) to left (apex). The form of the envelope and the phase relations of the traveling wave are approximately those given by von Békésy (1960). To the envelope is added, near its left end, the effect of the cochlear amplifier. A tone of 2000 Hz thereby adds a peak at about 14 mm from the (human) stapes. The peak corresponds to the "tip" of the tuning curve for CF = 2000 Hz. *Reprinted from Davis, H., 1983. An active process in cochlear mechanics. Hear. Res. 9, 79–90, with permission from Elsevier.*

1.3.2 The Cochlear Amplifier

Because of the high-frequency selectivity of the auditory system, Gold (1948) predicted that active feedback mechanisms must amplify passive BM movements induced by sound in a frequency-selective way. This active cochlear amplification process depends critically on OHCs, which are thought to act locally in the cochlea. When a pure tone stimulates a passive BM a resonance occurs at a unique location and activates the OHCs. These activated OHCs feed energy back into the system thereby enhancing the BM vibration. Because of saturation, the cochlear amplifier shows a compressive nonlinearity so that the lowest SPL sounds are amplified substantially more than high SPL ones (Müller and Gillespie, 2008). Hudspeth (2008) detailed this active process as characterized by amplification, frequency selectivity, compressive nonlinearity, and the generation of spontaneous otoacoustic emissions (SOAEs) (Fig. 1.4).

The OHC electromotile response is also nonlinear and works in a cycle-by-cycle mode up to a frequency of at least 70 kHz (Dallos and Fakler, 2002). Recently, a gene that is specifically expressed in OHCs was isolated and termed *Prestin* (Zheng et al., 2000). The action of prestin is orders of magnitude faster than that of any other cellular motor protein. Note that gene

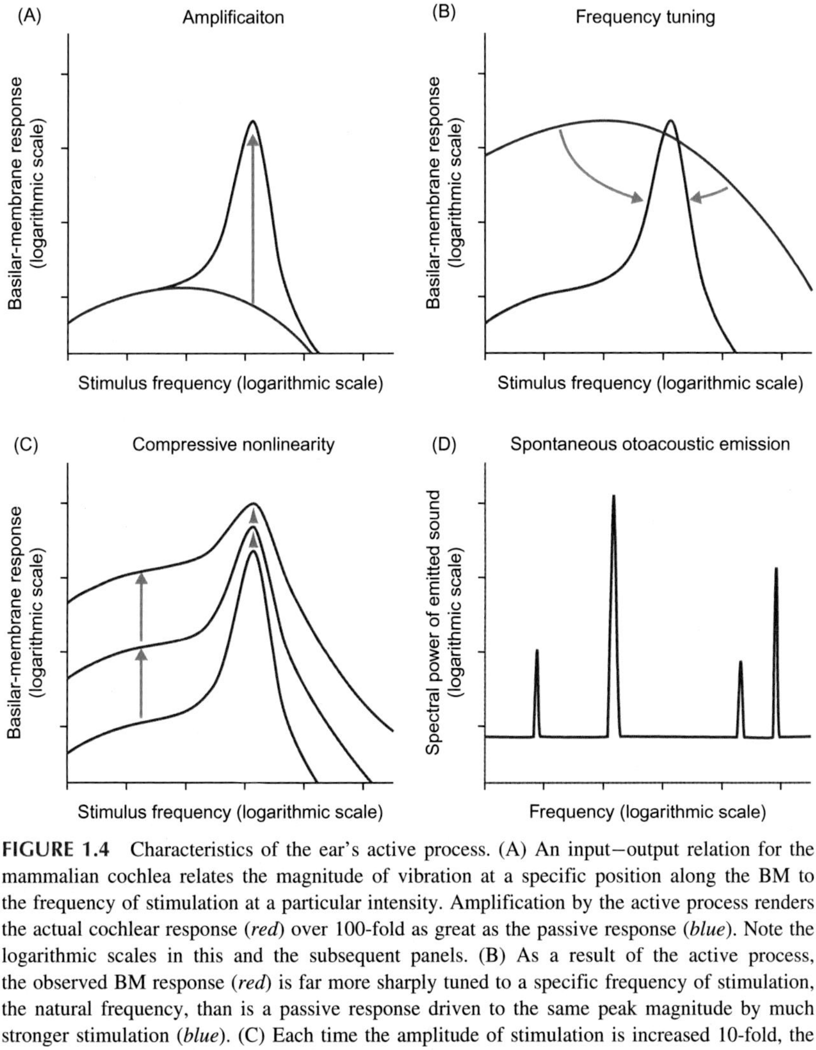

FIGURE 1.4 Characteristics of the ear's active process. (A) An input–output relation for the mammalian cochlea relates the magnitude of vibration at a specific position along the BM to the frequency of stimulation at a particular intensity. Amplification by the active process renders the actual cochlear response (*red*) over 100-fold as great as the passive response (*blue*). Note the logarithmic scales in this and the subsequent panels. (B) As a result of the active process, the observed BM response (*red*) is far more sharply tuned to a specific frequency of stimulation, the natural frequency, than is a passive response driven to the same peak magnitude by much stronger stimulation (*blue*). (C) Each time the amplitude of stimulation is increased 10-fold, the passive response distant from the natural frequency grows by an identical amount (*green* arrows). For the natural frequency at which the active process dominates, however, the maximal response of the BM increases by only $10^{1/3}$, a factor of about 2.15 (*orange* arrowheads). This compressive nonlinearity implies that the BM is far more sensitive than a passive system at low stimulus levels, but approaches the passive level of responsiveness as the active process saturates for loud sounds. (D) The fourth characteristic of the active process is SOAE, the unprovoked production of one or more pure tones by the ear in a very quiet environment. For humans and many other species, the emitted sounds differ between individuals and from ear to ear but are stable over months or even years. *Reprinted from Hudspeth, A.J., 2008. Making an effort to listen: mechanical amplification in the ear. Neuron 59, 530–545, with permission from Elsevier.*

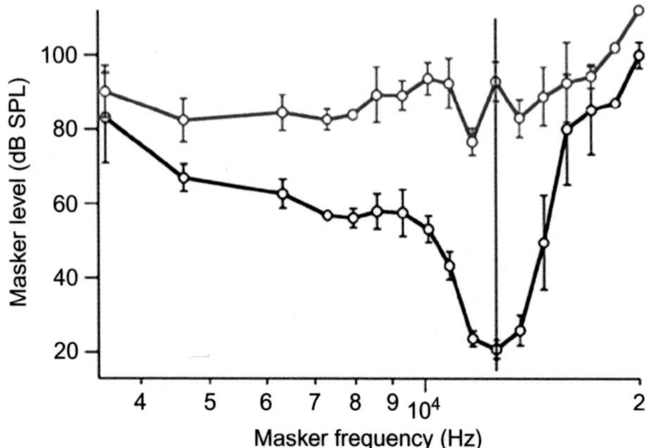

FIGURE 1.5 Average (±SD) compound action potential masking tuning curves from an RW electrode for KI and wild-type mice. Probe tone frequency: 12 kHz. *Reprinted from Dallos, P., Wu, X., Cheatham, M.A., Gao, J., Zheng, J., Anderson, C.T., et al., 2008. Prestin-based outer hair cell motility is necessary for mammalian cochlear amplification. Neuron 58, 333–339, with permission from Elsevier.*

names are commonly indicated in italics whereas the expressed protein, which may have the same name, is indicated in normal font. Prestin is required for normal auditory function (Fig. 1.5) because *prestin* knockout (KO), or knockin (KI), mice do not exhibit OHC electromotility (Liberman et al., 2002; Dallos et al., 2008) and thus do not show cochlear amplification. A gene KO is a genetic technique in which one of an organisms genes is made inoperative ("knocked out" of the organism). A gene KI refers to a genetic engineering method that involves the insertion of a protein coding a cDNA sequence at a particular locus in an organism's chromosome.

Functional consequences of loss of this nonlinear amplification process result in hearing loss, loudness recruitment, reduced frequency selectivity, and changes temporal processing. This manifests itself in hearing-impaired listeners as difficulties in speech understanding, especially in complex acoustic backgrounds (Oxenham and Bacon, 2003).

van der Heijden and Versteegh (2015) recently challenged the involvement of the cochlear amplifier in BM movement, as sketched above. They recorded the vibrations at adjacent positions on the BM in sensitive gerbil cochleas with a single-point laser vibrometer to measure the velocity of the BM. This measurement was converted in a putative power amplification by the action of the OHCs, and the local wave propagation on the BM. No local power amplification of soft sounds was evident, and this was combined with strong local attenuation of intense sounds. van der Heijden and Versteegh (2015) also reported that: "The waves slowed down abruptly when approaching their peak, causing an energy densification that quantitatively matched the amplitude peaking, similar to the growth of sea waves approaching the beach."

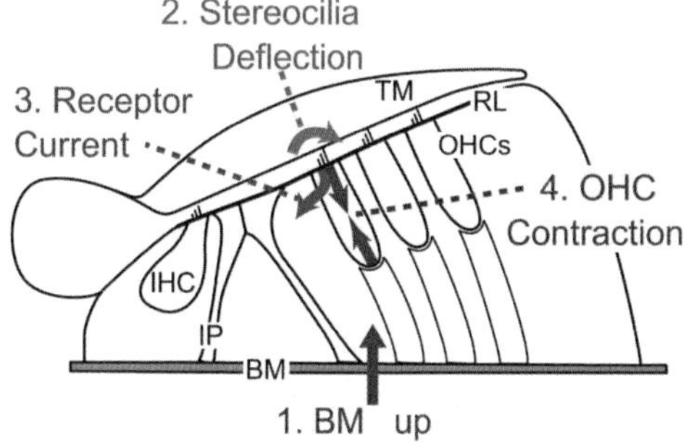

FIGURE 1.6 Steps in the cochlear amplification of BM motion for BM movement toward scala vestibuli. TM, tectorial membrane; RL, reticular lamina; IP, inner pillar. *Reprinted from Guinan, J.J., Salt, A., Cheatham, M.A., 2012. Progress in cochlear physiology after Békésy. Hear. Res. 293, 12–20, with permission from Elsevier.*

1.3.3 Mechanoelectrical Transduction

"The [BM] displacement is transferred to the hair bundles by means of the tectorial membrane, which contacts the OHC stereocilia and produces fluid movements that displace the IHC stereocilia. Movement of the hairs in the excitatory direction (i.e., toward the tallest row) depolarizes the hair cells whilst opposite deflections hyperpolarize them" (Furness and Hackney, 2008).

There are two hair cell types in the cochlea: IHCs and OHCs. The IHCs receive up to 95% of the auditory nerve's afferent innervation (Spoendlin and Schrott, 1988), but are fewer in number than the OHCs by a factor of 3–4 (He et al., 2006). As we have seen, the OHCs provide a frequency-dependent boost to the BM motion, which enhances the mechanical input to the IHCs, thereby promoting enhanced tuning and amplification. This occurs as follows as labeled in Fig. 1.6: (1) Pressure difference across the cochlear partition causes the BM to move up (purple arrow). (2) The upward BM movement causes rotation of the organ of Corti toward the modiolus and shear of the reticular lamina relative to the tectorial membrane that deflects OHC stereocilia in the excitatory direction (green arrow). (3) This stereocilia deflection opens mechanoelectrical transduction channels, which increases the receptor current driven into the OHC (blue arrow) by the potential difference between the endocochlear potential (+100 mV) and the OHC resting potential (−40 mV). This depolarizes the OHC. (4) OHC depolarization causes conformational changes in prestin molecules that induce a reduction in OHC length (red arrows). The OHC contraction pulls the BM upward toward the reticular lamina, which amplifies BM motion when the pull on the BM is in the correct phase. In contrast to OHCs that are displacement detectors, IHCs are sensitive to velocity of the fluid surrounding the stereocilia (Guinan et al., 2012).

1.3.4 Cochlear Microphonics and Summating Potentials

Both IHC and OHC generate receptor potentials in response to sound (Russell and Sellick, 1978; Dallos et al., 1982). It has long been known that

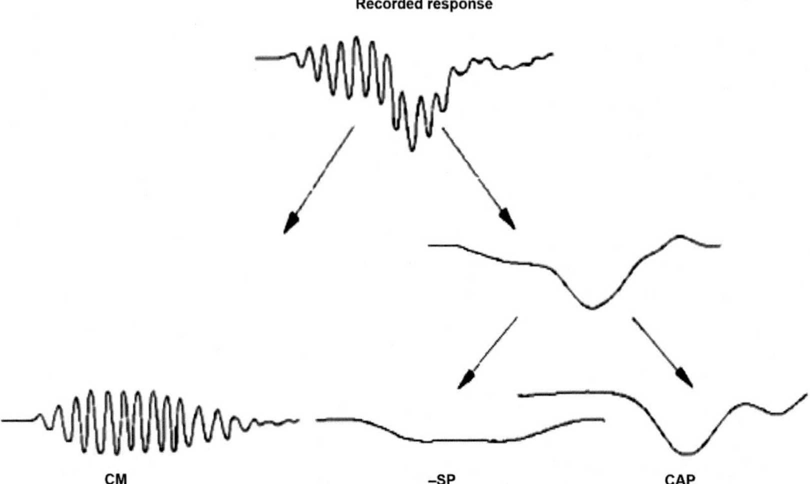

FIGURE 1.7 Sound-evoked gross potentials in the cochlea. In response to short tone bursts three stimulus-related potentials can be recorded from the cochlea. These potentials, CM, SP, and CAP, appear intermingled in the recorded response from the promontory. By presenting the stimulus alternately in phase and counter phase and averaging of the recorded response, a separation can be obtained between CM on the one hand and CAP and SP on the other. High pass filtering provides a separation between SP and CAP. This can also be obtained by increasing the repetition rate of the stimuli, which results in an adaptation of the CAP but leaves the SP unaltered. *From Eggermont, J.J., 1974. Basic principles for electrocochleography. Acta. Otolaryngol. Suppl. 316, 7–16 (Eggermont, 1974).*

population responses from the cochlea can be recorded at remote sites such as the round window, tympanic membrane, or even from the scalp and can be used clinically (Eggermont et al., 1974; Fig. 1.7). These responses are called the cochlear microphonic (CM) and the summating potential (SP). The CM is produced almost exclusively from OHC receptor currents and when recorded from the RW membrane is dominated by the responses of OHCs in the basal turn. The SP is a direct-current component resulting from the nonsymmetric depolarization–hyperpolarization response of the cochlea, which can be of positive or negative polarity, and is likely also generated dominantly by the OHCs (Russell, 2008). The compound action potential (CAP) is mixed in with the CM and SP and will be described in Section 1.4.3.

1.3.5 Otoacoustic Emissions

"Unlike other sensory receptor systems, the inner ear appears to generate signals of the same type as it is designed to receive. These sounds, called otoacoustic emissions (OAEs), have long been considered byproducts of the cochlear amplifier, the process that makes cochlear mechanics active by adding mechanical energy at the same frequency as a stimulus tone in a positive feedback process. This feature of the inner ear is one of the most important distinctions from other sensory receptors" (Siegel, 2008).

Kemp (1978) discovered that sound could evoke "echoes" from the ear. These echoes, called OAEs, result from the action of the cochlear amplifier. Guinan et al. (2012) described their generation as follows: "As a traveling wave moves apically [along the BM] it generates distortion due to cochlear nonlinearities (mostly from nonlinear characteristics of the OHC [mechanoelectrical transduction] channels, the same source that produces the nonlinear growth of BM motion), and encounters irregularities due to variations in cellular properties. As a result, some of this energy travels backwards in the cochlea and the middle ear to produce OAEs." Normal human ears generally exhibit SOAEs. SOAEs arise from multiple reflections of forward and backward traveling waves that are powered by cochlear amplification likely via OHC-stereocilia resonance (Shera, 2003). OAEs can be measured with a sensitive microphone in the ear canal and provide a noninvasive measure of cochlear amplification. There are two main types of OAEs in clinical use. Transient-evoked OAEs (TEOAEs) are evoked using a click stimulus. The evoked response from a click covers the frequency range up to around 4 kHz. Distortion product OAEs (DPOAEs) are evoked using a pair of primary tones f1 and f2 ($f1 < f2$) and with a frequency ratio $f2/f1 < 1.4$. In addition to the stimulus tones the spectrum of the ear canal sound contains harmonic and intermodulation distortion products at frequencies that are simple arithmetical combinations of the two tones. The most commonly measured DPOAE is at the frequency $2f1 - f2$ (Siegel, 2008). Recording of OAEs has become the main method for newborn and infant hearing screening (see chapter: Early Diagnosis and Prevention of Hearing Loss).

1.4 THE AUDITORY NERVE

1.4.1 Type I and Type II Nerve Fibers

The cell bodies of auditory afferent neurons in mammals form the spiral ganglion, which runs along the modiolar edge of the organ of Corti. The peripheral axons of type I afferents (also known as radial fibers) contact only a single IHC (Robertson, 1984). However, each mammalian IHC provides input to 5–30 type I afferents (depending on the species), allowing parallel processing of sound-induced activity (Rutherford and Roberts, 2008). Type I neurons constitute 90–95% of cochlear nerve afferents (Spoendlin, 1969; Liberman, 1982). Both the peripheral axons (also called dendrites) and the central axons as well as the cell bodies of a type I afferent neurons in mammals are myelinated. This increases conduction velocity, reduces temporal jitter, and decreases the probability of conduction failure across the cell body during high-frequency action-potential firing (Hossain et al., 2005).

A small population of afferent axons in the mammalian cochlea (type II) are unmyelinated (Liberman et al., 1990), and each type II axon synapses with many OHCs. They may be monitoring (like muscle spindles) the state of the motor aspects of the OHCs, but may not contribute to the perception of sound. Like the type I afferents, they may initiate action potentials near their distal tips (Hossain et al., 2005). OHC synapses typically release single neurotransmitter vesicles with low probability so that extensive summation is required to reach the relatively high action potential initiation threshold. Modeling suggests that neurotransmitter release from at least six OHCs is required to trigger an action potential in a type II neuron (Weisz et al., 2014).

Recently Flores et al. (2015) suggested that type II cochlear afferents might be involved in the detection of noise-induced tissue damage. They implied that this represents a novel form of sensation, termed auditory nociception, potentially related to "pain hyperacusis" (Tyler et al., 2014). Using immunoreactivity to c-Fos, Flores et al. (2015) recorded neuronal activation in the brainstem of Vglut3$^{-/-}$ mice, in which the type I afferents were silenced. In these deaf mice, they found responses to hair-cell-damaging noise, but not to non-traumatic noise, in cochlear nucleus neurons. This response originated in the cochlea. Flores et al. (2015) concluded that their findings "imply the existence of an alternative neuronal pathway from cochlea to brainstem that is activated by tissue-damaging noise and does not require glutamate release from IHCs." Corroboration of this idea that type II afferents are nociceptors comes from Liu et al. (2015), who showed that type II afferents are activated when OHCs are damaged. This response depends on purinergic receptors, binding ATP released from nearby supporting cells in response to hair cell damage. They found that selective activation of the metabotropic purinergic receptors increased type II afferent excitability, a potential mechanism for putative pain hyperacusis.

1.4.2 Type I Responses

Nearly all recordings of action potentials in mammalian ANFs are from axons of type I neurons, and nearly all type I neurons fire action potentials spontaneously. The spontaneous firing rate (SFR) ranges from less than 5 to approximately 100 spikes/s, irrespective of the ANFs CF (Kiang et al., 1965; Tsuji and Liberman, 1997). The SFRs are correlated with both axon diameter and ribbon synapse location in the IHCs (Liberman, 1980; Merchan-Perez and Liberman, 1996; Tsuji and Liberman, 1997). In these studies ANFs with high SFRs were found in spiral ganglion neurons (SGNs) with large diameter peripheral axons and contacted IHCs predominantly on the inner pillar face (cf. Fig. 1.6). Low- and intermediate-SFR fibers contact the modiolar face and have synapses with larger ribbons and more vesicles than synapses with high-SFR fibers. Because the 5–30 afferent synapses on each IHC display this range of characteristics, and because ANFs of the same CF differ in

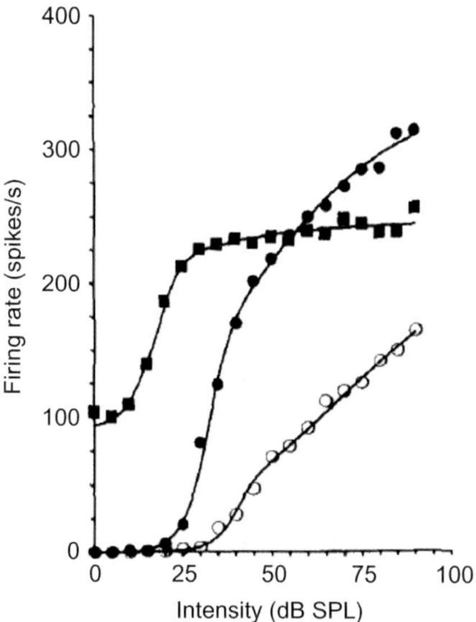

FIGURE 1.8 Typical examples of firing rate–intensity functions of the three different fiber types demonstrating the relationship between threshold, SFR, and rate–intensity function shape. CAP threshold in this frequency range was 25 dB SPL. (■) A fiber with threshold below CAP threshold, high spontaneous rate (84.4 spikes/s), and saturating rate–intensity function; (●) a fiber with threshold near CAP threshold, low spontaneous rate (0.2 spikes/s), and sloping-saturating rate–intensity function; (○) a fiber with threshold above CAP threshold, zero spontaneous rate, and straight rate–intensity function. *Reprinted from Müller, M., Robertson, D., Yates, G.K., 1991. Rate-versus-level functions of primary auditory nerve fibres: evidence for square law behaviour of all fibre categories in the guinea pig. Hear. Res. 55, 50–56 (Müller et al., 1991), with permission from Elsevier.*

SFR, it is likely that each IHC synapses with high-SFRs as well as medium- and low-SFRs ANFs (Rutherford and Roberts, 2008). High-SFR neurons have low thresholds and their driven firing rate saturates at low SPL. Medium- and low-SFR neurons have higher threshold and typically do not show saturating firing rates (Fig. 1.8).

1.4.3 Compound Action Potentials

The thresholds of the CAP for tone pips, as measured by electrocochleography (Fig. 1.7), are excellent indicators of hearing sensitivity and can be used to construct detailed audiograms (Eggermont, 1976). These can be used, in combination with auditory brainstem responses (ABRs), to determine the cochlear or retrocochlear origin of sensorineural hearing loss (Santarelli et al., 2009). Simultaneous and forward-masking of the CAPs can be used to estimate frequency-tuning curves in humans (Eggermont, 1977) that are very similar in characteristics to those in guinea pigs (Dallos and Cheatham, 1976; Fig. 1.5).

CM, SP, and CAP can be recorded from the promontory of the human cochlea (Appendix Fig. A2) by inserting (under local anesthesia) an electrode through the eardrum (Eggermont et al., 1974) or from the eardrum itself (Salomon and Elberling, 1971). The CAP is also represented by wave I of the ABR. However, the recorded potential is larger if the recording electrode is put closer to the source, for instance in the ear canal or better even on the wall of the cochlea. Namely, at the cochlear wall (promontory) the CAP to a 80 dB HL tone pip is typically 10–30 μV, at the ear drum one records 1–3 μV, in the ear canal maximally 1 μV, and at the earlobe or mastoid (as in ABR recording) about 0.1–0.3 μV (Eggermont et al., 1974). A detailed description of the transtympanic electrocochleography method and some basic findings can be found in the Appendix.

CAPs capture the synchronous activation of the ANFs. Low SPLs activate the high-SFR fibers, and increasing sound levels gradually activate medium- and then low-SFR fibers. Bourien et al. (2014) analyzed the contribution of the SFR types to the CAP 6 days after 30-min infusion of ouabain into the RW niche of Mongolian gerbils. Ouabain inhibits the plasma membrane sodium/potassium pump thereby preventing spiking in a dose-dependent fashion. CAP amplitude and threshold plotted against loss of ANFs revealed three ANF pools (Fig. 1.9): (1) a highly ouabain-sensitive pool with low SFR (high threshold), which did not contribute to either CAP threshold or amplitude, (2) a less sensitive pool constituting medium-SFR fibers, which only contributed to CAP amplitude, and (3) a ouabain-resistant pool with high-SFR fibers, required for CAP threshold and amplitude. Bourien et al. (2014) suggested that a low-SFR fiber loss leaves the CAP unaffected, potentially as a result of the delayed and broad first-spike latency distribution of these low-SFR fibers (Fig. 1.9). Bourien et al. (2014) also demonstrated that substantial ANF loss can coexist with normal hearing threshold and even unchanged CAP amplitude. These findings contrasts with the correlation of ABR-wave I amplitude loss at high stimulus levels in mice with loss of high threshold fibers after noise trauma in mice (Kujawa and Liberman, 2009).

1.5 RIBBON SYNAPSES

IHCs innervate type I nerve fibers via specialized ribbon synapses. The synaptic ribbon is a presynaptic electron-dense structure surrounded by neurotransmitter-filled vesicles and specialized for graded synaptic transmission (Sterling and Matthews, 2005; Moser et al., 2009; Glowatzki et al., 2008). Ribbon synapses can release precise changes in the amounts of neurotransmitter in response to very small changes in membrane potential. This allows both the exquisite following of the temporal aspects of the stimulus

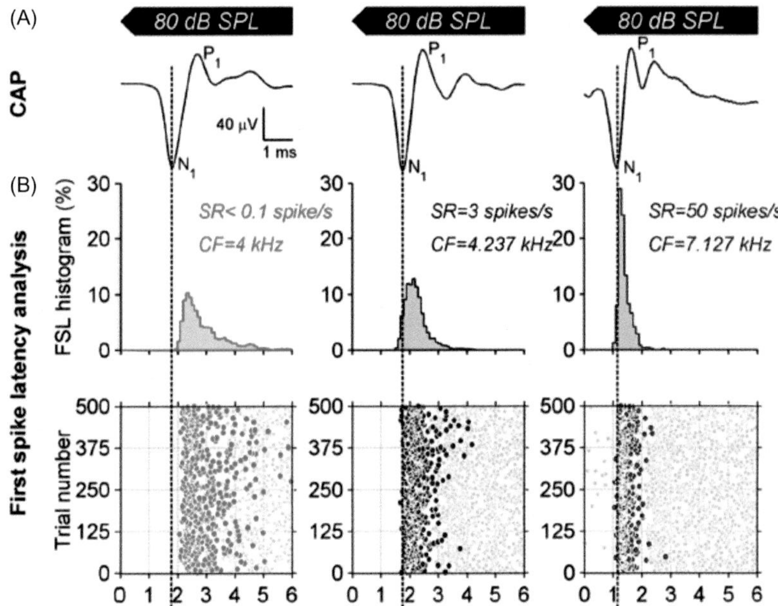

FIGURE 1.9 Contribution of single auditory nerve fibers to the CAP. (A) and (B) CAP (A) and single-fiber action potentials from gerbil auditory nerve (B) were simultaneously recorded and plotted together for low-SR (*green*), medium-SR (*blue*), and high-SR fibers (*red*). Vertical dashed lines show the CAP N1 latency. Stimulation was a tone burst presented at the CF of the fiber (1 ms rise/fall, 10 ms duration, 11 bursts/s, 500 presentations, 80 dB SPL, and alternating polarity). *Reprinted from Bourien, J., Tang, Y., Batrel, C., Huet, A., Lenoir, M., Ladrech, S., et al., 2014. Contribution of auditory nerve fibers to compound action potential of the auditory nerve. J. Neurophysiol. 112, 1025–1039.*

and releasing transmitter at high rates for long periods of time (Logiudice et al., 2009). High- and low-spontaneous release (SR) synapses coexist on the same IHC. Kantardzhieva et al. (2013) found that low-SR ribbons had more synaptic vesicles compared with those of high-SR synapses.

Khimich et al. (2005) showed that a lack of active-zone-anchored synaptic ribbons reduced the presynaptic readily releasable vesicle pool and impaired synchronous auditory signaling. Both exocytosis of the releasable vesicle pool and the number of synchronously activated SGNs co-varied with the number of anchored ribbons during development. Ribbon-deficient IHCs were still capable of sustained transmitter release. They concluded that ribbon-dependent synchronous release of multiple vesicles at the hair cell afferent synapse is essential for normal hearing. Ribbon synapse dysfunctions, termed auditory synaptopathies (see chapter: Types of Hearing Loss), impair the audibility of sounds to varying degrees but commonly affect neural encoding of acoustic temporal cues that are essential for speech comprehension (Moser et al., 2013). Clinical features of auditory synaptopathies are similar to those accompanying auditory neuropathy spectrum disorder, a group of genetic and acquired disorders of SGNs (see chapter: Types of Hearing Loss). Acquired synaptopathies result from noise-induced hearing loss because of excitotoxic synaptic damage and subsequent gradual neural degeneration (Kujawa and Liberman, 2009). Gradual loss of ribbon synapses potentially contributes to age-related hearing loss (Kujawa and Liberman, 2015; Fernandez et al., 2015).

1.6 THE CENTRAL AFFERENT SYSTEM

I will briefly review the structure and function of the afferent auditory pathways that are relevant for understanding hearing problems (based partially on Eggermont, 2001, 2014). Afferent pathways refer to bottom-up neural activity, i.e., starting with the spiking of the ANFs. Activity that originates in "higher" auditory centers and that typically modulates "lower" centers is carried by the efferents (Section 1.7).

1.6.1 Parallel Processing Between Cochlea and Inferior Colliculus

Parallel processing, starting at the cochlear nucleus as a result of the trifurcation of ANFs with outputs in the anteroventral cochlear nucleus (AVCN), posteroventral cochlear nucleus (PVCN), and dorsal cochlear nucleus (DCN), allows the initial segregation of sound localization (e.g., binaural processing) and sound identification (e.g., pitch processing) pathways (Fig. 1.10). The parallel processing results initially from different projections of the high-SFR and low- and medium-SFR fibers to neurons in the cochlear nucleus. ANFs with different SFR tend to project to different cell groups in the AVCN. According to Liberman (1991) the small cell cap was almost exclusively innervated by fibers with the highest acoustic thresholds. Within anterior AVCN, all SFR groups innervated spherical cells, whereas almost all multipolar cell innervation was only from high threshold fibers. Multipolar cell innervation in the posterior AVCN was from all SFR groups, whereas globular bushy cells (GBCs) were preferentially innervated by low threshold fibers.

The output of the ventral cochlear nucleus (VCN) largely follows the anatomical anterior and posterior subdivisions: AVCN GBC output is involved in sound localization and projects to the ipsilateral superior olivary complex (SOC) and via the MNTB to the contralateral SOC. The responses of spherical bushy cells (SBCs) to tones are sharply tuned and phase-locked for frequencies up to 3–5 kHz. As a consequence of this projection pattern, neurons in the medial superior olive (MSO) are activated in a systematic function of the location of sound in the horizontal plane. In contrast, the PVCN appears to be only involved in the identification of sound and its

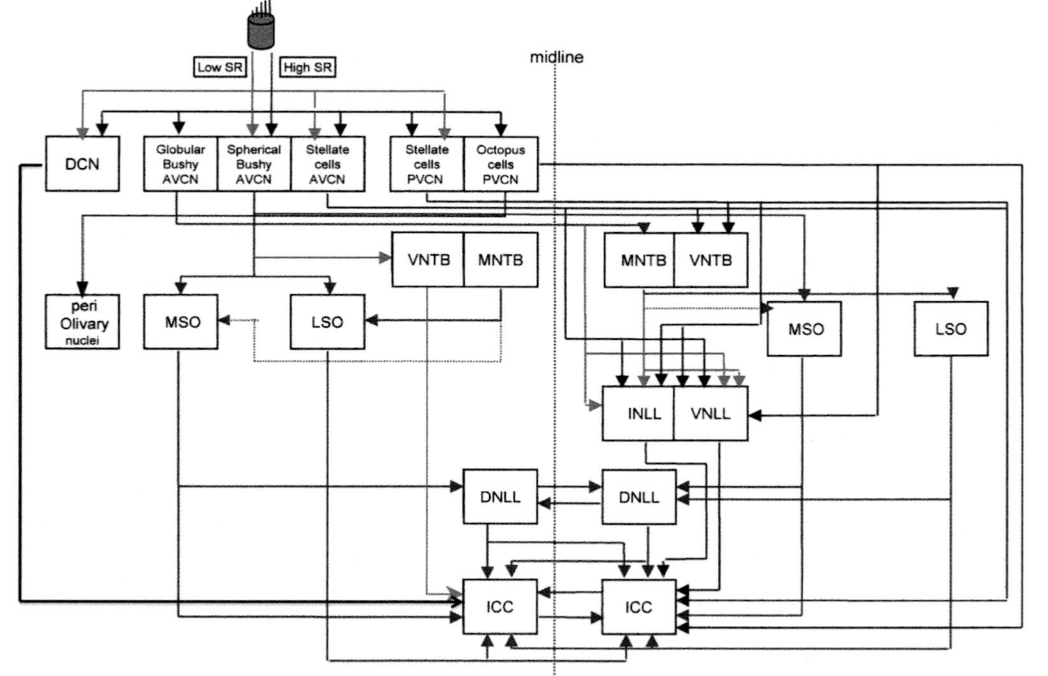

FIGURE 1.10 Simplified scheme of the afferent pathways in the auditory system up to the IC in the cat. Various levels of parallel processing and convergence are noted. First of all, low-SFR fibers have partially segregated projections from medium- and high-SFR fibers (here indicated by "high SR"). After the separate processing in the major cell groups of the VCN and DCN the pathways follow a distinct course for the sound localization processes (*red*) and the sound identification processes (*black*), or belong to both at some point on their course (*gray*). All pathways converge in the IC, however, only the ICC is shown. Not all known pathways are included. AVCN, anteroventral cochlear nucleus; DCN, dorsal cochlear nucleus; PVCN, posteroventral cochlear nucleus; MNTB, medial nucleus of the trapezoid body; VNTB, ventral nucleus of the trapezoid body; MSO, medial superior olive; LSO, lateral superior olive; DNLL, dorsal nucleus of the lateral lemniscus; VNLL, ventral nucleus of the lateral lemniscus; ICC, central nucleus of the inferior colliculus. *Reprinted from Eggermont, J.J., 2001. Between sound and perception: reviewing the search for a neural code. Hear. Res. 157, 1–42, with permission from Elsevier.*

output bypasses the SOC to project to the monaural nuclei of the lateral lemniscus (LL). The sound localization pathways are indicated with red lines in Fig. 1.10. The SBCs in the AVCN preserve and convey the timing information of ANFs bilaterally to the MSO (Smith et al., 1993; Oertel, 1999; Grothe and Koch, 2011; van der Heijden et al., 2013). The sound identification pathways (Fig. 1.10) are drawn as full-black lines (completely separate from the localization path) and gray lines (splitting of the localization path). Stellate cell output forms a major direct pathway from the VCN to the contralateral midbrain but also to the ipsilateral peri-olivary nuclei and to the contralateral ventral nucleus of the trapezoid body and the ventral nucleus of the lateral lemniscus (VNLL). The sharply frequency-tuned, tonic responses of stellate cells provide a firing-rate representation of the spectral content of sounds.

Both the monaural and binaural pathways from the auditory brainstem project to the central nucleus of the inferior colliculus (ICC). Neurons in the ICC project via the thalamus to the auditory forebrain. Thus, the cellular anatomy of the IC may provide the substrate for integration of the ascending pathways on single projection neurons (Winer and Schreiner, 2005). The operation of the IC consists largely of integrating inhibitory and excitatory inputs, which themselves have different temporal properties, and so interact to produce filters for temporal features of sound. The time constants of the filters suggest that they are relevant for the analysis of sound envelope, such as the estimation of the duration, the amplitude modulation rate, or the rate of frequency modulation (Covey and Casseday, 1999). The IC appears to act as the nexus of the auditory system, by sending activity to thalamus and thus to auditory cortex, as well as to more peripheral areas such as the cochlear nucleus via centrifugal connections (Gruters and Groh, 2012).

Most of the response types described for the CN are also found in the IC. Non-monotonic rate–intensity functions appear in the ICC, potentially reflecting similar non-monotonic input from the DCN (Aitkin, 1986). The ANFs and ICC are both tonotopically organized. However, whereas the auditory nerve has a smooth representation of CF, the ICC shows a step-wise progression of CF along its main topographic axis thought to reflect a framework for representation of psychophysical critical bands (Schreiner and Langner, 1997). Thus the ICC could comprise both analytic and integrative properties in the frequency domain. In addition, periodicity information in the cat may also be topographically organized orthogonal to the fine structure of the frequency representation (Langner and Schreiner, 1988). The co-localization, albeit along orthogonal axes, of frequency pitch and periodicity pitch in the ICC may potentially provide for a similar organization proposed for auditory cortex in the Mongolian gerbil (Schulze and Langner, 1997).

Recent work does not agree with these assumptions. Schnupp et al. (2015) made very detailed tonotopic and periodotopic maps in the IC of Mongolian gerbils, using pure tones and different periodic sounds, including click trains, sinusoidally amplitude modulated noise and iterated rippled noise. They found that while the tonotopic map exhibited a clear and highly reproducible gradient in all animals, periodotopic maps varied greatly across different types of periodic sound and from animal to animal. Furthermore, periodotopic gradients typically explained only about 10% of the variance in modulation tuning between recording sites. Thus, these findings do not corroborate the idea that the gerbil ICC might exhibit a periodotopic gradient, which runs anatomically orthogonal to the tonotopic axis.

1.6.2 Parallel Processing Between IC and Auditory Cortex

The output pathways of the IC in cats comprise: (1) a tonotopic, lemniscal, pathway that originates in the ICC (and ultimately in the cochlea), terminates in the ventral division of the medial geniculate body (MGBv), and continues to the primary auditory cortex (A1). (2) A non-tonotopic, lemniscal-adjunct, pathway (also called non-lemniscal pathway), that originates in the cortical (ICX) and paracentral subdivisions of the IC and targets the deep dorsal and ventrolateral nuclei of the MGB. This continues to non-tonotopic cortical areas. (3) A diffuse pathway, that originates from neurons throughout the IC and lateral tegmentum, sends axons to the medial division of the MGB and continues to layer I of all auditory cortical areas (Aitkin, 1986; Winer and Schreiner, 2005).

1.6.2.1 Splitting up the Lemniscal Pathway

Recent observations by Straka et al. (2013, 2014a,b) suggest that the lemniscal pathway may in addition be split into two parallel steams to the MGBv. They found evidence that, compared with neurons in caudal and medial regions within an isofrequency lamina of the ICC, neurons in rostral and lateral regions had lower thresholds and hence shorter first-spike latencies with less spiking jitter and shorter spike-burst duration. In addition local field potentials were larger and had shorter latencies. Thus, there were two distinct clusters of response features located in either the caudal and medial or the rostral and lateral parts of the ICC. Straka et al. (2014b) suggested also that some of these functional zones within the ICC may be maintained as two distinct sublemniscal pathways through the MGBv to the core auditory cortex. In cats and rats, neurons in the rostral parts of the MGBv project throughout the auditory cortex, including the A1, whereas those in the caudal MGBv project to core cortex regions primarily outside of A1 (Morel and Imig, 1987; Polley et al., 2007; Rodrigues-Dagaeff et al., 1989; Storace et al., 2010). These findings suggest that the rostral pathway is the main ascending auditory pathway, whereas the caudal pathway may serve a more modulatory role for sound processing (Straka et al., 2014a). This may be relevant for the selection of stimulation sites for the auditory midbrain implant (see chapter: Auditory Brainstem and Midbrain Implants).

1.6.3 Parallel Processing in Auditory Cortex

Auditory processing in primates recognizes three major regions (core, belt, parabelt), subdivided into 13 areas. The connections between areas are topographically ordered in a manner consistent with information flow along two major anatomical axes: core–belt–parabelt (Hackett et al., 1998a,b) and caudal–rostral (Hackett, 2011). Kaas and Hackett (2008) already described early auditory processing in humans as to be similar to those of monkeys. Woods et al. (2009) found that core and belt areas mapped the basic acoustic features of sounds while surrounding higher order parabelt regions were tuned to more abstract stimulus attributes. In an fMRI study, Langers (2014) showed multiple fields on the superior surface of the temporal lobe in both hemispheres, which were distinguishable on the basis of tonotopic gradient direction. They suggested that these were the human homologues of the core areas A1 and R in primates (Fig. 1.11). Saenz and Langers (2014) argued that the data in the literature are converging toward an interpretation that core fields A1 and R fold across the rostral and caudal banks of Heschl's gyrus, with tonotopic gradients laid out in a distinctive V-shaped manner. This suggested to them an organization largely homologous with nonhuman primates.

1.7 THE EFFERENT SYSTEM

1.7.1 Effects of Olivocochlear Bundle Activity

A group of nerve fibers, termed the crossed olivocochlear bundle (COCB), that project from the brainstem to the inner ear was first described by Rasmussen (1946). Nearly two decades later, Fex (1962) reported that single auditory nerve fibers were inhibited by stimulation of the COCB. Subsequently, Wiederholt and Kiang (1970) confirmed that COCB stimulation in general reduces the spike-discharge rate for tones at the CFs of the fibers. Realizing that the COCB could be activated by contralateral sound, Buño (1978) found that contralateral tonal stimulation at low sound intensities, but more than 20 dB above the nerve fiber threshold for ipsilateral activation, decreased the ipsilateral sound-evoked activity of about 25% of the units studied.

Two types of efferents innervate the cochlea: the efferents projecting underneath the IHCs and synapsing with the SGN dendrites, and those connecting directly the OHCs. These efferents originate from different areas in the SOC in the brainstem and run through the vestibular nerve. The efferents that connect the afferent type I fibers originate in the lateral superior olive.

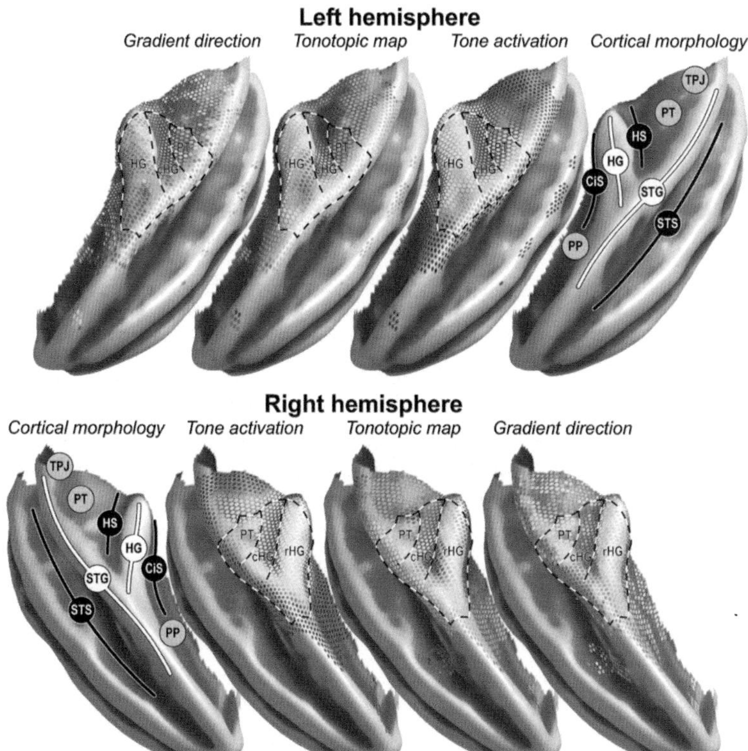

FIGURE 1.11 A three-dimensional rendering of key results on the group-average semi-inflated temporal lobe surface. Notable morphological features include the planum polare (PP), circular sulcus (CiS), Heschl's gyrus (HG), Heschl's sulcus (HS), planum temporale (PT), temporoparietal junction (TPJ), superior temporal gyrus (STG), and superior temporal sulcus (STS). The superimposed dashed lines delineate the approximate outline of two fields on rostral (rHG) and caudal (cHG) HG that could be clearly distinguished on the basis of the tonotopic organization. Evidence for at least one additional field labeled PT was additionally found further posteriorly. On the lateral side, adjacent to STS, the organization was ambiguous. *Reprinted from Langers, D.R., 2014, Assessment of tonotopically organised subdivisions in human auditory cortex using volumetric and surface-based cortical alignments. Hum. Brain Mapp. 35, 1544–1561, with permission from John Wiley & Sons, Inc.*

The OHCs' efferent neurons originate from the ventral nuclei of the trapezoid body. Most authors use the terms lateral and medial efferent systems to designate the efferents below the IHC and those of the OHCs. The lateral efferents, which represent about 50–65% of the olivocochlear bundle fibers, are unmyelinated and project toward the ipsilateral cochlea. The medial efferents are myelinated and reach the OHCs via the crossed and uncrossed components of the olivocochlear bundle. The crossed component predominates with 70–75% of the medial efferent innervation.

1.7.2 Recording From Efferent Neurons

Liberman and Brown (1986) were the first to record from efferent fibers of the olivocochlear bundle and observed that their firings were characterized by regular inter-spike intervals. Within the cat cochlea, efferent neurons branched profusely to innervate as many as 84 OHCs over as much as 2.8 mm of the organ of Corti. Efferent nerve fibers had frequency-tuning curves similar to those of primary afferents. Liberman (1988) found that SFR was weakly correlated with threshold among efferent neurons: Those with SFRs more than 1 spikes/s were generally more sensitive than spontaneously inactive fibers. The spontaneous discharge rate was dependent on stimulation history: Some units with zero SFR became spontaneously active after several minutes of continuous noise stimulation. Most efferent units appeared monaural (91%), with roughly two-thirds excited by ipsilateral stimuli and one-third by contralateral stimuli. Liberman (1988) also observed that increases in efferent discharge rate, both during and after noise stimulation, suggested that the efferent feedback to the periphery is especially high in a noisy environment. Thus, the presence of this feedback might increase the detectability of narrow-band stimuli embedded in noise and help protect the ear from adverse effects of acoustic overstimulation (Liberman, 1988).

1.7.3 Protective Effects of Efferent Activity

Puel et al. (1988) had originally reported that ipsilateral efferent nerve fibers protect against the effects of intense sound exposure. Accordingly, Warren and Liberman (1989a) found that contralaterally presented tones and broadband noise suppressed auditory nerve afferent responses to ipsilateral tones at their CF, but not to tones off CF. The suppression due to contralateral sound was not immediate but developed in 100–200 ms and also decayed over the same time span. A follow-up study (Liberman and Gao, 1995) found that results from temporary threshold shift (TTS) inducing experiments could not be extrapolated to permanent threshold shift (PTS) conditions, which are relevant to noise-induced hearing loss in humans. This implies that the protective effect of the efferent system is not present for traumatic noise exposures.

Here I will introduce pretrauma conditioning as a protective measure, because it may depend on the action of the efferent system (extracted from Eggermont, 2014). Canlon et al. (1988) could reduce the PTSs produced by traumatic noise by pre-exposing guinea pigs to a moderate level acoustic stimulus (1 kHz tone at 81 dB SPL presented continuously for 24 days). The sound pretreatment resulted in an approximately 20 dB reduction in the threshold shift relative to animals that were not pre-exposed and allowed complete recovery from the threshold shift after 2 months. After the 8-week recovery period, the control (not pre-exposed) group continued to show a threshold shift of 14 dB (at 0.5 kHz) to 35 dB (at 4 kHz). Within this recovery period none of the control animals returned to pre-exposure threshold values. Following up with the same paradigm, Canlon and Fransson (1995) showed that sound conditioning did not cause any significant functional or morphological alteration to the guinea pig cochlea. Traumatic noise exposure in the unconditioned group affected nearly 100% of the OHCs in a region centered at the 14-mm distance point from the RW. Following traumatic exposure, the sound-conditioned group showed a significantly less (~50%) OHC loss compared to the unconditioned group.

Kujawa and Liberman (1997) showed that total loss of the OCB significantly increased the noise-induced PTS, whereas loss of the COCB alone did not. They also found that the conditioning exposure in de-efferented animals increased the PTS from the traumatic exposure. In addition, animals undergoing sham surgery appeared protected whether or not they received the conditioning noise exposure. This suggests that conditioning-related protection may arise from a generalized stress response, which can be elicited by, e.g., noise exposure, brain surgery, or just handling. Kujawa and Liberman (1999) later suggested that sound conditioning changes the physiology of the OHCs, which are the peripheral targets of the olivocochlear fiber reflex. Darrow et al. (2007) studying the effects of selective removal of lateral olivocochlear fiber efferents again came to the conclusion that the lateral olivocochlear fibers modulate ANF excitability and thereby protect them from damage during acute acoustic trauma.

Summarizing, the action of the olivocochlear bundle protects the cochlear structures. Medial olivocochlear innervation of the OHCs prevents the loss of synaptic structure in IHCs through the inhibition of the cochlear amplification in noise- and age-related hearing loss. In parallel, lateral olivocochlear innervation governs the excitability of the ANFs to prevent glutamate-induced excitotoxicity and to promote synaptic repair (Nouvian et al., 2015). Effects of the olivocochlear bundle can be measured clinically using the acoustic reflex, also termed middle ear muscle reflex (MEMR; see chapter: Types of Hearing Loss).

1.7.4 Measuring Efferent Effects Using OAEs

Collet et al. (1990) were the first to investigate the possibility that contralateral auditory stimulation along medial efferent system pathways may alter active cochlear micromechanics and hence affect evoked OAEs in humans. They showed that low-intensity (>30 dB SPL) contralateral white noise reduced OAE amplitude. Their results suggested a way for functional exploration of the medial olivocochlear efferent system. Maison and Liberman (2000) measured the strength of a sound-evoked neuronal feedback pathway to the inner ear by examining DPOAEs from the normal ear. DPOAE amplitude decay was inversely correlated with the degree of hearing loss after subsequent noise exposure. Clinically, this is typically done using the standard tonal probe at 226 Hz and a stimulus tone at a frequency within a patient's "normal" audiometric range, a weakened MEMR can be seen in humans with noise-induced hearing loss, acoustic neuroma, and potentially auditory neuropathy (see chapter: Types of Hearing Loss; Valero et al., 2016).

1.7.5 Preventing Age-Related Synaptopathy?

Liberman et al. (2014) noted that moderate exposures (80–100 dB), causing only transient threshold elevation, can cause degeneration of the synapses connecting to spiral ganglion cell dendrites, without morphological loss of hair cells (Kujawa and Liberman, 2009; Maison et al., 2013). This primary neuronal degeneration appears, within hours postexposure, as a loss of synaptic terminals on IHCs (Robertson, 1983), likely from glutamate excitotoxicity (Pujol and Puel, 1999). Death of the spiral ganglion cells, appeared to be much slower, and continuing for months to years postexposure (Liberman and Kiang, 1978). Liberman et al. (2014) noted that this "hidden hearing loss" (Schaette and McAlpine, 2011) would likely be important in presbycusis as the aging human cochlea can show significant neural degeneration well before morphological loss of hair cells (Makary et al., 2011). We present more detail on the various deficits that comprise the auditory neuropathy spectrum in Chapter 5, Types of Hearing Loss.

1.8 SOUND LOCALIZATION

Localizing a sound source involves estimating azimuth (in the horizontal plane), elevation (in the vertical plane), and distance to the source. Most vertebrates can determine the horizontal location of a sound source by evaluating interaural phase/time differences (ITDs), interaural level differences (ILDs), and in the vertical plane by interaural spectral differences (Eggermont, 1988). According to the duplex theory of sound localization, localization of low- and high-frequency sounds largely depends on analysis of the ITD and ILD, respectively (Keating et al., 2014). Although first proposed more than a century ago (Rayleigh, 1907), the duplex theory provides a remarkably successful account of the way in which humans localize pure tones (Macpherson and Middlebrooks, 2002). ITD processing is effective at low frequencies due to the facts that phase locking by the action potentials of ANFs is prominent. In contrast, ILD is most pronounced for high-frequency sounds. This is because for head size larger than one-quarter of the sound wavelength (as a general rule), the head diffracts the sound and becomes an obstacle to sound propagation, thereby producing a head shadowing effect (Feng and Christensen-Dalsgard, 2008). For instance the diameter of the human head is about 17 cm, which is the wavelength of an approximately 2 kHz tone, so that frequencies below 500 Hz bypass the head without attenuation. Such low frequencies may only be localized based on phase differences between the two ears. The role of phase differences (equivalent to time differences) in localization of low-frequency tones was first reported by Lord Rayleigh (1907) in his Sedgwick lecture:

> *There could be no doubt but that relative intensities at the two ears play an important part in the localization of sound. Thus if a fork of whatever pitch be held close to one ear, it is heard much the louder by that ear, and is at once referred instinctively to that side of the head. It is impossible to doubt that this is a question of relative intensities. On the other hand, as we have seen, there are cases where this explanation breaks down. When a pure tone of low pitch is recognized as being on the right or the left, the only alternative to the intensity theory is to suppose that the judgement is founded upon the difference of* phases *at the two ears.*

The mechanisms by which the ITDs are used are still hotly debated. In an extensive review of the evolution and proposed mechanisms of sound localization, Grothe (2003) stated that: "Traditionally, it was thought that the underlying mechanism involved only coincidence detection of excitatory inputs from the two ears. However, recent findings have uncovered profound roles for synaptic inhibition in the processing of interaural time differences."

The current textbook view of ITD processing is still based on the model by Jeffress (1948). It has three fundamental assumptions (Grothe, 2003). First, the temporal pattern of an acoustic stimulus is preserved in the firing pattern of the excitatory projections to the ITD-encoding structure. Action potentials could be time-locked to the low-frequency carrier, to the stimulus onset, or to envelope changes. Second, the ITD-encoding neurons receive such time-locked inputs from both ears and fire maximally when the action potentials from the two sides arrive at the same time. Third, the excitatory inputs are arranged as delay lines that project to an array of neurons in the ITD detector that all respond to the same stimulus frequency (Fig. 1.12). Symmetrical axonal inputs from both sides (the neuron in the middle of the vertical array of coincidence detector neurons) would provide coincident inputs only when the stimulus is straight ahead, reaching both ears simultaneously. A neuron with a short delay line from the left, but a long delay line from the right ear (top neuron in the array) would respond maximally when an ITD compensates for the mismatch in delays from the two ears. So, it would respond maximally to stimuli in the right hemifield. In contrast, short delays from the right and long delays from the left ear would tune a neuron to respond maximally to stimuli in the left hemifield. A systematic arrangement of these delay lines (as shown in Fig. 1.12) would create a systematic map of best ITDs, and so a topographic map of azimuthal space.

The bird ITD-encoding system seems to have evolved in a way that closely matches Jeffress's predictions (Köppl and Carr, 2008). However, increasingly conflicting evidence began to raise doubts about the validity of the Jeffress model for the mammalian ITD encoder. The degree to which

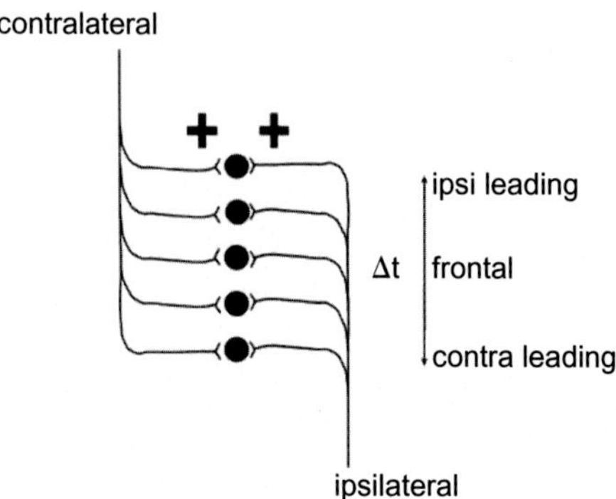

FIGURE 1.12 The Jeffress model of ITD detection. An array of coincidence detector neurons that only respond if binaural excitatory inputs arrive exactly simultaneously. Variation of axonal length of the inputs causes different traveling times of the neural activity to the coincidence detector neurons (the so-called "delay lines"). Consequently, for different azimuthal positions of a sound source (=different ITDs) coincidence of inputs will occur at different neurons. Thereby a systematic place code of azimuthal position could be created. *Reprinted from Grothe, B., 2000. The evolution of temporal processing in the medial superior olive, an auditory brainstem structure. Prog. Neurobiol. 61(6), 581–610 (Grothe, 2000), with permission from Elsevier.*

this requires only minor changes to the Jeffress model (Joris and Yin, 2007) or a complete overhaul thereof (Ashida and Carr, 2011; Grothe et al., 2010) is still debated. What is clear is the dominant role that time plays in sound localization in azimuth.

In the brainstem there is a tuned array of azimuths based on a population of neurons narrowly tuned to different ITDs and ILDs across the behaviorally relevant range, so that the source azimuth is represented by which neurons in the array are activated by stimulus azimuth, i.e., as a spatial map. Behaviorally, and in the cortex, there are in fact two general neurophysiological models of sound lateralization mechanisms, which may be active in humans and other mammals and in birds (Fig. 1.13). The second model needs only two neural channels, each broadly tuned to interaural cue values favoring one acoustic hemifield, so that, especially for sources near the midline, cue value and therefore source azimuth is encoded by the relative activation of the two neural populations (Phillips, 2008).

The data from various studies in auditory cortex, from rat (Yao et al., 2013) to cat (Eggermont and Mossop, 1998) to primates, favor a perceptual architecture for sound lateralization in auditory cortex based on the activity of two, hemifield-tuned azimuthal channels. Yao et al. (2015) showed in rat A1 strong responses in the contralateral hemifield spatial sensitivity, sharp cutoffs across the midline, and weak, sound-level-tolerant responses to ipsilateral sounds. This sensitivity results from computations in the lemniscal pathway, in which sharp, level-tolerant spatial sensitivity arises between ICC and the MGBv, projects to the forebrain to support perception of sound

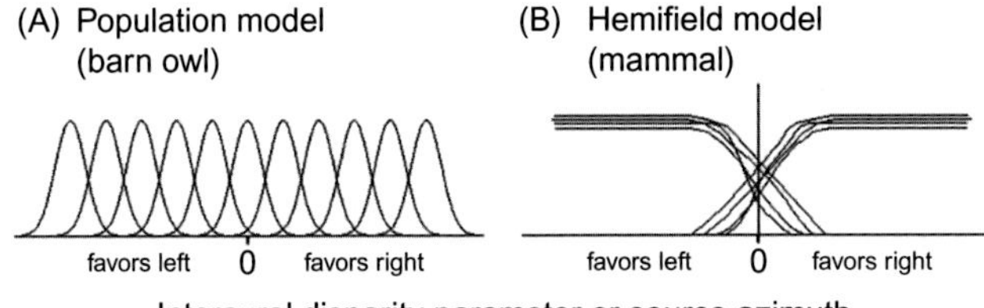

FIGURE 1.13 Schematic diagram illustrating the main features of two models of sound lateralization mechanisms in humans. In (A), which is derived largely from work in the barn owl, individual neurons are narrowly tuned to different values along an interaural disparity parameter, so that a stimulus disparity, and therefore azimuth, is indicated by which neurons of the array are active. In (B), which is derived largely from cortical recordings in mammals, there are only two "channels." Each is comprised of neurons having similar, though not identical, hemifield tuning to interaural disparity. Thus, source disparity, and therefore azimuth, is indicated by the relative rates of activation of the two channels. *Reprinted from Phillips, D.P., 2008. A perceptual architecture for sound lateralization in man. Hear. Res. 238, 124–132, with permission from Elsevier.*

location. In addition there is a tectal pathway, in which such sharp sensitivity arises from neural computations between ICC and brachium of the IC and projects to the superior colliculus. Cross-modal interaction in the superior colliculus (see chapter: Multisensory Processing), where visually induced saccadic eye movement is coded, could support the demonstrated saccadic eye movements to sounds (Bell et al., 2005).

1.9 SUMMARY

New findings about the structure and function of the auditory system continue to emerge. We describe here, besides the "classical" textbook knowledge on the action of the cochlea and central nervous system, recent insights into the effects of subclinical noise exposures on the ribbon synapses, and insights into the protective effects of the efferent system for the effects of these subclinical noise exposures. This protective effect does not seem to include traumatic noise exposure, albeit that conditioning with moderate level sound that presumably includes activity of the efferent system induces such protection. The advent of powerful imaging techniques applied to the human auditory cortex has allowed a comparison with the detailed functional knowledge of these areas in nonhuman primates and suggests very strong similarities between them.

REFERENCES

Aitkin, L., 1986. The Auditory Midbrain. Humana Press, Clifton, NJ.

Ashida, G., Carr, C.E., 2011. Sound localization: Jeffress and beyond. Curr. Opin. Neurobiol. 21, 745–751.

Bass, A.H., McKibben, J.R., 2003. Neural mechanisms and behaviors for acoustic communication in teleost fish. Prog. Neurobiol. 69, 1–26.

Bell, A.H., Meredith, M.A., Van Opstal, A.J., Munoz, D.P., 2005. Crossmodal integration in the primate superior colliculus underlying the preparation and initiation of saccadic eye movements. J. Neurophysiol. 93, 3659–3673.

Bourien, J., Tang, Y., Batrel, C., Huet, A., Lenoir, M., Ladrech, S., et al., 2014. Contribution of auditory nerve fibers to compound action potential of the auditory nerve. J. Neurophysiol. 112, 1025–1039.

Brownell, W.E., 1984. Microscopic observation of cochlear hair cell motility. Scan. Electron. Microsc. 3, 1401–1406.

Buño Jr., W., 1978. Auditory nerve fiber activity influenced by contralateral ear sound stimulation. Exp. Neurol. 59, 62–74.

Canlon, B., Fransson, A., 1995. Morphological and functional preservation of the outer hair cells from noise trauma by sound conditioning. Hear. Res. 84, 112–124.

Canlon, B., Borg, E., Flock, A., 1988. Protection against noise trauma by pre-exposure to a low level acoustic stimulus. Hear. Res. 34, 197–200.

Christie, K.,W., Sivan-Loukianova, E., Smith, W.,C., Aldrich, B.,T., Schon, M.,A., Roy, M.M., et al., 2013. Physiological, anatomical, and behavioral changes after acoustic trauma in *Drosophila melanogaster*. Proc. Natl. Acad. Sci. U. S. A. 110, 15449–15454.

Collet, L., Kemp, D.T., Veuillet, E., Duclaux, R., Moulin, A., Morgon, A., 1990. Effect of contralateral auditory stimuli on active cochlear micro-mechanical properties in human subjects. Hear. Res. 43, 251–261.

Conner, W.E., Corcoran, A.J., 2012. Sound strategies: the 65-million-year-old battle between bats and insects. Annu. Rev. Entomol. 57, 21–39.

Covey, E., Casseday, J.H., 1999. Timing in the auditory system of the bat. Annu. Rev. Physiol. 61, 457–476.

Dallos, P., 1973. The Auditory Periphery. Biophysics and Physiology. Academic Press, New York.

Dallos, P., Cheatham, M.A., 1976. Compound action potential (AP) tuning curves. J. Acoust. Soc. Am. 59, 591–597.

Dallos, P., Fakler, B., 2002. Prestin, a new type of motor protein. Nat. Rev. Mol. Cell. Biol. 3, 104–411.

Dallos, P., Santos-Sacchi, J., Flock, A., 1982. Intracellular recordings from cochlear outer hair cells. Science 218, 582–584.

Dallos, P., Wu, X., Cheatham, M.A., Gao, J., Zheng, J., Anderson, C.T., et al., 2008. Prestin-based outer hair cell motility is necessary for mammalian cochlear amplification. Neuron 58, 333–339.

Darrow, K.N., Maison, S.F., Liberman, M.C., 2007. Selective removal of lateral olivocochlear efferents increases vulnerability to acute acoustic injury. J. Neurophysiol. 97, 1775–1785.

Davis, H., 1983. An active process in cochlear mechanics. Hear. Res. 9, 79–90.

Eggermont, J.J., 1974. Basic principles for electrocochleography. Acta. Otolaryngol. Suppl. 316, 7–16.

Eggermont, J.J., 1976. Electrocochleography. In: Keidel, W.D., Neff, W.D. (Eds.), Handbook of Sensory Physiology. Springer-Verlag, New York, pp. 626–705.

Eggermont, J.J., 1977. Compound action potential tuning curves in normal and pathological human ears. J. Acoust. Soc. Am. 62, 1247–1251.

Eggermont, J.J., 1988. Mechanisms of sound localization in anurans. In: Fritzsch, B., Ryan, M., J., Wilczynski, W., Hetherington, T.E., Walkowiak, W. (Eds.), The Evolution of the Amphibian Auditory System. John Wiley & Sons, New York, pp. 307–336.

Eggermont, J.J., 2001. Between sound and perception: reviewing the search for a neural code. Hear. Res. 157, 1–42.

Eggermont, J.J., 2014. Noise and the Brain. Experience Dependent Developmental and Adult Plasticity. Academic Press, London.

Eggermont, J.J., Mossop, J.E., 1998. Azimuth coding in primary auditory cortex of the cat. I. Spike synchrony vs. spike count representations. J. Neurophysiol. 80, 2133–2150.

Eggermont, J.J., Odenthal, D.W., Schmidt, P.H., Spoor, A., 1974. Electrocochleography. Basic principles and clinical application. Acta Otolaryngol. Suppl. 316, 1–84.

Evans, E.F., Klinke, R., 1982. The effects of intracochlear and systemic furosemide on the properties of single cochlear nerve fibres in the cat. J. Physiol. 331, 409–427.

Fay, R.R., 1988. Hearing in Vertebrates. A Psychophysics Databook. Hill-Fay Associates, Winnetka, IL.

Feng, A.S., Christensen-Dalsgard, J., 2008. Interconnections between the ears in nonmammalian vertebrates. Chapter 3.13. The Senses. A Comprehensive Reference. Elsevier, pp. 217–224.

Fernandez, K.A., Jeffers, P.W.C., Lall, K., Liberman, M.C., Kujawa, S.G., 2015. Aging after noise exposure: acceleration of cochlear synaptopathy in "recovered" ears. J. Neurosci. 35, 7509–7520.

Fex, J., 1962. Auditory activity in centrifugal and centripetal cochlear fibers in cat. A study of a feedback system. Acta Physiol. Scand. 55 (Suppl. 189), 1–68.

Flores, E.N., Duggan, A., Madathany, T., Hogan, A.K., Márquez, F.G., Kumar, G., et al., 2015. A non-canonical pathway from cochlea to brain signals tissue-damaging noise. Curr. Biol. 25, 606–612.

Fonseca, P.J., Münch, D., Hennig, R.M., 2000. How cicadas interpret acoustic signals. Nature 405, 297–298.

Furness, D.N., Hackney, C.M., 2008. Molecular anatomy of receptor cells and organ of Corti. Chapter 3.06. The Senses. A Comprehensive Reference. Elsevier, pp. 108–137.

Garstang, M., 2004. Long-distance, low-frequency elephant communication. J. Comp. Physiol. A Neuroethol. Sens. Neural Behav. Physiol. 190, 791–805.

Glowatzki, E., Grant, L., Fuchs, P., 2008. Hair cell afferent synapses. Curr. Opin. Neurobiol. 18, 389–395.

Gold, T., 1948. Hearing. II. The physical basis of the action of the cochlea. Proc. Roy. Soc. B 135, 492–498.

Grothe, B., 2000. The evolution of temporal processing in the medial superior olive, an auditory brainstem structure. Prog. Neurobiol. 61, 581–610.

Grothe, B., 2003. New roles for synaptic inhibition in sound localization. Nat. Rev. Neurosci. 4, 540–550.

Grothe, B., Koch, U., 2011. Dynamics of binaural processing in the mammalian sound localization pathway—the role of GABA(B) receptors. Hear. Res. 279, 43–50.

Grothe, B., Pecka, M., McAlpine, D., 2010. Mechanisms of sound localization in mammals. Physiol. Rev. 90, 983–1012.

Gruters, K.G., Groh, J.M., 2012. Sounds and beyond: multisensory and other non-auditory signals in the inferior colliculus. Front. Neural Circuits 6, 96.

Guinan Jr., J.J., Salt, A., Cheatham, M.A., 2012. Progress in cochlear physiology after Békésy. Hear. Res. 293, 12–20.

Hackett, T.A., 2011. Information flow in the auditory cortical network. Hearing Research 271, 133–146.

Hackett, T.A., Stepniewska, I., Kaas, J.H., 1998a. Subdivisions of auditory cortex and ipsilateral cortical connections of the parabelt auditory cortex in macaque monkeys. J. Comp. Neurol. 394, 475–495.

Hackett, T.A., Stepniewska, I., Kaas, J.H., 1998b. Thalamocortical connections of the parabelt auditory cortex in macaque monkeys. J. Comp. Neurol. 400, 271–286.

He, D.Z., Zheng, J., Kalinec, F., Kakehata, S., Santos-Sacchi, J., 2006. Tuning in to the amazing outer hair cell: membrane wizardry with a twist and shout. J. Membr. Biol. 209, 119–134.

Heffner, H.E., Heffner, R.S., 2007. Hearing ranges of laboratory animals. J. Am. Assoc. Lab. Anim. Sci. 46, 20–22.

Hossain, W.A., Antic, S.D., Yang, Y., Rasband, M.N., Morest, D.K., 2005. Where is the spike generator of the cochlear nerve? Voltage-gated sodium channels in the mouse cochlea. J. Neurosci. 25, 6857–6868.

Hudspeth, A.J., 2008. Making an effort to listen: mechanical amplification in the ear. Neuron 59, 530–545.

Jeffress, L.A., 1948. A place theory of sound localization. J. Comp. Physiol. Psychol. 41, 35–39.

Joris, P.X., Yin, T.C.T., 2007. A matter of time: internal delays in binaural processing. Trends Neurosci. 30, 70–78.

Kaas, J.H., Hackett, T.A., 2008. The functional neuroanatomy of the auditory cortex. Chapter 3.45. The Senses. Elsevier, pp. 765–780.

Kantardzhieva, A., Liberman, M.C., Sewell, W.F., 2013. Quantitative analysis of ribbons, vesicles, and cisterns at the cat inner hair cell synapse: correlations with spontaneous rate. J. Comp. Neurol. 521, 3260–3271.

Keating, P., Nodal, F.R., King, A.J., 2014. Behavioural sensitivity to binaural spatial cues in ferrets: evidence for plasticity in the duplex theory of sound localization. Eur. J. Neurosci. 39, 197–206.

Kemp, D.T., 1978. Stimulated acoustic emissions from within the human auditory system. J. Acoust. Soc. Am. 64, 1386–1391.

Khimich, D., Nouvian, R., Pujol, R., Dieck, T.S., Egner, A., Gundelfinger, E.D., et al., 2005. Hair cell synaptic ribbons are essential for synchronous auditory signalling. Nature 434, 889–894.

Kiang, N.Y.S., Watanabe, T., Thomas, C., Clark, L.F., 1965. Discharge Patterns of Single Fibers in the Cat's Auditory Nerve. MIT Press, Cambridge, MA.

Köppl, C., Carr, C.E., 2008. Maps of interaural time difference in the chicken's brainstem nucleus laminaris. Biol. Cybern. 98, 541–555.

Kujawa, S.G., Liberman, M.C., 1997. Conditioning-related protection from acoustic injury: effects of chronic deefferentation and sham surgery. J. Neurophysiol. 78, 3095–3106.

Kujawa, S.G., Liberman, M.C., 1999. Long-term sound conditioning enhances cochlear sensitivity. J. Neurophysiol. 82, 863–873.

Kujawa, S.G., Liberman, M.C., 2009. Adding insult to injury: cochlear nerve degeneration after "temporary" noise-induced hearing loss. J. Neurosci. 29, 14077–14085.

Kujawa, S.G., Liberman, M.C., 2015. Synaptopathy in the noise-exposed and aging cochlea: primary neural degeneration in acquired sensorineural hearing loss. Hear. Res. 330, 191–199.

Langers, D.R., 2014. Assessment of tonotopically organised subdivisions in human auditory cortex using volumetric and surface-based cortical alignments. Hum. Brain Mapp. 35, 1544–1561.

Langner, G., Schreiner, C.E., 1988. Periodicity coding in the inferior colliculus of the cat. I. Neuronal mechanisms. J. Neurophysiol. 60, 1799–1822.

Liberman, M.C., 1980. Morphological differences among radial afferent fibers in the cat cochlea: an electron-microscopic study of serial sections. Hear. Res. 3, 45–63.

Liberman, M.C., 1982. Single-neuron labeling in the cat auditory nerve. Science 216, 1239–1241.

Liberman, M.C., 1988. Physiology of cochlear efferent and afferent neurons: direct comparisons in the same animal. Hear. Res. 34, 179–191.

Liberman, M.C., 1991. Central projections of auditory-nerve fibers of differing spontaneous firing rate. I. Anteroventral cochlear nucleus. J. Comp. Neurol. 313, 240–258.

Liberman, M.C., Brown, M.C., 1986. Physiology and anatomy of single olivocochlear neurons in the cat. Hear. Res. 24, 17–36.

Liberman, M.C., Gao, W.Y., 1995. Chronic cochlear de-efferentation and susceptibility to permanent acoustic injury. Hear. Res. 90, 158–168.

Liberman, M.C., Kiang, N.Y., 1978. Acoustic trauma in cats. Cochlear pathology and auditory-nerve activity. Acta Otolaryngol. Suppl. 358, 1–63.

Liberman, M.C., Dodds, L.W., Pierce, S., 1990. Afferent and efferent innervation of the cat cochlea: quantitative analysis with light and electron microscopy. J. Comp. Neurol. 301, 443–460.

Liberman, M.C., Gao, J., He, D.Z., Wu, X., Jia, S., Zuo, J., 2002. Prestin is required for electromotility of the outer hair cell and for the cochlear amplifier. Nature 419, 300–304.

Liberman, M.C., Liberman, L.D., Maison, S.F., 2014. Efferent feedback slows cochlear aging. J. Neurosci. 34, 4599–4607.

Liu, C., Glowatzki, E., Fuchs, P.A., 2015. Unmyelinated type II afferent neurons report cochlear damage. Proc. Natl. Acad. Sci. U. S. A. 112, 14723–14727.

Logiudice, L., Sterling, P., Matthews, G.G., 2009. Vesicle recycling at ribbon synapses in the finely branched axon terminals of mouse retinal bipolar neurons. Neuroscience 164, 1546–1556.

Macpherson, E.A., Middlebrooks, J.C., 2002. Listener weighting of cues for lateral angle: the duplex theory of sound localization revisited. J. Acoust. Soc. Am. 111, 2219–2236.

Maison, S.F., Liberman, M.C., 2000. Predicting vulnerability to acoustic injury with a noninvasive assay of olivocochlear reflex strength. J. Neurosci. 20, 4701–4707.

Maison, S.F., Usubuchi, H., Liberman, M.C., 2013. Efferent feedback minimizes cochlear neuropathy from moderate noise exposure. J. Neurosci. 33, 5542–5552.

Makary, C.A., Shin, J., Kujawa, S.G., Liberman, M.C., Merchant, S.N., 2011. Age-related primary cochlear neuronal degeneration in human temporal bones. J. Assoc. Res. Otolaryngol. 12, 711–717.

Manley, G.A., 2002. Evolution of structure and function of the hearing organ of lizards. J. Neurobiol. 53, 202–211.

Merchan-Perez, A., Liberman, M.C., 1996. Ultrastructural differences among afferent synapses on cochlear hair cells: correlations with spontaneous discharge rate. J. Comp. Neurol. 371, 208–221.

Merchant, S.N., Ravicz, M.E., Puria, S., Voss, S.E., Whittemore Jr., K.R., Peake, W.T., et al., 1997. Analysis of middle ear mechanics and application to diseased and reconstructed ears. Am. J. Otol. 18, 139–154.

Morel, A., Imig, T.J., 1987. Thalamic projections to fields A, AI, P, and VP in the cat auditory cortex. J. Comp. Neurol. 265, 119–144.

Moser, T., Neef, A., Khimich, D., 2009. Mechanisms underlying the temporal precision of sound coding at the inner hair cell ribbon synapse. J. Physiol. 576, 55–62.

Moser, T., Predoehl, F., Starr, A., 2013. Review of hair cell synapse defects in sensorineural hearing impairment. Otol. Neurotol. 34, 995–1004.

Müller, M., Robertson, D., Yates, G.K., 1991. Rate-versus-level functions of primary auditory nerve fibres: evidence for square law behaviour of all fibre categories in the guinea pig. Hear. Res. 55, 50–56.

Müller, U., Gillespie, P., 2008. Silencing the cochlear amplifier by immobilizing prestin. Neuron 58, 299–301.

Nouvian, R., Eybalin, M., Puel, J.-L., 2015. Cochlear efferents in developing adult and pathological conditions. Cell Tissue Res. 361, 301–309.

Oertel, D., 1999. The role of timing in the brain stem auditory nuclei of vertebrates. Annu. Rev. Physiol. 61, 497–519.

Oxenham, A.J., Bacon, S.P., 2003. Cochlear compression: perceptual measures and implications for normal and impaired hearing. Ear Hear. 24, 352–366.

Peake, W.T., Rosowski, J.J., Lynch III, T.J., 1992. Middle-ear transmission: acoustic versus ossicular coupling in cat and human. Hear. Res. 57, 245–268.

Phillips, D.P., 2008. A perceptual architecture for sound lateralization in man. Hear. Res. 238, 124–132.

Polley, D.B., Read, H.L., Storace, D.A., Merzenich, M.M., 2007. Multiparametric auditory receptive field organization across five cortical fields in the albino rat. J. Neurophysiol. 97, 3621–3638.

Popper, A.N., Fay, R.R., 2011. Rethinking sound detection by fishes. Hear. Res. 273, 25–36.

Puel, J.L., Bobbin, R.P., Fallon, M., 1988. The active process is affected first by intense sound exposure. Hear. Res. 37, 53–63.

Pujol, R., Puel, J.-L., 1999. Excitotoxicity, synaptic repair, and functional recovery in the mammalian cochlea: a review of recent findings. Ann. N. Y. Acad. Sci. 884, 249–254.

Puria, S., Steele, C.R., 2008. Mechano-acoustical transformations. Chapter 3.10. The Senses. A Comprehensive Reference. Elsevier, pp. 166–201.

Rasmussen, G.L., 1946. The olivary peduncle and other fibre projections of the superior olivary complex. J. Comp. Neurol. 84, 141–219.

Rayleigh, L., 1907. XII. On our perception of sound direction. Philos. Mag. Ser. 6 13 (74), 214–232.

Rhode, W.S., 1971. Observations of the vibration of the basilar membrane in squirrel monkeys using the Mössbauer technique. J. Acoust. Soc. Am. 49 (Suppl. 2), 1218.

Robertson, D., 1983. Functional significance of dendritic swelling after loud sounds in the guinea pig cochlea. Hear. Res. 9, 263–278.

Robertson, D., 1984. Horseradish peroxidase injection of physiologically characterized afferent and efferent neurones in the guinea pig spiral ganglion. Hear. Res. 15, 113–121.

Robles, L., Ruggero, M.A., 2001. Mechanics of the mammalian cochlea. Physiol. Rev. 81, 1305–1352.

Rodrigues-Dagaeff, C., Simm, G., De Ribaupierre, Y., Villa, A., De Ribaupierre, F., Rouiller, E. M., 1989. Functional organization of the ventral division of the medial geniculate body of the cat: evidence for a rostro-caudal gradient of response properties and cortical projections. Hear. Res. 39, 103–125.

Russell, I.J., 2008. Cochlear receptor potentials. Chapter 3.20. The Senses. A Comprehensive Reference. Elsevier, pp. 320–358.

Russell, I.J., Sellick, P.M., 1978. Intracellular studies of hair cells in the mammalian cochlea. J. Physiol. 284, 261–290.

Rutherford, M.A., Roberts, W.M., 2008. Afferent synaptic mechanisms. Chapter 3.22. The Senses. A Comprehensive Reference. Elsevier, pp. 366–395.

Saenz, M., Langers, D.R., 2014. Tonotopic mapping of human auditory cortex. Hear. Res. 307, 42–52.

Salomon, G., Elberling, C., 1971. Cochlear nerve potentials recorded from the ear canal in man. Acta Otolaryngol. 71, 319–325.

Santarelli, R., Del Castillo, I., Rodríguez-Ballesteros, M., et al., 2009. Abnormal cochlear potentials from deaf patients with mutations in the otoferlin gene. J. Assoc. Res. Otolaryngol. 10, 545–556.

Schaette, R., McAlpine, D., 2011. Tinnitus with a normal audiogram: physiological evidence for hidden hearing loss and computational model. J. Neurosci. 31, 13452–13457.

Schnupp, J.W.H., Garcia-Lazaro, J.A., Lesica, N.A., 2015. Periodotopy in the gerbil inferior colliculus: local clustering rather than a gradient map. Front. Neural Circuits 9, 37.

Schreiner, C.E., Langner, G., 1997. Laminar fine structure of frequency organization in auditory midbrain. Nature 388, 383–386.

Schulze, H., Langner, G., 1997. Representation of periodicity pitch in the primary auditory cortex of the Mongolian gerbil. Acta Otolaryngol. Suppl. 532, 89–95.

Shera, C.A., 2003. Mammalian spontaneous otoacoustic emissions are amplitude-stabilized cochlear standing waves. J. Acoust. Soc. Am. 114, 244–262.

Shorey, H.H., 1962. Nature of the sound produced by *Drosophila melanogaster* during courtship. Science 137, 677–678.

Siegel, J., 2008. Otoacoustic emissions. Chapter 3.15. The Senses. A Comprehensive Reference. Elsevier, pp. 237–261.

Smith, P.H., Joris, P.X., Yin, T.C., 1993. Projections of physiologically characterized spherical bushy cell axons from the cochlear nucleus of the cat: evidence for delay lines to the medial superior olive. J. Comp. Neurol. 331, 245–260.

Spoendlin, H., 1969. Innervation patterns in the organ of corti of the cat. Acta Otolaryngol. 67, 239–254.

Spoendlin, H., Schrott, A., 1988. The spiral ganglion and the innervation of the human organ of corti. Acta Otolaryngol. 105, 403–410.

Sterling, P., Matthews, G., 2005. Structure and function of ribbon synapses. Trends Neurosci. 28, 20–29.

Storace, D.A., Higgins, N.C., Read, H.L., 2010. Thalamic label patterns suggest primary and ventral auditory fields are distinct core regions. J. Comp. Neurol. 518, 1630–1646.

Straka, M.M., Schendel, D., Lim, H.H., 2013. Neural integration and enhancement from the inferior colliculus up to different layers of auditory cortex. J. Neurophysiol. 110, 1009–1020.

Straka, M.M., McMahon, M., Markovitz, C.D., Lim, H.H., 2014a. Effects of location and timing of co-activated neurons in the auditory midbrain on cortical activity: implications for a new central auditory prosthesis. J. Neural. Eng. 11, 046021.

Straka, M.M., Schendel, D., Lim, H.H., 2014b. Response features across the auditory midbrain reveal an organization consistent with a dual lemniscal pathway. J. Neurophysiol. 112, 981–998.

Tsuji, J., Liberman, M.C., 1997. Intracellular labeling of auditory nerve fibers in guinea pig: central and peripheral projections. J. Comp. Neurol. 381, 188–202.

Tyler, R.S., Pienkowski, M., Roncancio, E.R., Jun, H.J., Brozoski, T., Dauman, N., et al., 2014. A review of hyperacusis and future directions: part I. Definitions and manifestations. Am. J. Audiol. 23, 420–436.

Valero, M.D., Hancock, K.E., Liberman, M.C., 2016. The middle ear muscle reflex in the diagnosis of cochlear neuropathy. Hear. Res. 332, 29–38.

van der Heijden, M., Versteegh, C.P.C., 2015. Energy flux in the cochlea: evidence against power amplification of the traveling wave. J. Assoc. Res. Otolaryngol. 16, 581–597.

van der Heijden, M., Lorteije, J.A., Plauska, A., Roberts, M.T., Golding, N.L., Borst, J.G., 2013. Directional hearing by linear summation of binaural inputs at the medial superior olive. Neuron 78, 936–948.

von Békésy, G., 1960. Experiments in Hearing. McGraw-Hill, New York.

Warren III, E.H., Liberman, M.C., 1989. Effects of contralateral sound on auditory-nerve responses. I. Contributions of cochlear efferents. Hear. Res. 37, 89–104.

Weisz, C.J., Glowatzki, E., Fuchs, P.A., 2014. Excitability of type II cochlear afferents. J. Neurosci. 34, 2365–2373.

Wiederholt, M.L., Kiang, N.Y.S., 1970. Effects of electricstimulation of the crossed-olivocochlear bundle on single auditory-nerve fibers in the cat. J. Acoust. Soc. Am. 48, 950–965.

Winer, J.A., Schreiner, C.E. (Eds.), 2005. The Inferior Colliculus. Springer Science + Business Media, New York.

Woods, D.L., Stecker, G.C., Rinne, T., Herron, T.J., Cate, A.D., Yund, E.W., et al., 2009. Functional maps of human auditory cortex: effects of acoustic features and attention. PLoS One 4, e5183.

Yao, J.D., Bremen, P., Middlebrooks, J.C., 2013. Rat primary auditory cortex is tuned exclusively to the contralateral hemifield. J. Neurophysiol. 110, 2140–2151.

Yao, J.D., Bremen, P., Middlebrooks, J.C., 2015. Transformation of spatial selectivity along the ascending auditory pathway. J. Neurophysiol. 113, 3098–3111.

Zheng, J., Shen, W., He, D.Z., Long, K.B., Madison, L.D., Dallos, P., 2000. Prestin is the motor protein of cochlear outer hair cells. Nature 405, 149–155.

Chapter 2

Brain Plasticity and Perceptual Learning

The brain—young as well as adult—has the capacity to adapt in response to changes in the external environment, to changes in the sensory input due to peripheral injury such as hearing loss, and as a result of training (perceptual learning). Brain plasticity is a determining aspect of the potential success of hearing aids and cochlear implants (CIs). It basically allows the user to adjust to the more or less deformed way the environment and speech are presented to the hard of hearing or completely deaf persons. This learning may often be enhanced by training. Here, I will not present exhaustive details about human developmental plasticity, this is discussed in Chapter 8, Early Diagnosis and Prevention of Hearing Loss, and its relation to CI use is reviewed in Chapter 11, Cochlear Implants.

2.1 THE EXTERNAL ENVIRONMENT

Simply plugging the ears or exposing humans to low-level sound for 2 weeks is sufficient to induce reversible changes in loudness perception (Formby et al., 2003). In this study, normal hearing (NH) human volunteers were asked to wear either earplugs or a set of open-canal, in-the-ear speakers producing a low-level noise between 1 and 8 kHz with a peak level of 50 dB SPL at approximately 6 kHz. Earplugs or earphones were worn for at least 23 h/day for 2 weeks, and subjects performed loudness judgments on 500 and 2000 Hz tones before and after treatment. The noise-exposed subjects needed an additional 4–8 dB of sound level to match their pre-exposure loudness judgments. Conversely, subjects who wore earplugs needed up to 5–9 dB less sound level compared with their baseline judgments. Hearing thresholds were not affected by either treatment. The noise-exposed subjects showed no difference in posttreatment loudness judgments between 500 and 2000 Hz, despite the fact that the noise spectrum did not extend to 500 Hz. Two possible explanations for the loudness changes are that the auditory system undergoes physiological changes or that the listeners simply recalibrate their behavioral criteria (Formby et al., 2003). These physiological changes could occur in auditory cortex, and subsequently by cortical modulation of

Hearing Loss. DOI: http://dx.doi.org/10.1016/B978-0-12-805398-0.00002-5

processing at the brainstem level (Luo et al., 2008) could change the gain of the brainstem mechanisms (Eggermont, 2014).

An interesting variant on the Formby et al. (2003) procedures is monaural ear plugging, which as expected affects sound localization in azimuth. However, Van Wanrooij and Van Opstal (2007) found that the effect was minimal for low-level sound presented to the unplugged ear side. In contrast, they observed that at higher sound levels azimuth localization was highly perturbed. The plug thus creates interaural level differences (see chapter: Hearing Basics) that far exceed the normal physiological range provided by the head shadow. Yet, these erroneous cues were incorporated in forming the azimuth percept, giving rise to larger localization errors than for low-intensity stimuli for which the binaural cues were absent. Thus, listeners rely on monaural spectral cues for sound-source azimuth localization as soon as the binaural difference cues break down. Hofman et al. (1998) had provided another interesting variant on plastic phenomena in sound localization by changing the pinnae with molds. Localizing sounds in elevation was dramatically degraded immediately after pinna modification, however, accurate performance improved again over a time period of 6 weeks, after which the learning process stabilized. Immediately, after removing the molds, the subject's localization accuracy with undisturbed ears was still as high as before the start of the experiment. Apparently, the auditory system had acquired a new representation of the pinna transfer functions, without interfering with the old set.

2.1.1 Critical and Sensitive Periods

The human cochlea is fully developed by 24 weeks of gestation. A blink startle response can first be elicited (acoustically) at 24–25 weeks and is constantly present at 28 weeks. Hearing thresholds are 40 dB SPL at 27–28 weeks and reach the adult threshold of 13.5 dB SPL by 42 weeks of gestation (Birnholz and Benacerrah, 1983). Early born preterm children often end up in the neonatal intensive care unit (NICU). Quite often they show signs of auditory neuropathy and sensorineural hearing loss; however, even in case they do not, they may have other neurological problems from which they only very slowly recover (Marlow et al., 2005).

In an NICU, noise is continuously present in the confines of an incubator. A big issue is the so far largely unknown effect of such prolonged noise exposure in the NICU on the neonatal brain. Whereas it has been established that this does not cause audiometric hearing loss, it may still have profound effects on hearing, as animal studies suggest (Zhang et al., 2001; Chang and Merzenich, 2003). In neonatal and adult animals, band-pass noise exposure leads to contracting tonotopic regions surrounded by expanding tonotopic regions (Pienkowski and Eggermont, 2012). Potential extrapolations can be drawn that pertain to human auditory development. Several studies of long-term outcomes in NICU graduates mention speech and language problems (Stjernqvist and Svenningsen, 1999; Marlow et al., 2005; Kern and Gayraud, 2007). However, few studies have specifically linked speech and language problems with noise type and levels.

Both adult and critical period (CP) animals show plastic changes, i.e., a capacity for change in the structure and/or function of the nervous system, as a result of sensory experience following passive exposure to tonal or noise stimuli. A sensitive period is one, typically developmental, where brain plasticity results as a function of sensory experience. The CP, in general, is considered a time period within a sensitive period when the best neural representation of the environment is selected from among the many competing inputs that affect the maturing nervous system. The growth and function of lateral inhibitory circuits may be important for terminating the CP. The difficulty of this problem is highlighted by the fact that the closure of the early CP may be dependent on the input received (Zhang et al., 2001; Chang and Merzenich, 2003). Note that in adult rats exposed to continuous noise Zheng (2012) found a complete disappearance of tonotopic order, i.e., as if the rats had reentered a condition similar to the CP rats. Tonotopy, also called cochleotopy, an ordered spatial representation of frequency, is found along the so-called lemniscal pathway (see chapter: Hearing Basics; Fig. 2.1) from brainstem to auditory cortex. Only 5 of the 13 auditory cortical areas in cats or primates (including humans) are tonotopically organized (see chapter: Hearing Basics).

In neonatal animals pulsed noise stimulation, at a level not causing hearing loss, disrupts the tonotopic map and broadens frequency tuning (Zhang et al., 2002), whereas in adult animals (de Villers-Sidani et al., 2008) map changes do not occur but behavioral effects related to broader frequency tuning are evident. Depending on the precise CP day, tonal stimulation in CP animals either expands the region of single frequency stimulation (de Villers-Sidani et al., 2007) and up to an octave-wide region on either side, or contacts the region of multitone stimulation and expands the surrounding frequencies (Zhang et al., 2001). In adult animals, the stimulated region contracts regardless being stimulated with band-pass tonal or noise stimuli, whereas the bordering regions dramatically expand (Pienkowski and Eggermont, 2012). These changes in adults spontaneously recover, those in CP animals only in the case of continuous noise (Chang and Merzenich, 2003), which delayed closure of the CP. For the pulsed noise or tonal stimulation in CP animals spontaneous recovery does not occur. The relationship between map changes in A1 and behavior remains unclear (Pienkowski and Eggermont, 2011; Eggermont, 2013).

It is not exactly known whether there are similar CPs in human auditory development (Eggermont, 2013), but from the CI literature one may derive CPs for the necessity of auditory stimulation for binaural hearing (<2 years of unilateral hearing, i.e., in single-sided deafness (Ponton et al., 2001), or using one CI (Gordon et al., 2012)). Support for this plasticity and the

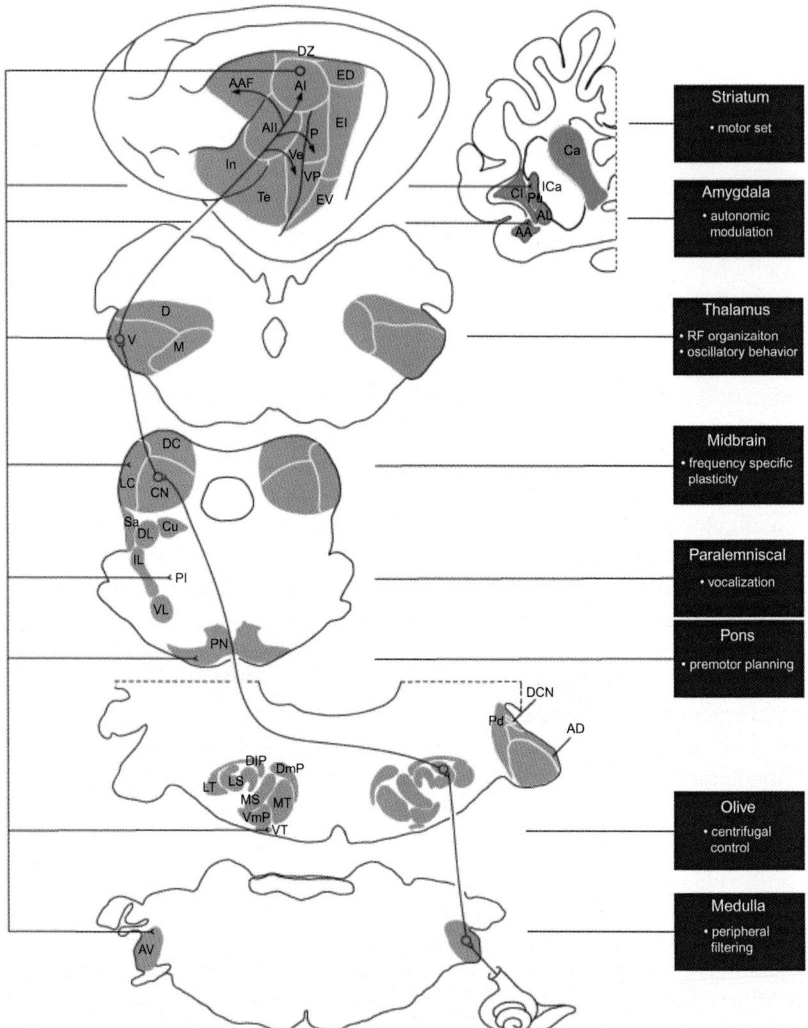

FIGURE 2.1 Ascending lemniscal (black) and descending (blue) projections in the central auditory system. AA, amygdala, anterior nucleus; AAF, anterior auditory field; AI, auditory cortex, primary area; AII, auditory cortex, secondary area; AD, dorsal cochlear nucleus, anterior part; AL, amygdala, lateral nucleus; Av, anteroventral cochlear nucleus; Ca, caudate nucleus; Cl, claustrum; CN, central nucleus of the inferior colliculus; Cu, cuneiform nucleus; D, dorsal nucleus of the medial geniculate body or dorsal; DC, dorsal cortex of the inferior colliculus; DCN, dorsal cochlear nucleus; DL, dorsal nucleus of the lateral lemniscus; DlP, dorsolateral periolivary nucleus; DmP, dorsomedial periolivary nucleus; DZ, dorsal auditory zone (suprasylvian fringe); ED, posterior ectosylvian gyrus, dorsal part; EI, posterior ectosylvian gyrus, intermediate part; EV, posterior ectosylvian gyrus, ventral part; IL, intermediate nucleus of the lateral lemniscus; In, insular cortex; ICa, internal capsule; LT, lateral nucleus of the trapezoid body; M, medial division MGB; MT, medial nucleus of the trapezoid body; PN, pontine nuclei; Pu, putamen; Sa, nucleus sagulum; Te, temporal cortex; V, pars lateralis of the ventral division MGB; Ve, auditory cortex, ventral area; VL, ventral nucleus of the lateral lemniscus; VmP, ventromedial periolivary nucleus; VP, auditory cortex, ventral posterior area; VT, ventral nucleus of the trapezoid body. *Reprinted from Winer, J.A., 2006. Decoding the auditory corticofugal systems. Hear. Res. 212, 1–8, with permission from Elsevier.*

positive effects of early intervention in congenitally unilaterally deaf white cats is provided in the work of Kral et al. (2013) and Tillein et al. (2016). Also the development of certain auditory evoked response components, i.e., N_1 does not develop after more than 3 years of deafness under the age of 6 (Ponton and Eggermont, 2001) suggests a CP. Normal language development in CI patients also suggests a CP (Svirsky et al., 2000). Conductive hearing loss in children is a major determinant of language delay and may potentially cause long-lasting deficits (see chapter: Causes of Acquired Hearing Loss).

2.2 LEARNING PARADIGMS

The main point of including animal data in the context of this chapter is to elucidate the various mechanisms underlying brain plasticity. Weinberger (1995) grouped these mechanisms in the following categories: habituation, sensitization, and conditioning. The first two are considered nonassociative learning; conditioning, including classical conditioning and operant conditioning, is a form of associative learning (Fig. 2.2). The following is also extensively discussed in the work of Eggermont (2014).

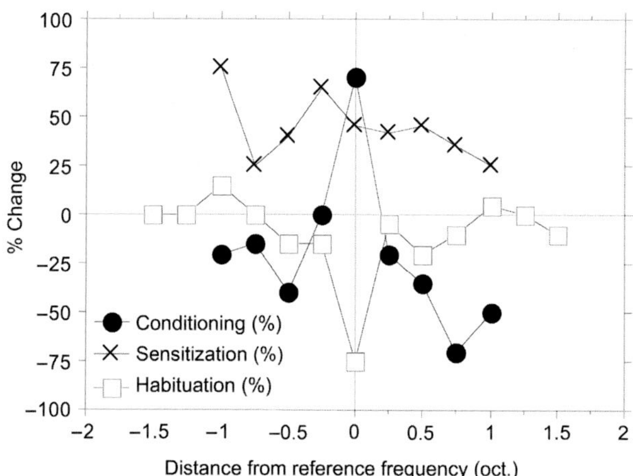

FIGURE 2.2 The effects of learning on the frequency tuning of neurons in A1. Normalized group difference functions show changes in response as a function of octave distance from the reference frequency, either CS frequency, best frequency, or repetition frequency. Conditioning (•) produces a specific increase in A1 response to the CS frequency with reduced responses at different frequencies. Sensitization (×) training produces a nonspecific increase in response across all frequencies (tone–shock unpaired). Repeated presentation (□) of the same tone alone (habituation) produces a specific decreased response at that frequency. *Data from Weinberger, N.M., 1995. Dynamic regulation of receptive fields and maps in the adult sensory cortex. Annu. Rev. Neurosci. 18, 129–158.*

2.2.1 Nonassociative Learning

Nonassociative learning is a change in a response to a stimulus that does not involve associating the presented stimulus with another stimulus or event such as reward or punishment. There are two forms: habituation and sensitization.

2.2.1.1 Habituation

Repeated stimulus presentation, without any other consequences, results in loss of attention and behavioral responses, and usually, in a reduction of the neural responses within relevant sensory cortex. This process of learning not to attend to such a stimulus is termed habituation. It differs from sensory adaptation and fatigue as habituation can occur at long interstimulus intervals, develops more rapidly with weaker stimulus intensity, and is highly specific to the parameters of the repeated stimulus (Fig. 2.2). Repeated sensory stimulation is widely used in studies of sensory cortex where both respond to decrements and increments, with modification of receptive field (RF) properties, have been reported. Thus, habituation is a form of adaptive behavior (or neuroplasticity) that is classified as nonassociative learning.

In the last few decades a phenomenon of relative decrease in neural activity of a frequently presented stimulus relative to the much larger activity produced by an infrequent stimulus was discovered (Näätänen, 1975; Näätänen et al., 1997). Under passive listening conditions the difference waveform between the long-latency auditory evoked potential (AEP) to the unexpected (also called deviant or odd-ball) sound and the AEP to the frequent sound was called the mismatch negativity (MMN). The MMN was interpreted as a reflection of neural information processing in the brain that would allow behavioral detection of a difference in, or discrimination of, the two sounds. At the single-unit level this phenomenon was later named "stimulus-specific adaptation" (SSA) and proposed as "a neural correlate of the MMN" (Ulanovsky et al., 2003; Dean et al., 2005; Pérez-Gonzales and Malmierca, 2014). An extensive discussion on SSA can be found in the work of Eggermont (2015).

2.2.1.2 Sensitization

Sensitization refers to the process by which a synapse becomes more efficient in its response to a stimulus. An example of sensitization is that of kindling (Valentine et al., 2004), where repeated stimulation of hippocampal or amygdala neurons (even when induced by repeated stimulation of primary auditory cortex) eventually leads to seizures in laboratory animals. Having been sensitized, very little stimulation is required to produce the seizures. Sensitization is also a nonassociative learning process because it involves increased responses with repeated presentations to a single stimulus. The increase of responsiveness can be more generalized than to the presented stimulus (Fig. 2.2).

2.2.2 Classical Conditioning

Classical conditioning involves two sequential stimuli, where the second stimulus is strong and biologically significant, e.g., food or a noxious stimulus. The first stimulus is referred to as the conditioned stimulus (CS), and the second stimulus as the unconditioned stimulus (US). It was thought that repeated pairings of the CS and US were necessary for conditioning to emerge; however, many conditioned responses (CRs) can be learned with a single trial as in fear conditioning and taste aversion learning (Rescorla, 1988). Fear conditioning, for example, using a tone paired with a shock, is the most commonly used model. A behaviorally neutral CS is followed by a nociceptive US. After a few pairings, animals and humans react to the CS with autonomic (change in heart rate, interruption of respiratory rhythm, increase in blood pressure) as well as somatic (e.g., freezing) fear-related CRs.

2.2.3 Instrumental or Operant Conditioning

In instrumental conditioning, the presentation of a sensory stimulus is contingent upon a behavioral response. For example, an animal might be required to press a bar (response) to receive food (stimulus), but the stimulus may be any sensory event, not merely food or water. In general, most behavior alters the sensory environment, placing subjects in a feedback loop with their environment. Instrumental conditioning can occur after sensory deafferentation (e.g., amputation of a digit, lesion in the retina, high-frequency hearing loss), when a subject's attempts to behaviorally compensate for its sensory deficit produce new relationships between behavior and its sensory responses. Converging evidence, obtained using techniques from auditory neurophysiology and the neurobiology of learning and memory, supports the idea that the primary auditory cortex acquires and retains specific memory traces about the behavioral significance of selected sounds. Stimulating the cholinergic system of the nucleus basalis in the forebrain or the ventral tegmental nucleus in the brainstem is sufficient to induce both specific memory traces and specific behavioral memory (Weinberger, 2004).

2.2.4 Receptive Field and Tonotopic Map Plasticity in Auditory Cortex

Recall that the tonotopically organized auditory pathway terminates in four or five tonotopically organized cortical areas (Fig. 2.1), of which the primary area A1 is the most studied. From auditory cortex, efferent fibers project to nearly all subcortical auditory areas and also to the amygdala and striatum.

Thalamic processes probably cannot fully account for plasticity in the auditory cortex during learning. The ventral medial geniculate body (MGBv) in the auditory thalamus is the lemniscal source of frequency-specific input to granular layers of the auditory cortex, but develops no plasticity to the CS during training (reviewed by Weinberger and Diamond, 1987) and only a very weak and highly transient RF plasticity after conditioning (Edeline, 1999). The magnocellular medial geniculate body (MGBm) provides non-lemniscal input to upper layers of the auditory cortex. MGBm cells do develop increased responses to the CS during training, and their RFs are retuned to favor the CS frequency (Edeline and Weinberger, 1992). However, their RFs are much more complex and broadly tuned than those of auditory cortical cells, so it seems unlikely that the highly frequency-specific cortical RF plasticity is simply projected from this nucleus, although this cannot yet be discounted.

The "receptive field–conditioning" approach was first used in the mid-1980s in two nonprimary auditory fields and ventral ectosylvian cortex, because it was assumed that A1 would be less plastic, based on dominant beliefs in auditory physiology at that time. The first such study involved classical fear conditioning in the adult guinea pig (Bakin and Weinberger, 1990). Subsequent studies revealed that auditory RF plasticity developed within five trials, was discriminative, exhibited consolidation, and lasted at least up to 8 weeks posttraining (Weinberger, 1998). Tuning shifts were generally assessed at stimulus levels used for training, i.e., well above threshold (e.g., 70 dB SPL), although even when trained at one level, they could develop across the range of 10–80 dB SPL. Tuning shifts to the CS frequency have also been found with rewarding brain stimulation as the US (Kisley and Gerstein, 2001). Auditory RF plasticity was highly frequency-specific, consistently exhibiting increased responses only at or near the CS frequency across subjects, with decreased responses to lower and higher frequencies (Fig. 2.2). CS-specific increased responses and tuning shifts were associative, as they required CS and US pairing. Habituation produces the opposite effect (Fig. 2.2), i.e., a specific decrease in response to the repeated stimulus, with little or no change in response to other frequencies (Condon and Weinberger, 1991).

Changes in spectral and temporal response properties of cortical neurons underlie many forms of use-dependent learning. The scale and direction of these changes appear to be determined by specific features of the behavioral tasks that evoke cortical plasticity. Extensive training in a frequency discrimination task leads to better discrimination ability and to an expansion of the cortical area responsive to the frequencies used during training (Recanzone et al., 1993) but see contrasting findings in the work of Brown et al. (2004). Furthermore, a study by Talwar and Gerstein (2001) shows that the parallels between training-related and artificially induced cortical plasticity may only be superficial. If the expansion of the cortical area is indeed responsible for the improved performance, then it should be possible to achieve improved performance through intracortical microstimulation (ICMS) instead of training.

Talwar and Gerstein's results, however, show that this is clearly not the case: Training and ICMS may both induce "representational plasticity," but while training improves performance, ICMS does not. Compare these results with the effects of frequency discrimination training reported by Recanzone et al. (1993). In these studies, owl monkeys were trained to discriminate a target frequency from slightly different frequencies. For correct target detection, the animal received a food reward, but for an incorrect response (a "false alarm") the animal was "punished" with a timeout—testing was suspended for a short while and the animal had to wait before it would be given the opportunity to try for another reward. Recanzone et al. (1993) also observed a sharpening of frequency tuning. Talwar and Gerstein's experiments may have led to a broadening of frequency tuning in neurons near the microstimulation site, so in this important respect the effect of ICMS may well be the opposite of that produced by training in a frequency discrimination task.

Recent findings by Pienkowski and Eggermont (2009) showed that tonotopic map changes based on cortical multi- or single-unit recordings are similar to those based on evoked local field potentials (LFPs) in primary auditory cortex, which reflect the output of the MGBv. LFPs from primary auditory cortex are, in essence, the intracortically recorded equivalent of the scalp recorded middle latency response. The LFP constitutes a collective property of a neuronal ensemble, i.e., is a measure of synchronous postsynaptic potential activity of a population of neurons. This activity may give rise to the spiking output of the cortical cells (Eggermont et al., 2011). In the unexposed control cat (Fig. 2.3, top), the tonotopic organization of AI is clear from both the SU and the LFP data. With little exception, lower-frequency octaves are represented progressively more posteriorly. The LFP data recorded after 6–12 weeks recovery from a passive 68 dB SPL exposure to an enhanced acoustic environment (EAE), in the range of 4–20 kHz, show dramatic changes in the tonotopic map, which again parallel the SU data (Fig. 2.3, bottom), suggesting that the EAE-induced changes are also present at the thalamic level. It is known that tonotopic map changes in adult cats that result from cochlear lesions are quite similar in the thalamus and primary cortex (Kamke et al., 2003; Rajan et al., 1993), but far less pronounced in the inferior colliculus (Irvine et al., 2003). The EAE-induced changes observed in the LFP data may thus originate in the thalamus, though it is also possible that they in turn are caused by corticofugal activity (Zhang and Suga, 2000; Zhang and Yan, 2008).

2.2.5 Environmental Enrichment

The standard definition of an enriched environment is "a combination of complex inanimate and social stimulation." This definition implies that the relevance of single contributing factors cannot be easily isolated but there are good reasons to assume that it is the interaction of factors that is an essential element of an enriched environment, not any single element that is hidden in

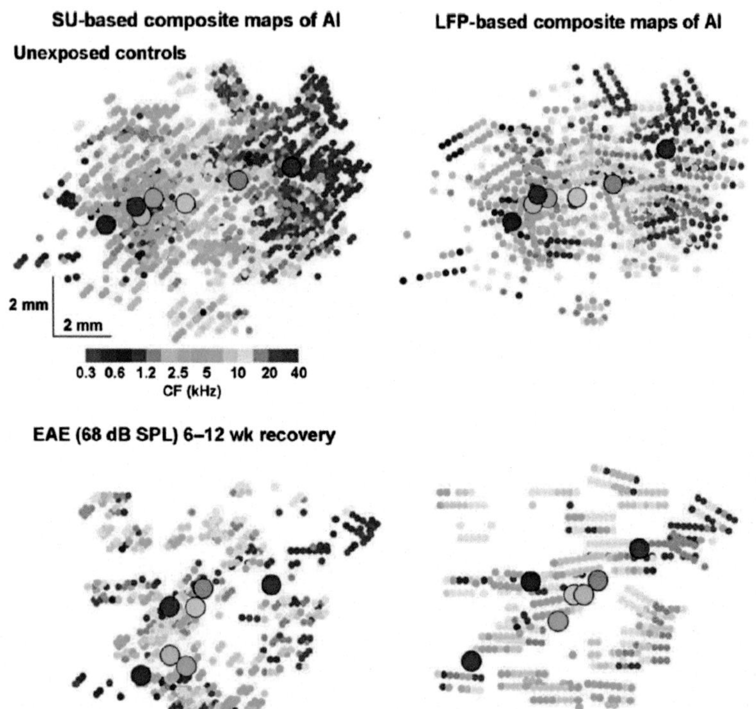

FIGURE 2.3 Composite CF maps of AI from unexposed controls (upper panels) and EAE-exposed animals with 6–12 weeks recovery in quiet (bottom panels), derived from SU (left-most panels) and LFP recordings (right-most panels). The individual cat maps were aligned on the position of the high-frequency (anterior) boundary of AI. The average 2D positions of each of the seven octaves from 312.5 Hz to 40 kHz are indicated by the large colored circles. *Reprinted from Pienkowski, M., Eggermont, J.J., 2009. Long-term, partially-reversible reorganization of frequency tuning in mature cat primary auditory cortex can be induced by passive exposure to moderate-level sounds. Hear. Res. 257, 24–40, with permission from Elsevier.*

the complexity. This may play a role in the NICU (see chapter: Early Diagnosis and Prevention of Hearing Loss). It is well known that adult rats that are handled briefly (15 min each day) for the first 3 weeks of life show markedly reduced responses to various stressful stimuli and better cognitive performance (van Praag et al., 2000). One of the most remarkable features of enrichment studies is that the changes in the brain can be detected even when the enriched experience, typically including tactile, visual, and auditory stimuli, is provided to an adult or aged animal. This finding underscores the possibility that experimental enrichment is a reversal of the impoverishment generally found in the laboratory setting (animals housed in small cages) rather than enrichment over a natural setting (van Praag et al., 2000).

In a study by Engineer et al. (2004) rats were housed in enriched or standard conditions. The standard condition consisted of one or two rats housed in hanging cages within an animal colony room. The enriched condition consisted of four to eight rats housed in a room with devices that generated different sounds when rats crossed a motion detector path, stepped on weight sensors, or passed through hanging bars. In addition, each rotation of the running wheel triggered a brief tone and light flash, and a CD player played 74 sounds, including tones, noise bursts, musical sequences, and other complex sounds, in random order. Some of these sounds were associated with delivery of a sugar reward. While the sounds in this enriched environment were more diverse and behaviorally relevant, rats in standard laboratory conditions heard approximately the same number of sounds each day. Evoked potentials from awake rats and extracellular spike recordings from anesthetized rats showed that enrichment dramatically increased the strength of auditory cortex responses. Cortical firing rates of both young and adult animals increased from exposure to an enriched environment and were reduced by exposure to an impoverished environment. Housing condition resulted in rapid remodeling of cortical responses in 2 weeks. Recordings made under anesthesia indicated that enrichment increased the number of neurons activated by any sound. This finding shows that the evoked potential plasticity documented in awake rats was not due to differences in behavioral state. Finally, enrichment made primary auditory cortex neurons more sensitive to low-level sounds and more frequency selective. It is interesting to contrast this with the effects of passively increasing or decreasing auditory inputs in humans (the experiments by Formby et al., 2003, discussed in Section 2.1). Here more sound exposure reduced the activity of the auditory brain, whereas less sound exposure increased the central gain. It is thus likely that the very different results obtained in an enriched acoustic environment rely on multimodal interaction (see chapter: Multisensory Processing). An alternative explanation is that "active" enrichment sensitizes the auditory system, whereas "passive" enrichment habituates it.

The general sensory environment has a role in "shaping" the organization of and processing in the auditory system as we have seen above. The potential for representational plasticity—alterations in central maps following long-lasting changes in peripheral input—is likely higher during the developmental period but it is nevertheless still present in adults. As described in the previous section, Pienkowski and Eggermont (2009) found that behaviorally irrelevant stimuli also induce plasticity potential via an initial process of fast habituation followed by extended changes in thalamic and/or intracortical inhibition (see also chapter: Causes of Acquired Hearing Loss).

2.3 PERCEPTUAL LEARNING

2.3.1 Bottom–Up Learning

Now we move to human learning. Perceptual learning can be defined as practice-induced (training) improvement in the ability to perform specific perceptual tasks. Perceptual learning has traditionally been portrayed as a bottom–up phenomenon that improves encoding or decoding of the trained stimulus by inducing changes in those processes that act as bottlenecks to performance (Dosher and Lu, 2005). Amitay et al. (2006) phrased it as: "Practice may not 'make perfect', but it can certainly improve skills and abilities, including the ability to detect or discriminate a variety of sensory stimuli."

Amitay et al. (2013) recently demonstrated that variations in the internal representation of identical input stimuli (1 kHz tones) may drive the decision process in an odd-one-out task. They found that learning occurred on the same "odd-one-out" frequency discrimination task in which the tones were physically identical (an impossible task; Amitay et al., 2006) and was no different in magnitude than learning produced by either adaptive training or training with a constant but very small frequency difference. This demonstrated that discrimination learning can occur in the absence of an external signal on which to base a discrimination decision, suggesting that these decisions are made based on variations produced by internal noise in the central nervous system (Amitay et al., 2014).

2.3.2 Top–Down Learning

Top–down learning involves attention, adapting to a new acoustic environment as a result of hearing loss, hearing aids, or CIs, and also improves speech understanding in noisy environments.

In contrast to a bottom–up process learning, the Reverse Hierarchy Theory (RHT) asserts that learning is a top–down guided process, which begins at high-level areas of the sensory system, and when these do not suffice, progresses backward to the input levels, which have a better signal-to-noise ratio (Ahissar and Hochstein, 2004). Bernstein et al. (2013) phrased it as: "According to RHT, perceptual learning is the access to and remapping of lower-level input representations to higher-level representations. To carry out the remapping, perceptual learning involves 'perception with scrutiny.' That is, a backward (corticofugal) search must be initiated to access the representational level of the information needed to carry out the perceptual task". Ahissar et al. (2009) noted that, for an auditory RHT, if perceptual learning is possible with a more complex stimulus, then specific benefits could be seen at both lower and higher level representations. Lower levels of cortical processing would be sufficient for improvements with specific frequency bands and temporal resolution. For instance, acoustic differences such as fine spectrotemporal acoustic features may be encoded at low levels of the auditory hierarchy, but not at its higher levels, which integrate across time and frequency, and form more abstract, spectrotemporally broader, categories (Nahum et al., 2008).

2.3.3 Extending the Reverse Hierarchy Theory

Behavioral studies indicate that synchronous presentation of audiovisual (AV) stimuli facilitates speech recognition in comparison to audio alone (Ross et al., 2007; Seitz et al., 2006; Shams and Seitz, 2008; Sumby and Pollack, 1954), and that the more reliable modality has the greater influence on the eventual perception (Alais and Burr, 2004). For example, Grahn et al. (2011) reported that an internal auditory rhythm representation may be triggered during visual rhythm perception, though not vice versa.

Proulx et al. (2014) expanded unisensory RHT learning to multisensory learning. They described it in Fig. 2.4 as follows: "Unisensory learning is shown as being modality specific, such that an auditory task (green) is supported by either low-level auditory areas for specific learning or high-level auditory areas for general learning (Ahissar, 2001). A visual task (red) exhibits similar activity in visual areas (Ahissar and Hochstein, 2004), again with low- and high-level areas defined in afferent terms, with low-level corresponding to primary sensory areas and high-level to association cortex." In Fig. 2.4B, it is

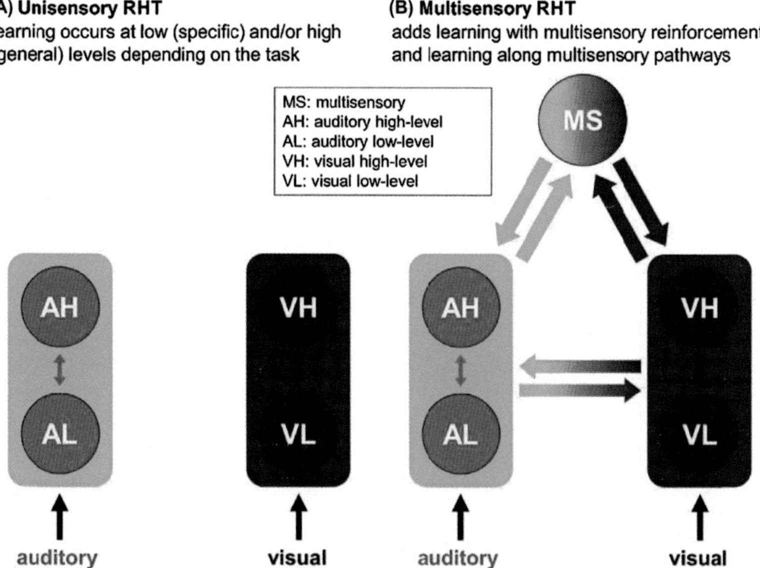

FIGURE 2.4 Figure depicting a unisensory and a multisensory RHT of perceptual learning. (A) Unisensory learning is shown as being modality specific, such that an auditory task (green) is supported by either low-level auditory areas for specific learning or high-level auditory areas for general learning. (B) Multisensory learning is shown to represent the same possible mechanisms for two different conditions: First, learning under multisensory stimulation can lead to correlated activity in higher-level multisensory areas; second, learning can progress from primary sensory areas to higher-level multisensory areas under complex unisensory stimulation. *Reprinted from Proulx, M.J., Brown, D.J., Pasqualotto, A., Meijer, P., 2014. Multisensory perceptual learning and sensory substitution. Neurosci. Biobehav. Rev. 41, 16–25, with permission from Elsevier.*

shown that learning under multisensory stimulation can lead to correlated neural activity in higher-level multisensory areas (Shams and Seitz, 2008). In addition, learning may progress from primary sensory areas to higher-level multisensory areas under complex unisensory stimulation. Feedback of this activity down the hierarchy then causes generalization across stimulus modalities when these higher-level, multisensory areas are implicated in learning either unisensory or multisensory tasks (Proulx et al., 2014).

Proulx et al. (2014) concluded that multisensory learning is more ecologically relevant than unisensory perceptual learning because human cognition evolved in a multisensory environment and our daily interaction with the external world is also multisensory.

2.4 AUDITORY TRAINING

2.4.1 Adults

Following chapter 7 from Eggermont (2014): Learning perceptual skills is characterized by rapid improvements in performance within the first hour of training (fast perceptual learning) followed by more gradual improvements that take place over several daily practice sessions (slow perceptual learning). Tremblay et al. (1998) were the first to demonstrate that training-associated changes in AEP activity, i.e., notably the P_2 component (cf. Fig. 8.1), occur at a preattentive level and may precede behavioral learning. This was demonstrated using the MMN, which increased with each testing day. Tremblay et al. (1998) results suggested that auditory training alters the neural activity that provides the necessary coding for speech-sound learning, that changes in neural activity occur rapidly during training, and that these changes are later integrated into functional behavior.

What are the neural substrates underlying these improvements in learning perceptual skills? Alain et al. (2007) recorded event-related potentials (ERPs) from the brain while listeners were presented with two phonetically different vowels. Listeners' ability to identify both vowels improved gradually during the first hour of testing and was paralleled by enhancements in an early evoked response component (Ta, latency $\sim$130 ms) localized in the right auditory cortex and a late evoked response component (T350, latency $\sim$350 ms) localized in the right anterior superior temporal gyrus and/or inferior prefrontal cortex (Fig. 2.5). To test whether or not the changes in ERP amplitude over the right temporal lobe (Ta and T350 components) were affected by prior task experience, ERPs recorded from the same participants were compared to those recorded 1 week later. Half of the participants received four 35-min daily practice sessions on the double-vowel task between the two ERP recording sessions, whereas the other half received no practice. For the trained group, as in the findings of Tremblay et al. (1998) there was a marked increase in P_2 amplitude after extended training.

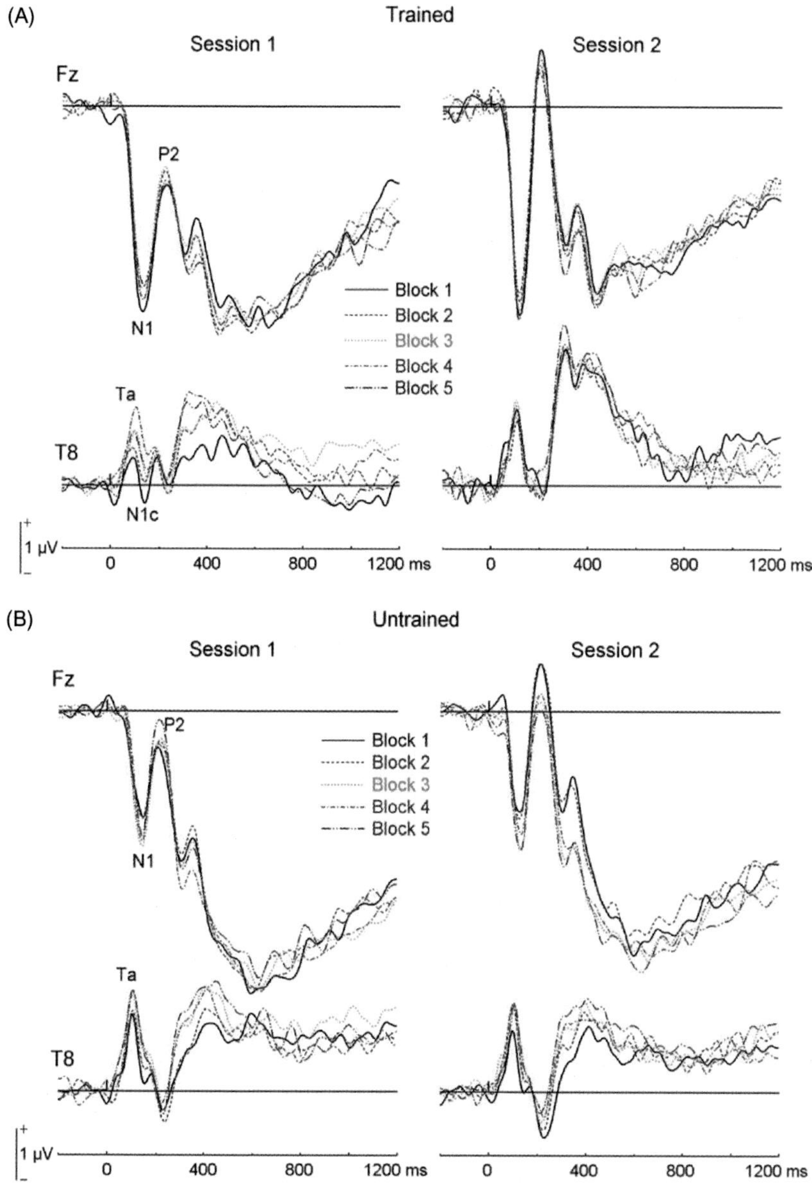

FIGURE 2.5 ERPs illustrate rapid learning. (A) Group mean ERPs recorded during the first and second ERP sessions in the trained group as a function of blocks of trials. (B) Group mean ERPs recorded during the first and second ERP sessions in the untrained group as a function of blocks of trials. Fz, midline frontal; T8, right temporal. *Reprinted from Alain, C., Snyder, J.S., He, Y., Reinke, K.S., 2007. Changes in auditory cortex parallel rapid perceptual learning. Cereb. Cortex 17, 1074–1084, by permission of Oxford University Press.*

However, for this trained group, there was no significant change in the response recorded from a right temporal electrode (T8) as a function of block (order within a session) for the 100–140 ms interval and around 350 ms after training (Session 2). The rapid enhancement in cortical responsiveness in the 100–140 ms poststimulus onset interval is thus modulated by prior experience, being present only in individuals who did not have the opportunity to practice the task in the preceding days—i.e., were not trained. These neuroplastic changes depended on listeners' attention and were preserved only if practice was continued. Familiarity with the task structure (procedural learning) was not sufficient. Alain et al. (2007) thus showed that the neuroplastic changes occurred rapidly within sessions, demonstrating the adaptability of human speech segregation mechanisms.

In a subsequent study Tremblay et al. (2010) found that enhanced P_2 activity reflected sound recognition that built over time as part of the learning experience. However, despite observing enhanced neural response patterns, especially when sound was paired with a listening task, enhanced P_2 amplitudes did not coincide with measurable improvement in perception. Tremblay et al. (2010) speculated that: "Repeated stimulus exposure primes the auditory system in a general way that is not stimulus specific and can be recalled following a long period of time. In contrast to exposure, training exercises shape the system such that the acoustic distinctions that make specific sounds relevant are reinforced, and perceptual gains can be made."

Is there a difference of training between activity in primary (A1) and secondary (A2) auditory cortex? This has been assessed by comparing effects in ERPs specific to these areas, such as the auditory steady-state response (ASSR) originating in A1 on Heschl's gyrus and N_1–P_2–N_2 generated by neurons in A2. Gander et al. (2010) studied the effect of auditory training on the 40 Hz ASSR. Repeated exposure to this stimulus advanced the phase of the ASSR (shortened the time delay between the 40 Hz response and stimulus waveforms). The phase advance appeared at the outset of the second of two sessions separated by 24–72 h, did not require active training, and was not accompanied by changes in ASSR amplitude over this time interval. Training for 10 sessions revealed further advances in ASSR phase and also an increase in ASSR amplitude, but the amplitude effect lagged that on phase and did not correlate with perceptual performance while the phase advance did. A control group trained for a single session showed a phase advance but no amplitude enhancement when tested 6 weeks later. In both experiments attention to auditory signals increased ASSR amplitude but had no effect on ASSR phase. This suggests a persistent form of neural plasticity expressed in the phase of ASSRs generated from the region of A1, which occurs either in A1 or in subcortical nuclei projecting to this region. This could represent a form of sensitization.

Gander et al. (2010) also showed the effects of attention on AEPs that are generated in A2 (Fig. 2.6). On day 1 the amplitude of N_1, P_2, and N_2

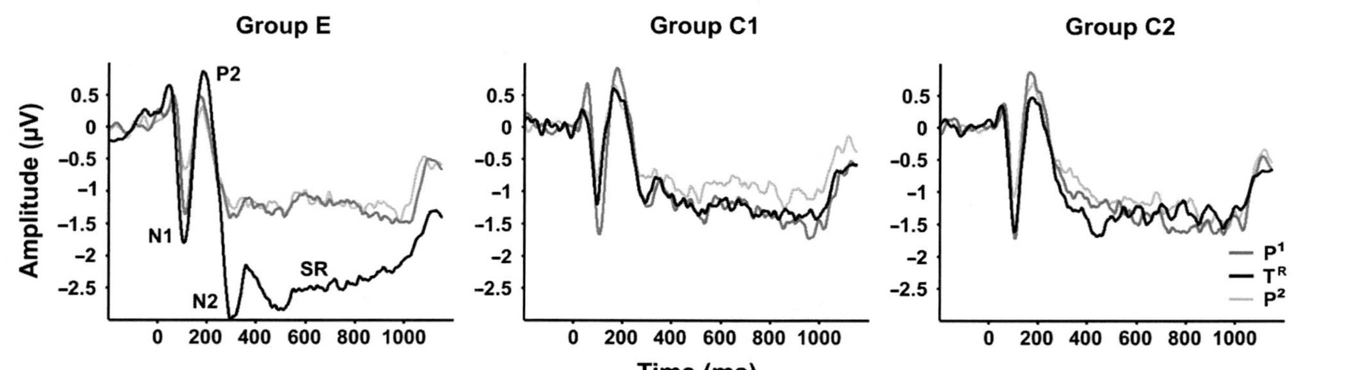

FIGURE 2.6 Fast modulation of transient responses by attention. N1, P2, N2, and SRs are identified in the left panel. Auditory attention was required in the T^R (task) stage (black lines) for the experimental group (E); P^1 (dark gray lines) and P^2 (light gray lines) are passive listening conditions. C1 and C2 are two control groups. *Reprinted from Gander, P.E., Bosnyak, D.J., Roberts, L.E., 2010. Acoustic experience but not attention modifies neural population phase expressed in human primary auditory cortex. Hear. Res. 269, 81–94, with permission from Elsevier.*

components and the auditory sustained response (SR) increased significantly from the passive stage (dark gray line, P^1) to the task (paying attention) stage (black line, T^R) in the training group (E) and returned to initial levels in the second passive stage (light gray line, P^2). In contrast, these responses tended to decrease over stages in the two control groups (C1 and C2), but only reaching significance for N_1. In addition, P_2 amplitude (but no other response) increased significantly between days 1 and 2 in Groups E and C1 (data not shown) with no prior evidence for an increase within day 1 and no effect of stage or group. Summarizing, effects of acoustic experience were detected in both A2 (P_2 amplitude) and A1 (ASSR phase) at the outset of the second session following an interval of 24–72 h, regardless of the conditions of task attention.

Auditory brain plasticity is typically considered to be mainly occurring in the thalamocortical circuit. As we have seen (Fig. 2.1) this circuit influences subcortical auditory structures via its efferent connections. In Eggermont (2012), I wrote "The investigation into the role of descending connections from cortex to the ICC of the big brown bat (Yan and Suga, 1998) was the start of a large series of investigations into cortical control of the auditory system. They found that electrical stimulation of cortex paired with a tone, with a frequency equal to the BF of the stimulation site, altered the frequency map in the ICC such that it enhanced the extent of the paired-tone frequency representation. In exploring the corticofugal effects to levels below the midbrain Luo et al. (2008) found that cortical activation increased the response magnitudes and shortened response latencies of CN neurons with BFs matched to the cortical stimulation site, whereas it decreased response magnitudes and lengthened response latencies of unmatched CN neurons. In addition, cortical activation shifted the frequency tunings of unmatched CN neurons toward those of the activated cortical neurons." Obviously, the auditory cortex apparently implements a long-range feedback mechanism to select or filter incoming signals from the ear. Yan and Suga (1998) suggested that the corticofugal system is involved in the long-term improvement and adjustment of subcortical auditory information processing. This suggests that subcortical structures will also show plastic changes.

Evidence for subcortical plasticity following pitch discrimination training was found in changes of the frequency-following response (FFR), a brainstem-generated evoked potential (Carcagno and Plack, 2011). Twenty-seven adult listeners were trained for 10 h on a pitch discrimination task using one of three different complex tone stimuli. One had a static pitch contour, one had a rising pitch contour, and one had a falling pitch contour. Trained participants showed significant improvements in pitch discrimination compared to the control group for all three trained stimuli. Also, the robustness of FFR neural phase locking to the sound envelope increased significantly more in trained participants compared to the control group for the static and rising contour, but not for the falling contour. Changes in FFR strength were partly specific for stimuli with the same pitch modulation (dynamic vs static) of the trained stimulus. Changes in FFR strength, however, were not specific for stimuli with the same pitch trajectory (rising vs falling) as the trained stimulus. These findings indicate that even relatively low-level processes in the mature auditory system are subject to experience-related change. Whether these changes are intrinsic to the brainstem or are the result of modulation by corticofugal projections (Luo et al., 2008) remains to be clarified.

2.4.2 Effects of Passive Exposure

Ross and Tremblay (2009) studied the effect of sound exposure on N_1 and P_2 responses during two experimental sessions on different days with young, middle-aged, and older participants who passively listened to speech—two versions of a /ba/ syllable—and a noise sound of the same duration. N_1 and P_2 are functionally distinct responses with P_2 sources located more anterior than N_1 in auditory cortices. P_2 also matures much faster than N_1 and likely is driven by the reticular activating system (Ponton et al., 2000, 2002). N_1 amplitudes decreased continuously during each recording session, but completely recovered between sessions. In contrast, P_2 amplitudes were fairly constant within a session but increased from the first to the second day of MEG recording. Whereas the modest N_1 amplitude decrease was independent of age, the larger amount of P_2 amplitude increase diminished with age (Fig. 2.7). Temporal dynamics of N_1 and P_2 amplitudes were interpreted as reflecting neuroplastic changes along different timescales (see also Tremblay et al., 2010, 2014).

Ross and Tremblay (2009) found that passive sound exposure resulted in a continuous decline in N_1 amplitude within experimental sessions followed by a recovery between sessions, as expected from short-term adaptation (Eggermont, 2015). In contrast, P_2 amplitude was relatively constant within a session but increased between first and second sessions taking place on different days. The increase of P_2 following passive sound exposure continued to last for several days. This emphasizes the importance of the P_2 response for studying effects of experience, memory, and learning (Tremblay and Kraus, 2002). Neuroplastic modifications of P_2 occurred in all age groups, indicating that brain functions are malleable throughout the lifespan. Because the effect size decreased with increasing age this suggests a reduced capacity for plastic reorganization in later life. Recently, P_2 was suggested as a biomarker for learning (Tremblay et al., 2014).

2.4.3 Auditory Training in Cochlear Implant Patients

Acclimatization is the notion that systematic changes in auditory performance take place as a result of changes in acoustic information available to

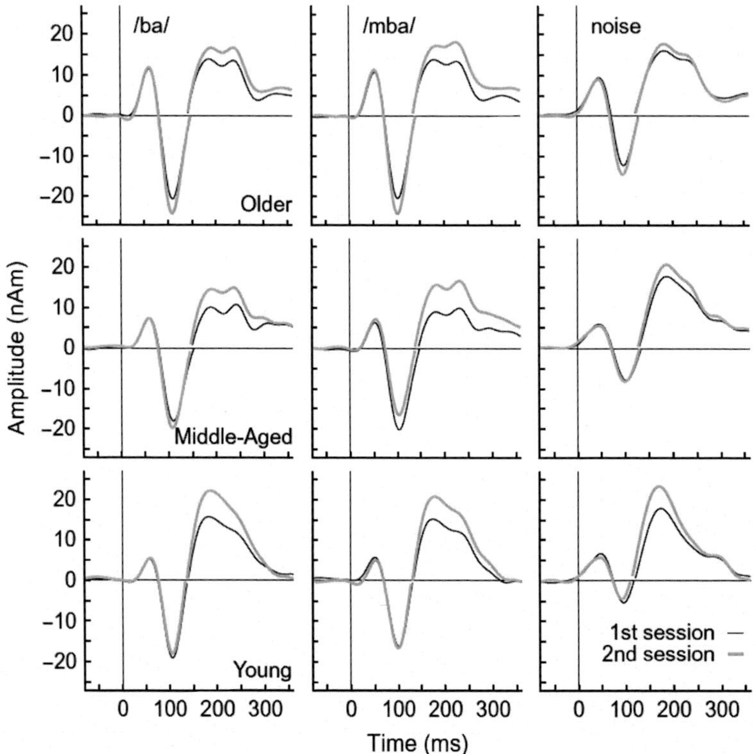

FIGURE 2.7 Changes in the response waveforms between first and second session. The waveforms were averaged across hemispheres and shown separately for age groups and stimulus types. The first part of the waveforms until the zero crossing between N_1 and P_2 wave was calculated based on the N_1 source model and the later part based on the P_2 source model. *Reprinted from Ross, B., Tremblay, K., 2009. Stimulus experience modifies auditory neuromagnetic responses in young and older listeners. Hear. Res. 248, 48–59, with permission from Elsevier.*

the listener, which is basically a bottom–up mechanism (see chapter: Hearing Aids, for a discussion of its role in hearing aids). There is support in favor of acclimatization among implant users. However, longitudinal studies suggest it can take up to 30–40 months following cochlear implantation for some individuals to make maximal use of their hearing prosthesis (reviewed by Tremblay and Moore, 2012). We have shown with cortical auditory evoked potential (CAEP) components P_1 and N_1 that the latency of P_1 depends on the "time in sound" (Ponton et al., 1996). We found that when CI users, regardless of their chronological age, wore their implant for 1 year (1-year time in sound), P_1 latencies resembled those of an NH 1-year-old infant. This finding suggests the development of the planum temporale, which generates the CAEP is arrested following deafness (Kral and Eggermont, 2007). When the CI is activated the maturation starts up again and continues from where it was at the time of onset of deafness. However, the longitudinal data of Ponton and Eggermont (2001) showed that a 3-year deprivation of sound under the age of 6 appears to be too long to develop the appropriate CAEPs even in young adulthood. Functional development in such cases may need top–down learning as facilitated by training programs.

Fu and Shannon (1999) demonstrated that speech performance of CI users (see chapter: Cochlear Implants) was highly dependent on a well-matched (relative to normal) frequency-to-electrode mapping. They found that a tonotopic shift in the frequency-to-electrode map resulted in immediate and significant drops in speech performance. In contrast, Rosen et al. (1999) observed a rapid accommodation by NH subjects listening to frequency-shifted speech via a 4-band noise vocoder. Fu et al. (2002) noted that the effects of learning also in experienced implant users who had to adapt to new stimulation patterns provided, i.e., by updated speech processors and speech processing strategies (see chapter: Cochlear Implants). Most of these subjects showed rapid improvement during the first 3–6 months, but little beyond 6 months (Shallop and McGinn-Brunelli, 1995; Dorman and Loizou, 1997; Pelizzone et al., 1999). In general, the learning patterns were similar for naive listeners adapting to the new sensory patterns of a CI or experienced users adapting to changes in the speech processor.

To assess whether moderate speech training can accelerate this learning process, Fu et al. (2005a) trained NH listeners with spectrally shifted speech via an eight-channel acoustic simulation of CI speech processing. Baseline vowel and consonant recognition was measured for both spectrally shifted and unshifted speech. Short daily training sessions were conducted over five consecutive days, using four different protocols. For the test-only protocol, Fu et al. (2005a) found no improvement over the 5-day period. Sentence training also provided little benefit for vowel recognition. "However, after five days of targeted phoneme training, subjects' recognition of spectrally shifted vowels significantly improved in most subjects. This improvement did not generalize to the spectrally unshifted vowel and consonant tokens, but improved with targeted vowel contrast training" (Fu et al., 2005a). In a parallel study, Fu et al. (2005b) evaluated the effect of targeted phonemic contrast on English speech recognition by 10 adult experienced CI users with limited speech recognition capabilities. Baseline (pretraining) multi-talker phoneme recognition performance was measured for at least 2 weeks or until performance asymptoted. After baseline measures were complete, participants trained at home using a program loaded onto their personal computers or loaner laptops. Participants were instructed to train at home 1 h per day, 5 days per week, for a period of 1 month or longer. During the training, production-based contrasts were targeted (i.e., second formant differences and duration); auditory and visual feedback was provided, allowing participants to compare their (incorrect) response to the correct response. Participants returned to the laboratory every 2 weeks for retesting (same tests as baseline measures). This showed that both vowel and consonant recognition significantly improved for all participants after training. Although performance significantly improved for all participants after 4 weeks or more of moderate training, for some participants, performance significantly improved after only a few hours of training, whereas others required a much longer time course.

Fu and Galvin (2007) integrating the findings from Fu et al. (2005a,b) found that in CI speech processing, patient performance is most strongly influenced by parameters that affect spectral cues. Acutely measured speech perception performance typically worsened with changes in frequency-to-electrode mapping, relative to the frequency allocation with which participants have prior experience (Fu and Shannon, 1999; Friesen et al., 1999). Fu and Galvin (2007) conclude that: "Even with years of experience with the device, performance remains below that of NH listeners, especially for difficult listening conditions (e.g., speech in noise) and listening tasks that require perception of spectrotemporal fine structure cues (e.g., music appreciation, and vocal emotion perception). Passive adaptation with current CI technology is clearly not enough. It is unclear whether 'active' auditory training may improve the degree of adaptation, or at least accelerate adaptation."

2.4.4 Auditory Learning in Children

Halliday et al. (2008) examined the effects of age on auditory learning, by giving 6- to 11-year-old children and adults approximately 1 h of training on a frequency discrimination task. They found that children on average had poorer frequency discrimination skills compared with adults at the outset of training, although performance improved with age. They showed that it was possible to induce auditory learning in children, even in those as young as 6 years of age. Halliday (2014) examined discrepancies in two studies (Moore et al., 2005; Halliday et al., 2012) that used the same training paradigm in mainstream children but came to opposite conclusions. She concluded that: "Differences in the randomization, blinding, experimenter familiarity and treatment of trained and control groups contributed to the different outcomes of the two studies. The results indicate that a plethora of factors can contribute to training effects and highlight the importance of well-designed randomized controlled trials in assessing the efficacy of a given intervention."

It is well known that individuals with NH experience a perceptual advantage for speech recognition in interrupted noise (Miller and Licklider, 1950) compared to continuous noise. In contrast, adults with hearing impairment (HI) and younger children with NH receive a minimal benefit (Festen and Plomp, 1990). To assess whether training HI children could alleviate this deficit, Sullivan et al. (2013) obtained speech scores in interrupted and continuous noise from pre-, post-, and 3 months posttraining from 24 children with moderate-to-severe hearing loss. The average improvement in SNR at

3 months posttraining was approximately 7.46 dB in interrupted noise and 7.13 dB in continuous noise for children who trained in interrupted noise for 7 h. Because a 1 dB improvement on the hearing-in-noise test (HINT) is equivalent to an 8.9% improvement in speech intelligibility (Nilsson et al., 1994), this corresponds to an approximate 63% improvement in word recognition based on the predicted 8.9% enhancement in speech intelligibility for every 1 dB change on the HINT.

Children who are diagnosed with auditory processing disorder (APD; see chapter: Hearing Problems) present an array of different symptoms that terms include auditory perception problems, auditory comprehension deficits, central auditory dysfunction, central deafness, and "word deafness." According to the National Institutes of Health, children with APD are described as being unable to recognize subtle differences between sounds in words, even though the sounds themselves are loud and clear. There is no known peripheral pathology to account for these difficulties, and no known treatment (Tremblay and Moore, 2012). The underlying auditory temporal processing hypothesis as underlying APD has been challenged and multimodal and cognitive factors including attention, motivation, and intelligence have been advanced as more likely causes (Moore et al., 2010). This has been extensively discussed in my book on "Auditory Temporal Processing and its Disorders" (Eggermont, 2015), where I came to the conclusion that a combination of developmental delay in temporal processing gives rise to cognitive processing deficits that persist following the maturation of the temporal processing mechanisms.

Recently, Loo et al. (2016) showed that: "A 12-week-long 5-day/week training with speech stimuli ranging from single words to complex sentences in the presence of competing stimuli under different conditions of spatial separation (thus resembling real-life listening conditions) led to improved speech-in-noise perception in tests that was reflected in improved functional listening in children with APD."

2.5 AV TRAINING

As we stated and elaborate upon in Chapter 3, Multisensory Processing postlingually deaf subjects learn the meaning of sounds after cochlear implantation by forming new associations between sounds and their sources (for a comprehensive review, see Butler and Lomber, 2013). Because CIs generate only coarse frequency responses, this imposes a dependency on visual cues, e.g., lip reading.

Bernstein et al. (2013) investigated how AV training might benefit or impede auditory perceptual learning of speech degraded by vocoding. The paired-associates (PA) AV training of one group of participants was compared with PA audio-only (A_O) training of another group. They found that PA training with A_O stimuli was reliably more effective than training with

AV stimuli. Bernstein et al. (2013) proposed that early AV speech integration can potentially impede auditory perceptual learning; but visual top—down access to relevant auditory features can promote auditory perceptual learning. Bernstein et al. (2013) noted that: "Vocoding has removed or distorted the basic auditory information that is typically mapped to phonetic features of natural speech. The phonetic feature level is inadequate to specify the phoneme category. But the visual speech information provides the needed phonetic information (Summerfield, 1987), the information is integrated, and the perceptual task is carried out at an immediate high-level of perception, as predicted by the reverse hierarchy theory." Bernstein et al. (2013) suggested that: "Top-down processing from visual speech representations may guide access to distinctive auditory features that can be remapped to phonetic features for novel speech transformations." Bernstein et al. (2013) found their results particularly relevant to training young cochlear implanted children who have not yet learned to read. They contrasted the need for multisensory information in children (see chapter: Multisensory Processing) with the use of orthographic representations or clear speech to guide perceptual learning literate NH adults (Davis et al., 2005; Hervais-Adelman et al., 2011).

2.6 MUSIC TRAINING

This section is partly based on chapter 9 from Eggermont (2014): Musical training has been related to language proficiency, spatial reasoning, and mathematical performance (Hannon and Trainor, 2007) and enhanced verbal memory (Chan et al., 1998). These results also showed that rhythm was significantly related to reading ability and, to a lesser degree, spelling ability. Anvari et al. (2002) tested 50 four-year olds and 50 five-year olds on a battery of musical and linguistic tests and showed that musical ability predicts early reading ability. Thus, music perception appears to engage auditory mechanisms related to reading that only partially overlap with those related to phonological awareness, suggesting that both linguistic and nonlinguistic general auditory mechanisms are involved in reading. To investigate the effect of musical experience on the neural representation of speech in noise, Parbery-Clark et al. (2009) compared subcortical neurophysiological responses to speech in quiet and noise in a group of highly trained musicians and nonmusician controls. Musicians, as measured by the brainstem generated FFR, were found to have a more robust subcortical representation of the acoustic stimulus in the presence of noise. Specifically, musicians demonstrated faster neural timing, enhanced representation of speech harmonics, and less degraded response morphology in noise. Neural measures were associated with better behavioral performance on the HINT for which musicians outperformed the nonmusician controls, suggesting that musical experience limits the negative effects of competing background noise. The same group (Strait et al., 2012) showed that the musicians' enhancement for the perception and neural encoding of speech in noise arises early in life, with more years of training relating with more robust speech processing in children aged 7–13 years. Musicians and nonmusicians did not differ on tests of visual working memory and attention. Herholz and Zatorre (2012) stated that: "Functional and structural changes due to musical experience take place at various stages of the auditory pathway, from the brainstem (Wong et al., 2007), to primary and surrounding auditory cortices (Bermudez et al., 2009; Schneider et al., 2002), to areas involved in higher-order auditory cognition (Lappe et al., 2011)."

These various approaches suggest that musical training has a beneficial impact on speech processing (e.g., hearing of speech in noise and prosody perception). Patel (2014) offered a conceptual framework based on the "expanded OPERA hypothesis" for understanding such effects based on mechanisms of neural plasticity. The OPERA hypothesis is that musical training drives adaptive plasticity in speech processing networks when five conditions are met. These are: (1) Anatomical *O*verlap in brain networks that process acoustic features used in both music and speech (e.g., waveform periodicity, amplitude envelope). (2) Music places higher demands on these shared networks than does speech, in terms of the *P*recision of processing. (3) Musical activities that engage this network elicit strong positive *E*motion, (4) Musical activities that engage this network are frequently *R*epeated. (5) Musical activities that engage this network are associated with focused *A*ttention. Under these conditions, and because speech shares these networks with music, neural plasticity drives the networks in question to function with higher precision than needed for ordinary speech communication. Patel et al. have started to apply these ideas to the training of nonmusical CI users.

But does music training in practice transfer to verbal skills? To assess this important aspect, Gordon et al. (2015) conducted a critical review of the literature built around key criteria needed to test the direct transfer hypothesis, including: (1) inclusion of music training versus control groups; (2) inclusion of pre- versus post-comparison measures, and (3) indication that reading instruction was held constant across groups. Thirteen studies, out of 178 reviewed at the abstract level, were identified (encompassing 901 subjects). They reported that: "Results supported the hypothesis that music training leads to gains in phonological awareness skills. The effect isolated by contrasting gains in music training vs. gains in control was small relative to the large variance in these skills ($d = 0.2$). Interestingly, analyses revealed that transfer effects for rhyming skills tended to grow stronger with increased hours of training. In contrast, no significant aggregate transfer effect emerged for reading fluency measures, despite some studies reporting large training effects."

2.7 TRAINING BY PLAYING ACTION VIDEO GAMES

Finally, I mention a potentially new tool that might be used in training certain aspects of hearing in difficult environments, namely, action video games. Immersion in video games is just another type of experiencing an enriched environment. It is different from perceptual learning, where the trainee typically learns the best template for the trained task, whereas action video game players (AVGPs) learn to find the best template as they are faced with new visual stimuli and new environments (Bavelier et al., 2010). As video gaming also involves rewards such as advancing to more difficult levels, it could result in a form of associative learning. Bavelier et al. (2010) stated that, if considered as a training paradigm, gaming differs from more standard methods on several dimensions. "(1) Gaming tends to be more varied in the skills it requires than standard training, which typically focuses on just one aspect of performance, as exemplified in the field of perceptual learning. (2) Unlike standard training paradigms, gaming is an activity that is highly engrossing and also extremely rewarding. Reward is likely to engage dopamine and possibly opiates and other neuromodulators. (3) Action games constantly require divided attention and its efficient reallocation as task demands change, most likely engaging neuromodulatory systems such as acetylcholine and dopamine, which are also known to enhance sensory processing and brain plasticity" (Bavelier et al., 2010). Importantly, several different aspects of attention are improved following action game play, including selective attention, divided attention, and sustained attention (Bavelier et al., 2012). It has also become clear that AVGPs "may not be endowed with better vision, better task-switching abilities, or better attentional tracking abilities per se, but instead they benefit from having the ability to quickly learn on the fly the diagnostic features of the tasks they face so as to readily excel on them" (Bejjanki et al., 2014).

2.8 SUMMARY

The brain—young as well as adult—has the capacity to change in response to changes in the external environment, to changes in the sensory input due to peripheral injury such as hearing loss, and as a result of training. Both adult and CP animals show plastic changes in auditory cortex following passive exposure to tonal or noise stimuli. A big issue is the so far largely unknown effect of such prolonged noise exposure in the NICU on the neonatal brain. Changes in a response to a stimulus that does not involve associating the presented stimulus with another stimulus or event such as reward or punishment occur via habituation and sensitization. An example is environmental enrichment where the changes in the brain can be detected even when the enriched experience, typically including tactile, visual, and auditory stimuli, is provided to an animal. Recent experiments with passive, moderate-level noise exposure demonstrated bottom–up driven cortical plasticity in both adult and juvenile animals. Perceptual learning, defined as practice-induced (training) improvement in the ability to perform specific perceptual tasks, has traditionally been portrayed as a bottom–up phenomenon that improves encoding or decoding of the trained stimulus by inducing changes in those processes that act as bottlenecks to performance. We also discuss the RHT that asserts that learning is a top–down guided process, which begins at high-level cortical areas of the sensory system. When this does not suffice, the process progresses backward to the input levels, which have a better signal-to-noise ratio. Auditory training alters the neural activity that codes for speech-sound learning. Its use in improving speech understanding in CI users is described, together with the specific problems of auditory training in children. AV training plays an important role in both. Music training presents another bridge to multisensory learning that may have benefits for CI users. Video gaming could be a new tool that might be used in training certain aspects of hearing in noisy environments, enhanced use of hearing aids and CIs.

REFERENCES

Ahissar, M., 2001. Perceptual training: a tool for both modifying the brain and exploring it. Proc. Natl. Acad. Sci. U. S. A. 98, 11842–11843.

Ahissar, M., Hochstein, S., 2004. The reverse hierarchy theory of visual perceptual learning. Trends Cogn. Sci. 8, 457–464.

Ahissar, M., Nahum, M., Nelken, I., Hochstein, S., 2009. Reverse hierarchies and sensory learning. Philos. Trans. R. Soc. Lond. B Biol. Sci. 364, 285–299.

Alain, C., Snyder, J.S., He, Y., Reinke, K.S., 2007. Changes in auditory cortex parallel rapid perceptual learning. Cereb. Cortex 17, 1074–1084.

Alais, D., Burr, D., 2004. The ventriloquist effect results from near-optimal bimodal integration. Curr. Biol. 14 (3), 257–262.

Amitay, S., Irwin, A., Moore, D.R., 2006. Discrimination learning induced by training with identical stimuli. Nat. Neurosci. 9, 1446–1448.

Amitay, S., Guiraud, J., Sohoglu, E., Zobay, O., Edmonds, B.A., Zhang, Y.X., et al., 2013. Human decision making based on variations in internal noise: an EEG study. PLoS One 8, e68928.

Amitay, S., Zhang, Y.-X., Jones, P.R., Moore, D.R., 2014. Perceptual training: top to bottom. Vision Res. 99, 69–77.

Anvari, S., Trainor, L.J., Woodside, J., Levy, B.A., 2002. Relations among musical skills, phonological processing, and early reading ability in preschool children. J. Exp. Child Psychol. 83, 111–130.

Bakin, J.S., Weinberger, N.M., 1990. Classical conditioning induces CS specific receptive field plasticity in the auditory cortex of the guinea pig. Brain Res. 536, 271–286.

Bavelier, D., Levi, D.M., Li, R.W., Dan, Y., Hensch, T.K., 2010. Removing brakes on adult brain plasticity: from molecular to behavioral interventions. J. Neurosci. 30, 14964–14971.

Bavelier, D., Green, C.S., Pouget, A., Schrater, P., 2012. Brain plasticity through the life span: learning to learn and action video games. Annu. Rev. Neurosci. 35, 391–416.

Bejjanki, V.R., Zhang, R., Li, R., Pouget, A., Green, C.S., Lue, Z.-L., et al., 2014. Action video game play facilitates the development of better perceptual templates. Proc. Natl. Acad. Sci. U. S. A. 111, 16961–16966.

Bermudez, P., Evans, A.C., Lerch, J.P., Zatorre, R.J., 2009. Neuro-anatomical correlates of musicianship as revealed by cortical thickness and voxel-based morphometry. Cereb. Cortex 19, 1583–1596.

Bernstein, L.E., Auer, E.T., Eberhardt, S.P., Jiang, J., 2013. Auditory perceptual learning for speech perception can be enhanced by audiovisual training. Front. Neurosci. 7, 34.

Birnholz, J.C., Benacerrah, B.R., 1983. The development of human fetal hearing. Science 222, 516–518.

Brown, M., Irvine, D.R.F., Park, V.N., 2004. Perceptual learning on an auditory discrimination task by cats: association with changes in primary auditory cortex. Cereb. Cortex 14, 952–965.

Butler, B.E., Lomber, S.G., 2013. Functional and structural changes throughout the auditory system following congenital and early-onset deafness: implications for hearing restoration. Front. Syst. Neurosci. 7, 92.

Carcagno, S., Plack, C.J., 2011. Subcortical plasticity following perceptual learning in a pitch discrimination task. J. Assoc. Res. Otolaryngol. 12, 89–100.

Chan, A.S., Ho, Y.C., Cheung, M.C., 1998. Music training improves verbal memory. Nature 396, 128.

Chang, E.F., Merzenich, M.M., 2003. Environmental noise retards auditory cortical development. Science 300, 498–502.

Condon, C.D., Weinberger, N.M., 1991. Habituation produces frequency-specific plasticity of receptive fields in the auditory cortex. Behav. Neurosci. 105, 416–430.

Davis, M.H., Johnsrude, I.S., Hervais-Adelman, A., Taylor, K., McGettigan, C., 2005. Lexical information drives perceptual learning of distorted speech: evidence from the comprehension of noise-vocoded sentences. J. Exp. Psychol. Gen. 134, 222–241.

Dean, I., Harper, N.S., McAlpine, D., 2005. Neural population coding of sound level adapts to stimulus statistics. Nat. Neurosci. 8, 1684–1689.

de Villers-Sidani, E., Chang, E.F., Bao, S., Merzenich, M.M., 2007. Critical period window for spectral tuning defined in the primary auditory cortex (A1) in the rat. J. Neurosci. 27, 180–189.

de Villers-Sidani, E., Simpson, K.L., Lu, Y.F., Lin, R.C., Merzenich, M.M., 2008. Manipulating critical period closure across different sectors of the primary auditory cortex. Nat. Neurosci. 11, 957–965.

Dorman, M.F., Loizou, P.C., 1997. Changes in speech intelligibility as a function of time and signal processing strategy for an Ineraid patient fitted with continuous interleaved sampling (CIS) processors. Ear Hear. 18, 147–155.

Dosher, B.A., Lu, Z.L., 2005. Perceptual learning in clear displays optimizes perceptual expertise: learning the limiting process. Proc. Natl. Acad. Sci. U. S. A. 102, 5286–5290.

Edeline, J.-M., 1999. Learning-induced physiological plasticity in the thalamo-cortical sensory systems: a critical evaluation of receptive field plasticity, map changes and their potential mechanisms. Prog. Neurobiol. 57, 165–224.

Edeline, J.-M., Weinberger, N.M., 1992. Associative retuning in the thalamic source of input to the amygdala and auditory cortex: receptive field plasticity in the medial division of the medial geniculate body. Behav. Neurosci. 106, 81–105.

Eggermont, J.J., 2012. The Neuroscience of Tinnitus. Oxford University Press, Oxford, UK.

Eggermont, J.J., 2013. On the similarities and differences of non-traumatic sound exposure during the critical period and in adulthood. Front. Syst. Neurosci. 7 (12), 1–12.

Eggermont, J.J., 2014. Noise and the Brain. Experience Dependent Developmental and Adult Plasticity. Academic Press, London.

Eggermont, J.J., 2015. Auditory Temporal Processing and its Disorders. Oxford University Press, Oxford, UK.

Eggermont, J.J., Munguia, R., Pienkowski, M., Greg Shaw, G., 2011. Comparison of LFP-based and spike-based spectro-temporal receptive fields and neural synchrony in cat primary auditory cortex. PLoS One 6 (5), e20046.

Engineer, N.D., Percaccio, C.R., Pandya, P.K., Moucha, R., Rathbun, D.L., Kilgard, M.P., 2004. Environmental enrichment improves response strength, threshold, selectivity, and latency of auditory cortex neurons. J. Neurophysiol. 92, 73–82.

Festen, J.M., Plomp, R., 1990. Effects of fluctuating noise and interfering speech on the speech-reception threshold for impaired and normal hearing. J. Acoust. Soc. Am. 88, 1725–1736.

Formby, C., Sherlock, L., Gold, S.L., 2003. Adaptive plasticity of loudness induced by chronic attenuation and enhancement of the acoustic background. J. Acoust. Soc. Am. 114, 55–58.

Friesen, L.M., Shannon, R.V., Slattery III, W.H., 1999. The effect of frequency allocation on phoneme recognition with the nucleus 22 cochlear implant. Am. J. Otol. 20, 729–734.

Fu, Q.J., Galvin, J.J., 2007. Perceptual learning and auditory training in cochlear implant recipients. Trends Amplif. 11, 193–205.

Fu, Q.J., Shannon, R.V., 1999. Recognition of spectrally degraded and frequency shifted vowels in acoustic and electric hearing. J. Acoust. Soc. Am. 105, 1889–1900.

Fu, Q.J., Shannon, R.V., Galvin, J.J., 2002. Perceptual learning following changes in the frequency-to-electrode assignment with the Nucleus-22 cochlear implant. J. Acoust. Soc. Am. 112, 1664–1674.

Fu, Q.J., Nogaki, G., Galvin III, J.J., 2005a. Auditory training with spectrally shifted speech: an implication for cochlear implant users' auditory rehabilitation. J. Assoc. Res. Otolaryngol. 6, 180–189.

Fu, Q.J., Nogaki, G., Galvin III, J.J., 2005b. Moderate auditory training can improve speech performance of adult cochlear implant patients. Acoust. Res. Lett. Online 6, 106–111.

Gander, P.E., Bosnyak, D.J., Roberts, L.E., 2010. Acoustic experience but not attention modifies neural population phase expressed in human primary auditory cortex. Hear. Res. 269, 81–94.

Gordon, K.A., Salloum, C., Toor, G.S., van Hoesel, R., Papsin, B.C., 2012. Binaural interactions develop in the auditory brainstem of children who are deaf: effects of place and level of bilateral electrical stimulation. J. Neurosci. 32, 4212–4223.

Gordon, R.L., Fehd, H.M., McCandliss, B.D., 2015. Does music training enhance literacy skills? A meta-analysis. Front. Psychol. 6, 1777.

Grahn, J.A., Henry, M.J., McAuley, J.D., 2011. FMRI investigation of cross-modal inter-actions in beat perception: audition primes vision, but not vice versa. Neuroimage 54 (2), 1231–1243.

Halliday, L.F., 2014. A tale of two studies on auditory training in children: a response to the claim that 'discrimination training of phonemic contrasts enhances phonological processing in mainstream school children' by Moore, Rosenberg and Coleman (2005). Dyslexia 20, 101–118.

Halliday, L.F., Taylor, J.L., Edmondson-Jones, A.M., Moore, D.R., 2008. Frequency discrimination learning in children. J. Acoust. Soc. Am. 123, 4393–4402.

Halliday, L.F., Taylor, J.L., Millward, K.E., Moore, D.R., 2012. Lack of generalization of auditory learning in typically developing children. J. Speech Lang. Hear. Res. 55, 168–181.

Hannon, E.E., Trainor, L.J., 2007. Music acquisition: effects of enculturation and formal training on development. Trends Cogn. Sci. 11, 466–472.

Herholz, S.C., Zatorre, R.J., 2012. Musical training as a framework for brain plasticity: behavior, function, and structure. Neuron 76, 486–502.

Hervais-Adelman, A., Davis, M.H., Johnsrude, I.S., Taylor, K.J., Carlyon, R.P., 2011. Generalization of perceptual learning of vocoded speech. J. Exp. Psychol. Hum. Percept. Perform. 37, 293–295.

Hofman, P., Van Riswick, J.G.A., Van Opstal, A.J., 1998. Relearning sound localization with new ears. Nat. Neurosci. 1 (5), 417–421.

Irvine, D.R., Rajan, R., Smith, S., 2003. Effects of restricted cochlear lesions in adult cats on the frequency organization of the inferior colliculus. J. Comp. Neurol. 467, 354–374.

Kamke, M.R., Brown, M., Irvine, D.R., 2003. Plasticity in the tonotopic organization of the medial geniculate body in adult cats following restricted unilateral cochlear lesions. J. Comp. Neurol. 459, 355–367.

Kern, S., Gayraud, F., 2007. Influence of preterm birth on early lexical and grammatical acquisition. First Lang. 27, 159–173.

Kisley, M.A., Gerstein, G.L., 2001. Daily variation and appetitive conditioning-induced plasticity of auditory cortex receptive welds. Eur. J. Neurosci. 13, 1993–2003.

Kral, A., Eggermont, J.J., 2007. What's to lose and what's to learn: development under auditory deprivation, cochlear implants and limits of cortical plasticity. Brain Res. Rev. 56, 259–269.

Kral, A., Hubka, P., Heid, S., Tillein, J., 2013. Single-sided deafness leads to unilateral aural preference within an early sensitive period. Brain 136, 180–193.

Lappe, C., Trainor, L.J., Herholz, S.C., Pantev, C., 2011. Cortical plasticity by short-term multimodal musical rhythm training. PLoS One 6, e21493.

Loo, J.H., Rosen, S., Bamiou, D.-E., 2016. Auditory training effects on the listening skills of children with auditory processing disorder. Ear Hear 37 (1), 38–47.

Luo, F., Wang, Q., Kashani, A., Yan, J., 2008. Corticofugal modulation of initial sound processing in the brain. J. Neurosci. 28, 11615–11621.

Marlow, N., Wolke, D., Bracewell, M., Samara, M., 2005. Neurologic and developmental disability at 6 years of age following extremely preterm birth. N. Engl. J. Med. 352, 9–19.

Miller, G.A., Licklider, J., 1950. The intelligibility of interrupted speech. J. Acoust. Soc. Am. 22, 167–173.

Moore, D.R., Rosenberg, J.F., Coleman, J.S., 2005. Discrimination training of phonemic contrasts enhances phonological processing in mainstream school children. Brain Lang. 94, 72–85.

Moore, D.R., Ferguson, M.A., Edmondson-Jones, A.M., Ratib, S., Riley, A., 2010. Nature of auditory processing disorder in children. Pediatrics 126, 382–390.

Näätänen, R., 1975. Selective attention and evoked potentials in humans—a critical review. Biol. Psychol. 2, 237–307.

Näätänen, R., Lehtokoski, A., Lennes, M., Cheour, M., Huotilainen, M., Livonen, A., et al., 1997. Language-specific phoneme representations revealed by electric and magnetic brain responses. Nature 385, 432–434.

Nahum, M., Nelken, I., Ahissar, M., 2008. Low-level information and high-level perception: the case of speech in noise. PLoS Biol. 6, e126.

Nilsson, M., Soli, S.D., Sullivan, J.A., 1994. Development of the hearing in noise test for the measurement of speech reception thresholds in quiet and in noise. J. Acoust. Soc. Am. 95, 1085–1099.

Parbery-Clark, A., Skoe, E., Kraus, N., 2009. Musical experience limits the degradative effects of background noise on the neural processing of sound. J. Neurosci. 29, 14100–14107.

Patel, A.D., 2014. Can nonlinguistic musical training change the way the brain processes speech? The expanded OPERA hypothesis. Hear. Res. 308, 98–108.

Pelizzone, M., Cosendai, G., Tinembart, J., 1999. Within-patient longitudinal speech reception measures with continuous interleaved sampling processors for Ineraid implanted subjects. Ear Hear. 20, 228–237.

Pérez-González, D., Malmierca, M.S., 2014. Adaptation in the auditory system: an overview. Front. Integr. Neurosci. 8, 19.

Pienkowski, M., Eggermont, J.J., 2009. Long-term, partially-reversible reorganization of frequency tuning in mature cat primary auditory cortex can be induced by passive exposure to moderate-level sounds. Hear. Res. 257, 24–40.

Pienkowski, M., Eggermont, J.J., 2011. Cortical tonotopic map plasticity and behavior. Neurosci. Biobehav. Rev. 35, 2117–2128.

Pienkowski, M., Eggermont, J.J., 2012. Reversible long-term changes in auditory processing in mature auditory cortex in the absence of hearing loss induced by passive, moderate-level sound exposure. Ear Hear. 33, 305–314.

Ponton, C.W., Eggermont, J.J., 2001. Of kittens and kids. Altered cortical maturation following profound deafness and cochlear implant use. Audiol. Neurootol. 6, 363–380.

Ponton, C.W., Don, M., Eggermont, J.J., Waring, M.D., Kwong, B., Masuda, A., 1996. Plasticity of the auditory system in children after long periods of complete deafness. Neuroreport 8, 61–65.

Ponton, C.W., Eggermont, J.J., Kwong, B., Don, M., 2000. Maturation of human central auditory system activity: evidence from multi-channel evoked potentials. Clin. Neurophysiol. 111, 220–236.

Ponton, C.W., Vasama, J.P., Tremblay, K., Khosla, D., Kwong, B., Don, M., 2001. Plasticity in the adult human central auditory system: evidence from late-onset profound unilateral deafness. Hear. Res. 154, 32–44.

Ponton, C.W., Eggermont, J.J., Khosla, D., Kwong, B., Don, M., 2002. Maturation of human central auditory system activity: separating auditory evoked potentials by dipole source modeling. Clin. Neurophysiol. 113, 407–420.

Proulx, M.J., Brown, D.J., Pasqualotto, A., Meijer, P., 2014. Multisensory perceptual learning and sensory substitution. Neurosci. Biobehav. Rev. 41, 16–25.

Rajan, R., Irvine, D.R., Wise, L.Z., Heil, P., 1993. Effect of unilateral partial cochlear lesions in adult cats on the representation of lesioned and unlesioned cochleas in primary auditory cortex. J. Comp. Neurol. 338, 17–49.

Recanzone, G.H., Schreiner, C.E., Merzenich, M.M., 1993. Plasticity in the frequency representation of primary auditory cortex following discrimination training in adult owl monkeys. J. Neurosci. 13, 87–103.

Rescorla, R.A., 1988. Behavioral studies of Pavlovian conditioning. Annu. Rev. Neurosci. 11, 329–352.

Rosen, S., Faulkner, A., Wilkinson, L., 1999. Adaptation by normal listeners to upward spectral shifts of speech: implications for cochlear implants. J. Acoust. Soc. Am. 106, 3629–3636.

Ross, B., Tremblay, K., 2009. Stimulus experience modifies auditory neuromagnetic responses in young and older listeners. Hear. Res. 248, 48–59.

Ross, L.A., Saint-Amour, D., Leavitt, V.M., Javitt, D.C., Foxe, J.J., 2007. Do you see what I am saying? Exploring visual enhancement of speech comprehension in noisy environments. Cereb. Cortex 17 (5), 1147–1153.

Schneider, P., Scherg, M., Dosh, H.G., Specht, H.J., Gutschalk, A., Rupp, A., 2002. Morphology of Heschl's gyrus reflects enhanced activation in the auditory cortex of musicians. Nat. Neurosci. 5, 688–694.

Seitz, A.R., Kim, R., Shams, L., 2006. Sound facilitates visual learning. Curr. Biol. 16 (14), 1422–1427.

Shallop, J.K., McGinn-Brunelli, T., 1995. Speech recognition performance over time with the Spectra 22 speech processor. Ann. Otol. Rhinol. Laryngol. Suppl. 166, 306–307.

Shams, L., Seitz, A.R., 2008. Benefits of multisensory learning. Trends Cogn. Sci. 12, 411–417.

Stjernqvist, K., Svenningsen, W., 1999. Ten year follow up of children born before 29 gestational weeks: health, cognitive development, behaviour and school achievement. Acta Paediatr. 88, 557–562.

Strait, D.L., Parbery-Clark, A., Hittner, E., Kraus, N., 2012. Musical training during early childhood enhances the neural encoding of speech in noise. Brain Lang. 123, 191–201.

Sullivan, J.R., Thibodeau, L.M., Assmann, P.F., 2013. Auditory training of speech recognition with interrupted and continuous noise maskers by children with hearing impairment. J. Acoust. Soc. Am. 133, 495–501.

Sumby, W.H., Pollack, I., 1954. Visual contribution to speech intelligibility in noise. J. Acoust. Soc. Am. 26 (2), 212–215.

Summerfield, A.Q., 1987. Some preliminaries to a comprehensive account of audio-visual speech perception. In: Dodd, B., Campbell, R. (Eds.), Hearing by Eye: The Psychology of Lip-Reading. Lawrence Erlbaum Associates, Inc, London, pp. 3–52.

Svirsky, M.A., Robbins, A.M., Kirk, K.I., Pisoni, D.B., Miyamoto, R.T., 2000. Language development in profoundly deaf children with cochlear implants. Psychol. Sci. 11, 153–158.

Talwar, S.K., Gerstein, G.L., 2001. Reorganization in awake rat auditory cortex by local microstimulation and its effect on frequency-discrimination behavior. J. Neurophysiol. 86, 1555–1572.

Tillein, J., Hubka, P., Kral, A., 2016. Monaural congenital deafness affects aural dominance and degrades binaural processing. Cereb. Cortex 26, 1762–1777.

Tremblay, K.L., Kraus, N., 2002. Auditory training induces asymmetrical changes in cortical neural activity. J. Speech Lang. Hear. Res. 45, 564–572.

Tremblay, K., Moore, D.R., 2012. Current issues in auditory plasticity and auditory training. Transl. Perspect. Aud. Neurosci. 165–189, Book 3.

Tremblay, K.L., Kraus, N., McGee, T., 1998. The time course of auditory perceptual learning: neurophysiological changes during speech-sound training. Neuroreport 9, 3557–3560.

Tremblay, K.L., Inoue, K., McClannahan, K., Ross, B., 2010. Repeated stimulus exposure alters the way sound is encoded in the human brain. PLoS One 5 (4), e10283.

Tremblay, K.L., Ross, B., Inoue, K., McClannahan, K., Collet, G., 2014. Is the auditory evoked P2 response a biomarker of learning? Front. Syst. Neurosci. 8, 28.

Ulanovsky, N., Las, L., Nelken, I., 2003. Processing of low-probability sounds by cortical neurons. Nat. Neurosci. 6, 391–398.

Valentine, P.A., Teskey, G.C., Eggermont, J.J., 2004. Kindling changes burst firing, neural synchrony and tonotopic organization of cat primary auditory cortex. Cereb. Cortex 14, 827–839.

van Praag, H., Kempermann, G., Gage, F.H., 2000. Neural consequences of environmental enrichment. Nat. Rev. Neurosci. 1, 191–198.

Van Wanrooij, M.M., Van Opstal, A.J., 2007. Sound localization under perturbed binaural hearing. J. Neurophysiol. 97, 715–726.

Weinberger, N.M., 1995. Dynamic regulation of receptive fields and maps in the adult sensory cortex. Annu. Rev. Neurosci. 18, 129–158.

Weinberger, N.M., 1998. Physiological memory in primary auditory cortex: characteristics and mechanisms. Neurobiol. Learn. Mem. 70, 226–251.

Weinberger, N.M., 2004. Specific long-term memory traces in primary auditory cortex. Nat. Rev. Neurosci. 5, 279–290.

Weinberger, N.M., Diamond, D.M., 1987. Physiological plasticity in auditory cortex: rapid induction by learning. Prog. Neurobiol. 29, 1–55.

Winer, J.A., 2006. Decoding the auditory corticofugal systems. Hear. Res. 212, 1–8.

Wong, P.C., Skoe, E., Russo, N.M., Dees, T., Kraus, N., 2007. Musical experience shapes human brainstem encoding of linguistic pitch patterns. Nat. Neurosci. 10, 420–422.

Yan, W., Suga, N., 1998. Corticofugal modulation of the midbrain frequency map in the bat auditory system. Nat. Neurosci. 1, 54–58.

Zhang, L.I., Bao, S., Merzenich, M.M., 2001. Persistent and specific influences of early acoustic environments on primary auditory cortex. Nat. Neurosci. 4, 1123–1130.

Zhang, L.I., Bao, S.W., Merzenich, M.M., 2002. Disruption of primary auditory cortex by synchronous auditory inputs during a critical period. Proc. Natl. Acad. Sci. U.S.A. 99, 2309–2314.

Zhang, Y., Suga, N., 2000. Modulation of responses and frequency tuning of thalamic and collicular neurons by cortical activation in mustached bats. J. Neurophysiol. 84, 325–333.

Zhang, Y., Yan, J., 2008. Corticothalamic feedback for sound-specific plasticity of auditory thalamic neurons elicited by tones paired with basal forebrain stimulation. Cereb. Cortex 18, 1521–1528.

Zheng, W., 2012. Auditory map reorganization and pitch discrimination in adult rats chronically exposed to low-level ambient noise. Front. Syst. Neurosci. 6, 65.

Chapter 3

Multisensory Processing

3.1 MULTIMODAL AUDITORY CORTICAL AREAS

Examples in Chapter 2, Brain Plasticity and Perceptual Learning show that perceptual learning is more efficient when the training is multisensorial. Here we will review that combining the information from different senses is essential for successful interaction with real-life situations. It is often believed that this integration occurs at later processing stages and mostly in higher association cortices (Ghazanfar and Schroeder, 2006), whereas other studies suggest that sensory convergence may occur in primary sensory cortex (Bizley et al., 2007; Ghazanfar et al., 2005; Kayser et al., 2007). However, the point of convergence may be even subcortical (Shore, 2005). Gruters and Groh (2012) noted that: "Converging anatomical and physiological evidence indicates that cells within the (inferior colliculus) IC are sensitive to visual, oculomotor, eye position, and somatosensory information as well as to signals relating to behavioral context and reward. . . . The presence of non-auditory signals throughout all subdivisions of the IC—including both ascending and descending regions—provides a point of entry for these signals to reach auditory processing at all stages from brainstem to cortex."

In this chapter, we initially limit ourselves to the neocortex, but pick up the subcortical areas in Section 3.3 by reviewing somato-auditory interactions along the entire auditory pathway.

3.1.1 Animal Data

The auditory cortex of nonhuman primates consists of 13 interconnected areas distributed across three major regions on the superior temporal gyrus: core, belt, and parabelt. Hackett et al. (2007) traced both corticocortical and thalamocortical connections in marmosets and macaques. In addition to those with core auditory cortex, the cortical connections of the belt areas CM and CL included somatosensory (retroinsular, Ri) and multisensory areas (temporal parietal). Thalamic inputs included the medial geniculate complex and several multisensory nuclei (supra-geniculate, posterior, limitans, medial pulvinar), but not the ventroposterior complex. The core (A1, R) and rostromedial areas of auditory cortex have only sparse multisensory

Hearing Loss. DOI: http://dx.doi.org/10.1016/B978-0-12-805398-0.00003-7

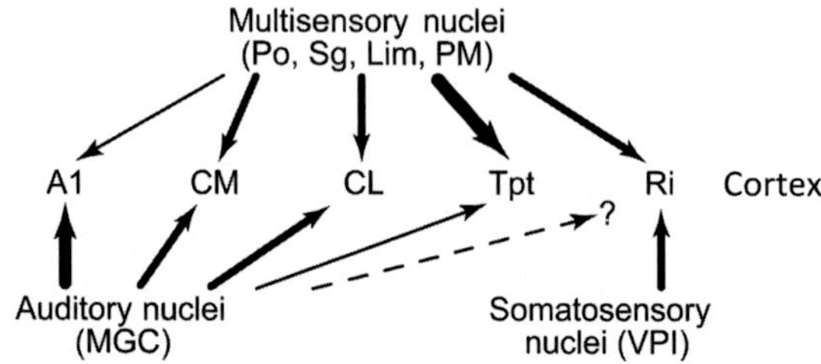

FIGURE 3.1 Summary of thalamocortical inputs to A1, CM, CL, Tpt, and Ri in this study. Heavier arrows indicate denser projections. Auditory areas receive the densest inputs from the MGC and variable projections from the multisensory nuclei. Tpt receives the densest multisensory inputs and modest auditory projections. Ri has uncertain auditory inputs and stronger inputs from multisensory and somatosensory nuclei. A1, primary auditory cortex; CL, caudolateral belt area; CM, caudomedial belt area; MGC, medial geniculate complex; Lim, limitans nucleus; PM, medial pulvinar; Po, posterior nucleus; Ri, retroinsular area; Sg, suprageniculate nucleus; Tpt, temporal pariotemporal are; VPI, ventroposterior nucleus, inferior division. *Reprinted from Hackett, T.A., De La Mothe, L.A., Ulbert, I., Karmos, G., Smiley, J., Schroeder, C.E., 2007. Multisensory convergence in auditory cortex, II. Thalamocortical connections of the caudal superior temporal plane. J. Comp. Neurol. 502, 924–952, with permission from John Wiley & Sons, Inc.*

connections (Fig. 3.1). Hackett et al. (2007) found that the Ri area is one of the principle sources of somatosensory input to the caudal belt, while multisensory regions of cortex and thalamus may also contribute. Recent studies suggest that sensory convergence can already occur in primary sensory cortices. A good example for early convergence appears to be the auditory cortex, for which auditory-evoked activity can be modulated by visual and tactile stimulation. Kayser et al. (2007) found that both primary (core) and nonprimary (belt) auditory fields in monkeys can be activated by the mere presentation of visual scenes. Audiovisual (AV) convergence was restricted to caudal fields; prominently the primary auditory cortex and belt fields and continued in the auditory parabelt and the superior temporal sulcus (STS) (Fig. 3.1). The same fields exhibited enhancement of auditory activation by visual stimulation and showed stronger enhancement for less effective stimuli, two characteristics of sensory integration. Kayser et al. (2007) found basically the same results in awake and anesthetized monkeys, indicating that attention does not have an effect. More extensive information can be found in the work of Eggermont (2015).

3.1.2 Human Findings

Diaconescu et al. (2011) noted that perceptual objects often have both a visual and an auditory component, i.e., they can be seen and heard, and this information arrives in cortex simultaneously through distinct sensory channels. The cross-modal object features are linked by reference to the primary sensory cortices. The binding of familiar, though continued exposure, auditory and visual components is referred to as semantic, multisensory integration. Diaconescu et al. (2011) recorded spatiotemporal patterns underlying multisensory processing at multiple cortical stages using magnetoencephalography recordings of meaningful related cross-modal and unimodal stimuli. Already for latencies of 100 ms after stimulus onset, posterior parietal brain regions responded preferentially to cross-modal stimuli irrespective of task instructions or relatedness between the auditory and visual components. Okada et al. (2013) recalled that: "In audiovisual speech studies, neural activity is consistently reported in posterior superior temporal sulcus (pSTS) and this site has been implicated in multimodal integration." Okada et al. (2013) then used fMRI to investigate how visual speech influences activity in auditory cortex above and beyond its response to auditory speech. Subjects were presented with auditory speech with and without congruent visual input. It appeared that congruent visual speech increased the BOLD activity in auditory cortex, indicating early multisensory processing.

3.1.3 Hearing Loss Affects Multisensory Representation in Animals

Cross-modal reorganization may occur following damage to mature sensory systems (Rebillard et al., 1977; Shepherd et al., 1999; Allman et al., 2009; Park et al., 2010) and may be the neural substrate that allows a compensatory visual function. Lomber et al. (2010) tested this hypothesis using a battery of visual psychophysical tasks and found that congenitally deaf cats, compared with hearing cats, have superior localization in the peripheral field and lower visual movement detection thresholds. In the deaf cats, selective reversible deactivation of posterior auditory cortex (PAF) by cooling eliminated superior visual localization abilities, whereas deactivation of the dorsal auditory cortex (DZ) eliminated superior visual motion detection. Thus, the different perceptual visual improvements were dependent on specific and different subregions of auditory cortex. Lomber et al. (2010)'s data suggested that: "The improved localization of visual stimuli in deaf cats was eliminated by deactivating area PAF, whereas the enhanced sensitivity to visual motion was blocked by disabling area DZ. Because neither cortical area influenced visual processing in hearing cats, these data indicate both that cross-modal reorganization occurred in the PAF and DZ and that the reorganization was functional and highly specific" (Fig. 3.2).

Extending the results on field DZ, Kok et al. (2014) found that: "Overall, the pattern of cortical projections to DZ was similar in both hearing and deafened animals. However, there was a progressive increase in projection strength among hearing and late- and early-deafened cats from an extrastriate visual cortical region known to be involved in the processing of visual motion, the posterolateral lateral suprasylvian area (PLLS)." This suggested that the increase in projection strength from PLLS is larger for early-deafened than in late-deafened animals.

Task	PAF Deactivation	DZ Deactivation
Visual localization in the peripheral field	Deficit	No deficit
Movement detection	No deficit	Deficit

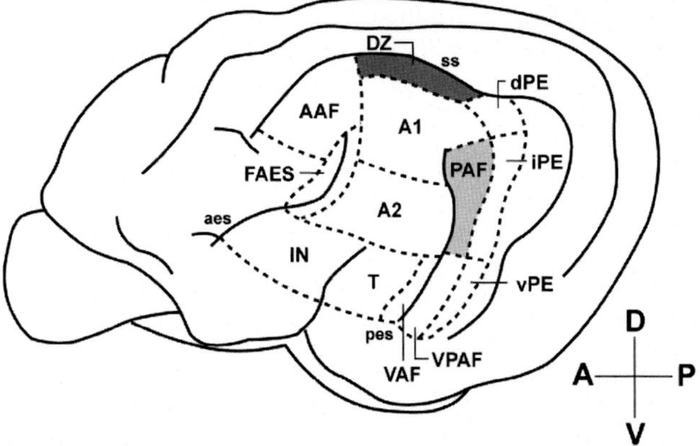

FIGURE 3.2 Summary diagram illustrating the double-dissociation of visual functions in auditory cortex of the deaf cat. Bilateral deactivation of PAF, but not DZ, resulted in the loss of enhanced visual localization in the far periphery. On the other hand, bilateral deactivation of DZ, but not PAF, resulted in higher movement detection thresholds. The lower panel shows a lateral view of the cat cerebrum highlighting the locations of PAF and DZ. PAF, posterior auditory field; DZ, dorsal zone. *Reprinted from Lomber, S.G., Meredithm M.A., Kralm A., 2010. Cross-modal plasticity in specific auditory cortices underlies visual compensations in the deaf. Nat. Neurosci. 13, 1421–1427, with permission from Macmillan Publishers Ltd.*

For more anterior auditory cortical areas, Meredith et al. (2011) found that in hearing cats, the cortical auditory field of the anterior ectosylvian sulcus (FAES; Fig. 3.2) is largely responsive to acoustic stimulation and its unilateral deactivation results in profound contralateral acoustic orienting deficits. They also found that recordings in the FAES of early-deafened adults revealed robust responses to visual stimulation in the contralateral visual field. A second group of early-deafened cats was trained to localize visual targets in a perimetry array. In these animals, cooling loops were surgically placed on the FAES to reversibly deactivate the region, which resulted in substantial contralateral visual orienting deficits. Meredith et al. (2011) found that "crossmodal plasticity can substitute one sensory modality for another while maintaining the functional repertoire of the reorganized region." Meredith and Lomber (2011) then looked at the effects of deafening on the anterior auditory field (AAF) and observed that neurons in early-deafened AAF could not be activated by auditory stimulation. Instead, the majority (78%) was activated by somatosensory cues, while fewer were driven by visual stimulation (44%). These results indicated to them that, "following postnatal deafness, both somatosensory and visual modalities participate in crossmodal reinnervation of the AAF." In a study investigating multisensory projections to AAF following early- and late-onset deafness, Wong et al. (2015) injected a retrograde tracer in AAF, which in early-deaf cats, resulted in ipsilateral neuronal labeling in visual and somatosensory cortices increased by 329% and 101%, respectively, whereas, labeling in auditory areas was reduced by 36%. Less marked differences were observed in late-deaf cats. Conserved thalamocortical connectivity, following early- and late-onset deafness, suggested that thalamic inputs to AAF do not depend on acoustic experience. However, corticocortical connectivity following early-onset deafness changed considerably thereby demonstrating the importance for cortical development of early acoustic experience.

3.1.4 Human Findings Following Sensory Deprivation

Demonstration of experience-dependent plasticity has been provided by studies of sensory-deprived individuals (e.g., blind or deaf), showing that brain regions deprived of their natural inputs change their sensory tuning to support the processing of inputs coming from the spared senses (Fig. 3.3).

Despite the massively different cortical activation in early blind subjects, Bavelier and Neville (2002) early on had stated in a review paper: "Reports that

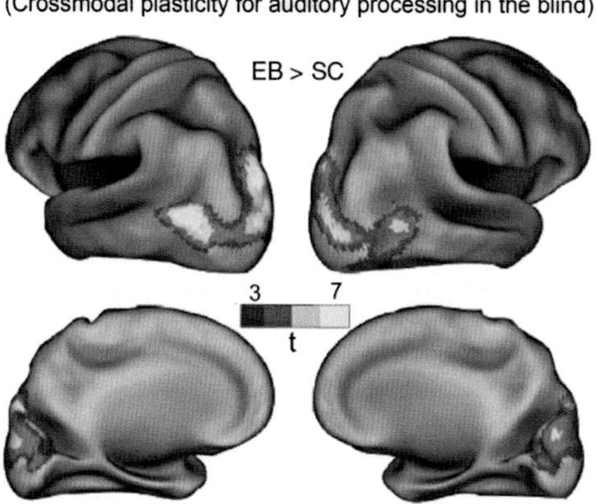

FIGURE 3.3 Example of the massive activation elicited by sounds in the occipital cortex of blind adults, based on data from Collignon et al. (2011): It depicts the activation obtained when contrasting early-blind individuals (EB) versus sighted controls (SC) when both groups of participants were exposed to auditory stimuli only. *Reprinted from Heimler, B., Weisz, N., Collignon, O., 2014. Revisiting the adaptive and maladaptive effects of crossmodal plasticity. Neuroscience 283, 44–63, with permission from Elsevier.*

visual areas V1 and V2 (the primary and secondary visual cortices, respectively) are recruited during auditory language processing in post-lingually deaf individuals after they receive cochlear implants (CIs) also indicate a link between plastic changes in the spared modality and deprivation of the auditory system." According to Giraud et al. (2001) this finding seems to reflect the greater reliance of CI users on visual cues during the processing of oral language, rather than plasticity caused by deafness per se. Bavelier and Neville (2002) noted that: "Psychophysical thresholds for visual contrast sensitivity, visual flicker, brightness discrimination, direction of motion and motion velocity are similar in deaf and hearing individuals." This agrees with an early study (Starlinger and Niemeyer, 1981) based on standard audiometry and tactile thresholds that did not show any differences between blind and sighted individuals.

Bavelier and Neville (2002) found that: "Individuals who became blind early in life, but not those who lost their sight later, can process sounds faster, localize sounds more accurately and have sharper auditory spatial tuning—as measured both behaviourally and using event-related potentials (ERPs)—than sighted individuals (Roder et al., 1999; Lessard et al., 1998)." This is supported by the finding that auditory and somatosensory ERPs over posterior cortical areas are larger and the processing is faster in blind than in sighted subjects, indicating that these areas are recruited by the remaining modalities (Calvert, 2001). This type of compensation might be mediated by enhanced recruitment of multimodal areas of cortex by the remaining modalities (Fig. 3.3). Not surprisingly, the areas that show reorganization after sensory deprivation seem to be part of the cortical network that mediates crossmodal processing in normally sighted, hearing individuals (Calvert, 2001).

Karns et al. (2012) reported that in the deaf Heschl's gyrus, the site of human primary auditory cortex, fMRI signal change was greater for somatosensory and bimodal stimuli than found in hearing participants. Particularly, visual responses in Heschl's gyrus, which were larger in deaf than hearing persons, were smaller than those elicited by somatosensory stimulation. However, in the superior temporal cortex the visual response was comparable to the somatosensory one. The same research group (Scott et al., 2014) using individually defined primary auditory cortex areas found that the fMRI signal change for more peripheral stimuli in the visual field was larger than for perifoveal stimuli in deaf. In contrast, there was no difference in hearing people. In addition, auditory cortex in the deaf contains significantly less white matter and larger gray matter–white matter ratios than in hearing participants (Emmorey et al., 2003; Shibata, 2007).

In a recent review paper, Heimler et al. (2014) reported that: "Buckley and Tobey (2011) recorded electrophysiological responses of early-deaf cochlear implant (CI) recipients elicited by visual motion. ... The authors observed a negative correlation between the strength of crossmodal recruitment and scores to linguistic tests, thus ultimately suggesting that crossmodal takeover interferes with proper language recovery." For an excellent

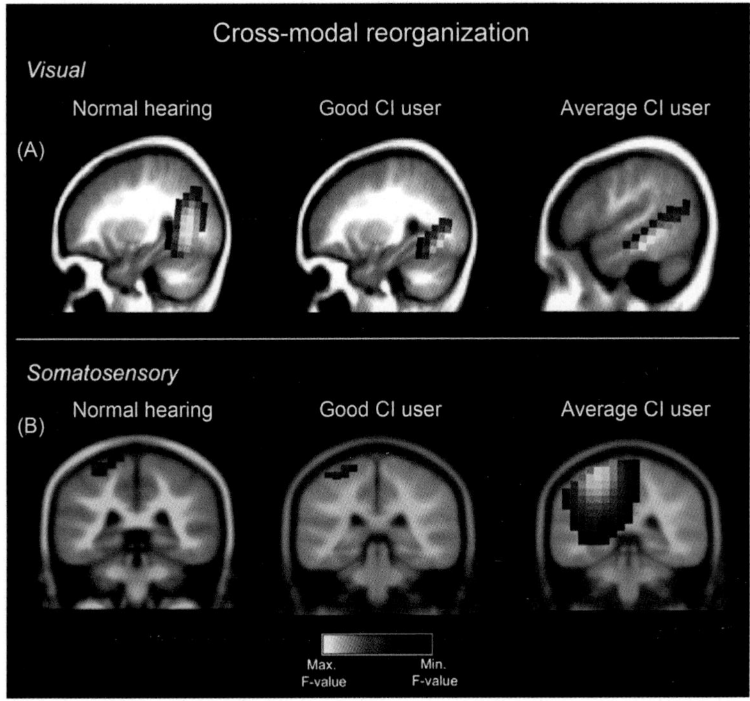

FIGURE 3.4 (A) Visual cross-modal reorganization in children with CIs. Visual gradient stimulation was presented to a child with normal hearing and two children with CIs. Current density reconstructions of the cortical visual P2 component computed via sLORETA show activated regions as illustrated on sagittal MRI slices. Yellow regions reflect maximal cortical activation, while brown/black regions reflect the areas of least activation. Left panel: A 10-year-old child with normal hearing shows activation of higher-order occipital cortices in response to visual stimuli. Middle panel: An 8-year-old cochlear implanted child with a speech perception score of 96% on the Lexical Neighborhood Test shows similar activation of higher-order visual areas, such as middle occipital gyrus, fusiform gyrus, and lingual gyrus. Right panel: In contrast, a 7-year-old cochlear implanted child with a speech perception score of 67% on the Multisyllabic Lexical Neighborhood Test shows activation of occipital areas and superior temporal gyrus and medial temporal gyrus. (B) Somatosensory cross-modal reorganization in children with CIs. Vibrotactile stimulation of the right index finger was presented to a child with normal hearing and two children with CIs. Current density reconstructions of the cortical somatosensory N70 component computed via sLORETA show activated regions as illustrated in coronal MRI slices. Left panel: A normal hearing 7-year-old child shows activation of somatosensory cortex in the postcentral gyrus. Middle panel: A 13-year-old cochlear implanted child with a speech perception score of 94% on the Consonant Nucleus Consonant (CNC) test shows similar activation of somatosensory cortex in postcentral gyrus. Right panel: In contrast, a 15-year-old cochlear implanted child who showed average performance on the CNC speech perception test (76%) exhibited activation of the somatosensory cortex, superior and transverse temporal gyri, and parietal cortex. *Reprinted from Sharma, A., Campbell, J., Cardon, G., 2015. Developmental and cross-modal plasticity in deafness: evidence from the P1 and N1 event related potentials in cochlear implanted children. Int. J. Psychophysiol. 95, 135–144, with permission from Elsevier.*

illustration of this see Sharma et al. (2015) (Fig. 3.4). Sandmann et al. (2012) found smaller visual-evoked P100 amplitudes and reduced visual cortex activation in CI users compared with normal hearing listeners. This suggests a visual takeover of the auditory cortex, and that such cross-modal plasticity may be one of the main sources of the high variability observed in CI outcomes. Heimler et al. (2014) suggested that: "Crossmodal plasticity is ultimately and unavoidably maladaptive for optimal auditory recovery and that its presence should be considered as a negative predictor of successful auditory restoration through cochlear implantation." This somewhat pessimistic statement has been weakened by recent animal studies using CIs in congenital deaf white cats, where Land et al. (2016) concluded that: "Cross-modal reorganization was less detrimental for neurosensory restoration than previously thought."

3.2 AV INTERACTION IN HUMANS

3.2.1 The McGurk Effect

McGurk and MacDonald (1976) reported a powerful multisensory illusion occurring with AV speech. The illusion has been called the McGurk effect. The best known case of the McGurk effect is when dubbing a voice saying /b/ onto a face articulating /g/ results in hearing /d/. This is called the fusion effect since the percept differs from the acoustic and visual components. Tiippana (2014) suggested that the McGurk effect should be defined as a categorical change in auditory perception induced by incongruent visual speech, resulting in a single percept of hearing something different than what the voice is saying. Second, when interpreting the McGurk effect, it is crucial to take into account the perception of the unisensory acoustic and visual stimulus components. There is a high degree of individual variability in how frequently the illusion is perceived: Some individuals almost always perceive the McGurk effect, while others rarely do (Mallick et al., 2015). Frequent perceivers of the McGurk effect were also more likely to fixate the mouth of the talker, and there was a significant correlation between McGurk frequency and mouth looking time (Gurler et al., 2015). To which extent visual input influences the percept depends on how coherent and reliable information each modality provides. Coherent information is integrated and weighted, e.g., according to the reliability of each modality, which is reflected in unisensory discriminability. Tiippana (2014) concluded: "When integration of auditory and visual information takes place, it results in a unified percept, without access to the individual components that contributed to the percept."

Nath and Beauchamp (2012) used fMRI to measure brain activity as McGurk perceivers and nonperceivers (susceptibility) were presented with congruent AV syllables, McGurk AV syllables, and non-McGurk incongruent syllables (stimulus condition). They found that the inferior frontal gyrus showed greater responses for incongruent stimuli with no difference between perceivers and nonperceivers. In contrast, the left auditory cortex showed a greater response in perceivers but was not affected by stimulus condition. Only one brain region, the left STS, showed a significant effect of both susceptibility and stimulus condition. The strength of the STS response was positively correlated with the likelihood of perceiving the McGurk effect.

In an event-related 7 Tesla fMRI study, Szycik et al. (2012) used three naturally spoken syllable pairs with matching AV information and one syllable pair designed to elicit the McGurk illusion. Successful fusion of AV speech appeared to be related to activity within the STS of both hemispheres. However, Erickson et al. (2014) found that the left pSTS exhibited significantly greater fMRI signal for congruent AV speech than for both auditory only (A_O) and visual only (V_O) trials. For McGurk speech, two clusters in the left posterior superior temporal gyrus (pSTG), just posterior to Heschl's gyrus or on its border, exhibited greater fMRI signal than both A_O and V_O trials. Erickson et al. (2014) proposed that some brain areas, such as left pSTS may be more critical for the *integration* of AV speech. Other areas, such as left pSTG, may generate the "corrected" or merged percept arising from conflicting auditory and visual cues (i.e., as in the McGurk effect).

3.2.2 Lip Reading

Studies in AV speech perception have shown that visual speech cues in the talker's face have a facilitating effect in terms of accuracy across a wide range of auditory signal-to-noise ratios. In a seminal study, Sumby and Pollack (1954) used speech intelligibility tests with, and without, supplementary visual observation of the speaker's facial and lip movements. They found that the visual contribution to speech intelligibility was independent of the speech-to-noise ratio under test. However, they found that "since there is a much greater opportunity for the visual contribution at low speech-to-noise ratios (Fig. 3.5), its absolute contribution can be exploited most profitably under these conditions."

Grant and Seitz (2000) investigated whether visible movements of the speech articulators could be used to improve the detection of speech in noise (Summerfield, 1987), thus demonstrating an influence of speech reading on the ability to detect, rather than recognize, speech. Grant and Seitz (2000) found that "when the speech-read sentence matched the target sentence, average detection thresholds improved by about 1.6 dB relative to the auditory condition." This suggested to Grant and Seitz (2000) that "visual cues derived from the dynamic movements of the fact during speech production interact with time-aligned auditory cues to enhance sensitivity in auditory detection." Bernstein et al. (2002) investigated in hearing adults whether visual speech perception depends on processing by the primary auditory cortex. Using fMRI, a pulse tone was presented and contrasted with gradient noise. During the same session, a silent video of a talker saying isolated

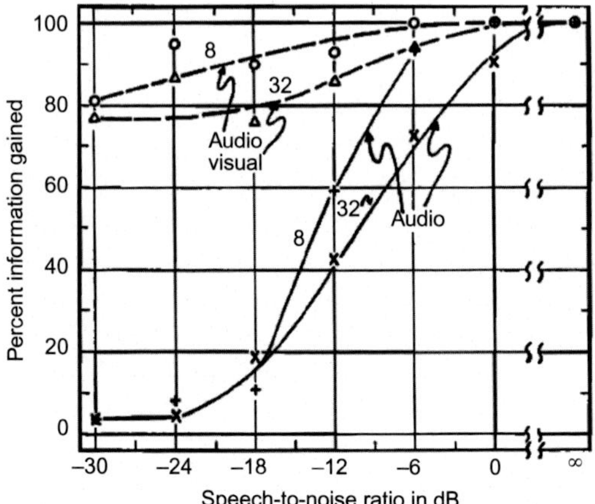

FIGURE 3.5 The information transmitted, relative to the information presented, as a function of the speech-to-noise ratio under test. The parameter on the curves is the size of the vocabulary (spondee words) under test. Upper curves are for auditory and visual presentation. Lower curves are for auditory presentation alone. Each point represents the results for 50 observations, pooled over subjects, for each of the words. *Reprinted from Sumby, W.H., Pollack, I., 1954. Visual contribution to speech intelligibility in noise. J. Acoust. Soc. Am. 26, 212–215, with permission from Acoustic Society of America.*

words was presented contrasted with a still face. They reported that: "Visual speech activated the superior temporal gyrus anterior, posterior, and lateral to the primary auditory cortex, but not the region of the primary auditory cortex. These results suggest that visual speech perception is not critically dependent on the region of primary auditory cortex."

3.2.3 Audio-visual Interaction in Development and Aging

3.2.3.1 Children

By varying signal-to-noise ratios, Ross et al. (2011) found that children benefited significantly less from observing visual articulations and displayed considerably less AV enhancement than adults. This suggested that improvement in the ability to recognize speech in noise and in AV integration during speech perception continues quite late into the childhood years. The implication is that a considerable amount of multisensory learning has to be achieved during the later school years.

Does visual speech fills in nonintact auditory speech (excised consonant onsets) in typically developing children from 4 to 14 years of age? Jerger et al. (2014) found that easy visual speech cues provided greater filling in than difficult cues. Only older children benefited from difficult visual speech cues, whereas all children benefited from easy visual speech cues, although 4- and 5-year-olds did not benefit as much as older children. To explore task demands, they compared results of this filling-in task with those of the McGurk task. Jerger et al. (2014) found that the influence of visual speech was associated with age and vocabulary abilities for the visual speech fill-in effect and was also associated with speech reading skills for the McGurk effect. According to the authors, this suggested that children perceive a speaker's utterance rather than the auditory stimulus per se.

Nath et al. (2011) used fMRI in 17 children aged 6–12 years to measure brain responses to the following three AV stimulus categories: McGurk incongruent, non-McGurk incongruent, and congruent syllables. Two separate analyses, one using independent functional localizers and another using whole-brain voxel-based regression, showed differences in the left STS between perceivers and nonperceivers. Nath et al. (2011) found that the STS of McGurk perceivers responded significantly more than that of nonperceivers to McGurk syllables, but not to other stimuli, and perceivers' hemodynamic responses in the STS were significantly prolonged. These results suggest that the STS is an important source of interindividual variability in children's AV speech perception, as it is in adults.

3.2.3.2 The Elderly

Tye-Murray et al. (2007) compared 53 adults with normal hearing and 24 adults with mild to moderate hearing impairment (all above age 65) on the performance of auditory-only, visual-only, and AV speech perception, using consonants, words, and sentences as stimuli. All testing was conducted in the presence of multitalker background babble. In general, the two groups of participants performed similarly on measures of V_O and AV speech perception. Tye-Murray et al. (2007) concluded that: "Despite increased reliance on visual speech information, older adults who have hearing impairment do not exhibit better V_O-speech perception or AV integration than age-matched individuals who have normal hearing."

Musacchia et al. (2009) examined the impact of hearing loss on AV processing in the aging population. They compared AV processing between 12 older adults with normal hearing and 12 older adults with mild to moderate sensorineural hearing loss. They recorded cortical-evoked potentials that were elicited by watching and listening to recordings of a speaker saying the syllable /bi/. Stimuli were presented in three conditions: when hearing the syllable /bi/ (A_O), when viewing a person say /bi/ (V_O), and when seeing and hearing the syllables simultaneously (AV). They found that in the AV condition, the normal hearing group showed a clear and consistent decrease in P_1 and N_1 latencies as well as a reduction in P_1 and N_1 amplitude compared with the sum of the unimodal components (auditory–visual). These integration effects were absent or less consistent in participants with hearing loss. Musacchia et al. (2009) suggested that these results demonstrate that hearing loss has a deleterious effect on how older adults combine what they see and hear.

3.2.4 Role of Audio-visual Interaction in Cochlear Implant Use

Post-lingually deaf subjects learn the meaning of sounds after cochlear implantation by forming new associations between sounds and their sources (for a comprehensive review, see Butler and Lomber, 2013). Because CIs generate only coarse frequency responses, this imposes a dependency on visual cues, e.g., lip reading. Giraud et al. (2001) hypothesized that cross-modal facilitation results from engagement of the visual cortex by purely auditory tasks. In fMRI recordings they showed recruitment of early visual cortex (V1/V2) when CI users listen to sounds with eyes closed. Blood flow in visual cortex increased for both noise and speech as a function of time postimplantation, and reflecting experience-dependent adaptations in the postimplant phase. Giraud and Truy (2002) found that both in normal hearing and cochlear implantees visual cortical areas participate in semantic processing of speech sounds. In CI patients speech activates the mid-fusiform gyrus in the vicinity of the so-called face area. This may be the correlate of increased usage of lip reading in CI patients.

Rouger et al. (2007) analyzed the longitudinal postimplantation evolution of word recognition in a large sample of CI users in unimodal (visual or auditory) and visuoauditory conditions. They found that, "despite considerable recovery of auditory performance during the first year post-implantation, CI patients maintain a much higher level of word recognition in speech-reading conditions compared with normally hearing subjects, even several years after implantation. ... This better performance is not only due to greater speech-reading performance, but, most importantly, also due to a greater capacity to integrate visual input with the distorted speech signal. [These] results suggest that these behavioral changes in CI users might be mediated by a reorganization of the cortical network involved in speech recognition that favors a more specific involvement of visual areas." Rouger et al. (2012) revealed that the neuroplasticity after cochlear implantation involved not only auditory but also visual and AV speech processing networks (Fig. 3.6). In addition, their findings suggested that the functional reorganizations in the auditory and AV speech processing networks, but also the visual speech-processing network, allow "a low-level integration of AV information early after the beginning of auditory stimulation through the implant, thus facilitating auditory-matching processes during speech reading."

Stropahl et al. (2015) found that lip-reading skills were significantly enhanced in the CI group compared to a normal hearing group, particularly after a longer duration of deafness. In contrast, face recognition was not significantly different between the two groups. In addition, auditory cortex activation in CI users was positively correlated with face recognition abilities. This confirms cross-modal reorganization for meaningful visual stimuli in CI users (see also Section 3.1.3).

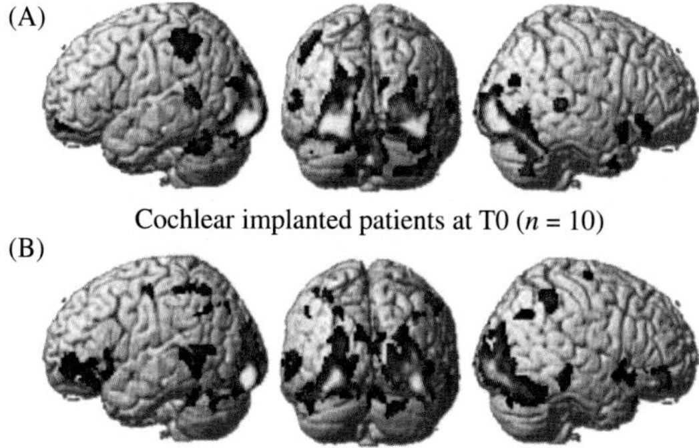

FIGURE 3.6 Brain activation patterns during speech reading. (A) CI patients at T0 ($n = 9$). (B) CI patients at T1 ($n = 8$). Speech reading elicits auditory activations in CI patients. The first PET session (T0) was performed as early as possible after the implant onset. The second PET session (T1) was performed as soon as the patient's auditory speech performance (as measured by the speech therapist) had reached a recognition level of 60% or above, or after a maximum of 1 year after the implant was activated. *Reprinted from Rouger, J., Lagleyre, S., Démonet, J.F., Fraysse, B., Deguine, O., Barone, P., 2012. Evolution of crossmodal reorganization of the voice area in cochlear-implanted deaf patients. Hum. Brain Mapp. 33, 1929–1940, with permission from John Wiley & Sons, Inc.*

3.3 AUDITORY–SOMATOSENSORY INTERACTION

The neural processes for hearing perception are modulated by vision and somatosensation. Especially the somato-auditory integration occurs at every level of the ascending auditory pathway: the cochlear nucleus (CN), IC, medial geniculate body (MGB), and the auditory cortex (Fig. 3.7). Somatic and auditory signals are potentially important for localizing sound sources and attending to salient stimuli. Somato-auditory integration distinguishes environmental from self-generated sounds and aids in the perception and generation of communication sounds (Gruters and Groh, 2012). The IC functions as the nexus of a highly interconnected sensory, motor, and cognitive network dedicated to synthesizing a higher-order auditory percept rather than simply reporting patterns of air pressure detected by the cochlea. The ICs afferents pass auditory information "upward" to thalamus, auditory cortex, and beyond, but also receive centrifugal connections and pass them on to more peripheral areas of the auditory pathway such as the CN (Gruters and Groh, 2012).

The following parts in this section are largely based on the comprehensive review by Wu et al. (2015).

3.3.1 The Dorsal Cochlear Nucleus

The CN is the first central nervous system station in the auditory system that integrates multisensory information. Specific sensory projection neurons to

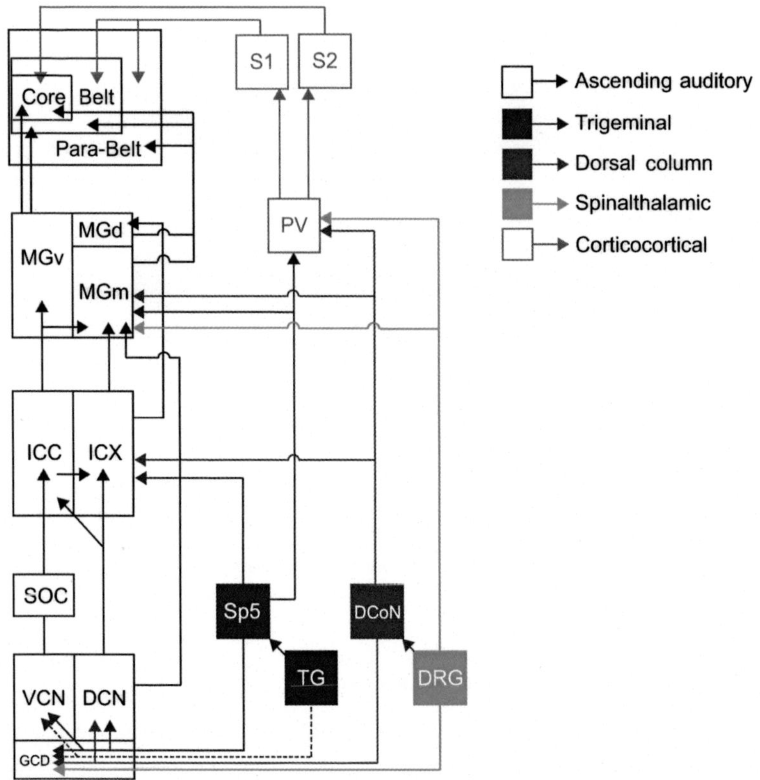

FIGURE 3.7 Multisensory integration in the central auditory pathway. Auditory pathways begin in VCN/DCN, ventral/dorsal cochlear nucleus. The GCD receives a majority of inputs from the somatosensory system (or marginal cell area of VCN). Somatosensory input is thus "already processed" as it traverses the central auditory pathway. Further inputs occur at each separate location. SOC, superior olivary complex; ICC/ICX, central/external nucleus of IC; TG, trigeminal ganglion; Sp5, spinal trigeminal nucleus; DRG, dorsal root ganglion; PV, posterior ventral nucleus of thalamus; S1/S2, primary/secondary somatosensory cortex. *From Wu, C., Stefanescu, R.A., Martel, D.T., Shore, S.E., 2015. Listening to another sense: somatosensory integration in the auditory system. Cell Tissue Res. 361, 233–250, with permission of Springer.*

the CN originate in the trigeminal ganglion and the spinal trigeminal nucleus (Sp5), dorsal root ganglion and dorsal column nuclei (DCoN), saccule, and vestibular nucleus (Wu et al., 2015). Most of the projections from nonauditory sensory ganglia and brainstem nuclei terminate in the CN granular cell domain (GCD; Fig. 3.7). The trigeminal ganglia directly innervate neurons in the cochlea, middle ear, shell area of the VCN including the GCD and the fusiform cell layer of the DCN. Their synapses contain small, spherical vesicles, indicating excitatory transmission (Shore, 2005). The interpolar and caudal Sp5 subnuclei project to many of the same CN regions as the trigeminal ganglion. These nuclei primarily relay pressure and proprioceptive information from the jaw, face, and scalp (Wu et al., 2015). The DCN processes externally generate auditory stimuli partly by comparing them with the sounds produced by an animal's own movements (Roberts and Portfors, 2008). A cerebellar-like circuitry, consisting of a principal output cell, the fusiform cell, and several inhibitory interneurons, assist in the suppression of self-generated sounds.

3.3.2 The Inferior Colliculus

Separate processing streams of auditory inputs from the brainstem converge in the IC (Fig. 3.7). Major projections are again received from somatosensory centers, Sp5 and DCoN, primary somatosensory cortex, as well as visual inputs from the retinal ganglion, SC and V1 (Gruters and Groh, 2012; Wu et al., 2015). The superior colliculus (SC) is primarily a visual structure but also contains an auditory space map. SC neurons in specific regions respond to acoustic stimuli from different spatial coordinates (Middlebrooks and Knudsen, 1984; Palmer and King, 1982). The SC also contains visual and somatosensory maps (Meredith and Stein, 1986). This allows the convergence of multisensory spatial representations serves higher functions such as orientation and localization. The dorsal part of the IC (ICX) also contains an auditory space map (Binns et al., 1992) as well as a somatotopic map (Aitkin et al., 1981). The presence of these two sensory maps suggests the ICX as a site of auditory and somatosensory spatial integration (Wu et al., 2015).

3.3.3 The Auditory Thalamus and Cortex

The auditory thalamus (MGB) contains ventral, dorsal, and medial subdivisions. The ventral MGB (MGBv) receives inputs only from the ICC. Its neurons respond with short latencies and comprise a tonotopic map. In contrast, the dorsal MGB (MGBd) neurons respond only weakly to auditory stimuli; their inputs are exclusively derived from the ventral medial edge of ICX. The magnocellular region of the MGB (MGBm) is the multisensory division of the auditory thalamus. It is analogous to ICX with broader frequency-tuning properties and receives nonauditory inputs that are derived from ICX, and a direct pathway from the DCN and the GCD region that bypass the IC (Schofield et al., 2014). Multisensory inputs to the MGBm include somatosensory afferents from the spinal-thalamic, dorsal column, and the trigeminal pathways, as well as visual afferents from SC. Multisensory responses in MGBm have been documented (Wu et al., 2015).

In the auditory cortex, as in subcortical structures, multisensory integration is mediated by neurons independently activated by more than one sensory input. In addition, neurons from a specific cortical area may be activated by a single modality but their responses are significantly modulated, i.e., enhanced or suppressed, by the input of a second modality. Basura et al. (2012)

investigated the effects of multisensory plasticity inducing stimulation on the somato-auditory integration of simultaneously recorded neurons in DCN and A1. Immediate (bimodal response) and long-lasting (bimodal plasticity) effects of Sp5-tone stimulation were facilitation or suppression of tone-evoked firing rates in DCN and A1. These facilitation or suppression effects lasted for up to an hour.

Nava et al. (2014) tested congenitally and late-deaf CI recipients, age-matched with two groups of hearing controls, on an audio-tactile redundancy paradigm, by measuring reaction times to unimodal and cross-modal redundant stimuli. This showed that both congenitally and late-deaf CI recipients were able to integrate audio-tactile stimuli. This suggested to Nava et al. (2014) that congenital and acquired deafness does not prevent the development and recovery of basic multisensory processing.

Interactions between the auditory and somatosensory system play an important role in enhancing human experience during dynamic contact between the hands and the environment (Kuchenbuch et al., 2014). They found that cross-modal cortical activity mediates preferential responses of the cortical area processing the more salient stimuli and inhibition of the cortical activity in the area processing the less salient stimuli. In addition, vibrotactile stimuli and tactile pulses without vibration activate the auditory belt area in animals and humans (Fu et al., 2003; Schürmann et al., 2006) with supra-additive integration (Kayser et al., 2005).

3.4 SUMMARY

Perceptual learning is more efficient when the training is multisensorial. Combining the information from different senses is essential for successful interaction with real-life situations. Converging anatomical and physiological evidence indicates that cells within the auditory midbrain are sensitive to visual, oculomotor, eye position, and somatosensory information as well as to signals relating to behavioral context and reward. This may influence the multisensory interaction in cortical areas, where visual and somatosensory stimuli modulate auditory processing. Following hearing loss, the auditory cortex may become responsive to visual and somatosensory stimuli, such that they may interfere with the functioning of CIs. More positive interactions are related to lip reading, which facilitates speech understanding in noisy environments and aids the CI user in speech understanding. Aging reduces this beneficiary effect of AV interaction.

REFERENCES

Aitkin, L.M., Kenyon, C.E., Philpott, P., 1981. The representation of the auditory and somatosensory systems in the external nucleus of the cat inferior colliculus. J. Comp. Neurol. 196, 25–40.

Allman, B.L., Keniston, L.P., Meredith, M.A., 2009. Adult deafness induces somatosensory conversion of ferret auditory cortex. Proc. Natl. Acad. Sci. U. S. A. 106, 5925–5930.

Basura, G.J., Koehler, S.D., Shore, S.E., 2012. Multi-sensory integration in brainstem and auditory cortex. Brain Res. 1485, 95–107.

Bavelier, D., Neville, H.J., 2002. Cross-modal plasticity: where and how? Nat. Rev. Neurosci. 3, 443–452.

Bernstein, L.E., Auer Jr., E.T., Moore, J.K., Ponton, C.W., Don, M., Singh, M., 2002. Visual speech perception without primary auditory cortex activation. Neuroreport 13, 311–315.

Binns, K.E., Grant, S., Withington, D.J., Keating, M.J., 1992. A topographic representation of auditory space in the external nucleus of the inferior colliculus of the guinea-pig. Brain Res. 589, 231–242.

Bizley, J.K., Nodal, F.R., Bajo, V.M., Nelken, I., King, A.J., 2007. Physiological and anatomical evidence for multisensory interactions in auditory cortex. Cereb. Cortex 17, 2172–2189.

Buckley, K.A., Tobey, E.A., 2011. Cross-modal plasticity and speech perception in pre- and postlingually deaf cochlear implant users. Ear Hear. 32, 2–15.

Butler, B.E., Lomber, S.G., 2013. Functional and structural changes throughout the auditory system following congenital and early-onset deafness: implications for hearing restoration. Front. Syst. Neurosci. 7, 92.

Calvert, G., 2001. Crossmodal processing in the human brain: insights from functional neuroimaging studies. Cereb. Cortex 11, 1110–1123.

Collignon, O., Vandewalle, G., Voss, P., Albouy, G., Charbonneau, G., Lassonde, M., et al., 2011. Functional specialization for auditory-spatial processing in the occipital cortex of congenitally blind humans. Proc. Natl. Acad. Sci. U. S. A. 108, 4435–4440.

Diaconescu, A.O., Alain, C., McIntosh, A.R., 2011. The co-occurrence of multisensory facilitation and cross-modal conflict in the human brain. J. Neurophysiol. 106, 2896–2909.

Eggermont, J.J., 2015. Auditory Temporal Processing and its Disorders.. Oxford University Press, Oxford, UK.

Emmorey, K., Allen, J., Bruss, J., Schenker, N., Damasio, H., 2003. A morpho-metric analysis of auditory brain regions in congenitally deaf adults. Proc. Natl. Acad. Sci. U. S. A. 100, 10049–10054.

Erickson, L.C., Zielinski, B.A., Zielinski, J.E.V., Liu, G., Turkeltaub, P.E., Leaver, A.M., et al., 2014. Distinct cortical locations for integration of audiovisual speech and the McGurk effect. Front. Psychol. 5, 534.

Fu, K.M., Johnston, T.A., Shah, A.S., Arnold, L., Smiley, J., Hackett, T.A., et al., 2003. Auditory cortical neurons respond to somatosensory stimulation. J. Neurosci. 23, 7510–7515.

Ghazanfar, A.A., Schroeder, C.E., 2006. Is neocortex essentially multisensory? Trends Cogn. Sci. 10, 278–285.

Ghazanfar, A.A., Maier, J.X., Hoffman, K.L., Logothetis, N.K., 2005. Multisensory integration of dynamic faces and voices in rhesus monkey auditory cortex. J. Neurosci. 25, 5004–5012.

Giraud, A.-L., Truy, E., 2002. The contribution of visual areas to speech comprehension: a PET study in cochlear implants patients and normal-hearing subjects. Neuropsychologia 40, 1562–1569.

Giraud, A.-L., Price, C.J., Graham, J.M., Frackowiak, R.S.J., 2001. Cross-modal plasticity underpins recovery of human language comprehension after auditory re-afferentation by cochlear implants. Neuron 30, 657–663.

Grant, K.W., Seitz, P.F., 2000. The use of visible speech cues for improving auditory detection of spoken sentences. J. Acoust. Soc. Am. 108, 1197–1208.

Gruters, K.G., Groh, J.M., 2012. Sounds and beyond: multisensory and other non-auditory signals in the inferior colliculus. Front. Neural Circuits 6, 96.

Gurler, D., Doyle, N., Walker, E., Magnotti, J., Beauchamp, M., 2015. A link between individual differences in multisensory speech perception and eye movements. Atten. Percept. Psychophys. 77, 1333–1341.

Hackett, T.A., De La Mothe, L.A., Ulbert, I., Karmos, G., Smiley, J., Schroeder, C.E., 2007. Multisensory convergence in auditory cortex, II. Thalamocortical connections of the caudal superior temporal plane. J. Comp. Neurol. 502, 924–952.

Heimler, B., Weisz, N., Collignon, O., 2014. Revisiting the adaptive and maladaptive effects of crossmodal plasticity. Neuroscience 283, 44–63.

Jerger, S., Damian, M.F., Tye-Murray, N., Abdi, H., 2014. Children use visual speech to compensate for non-intact auditory speech. J. Exp. Child. Psychol. 126, 295–312.

Karns, C.M., Dow, M.W., Neville, H.J., 2012. Altered cross-modal processing in the primary auditory cortex of congenitally deaf adults: a visual-somatosensory fMRI study with a double-flash illusion. J. Neurosci. 32, 9626–9638.

Kayser, C., Petkov, C.I., Augath, M., Logothetis, N.K., 2005. Integration of touch and sound in auditory cortex. Neuron 48, 373–384.

Kayser, C., Petkov, C.I., Augath, M., Logothetis, N.K., 2007. Functional imaging reveals visual modulation of specific fields in auditory cortex. J. Neurosci. 27, 1824–1835.

Kok, M.A., Chabot, N., Lomber, S.G., 2014. Cross-modal reorganization of cortical afferents to dorsal auditory cortex following early- and late-onset deafness. J. Comp. Neurol. 522, 654–675.

Kuchenbuch, A., Paraskevopoulos, E., Herholz, S.C., Pantev, C., 2014. Audio-tactile integration and the influence of musical training. PLoS One 9, e85743.

Land, R., Baumhoff, P., Tillein, J., Lomber, S.G., Hubka, P., Kral, A., 2016. Crossmodal plasticity in higher-order auditory cortex of congenitally deaf cats does not limit auditory responsiveness to cochlear implants. J. Neurosci. 36 (23), 6175–6185.

Lessard, N., Paré, M., Lepore, F., Lassonde, M., 1998. Early-blind human subjects localize sound sources better than sighted subjects. Nature 395, 278–280.

Lomber, S.G., Meredith, M.A., Kral, A., 2010. Cross-modal plasticity in specific auditory cortices underlies visual compensations in the deaf. Nat. Neurosci. 13, 1421–1427.

Mallick, D.B., Magnotti, J.F., Beauchamp, M.S., 2015. Variability and stability in the McGurk effect: contributions of participants, stimuli, time, and response type. Psychon. Bull. Rev. 22, 1299–1307.

McGurk, H., MacDonald, J., 1976. Hearing lips and seeing voices. Nature 264, 746–748.

Meredith, M.A., Lomber, S.G., 2011. Somatosensory and visual crossmodal plasticity in the anterior auditory field of early-deaf cats. Hear. Res. 280, 38–47.

Meredith, M.A., Stein, B.E., 1986. Visual, auditory, and somatosensory convergence on cells in superior colliculus results in multisensory integration. J. Neurophysiol. 56, 640–662.

Meredith, M.A., Kryklywy, J., McMillan, A.J., Malhotra, S., Lum-Tai, R., Lomber, S.G., 2011. Crossmodal reorganization in the early deaf switches sensory, but not behavioral roles of auditory cortex. Proc. Natl. Acad. Sci. U. S. A. 108, 8856–8861.

Middlebrooks, J.C., Knudsen, E.I., 1984. A neural code for auditory space in the cat's superior colliculus. J. Neurosci. 4, 2621–2634.

Musacchia, G., Arum, L., Nicol, T., Garstecki, D., Kraus, N., 2009. Audiovisual deficits in older adults with hearing loss: biological evidence. Ear Hear. 30, 505–514.

Nath, A.R., Beauchamp, M.S., 2012. A neural basis for interindividual differences in the McGurk effect, a multisensory speech illusion. Neuroimage 59, 781–787.

Nath, A.R., Fava, E.E., Beauchamp, M.S., 2011. Neural correlates of interindividual differences in children's audiovisual speech perception. J. Neurosci. 31, 13963–13971.

Nava, E., Bottari, D., Villwock, A., Fengler, I., Büchner, A., Lenarz, T., 2014. Audio-tactile integration in congenitally and late deaf cochlear implant users. PLoS One 9 (6), e99606.

Okada, K., Venezia, J.H., Matchin, W., Saberi, K., Hickok, G., 2013. An fMRI study of audiovisual speech perception reveals multisensory interactions in auditory cortex. PLoS One 8, e68959.

Palmer, A.R., King, A.J., 1982. The representation of auditory space in the mammalian superior colliculus. Nature 299, 248–249.

Park, M.H., Lee, H.J., Kim, J.S., Lee, J.S., Lee, D.S., Oh, S.H., 2010. Cross-modal and compensatory plasticity in adult deafened cats: a longitudinal PET study. Brain Res. 1354, 85–90.

Rebillard, G., Carlier, E., Rebillard, M., Pujol, R., 1977. Enhancement of visual responses on the primary auditory cortex of the cat after an early destruction of cochlear receptors. Brain Res. 129, 162–164.

Roberts, P.D., Portfors, C.V., 2008. Design principles of sensory processing in cerebellum-like structures. Early stage processing of electrosensory and auditory objects. Biol. Cybern. 98, 491–507.

Roder, B., Teder-Sälejärvi, W., Sterr, A., Rösler, F., Hillyard, S.A., Neville, H.J., 1999. Improved auditory spatial tuning in blind humans. Nature 400, 162–166.

Ross, L., Molholm, S., Blanco, D., Gomez-Ramirez, M., Saint-Amour, D., Foxe, J., 2011. The development of multisensory speech perception continues into the late childhood years. Eur. J. Neurosci. 33, 2329–2337.

Rouger, J., Lagleyre, S., Fraysse, B., Deneve, S., Deguine, O., Barone, P., 2007. Evidence that cochlear implanted deaf patients are better multisensory integrators. Proc. Natl. Acad. Sci. U. S. A. 104, 7295–7300.

Rouger, J., Lagleyre, S., Démonet, J.F., Fraysse, B., Deguine, O., Barone, P., 2012. Evolution of crossmodal reorganization of the voice area in cochlear-implanted deaf patients. Hum. Brain Mapp. 33, 1929–1940.

Sandmann, P., Dillier, N., Eichele, T., Meyer, M., Kegel, A., Pascual-Marqui, R.D., et al., 2012. Visual activation of auditory cortex reflects maladaptive plasticity in cochlear implant users. Brain 135, 555–568.

Schofield, B.R., Motts, S.D., Mellott, J.G., Foster, N.L., 2014. Projections from the dorsal and ventral cochlear nuclei to the medial geniculate body. Front. Neuroanat. 8, 10.

Schürmann, M., Caetano, G., Hlushchuk, Y., Jousmaki, V., Hari, R., 2006. Touch activates human auditory cortex. Neuroimage 30, 1325–1331.

Scott, G.D., Karns, C.M., Dow, M.W., Stevens, C., Neville, H.J., 2014. Enhanced peripheral visual processing in congenitally deaf humans is supported by multiple brain regions, including primary auditory cortex. Front. Hum. Neurosci. 8, 177.

Sharma, A., Campbell, J., Cardon, G., 2015. Developmental and cross-modal plasticity in deafness: evidence from the P1 and N1 event related potentials in cochlear implanted children. Int. J. Psychophysiol. 95, 135–144.

Shepherd, R.K., Baxi, J.H., Hardie, N.A., 1999. Response of inferior colliculus neurons to electrical stimulation of the auditory nerve in neonatally deafened cats. J. Neurophysiol. 82, 1363–1380.

Shibata, D., 2007. Differences in brain structure in deaf persons on MR imaging studied with voxel-based morphometry. Am. J. Neuroradiol. 28, 243–249.

Shore, S.E., 2005. Multisensory integration in the dorsal cochlear nucleus: unit responses to acoustic and trigeminal ganglion stimulation. Eur. J. Neurosci. 21, 3334–3348.

Starlinger, I., Niemeyer, W., 1981. Do the blind hear better? Investigations on auditory processing in congenital or early acquired blindness. I. Peripheral functions. Audiology 20, 503–509.

Stropahl, M., Plotz, K., Schönfeld, R., Lenarz, T., Sandmann, P., Yovel, G., et al., 2015. Cross-modal reorganization in cochlear implant users: auditory cortex contributes to visual face processing. Neuroimage 121, 159–170.

Sumby, W.H., Pollack, I., 1954. Visual contribution to speech intelligibility in noise. J. Acoust. Soc. Am. 26, 212–215.

Summerfield, Q., 1987. Some preliminaries to a comprehensive account of audio-visual speech perception. In: Dodd, B., Campbell, R. (Eds.), Hearing by Eye: The Psychology of Lip-Reading. Lawrence Erlbaum, Hillsdale, NJ.

Szycik, G.R., Stadler, J., Tempelmann, C., Münte, T.F., 2012. Examining the McGurk illusion using high-field 7 Tesla functional MRI. Front. Hum. Neurosci. 6, 95.

Tiippana, K., 2014. What is the McGurk effect? Front. Psychol. 5, 725.

Tye-Murray, N., Sommers, M.S., Spehar, B., 2007. Audiovisual integration and lipreading abilities of older adults with normal and impaired hearing. Ear Hear. 28, 656–668.

Wong, C., Chabot, N., Kok, M.A., Lomber, S.G., 2015. Amplified somatosensory and visual cortical projections to a core auditory area, the anterior auditory field, following early- and late-onset deafness. J. Comp. Neurol. 523, 1925–1947.

Wu, C., Stefanescu, R.A., Martel, D.T., Shore, S.E., 2015. Listening to another sense: somato sensory integration in the auditory system. Cell Tissue Res. 361, 233–250.

Part II

The Problem

Chapter 4

Hearing Problems

Imagine that your hearing sensitivity for pure tones is exquisite—not affected by frequent exposure to loud music or other noises—but that you have great problems in understanding speech even in a quiet environment. This occurs if you have a temporal processing disorder. Although hearing loss (HL) is in the ear, hearing problems originate in the brain. We often take hearing for granted; not realizing what good hearing allows us to do. Without hearing, communication with our fellow humans largely disappears. Substitutes for loss of hearing are sign language, which replaces hearing with vision, and cochlear implants, which restore hearing to a large extent. For hard of hearing persons, amplification with hearing aids restores the sense of sound but does not generally result in normal perception, except if the HL is of the conductive type. Here we will identify some effects of noise exposure on the central nervous system (CNS). In addition, we will present some annoying side effects of HL: a primary one being tinnitus, but an even more disturbing one is an increased probability of dementia.

4.1 THE VARIOUS CONSEQUENCES OF NOISE EXPOSURE

The commonly accepted level duration of exposure trade-off is shown in Fig. 4.1 together with equivalent exposure levels for permanent threshold shift (PTS), temporary threshold shift (TTS), and "assumed safe" conditions. It should be noted that "safe" means a very low likelihood of cochlear damage, and that all these conditions are based on 8 h exposure per day. The region indicated by "?" does not fulfill that criterion, and our own results have shown that there are specifically long-lasting effects on the auditory thalamus and cortex (Section 4.5). Nevertheless, there were no HLs, not even temporary, and no outer hair cell (OHC) abnormalities. So according to these criteria the time–intensity trading "red line" in Fig. 4.1, demarcating "safe" from "threatening," continues in the "?" region.

The vertical axis in Fig. 4.1 is in dB(A), where 0 dB(A) is the minimal audible sound intensity at each frequency in people with normal hearing. dB (A) and dB SPL are very close (±5 dB) between 0.5 and 6 kHz (Fig. 4.2). Regulations on admissible sound levels do not generally apply to recreational

Hearing Loss. DOI: http://dx.doi.org/10.1016/B978-0-12-805398-0.00004-9

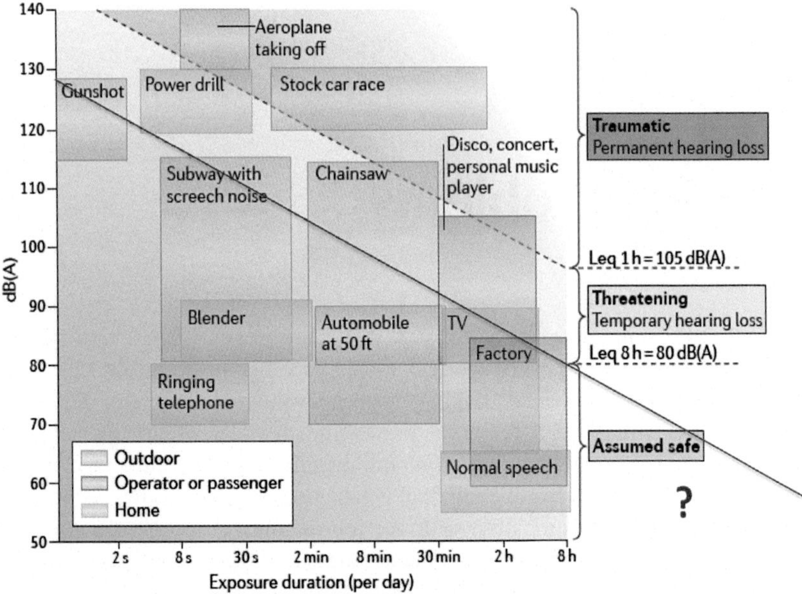

FIGURE 4.1 Typical range of common noise exposure durations and levels for a day. Such exposure profiles give an approximate range of equivalent sound level measure (Leq) for each noise source. A noise level of Leq 1 h = 105 dB(A) is currently accepted as traumatic—that is, inducing a permanent hair cell loss—and most studies assume that up to Leq 8 h = 80 dB(A) is safe for the auditory system over a lifetime. "A-weighting is the most commonly used of a family of curves defined in the International standard IEC 61672:2003 for the measurement of sound levels. The integrated energy below the weighting curve is typically indicated as dB(A). The A-weighting curve, which is basically the inverse of the human threshold as a function of frequency, is mandated for the measurement of environmental noise and industrial noise, as well as when assessing potential hearing damage and other noise health effects at all sound levels" (Eggermont, 2014). Assumed safe means that there is no structural damage to the cochlea and no loss of hearing sensitivity but could mean that there are potentially profound changes to the auditory cortex, which could lead to communication problems. Between 80 and 105 dB(A) is an area that is structurally threatening for the auditory system, in the sense that associated temporary HL for a few hours or days following the exposure is reported. The red question mark in the bottom right signals the putative safe zone for uninterrupted low-level exposure with long duration. *Modified from Gourévitch B., Edeline J.-M., Occelli F., Eggermont J.J., 2014. Is the din really harmless? Experience-related neural plasticity for non-traumatic noise levels. Nat. Rev. Neurosci. 15, 483–491.*

areas such as bars, sports, and concert venues (with some countries, such as the United Kingdom, being an exception), and if they do apply then typically it is only to what is audible outside the venue. While ear protection inside these establishments is advisable, it is generally not complied with. For instance, Cranston et al. (2013) identified possible hazardous noise sources in two hockey arenas, namely, enhanced reverberant conditions from enclosed environment, loud public address systems, and design and building

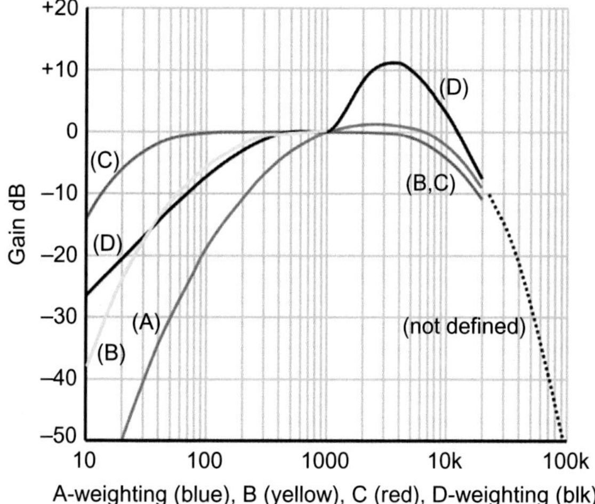

FIGURE 4.2 Filter curves (weightings) for sound level measurements (left). The filter gain is plotted as a function of sound frequency on double logarithmic scales. Sound levels with indications of sources and approximate loudness (right). A-weighting (blue) is the most commonly used of a family of curves defined in the International standard IEC 61672:2003 for the measurement of sound levels. The integrated energy below the weighting curve is typically indicated as dB(A). A-weighting, which is basically the inverse of the human threshold as a function of frequency, is mandated for the measurement of environmental noise and industrial noise, as well as when assessing potential hearing damage and other noise health effects at all sound levels. For measuring low-frequency (infra) sounds the C-weighting (red) is better. The B and D weightings are hardly ever used. *Modified from Eggermont, J.J., 2014. Noise and the Brain. Experience Dependent Developmental and Adult Plasticity, Academic Press, London.*

materials. They applied personal dosimeters worn by both workers and fans and area noise monitors to evaluate exposure levels in both venues. Noise monitor measurements for both venues recorded peak levels of 105–124 dB(A) with equivalent continuous SPLs 81–96 dB(A), and 110–117 dB(A) with equivalent continuous SPLs 85–97 dB(A). Taking 80 dB(A) and 8 h (Leq 8 h) exposure as the industrial norm, and using an exchange of 3 dB (the National Institute of Occupational Safety & Health, NIOSH norm) for time doubling or halving, would then result in an allowable exposure time at 100 dB of about 5 min/day (Fig. 4.1). Leq is a measure of the total sound energy averaged over the duration of the observation period (e.g., 8 h). Formally, this is $20\log_{10}$ of the ratio of a root mean square dB(A)-weighted sound pressure during the stated time interval to the reference sound pressure, divided by the exposure duration (Fig. 4.2). This gives a single value of sound level for any desired duration based on the amount of sound energy contained in the time-varying sound. Note that the use of Leq is often discouraged for very long durations (from months to years) a region indicated in Fig. 4.1 with the red question mark.

TABLE 4.1 Noise Exposure Phenomenology

	PTS noise	TTS noise	"Safe" noise–long duration
Hearing loss	Yes	No	No
ABR threshold shift	Yes	No	No
ABR wave I amplitude	Decreased	Decreased	Normal
Otoacoustic emission loss	Yes	No	No
Central gain change	Yes	Yes	Yes
SFR	Increased	Increased	Decreased/increased
Xcorr	Increased	?	Increased
Tonotopic map	Reorganized	?	Reorganized
Speech-in-noise deficit	Yes	Yes	Yes
OHC	Damage	Normal	Normal
IHC	Damage likely	Modiolar ribbon synapses	Normal
ANFs	Loss in HL range	Loss of high-threshold ANFs	Normal
Central damage	Yes	Yes	Yes

ABR, auditory brainstem response; ANFs, auditory nerve fibers; IHC, inner hair cell; SFR, spontaneous firing rate.

Depending on the exposure level and duration, several consequences result for the auditory system. They all share CNS changes, which may or may not be structural. These consequences are shown in Table 4.1 and further explained in the text.

4.1.1 Structural Changes in the Auditory Periphery

The effect of loud sound on the cochlea depends on the type used. Impulse and impact noises, such as gunfire, are characterized by high intensity and short duration and may produce immediate mechanical alterations to the cochlea. Continuous exposure at moderate to severe noise levels typically produces more subtle changes. Narrow-band noise exposure at approximately 110 dB SPL for 2 h resulted after more than 4 months postexposure in about 40 dB threshold elevation in cat ears and was correlated with loss or damage to hair cells. The orderliness of the stereocilia, on both IHC and OHC, showed the closest correlation with the thresholds of firing of ANFs (Liberman and Beil, 1979).

Scanning electron microscope examination of guinea pig cochleas immediately after a 1 h exposure to a pure tone ranging from 96 to 129 dB SPL showed little hair cell loss but widespread damage to the stereocilia (Robertson and Johnstone, 1980). The stereocilia are the locus of the mechanically sensitive ion channels that mediate transduction (Hudspeth, 2013; see chapter: Hearing Basics). In general, the OHCs are more vulnerable to noise trauma than the IHCs, regardless of the type of noise. This susceptibility may be an inherent property of the OHC biochemistry (Saunders et al., 1985) since there is a relation between the pathology seen with noise exposures and the vulnerability of OHCs to ototoxic drugs (Slepecky, 1986). As stereocilia damage progresses from disarray to fusion or loss, interfering with the patency of the transduction channels, the ANF firing thresholds increase. Subsequent cellular impairment involves protein-, lipid-, and glucose synthesis needed for cell repair and survival, and this results in permanent cell injury or cell death, leading to a permanent HL (Lim, 1986). Noise exposure also induces neurotoxicity caused by excessive release of glutamate by the IHCs, resulting in the influx of large and toxic quantities of Ca^{2+} ions in the postsynaptic area, and disruption of the synapse. This synaptic damage may extend well beyond regions with hair cell damage and likely can be attributed to excessive activation of those physically nondamaged hair cells. Glutamate neurotoxicity can recover in about 1 week after the trauma (Fig. 4.3; Puel et al., 1997; Nouvian et al., 2015).

In a mouse model, acoustic trauma causes retrograde degeneration of ANFs. It could be secondary when it follows IHC loss or primary as a result

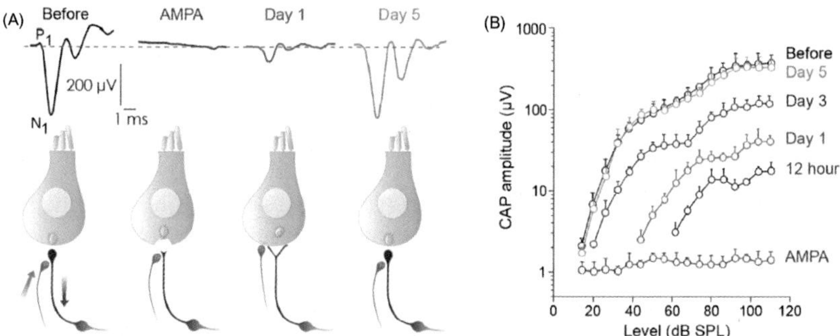

FIGURE 4.3 Synaptic rearrangement after α-amino-3-hydroxy-5-methyl-4-isoxazolepropionic acid (AMPA)-induced neurotoxicity. (A) The compound action potential (CAP), which reflects the synchronous activation of the ANFs, is completely abolished following intracochlear 200 μM AMPA perfusion and recovers during the next days (top). Schematic representation of the innervation changes during neurotoxic injury in the cochlea. Swelling of the afferent terminals caused by AMPA application accounts for the loss of CAP. During the repair of the afferent fibers terminals, efferent terminals connect in a transient fashion the hair cells (bottom). (B) CAP amplitude plots against the intensity of the sound stimulation (probed at 8 kHz). Note that the CAP amplitude recovers completely within 5 days after the neurotoxic injury. *From Nouvian, R., Eybalin, M., Puel, J-L., 2015. Cochlear efferents in developing adult and pathological conditions. Cell Tissue Res. 361, 301–309, with permission of Springer.*

of damage of specific ribbon synapses in the IHCs (Kujawa and Liberman, 2015). The first affected synapses are those with high-threshold, low-SFR, nerve fibers. The ANF degeneration starts when the peripheral dendrites to the IHCs are irreversibly damaged (Spoendlin, 1976). The integrity of its peripheral dendrite, and likely also the ribbon synapse, seems to be essential for the neuron to survive (Kujawa and Liberman, 2009, 2015). Without the peripheral dendrite synapsing on the IHCs there is no action potential activity in the ANFs and that may be required for their viability. ANFs with elevated thresholds at the characteristic frequency as a result of noise exposure in general have frequency-tuning curves with an abnormal shape (Liberman and Kiang, 1978; Robertson, 1982; Salvi et al., 1983). A strong correlation was found between cochlear regions with fused IHC stereocilia and ANF regions with V-shaped tuning curves and a reduction in SFR. Selective damage to the OHCs did not reduce the SFR, indicating that SFR in ANFs only depends on spontaneous transmitter release by the IHCs.

4.1.2 Central Effects of Permanent Threshold Shifts

The noise level and exposure durations that produce a PTS and structural changes are more extensively discussed in Chapter 6, Causes of Acquired Hearing Loss. Besides inducing a loss of sensitivity for sound, as reflected in the audiogram, there are also changes in the CNS. I reiterate here (Eggermont, 2014) that most studies on PTS in the central auditory system have been based on recordings from the brainstem (medulla, olive, pons) and midbrain. Adult chinchillas exposed to an octave-band noise centered at 4 kHz for 105 min at a level of 108 dB SPL (Morest et al., 1998) showed besides hair cell damage also partial deafferentation of the ipsilateral cochlear nucleus (CN). Most of this loss resulted subsequently to the degeneration of the ANFs. New growth of axons and axonal endings was observed in the CN following the deafferentation. Sixteen days following this exposure, the chinchillas showed transient axonal degeneration in the dorsal CN that was no longer visible at longer survival times. Meanwhile, ANF degeneration continued to extend basally in the cochlea, and 2 weeks to 2 months later was followed by spread of axonal degeneration into the corresponding high-frequency region of the ventral cochlear nucleus (VCN). Following a 3 h exposure to the same sound, there was degeneration of hair cells and ANFs (Muly et al., 2002). Secondary synaptic degeneration in the VCN was visible 1−16 weeks after the trauma. After several months, however, all these changes reversed and eventually the ANF endings again showed a normal appearance (Kim et al., 2004a). For periods of 6 and 8 months after this single exposure they observed a chronic, continuing process of neurodegeneration involving excitatory and inhibitory synaptic endings. This was accompanied by newly formed synaptic endings, which repopulated some of the sites vacated previously by axosomatic endings on globular bushy cells in the anterior part of the VCN. Noise-induced hearing loss (NIHL) thus resembles in part a neurodegenerative disease with the capacity for synaptic reorganization within the CN (Kim et al., 2004c). After noise exposure and recovery for up to 32 weeks, neuronal cell bodies lost both excitatory and inhibitory endings at first and later recovered a full complement of excitatory but not inhibitory terminals (Kim et al., 2004b). In addition to changes in the CN, there was transsynaptic terminal axonal degeneration in nuclei of the superior olivary complex, lateral lemniscus, and inferior colliculus (IC) (Morest et al., 1997; Kim et al., 1997). A consequence of these subcortical structural changes may be cortical tonotopic map reorganization. Melcher et al. (2013) have shown that a high-frequency HL in humans causes atrophy in prefrontal cortex. Typically, high-frequency HL is also accompanied by increased SFR from the CN to auditory cortex and includes increased neural synchrony, and tonotopic map reorganization in primary auditory cortex (Noreña and Eggermont, 2003, 2006).

4.1.3 Central Effects of Temporary Threshold Shifts

Lower level exposures often lead to TTSs (Fig. 4.1). Using confocal imaging of the inner ear in the mouse, Kujawa and Liberman (2009) showed that acoustic overexposure for 2 h with an 8–16 kHz band of noise at 100 dB SPL caused a moderate, but completely reversible, threshold elevation as measured by ABR. They observed no permanent changes in the otoacoustic emissions, which indicated that the exposure left OHCs, and therefore likely the less susceptible IHCs as well, intact. Despite the normal appearance of the cochlea and normal ABR thresholds there was an acute loss of the ribbon synapses located on the medial side of the IHC that are connected to high-threshold, low-SFR ANFs. This corresponded to the loss of suprathreshold ABR wave-1 amplitude for frequencies that showed a TTS. After some delay, this was followed by a progressive diffuse degeneration of the auditory nerve for these TTS frequencies. Likely consequences following TTS in humans, where clinical audiograms are normal, are difficulties understanding speech particularly in background noise, because of the loss of the high-threshold, low-SFR fiber activity, which is difficult to saturate by noise.

Recently, Wang and Ren (2012) showed that repeated TTS noise exposures (also 8–16 kHz noise presented at 100 dB SPL for 2 h) in initially 4-week-old CBA/CaJ mice do affect permanent hearing thresholds (HTs). Although ABR thresholds recovered fully in once- and twice-exposed animals, the growth function of ABR wave-1 amplitude (CAP activity) was significantly reduced at high SPL. However, a third dose of the same noise exposure resulted in a PTS. The pattern of PTS resembled that of age-related HL, i.e., high-frequency hearing impairment. Wang and Ren (2012) found that threshold elevation at the frequency locus matched with synaptic ribbon loss in the IHCs. They suggested that accumulation of afferent synapse damage over time with recurrent noise exposures could be a major contributor to age-related hearing impairment. Recently, Hickox and Liberman (2014) showed that stimulation at 94 dB SPL (2 h with an 8–16 kHz band of noise) did not result in significant synaptic ribbon loss. However, it cannot be excluded that repeated exposure at this level could cause ribbon loss, or even PTS.

4.1.4 Central Effects of Noise Exposure Without Threshold Shifts

Noreña et al. (2006) reported that at least 4 months of passive exposure of adult cats to a 4–20 kHz tone pip ensemble presented at 80 dB SPL, termed an enhanced acoustic environment (EAE), decreased the responsiveness of primary auditory cortex to sound frequencies in the exposure band, and increased the responsiveness to frequencies at the outer edges of the band. The reduced responsiveness to the EAE frequencies may be partially the result of habituation (see chapter: Brain Plasticity and Perceptual Learning). The increased responsiveness to frequencies above and below the EAE frequency band may result from reduced lateral inhibition provided by the neurons responding to the EAE frequencies. No HL as measured by ABR was found. Recordings of the SFR in control and EAE cats were split into three groups according to the location of the recording electrodes along the posteroanterior axis: posterior (<10% of the posteroanterior sulci distance), middle (10–70% of the posteroanterior sulci distance), and anterior (>70% of the posteroanterior sulci distance). These locations correspond to regions with CFs of less than 4, 4–20, and more than 20 kHz, respectively. Noreña et al. (2006) found that the "SFR was not significantly different between control and EAE cats for recordings with CFs normally corresponding to the EAE spectrum (middle area). On the other hand, the SFR was significantly increased for recordings normally corresponding to CFs of less than 5 kHz (posterior area) and more than 20 kHz (anterior area)." During 15 min periods of silence, Noreña et al. (2006) also recorded the spontaneous multi-unit activity (MUA) from pairs of electrodes and then calculated the neural synchrony (peak cross-correlation coefficient; Eggermont, 1992) between them. Noreña et al. (2006) found that the synchrony strength was lower in control cats (Fig. 4.4A) than in EAE cats (Fig. 4.4B) in posterior, middle, and anterior areas. Even the local neural synchrony (i.e., for electrode distances of ≤1 mm) was significantly increased in EAE cats as compared with control cats throughout the entire mapped area. In addition, in EAE cats, the neural synchrony was significantly enhanced in the posterior and anterior areas of A1 as compared with the middle area.

Pienkowski and Eggermont (2009) subsequently explored the effects of "safe" noise exposures, i.e., considerably below the 80 dB(A), 8 h equivalent, but with longer durations (in the "?" region of Fig. 4.1). They demonstrated qualitatively similar plastic changes for a 6-week EAE exposure at a level of

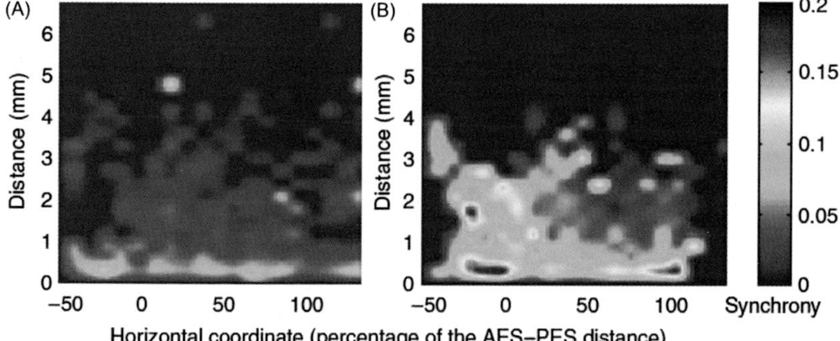

FIGURE 4.4 Neural synchrony maps in A1. (A and B) Synchrony, defined here as the peak strength of the corrected cross-correlogram, as a function of the position of the recording electrode along the posteroanterior axis (abscissa) and the distance between the two electrodes involved in the calculation of synchrony (ordinate) in (A) control and (B) EAE cats. For each electrode pair, positions along the posteroanterior axis are plotted. Colored bar—strength of neural synchrony. In control cats, the strongest synchrony was found between neighboring electrodes in the array and most correlations occurred locally. Note the increased synchrony in EAE cats as compared with control cats, especially for larger distances between electrodes. This probably signifies the stronger connections over large distances (i.e., into the reorganized region) made by horizontal fibers. In these cats, the range of strong correlations is much larger, especially in the −50 to 50% region, which reflects the entire area with characteristic frequencies of 5 kHz but also a substantial part of the 5–20 kHz area. In addition, the area with CFs of more than 20 kHz (70–125%) also showed strongly increased neural synchrony. AES, anterior ectosylvian sulcus; PES, posterior ectosylvian sulcus. *From Noreña, A.J., Gourévitch, B., Aizawa, N., Eggermont, J.J., 2006. Enriched acoustic environment disrupts frequency representation in cat auditory cortex. Nat. Neurosci. 9, 932–939.*

68 dB SPL. Again, no peripheral HL was induced by the exposure, and the resulting reorganization of the A1 tonotopic map occurring mostly during the 8- to 12-week recovery period (and much less during the exposure) resembled that following a restricted lesion of the sensory epithelium. They concluded that exposure-induced effects were present in the thalamus, as deduced from changes in short-latency, sound-evoked local field potentials (LFPs), but were further modified in A1. In contrast to the exposure at 80 dB SPl for at least 4 months (Noreña et al., 2006), Munguia et al. (2013) found that the SFR was not significantly changed for the EAE frequency range in exposed cats, but was significantly enhanced in the non-EAE frequency ranges as compared with controls in the same frequency range. The SFR in the EAE range was also significantly lower than in the non-EAE range in exposed cats (Fig. 4.5A).

Pienkowski and Eggermont (2009) also computed correlations in spontaneous spike firing between MUA pairs recorded on separate electrodes in AI. Averaged peak correlation coefficients are plotted in Fig. 4.5B as a function of the interelectrode distance, pooled in half-octave distance bins; data from control cats (2982 pairs) are shown as solid lines, data from cats exposed at

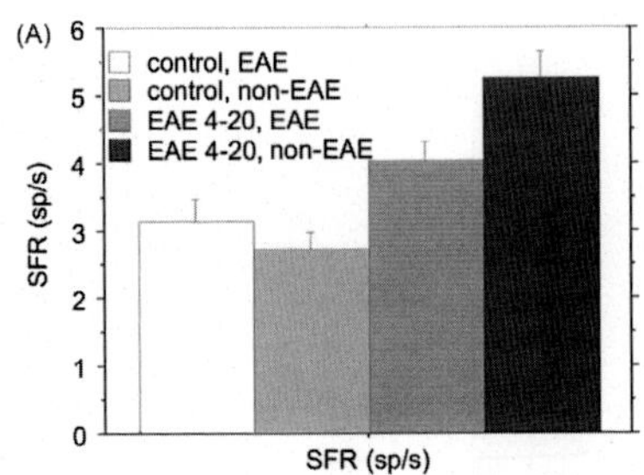

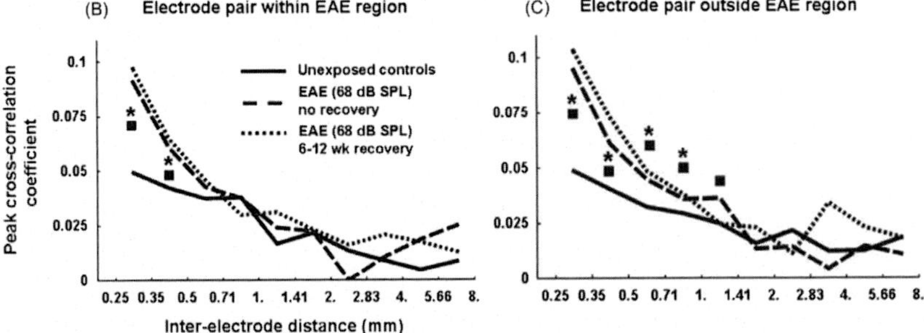

FIGURE 4.5 (A) Averaged SFR for EAE and non-EAE frequency ranges in control and exposed cats. (B, C) Averaged peak cross-correlation coefficients for spontaneous firing between multiple single-unit pairs in A1, plotted as a function of interelectrode distances (pooled in half-octave bins), for a group of unexposed control cats (solid curve), for cats exposed at 68 dB SPL and allowed no recovery (dashed curve), and for cats exposed at 68 dB SPL and allowed 6–12 weeks of recovery (dotted curve). Boxes and asterisks indicate significant differences in synchrony from control levels, at least at $p < 0.05$, for the no recovery and the 8- to 12-week groups, respectively. Results are plotted separately for electrode pairs within the EAE region (B) and for pairs located those outside the EAE region (C). *Reprinted from (A) Munguia, R., Pienkowski, M., Eggermont, J.J., 2013. Spontaneous firing rate changes in cat primary auditory cortex following long-term exposure to non-traumatic noise. Tinnitus without hearing loss? Neurosci. Lett. 546, 46–50, with permission from Elsevier; (B) Pienkowski, M., Eggermont, J.J., 2009. Long-term, partially-reversible reorganization of frequency tuning in mature cat primary auditory cortex can be induced by passive exposure to moderate-level sounds. Hear. Res. 257, 24–40, with permission from Elsevier.*

68 dB SPL and allowed no recovery (401 pairs) are shown as dashed lines, and data from cats exposed at 68 dB SPL and allowed 8–12 weeks of recovery (1250 pairs) are shown as dotted lines. As is usually observed (e.g., Eggermont, 2006), there was a decrease in the synchrony of spontaneous firing with increasing interelectrode distance in all three groups. In the shortest distance bin (0.25–0.35 mm), corresponding to the distance between nearest neighbors on the electrode arrays, the synchrony measured in both the no recovery and the 8- to 12-week recovery groups was approximately double that measured in controls ($p < 10^{-8}$). However, increased synchrony was observed only in the outside-EAE regions up to distances of approximately

2 mm in both exposed groups (Fig. 4.5C; boxes and asterisks indicate significant differences from control data, at least at $p < 0.05$, for the no recovery and 8- to 12-week groups, respectively). Beyond 2 mm, the data were noisier (fewer pairs) and not statistically different. Importantly, the increase in synchrony showed no sign of reverting to normal 12 weeks after the cessation of exposure in either the EAE or outside-EAE regions. If one considers increased SFR and synchrony as substrates of tinnitus, one cannot escape the notion that long-term exposure to nontraumatic noise can cause tinnitus.

The effect was similar but significantly weaker following an intermittent exposure (12 h-on/12 h-off) of the same type, duration, and intensity (Pienkowski and Eggermont, 2010). Pienkowski and Eggermont (2011)'s "intent with the intermittent exposure (12 h-on/12 h-off) was to simulate the alteration of noisy-work/quiet-rest environments, albeit at substantially lower intensity levels (~68 dB SPL) than are presently considered harmful to human hearing (~85 dB(A) for 8 h/day; OSHA, Standard 1926.52). The current standards for occupational noise exposure are apparently intended to just prevent permanent absolute threshold shifts (Kryter et al., 1966; May, 2000), though it has long been suspected that problems such as poor speech intelligibility in noise, as well as tinnitus, could arise after long-term exposure at lower intensities (Kjellberg, 1990; Kujala et al., 2004; Brattico et al., 2005). It may be that cortical response suppression and eventual tonotopic map reorganization, as observed in our studies, represent the neurophysiological underpinnings of such problems."

Do these surprising findings have a bearing on potential changes in the brain occurring from daily exposure to occupational noise? Pienkowski and Eggermont (2010) found qualitatively similar effects of passive exposure occurred when the EAE presentation was limited to 12 h/day. Compared to continuous exposure at the same SPL and over a similar duration (6–12 weeks), this intermittent exposure produced a smaller decrease in AI-neuron firing rate and LFP activity in response to sound frequencies in the exposure range, and an increase in LFP amplitude only for frequencies above the exposure range. As expected at these moderate exposure levels, cortical changes occurred in the absence of concomitant HL (i.e., absolute ABR-threshold shifts). Since there was some overlap in the amount of change in neural activity between the present intermittently exposed group and the continuously exposed group, it is expected that recovery from the effects of the intermittent exposure would also take a long time.

4.2 SOUND LOCALIZATION PROBLEMS

Animal data on sound localization with normal hearing have been presented in Chapter 1, Hearing Basics. Here we first present electrophysiological and psychoacoustical data from normal hearing humans and then proceed to the effects of HL and aging.

4.2.1 Findings in Normal Hearing Humans

Using magneto-encephalography (MEG) in humans to record long-latency cortical potentials, Salminen et al. (2009) found that sounds originating from locations within the same hemifield resulted in the same N_{1m}, (the magnetic equivalent of the N_1; cf. Fig. 8.1) activity distribution regardless of the spatial separation between the sound sources. When sounds were presented from locations in opposite hemifields it resulted in different N_{1m} distributions, suggesting activation of separate groups of neurons. These results are consistent with a rate code of spatial location formed by two opponent populations, one tuned to locations in the left and the other to those in the right.

In an EEG study, also aimed at distinguishing these two dominant models of sound localization, Magezi and Krumbholz (2010) found that the N_1 and P_2 amplitudes were highly consistent with the predictions of the opponent-channel model. They concluded that the majority of interaural time difference (ITD)-sensitive neurons—underlying the N_1 and P_2—in each hemisphere are tuned to ITDs from the contralateral hemifield. Edmonds and Krumbholz (2014) then measured the P_1−N_1−P_2 complex in response to changes in sound lateralization elicited by perceptually matched changes in ITD and/or interaural level difference (ILD). The P_1−N_1−P_2 responses to the ITD and ILD changes showed different morphology, however, they originated from overlapping areas of the cortex and showed clear evidence for functional coupling. Edmonds and Krumbholz (2014) suggested that in auditory cortex sound laterality is coded in an integrative way, but that independent information about ITD and ILD cues is preserved.

Evidence for a third, frontal, channel had been found in psychoacoustic ITD studies with low-frequency tones (Dingle et al., 2010) and ILD studies with both high- and low-frequency tones (Dingle et al., 2012). Dingle et al. (2013) then used a selective adaptation paradigm in combination with transposed tones to further probe for the existence of three (left, right, and midline) perceptual channels for sound source azimuth based on ITDs at high frequencies. They reported little evidence for a midline ITD channel at high frequencies.

4.2.2 Hearing Loss and Sound Localization

Normal hearing listeners can localize well in both the horizontal and vertical planes based on the binaural difference cues, ILDs and ITDs. The smallest error for broadband sounds in a study by Makous and Middlebrooks (1990) was 2° in the front and 9° at 60° azimuth in a two-dimensional head pointing task. However, controversy exists regarding how well hearing-impaired listeners can localize sound. Vertical localization and front-back discrimination rely on high-frequency spectral cues (>5 kHz) that are created by reflection and diffraction of sound by the external ear and, in particular, the pinna.

These cues are often referred to as "monaural spectral cues" since the analysis depends only on a signal being present in one ear, although the spectral cues would be available to both ears. Listeners with high-frequency HL have more difficulty than normal hearing listeners when tested in the sagittal plane and with elevation (Simon, 2005).

Binaural performance can vary markedly across subjects. Subjects with unilateral or asymmetric losses tend to show larger than normal thresholds for detecting ITDs and ILDs. Subjects with symmetrical losses sometimes show normal or near-normal localization for broadband noise stimuli. However, they often show impaired performance for narrowband stimuli (Moore, 1996). He mentioned that Bronkhorst and Plomp (1989) carried out speech-in-noise experiments using 17 subjects with mild to moderate symmetrical HLs and 17 subjects with mild to moderate asymmetrical losses. For speech and noise presented at 0° azimuth, the speech reception thresholds (SRTs) were approximately 2.5 dB higher than found for normally hearing subjects. However, when the noise azimuth was changed to 90° and the speech was still presented at 0°, the hearing-impaired subjects showed SRTs that were 5.1–7.6 dB higher than normal. In summary, hearing-impaired subjects tested under conditions where speech and noise are spatially separated, perform more poorly, relative to normal, than when the speech and noise come from the same position in space. The disadvantage appears to arise mainly from the inaudibility of high frequencies in the ear at which the speech-to-noise ratio is highest (Bronkhorst and Plomp, 1989).

Moore (1996) described potential origins of these localization deficits: (1) The relatively low SL of the stimuli because ITD discrimination in normally hearing subjects worsens markedly below about 20 dB SL; (2) differences in travel time or phase of spike initiation between the two ears; (3) abnormalities in phase locking. Abnormalities in ITD discrimination may occur as well, due to: (1) relatively low SL of the stimuli, (2) abnormal intensity coding, and (3) differences in intensity coding at the two ears. Some people with cochlear damage have essentially no ability to use spectral cues provided by pinna transformations to assess sound elevation.

4.2.3 Aging and Sound Localization

Compared with young human adults, older adults typically localize sound sources less accurately. Dobreva et al. (2011) reported that: "Findings underscore the distinct neural processing of the auditory spatial cues in sound localization and their selective deterioration with advancing age." In contrast, Otte et al. (2013) "found similar localization abilities in azimuth for all listeners, including the older adults with high-frequency hearing loss." According to a review by Briley and Summerfield (2014): "This contradiction could reflect stimulus or task differences between the two studies, or it could reflect differences in the participant groups." With respect to age, in the Dobreva et al. (2011) study the older adults were 70−81 years of age, whereas in the Otte et al. (2013) study they were 63−80 years of age. High-frequency HLs also began on average at a lower frequency in the Dobreva et al. (2011) population than in the Otte et al. (2013) group. Thus, it could be HL as well as age that underlie the localization problems.

4.3 THE COCKTAIL PARTY, WHERE IDENTIFICATION AND LOCALIZATION COME TOGETHER

> *Animals often use acoustic signals to communicate in groups or social aggregations in which multiple individuals signal within a receiver's hearing range. Consequently, receivers face challenges related to acoustic interference and auditory masking that are not unlike the human cocktail party problem, which refers to the problem of perceiving speech in noisy social settings. (Bee and Micheyl, 2008)*

One well-known example of difficulties that humans face in auditory scene analysis (Bregman, 1990) is described as the "cocktail party problem" (Cherry, 1953). This refers to the difficulty that even normal hearing persons sometimes have to understand speech in noisy social settings. Cherry (1953) studied how one can distinguish one speaker among a multitude of others in a crowded noisy scene such as a cocktail party. He asked: "On what logical principles could one design a machine whose reaction, in response to speech stimuli, would be analogous to that of a human being? How could it separate one of two simultaneous spoken messages?" McDermott (2009) described two distinct challenges for a listener in a "cocktail party" situation. The first is the problem of sound segregation. The auditory system must derive the properties of individual sounds from the mixture entering the ears, i.e., through a binaural filter to assess spatial proximity. The second challenge is that of directing attention to the sound source of interest while ignoring the others, i.e., an attention filter, and of switching attention between sources, as when intermittently following two conversations. Most of our cognitive processes can operate only on one thing at a time, so we typically select a particular sound source on which to focus. This process affects sound segregation, which is biased by what we attend to.

In a recent study, Bidet-Caulet et al. (2008) used 21 and 29 Hz amplitude-modulated sounds and recorded neural responses directly from the auditory cortex of epileptic patients. They found that different mechanisms were involved in the segregation or grouping of overlapping components as a function of the acoustic context. In their study, sound onset asynchrony was manipulated to induce the segregation or grouping of two concurrent components. This was done by starting the 21 Hz component 800 ms before the 29 Hz component (pitch continuity of the 21 Hz component), resulting in the perception of two streams. When the 21 and 29 Hz components were started synchronously this led to the percept of one sound. According to the authors, synchronization of transient responses could account for grouping of overlapping auditory components.

Considering the effects of HL on spatial filtering, Glyde et al. (2013) recognized that the process of understanding speech in simultaneous background noise involves separating the acoustic information into discrete streams. However, they found no significant relationship between spatial-processing ability and aging. In addition, no significant relationship was found between cognitive ability and spatial processing.

Recently, I summarized the cocktail party problem as: "It is amazing that in a noisy, multiple-people-talking environment, listeners with normal hearing can still recognize and understand the attended speech and simultaneously ignore background noise and irrelevant speech stimuli. ... Temporal structure has a key role in stream segregation. Timing synchrony of frequency partials allow fusion into a more complex sound, and if the frequency partials are harmonic the fusion is more likely. In contrast, timing asynchrony is a major element to distinguish one stream from two streams. Animal experiments have highlighted the role of temporal aspects by comparing behavioral data and recordings from the forebrain in the same species. Feature dependent forward suppression in auditory cortex may underlie streaming. Modeling studies suggest that stream formation depends primarily on temporal coherence between responses that encode various features of a sound source. Furthermore, it is postulated that only when attention is directed towards a particular feature (e.g. pitch) do all other temporally coherent features of that source (e.g. timbre and location) become bound together as a stream that is segregated from the incoherent features of other sources" (Eggermont, 2015).

4.4 OTHER CONSEQUENCES OF HEARING LOSS

4.4.1 Hyperacusis

Hyperacusis reflects an unusual hypersensitivity or discomfort induced by exposure to sound. To dissociate mechanisms underlying hyperacusis and tinnitus, Sheldrake et al. (2015) measured audiograms and loudness discomfort levels (LDLs) in 381 patients with a primary complaint of hyperacusis for the full standard audiometric frequency range from 0.125 to 8 kHz. On average, patients had mild high-frequency HL, but more than a third of the tested ears had normal HTs, i.e., $\leq$20 dB HL. LDLs were found to be significantly decreased compared to a normal hearing reference group, with average values around 85 dB HL across the frequency range. Sheldrake et al. (2015) found that LDLs tended to be higher at frequencies where HL was present ($r = 0.36$), suggesting that hyperacusis is unlikely to be caused by HT increase, in contrast to tinnitus for which HL is a main trigger. They also found that LDLs were decreased for all audiometric frequencies, regardless of the pattern or degree of HL. This suggested to Sheldrake et al. (2015) that hyperacusis might be due to a generalized increase in auditory gain, whereas tinnitus was thought to result from neuroplastic changes in a restricted frequency range. Thus, tinnitus and hyperacusis might not share a common mechanism.

The following description of potential mechanisms for hyperacusis is based on a more extensive exposition in chapter 3 of "The Neuroscience of Tinnitus" (Eggermont, 2012), and information published in more recent years is added.

4.4.1.1 Peripheral Aspects

Some peripheral hearing impairments give rise to oversensitivity for loud sounds (Baguley, 2003). For instance, hyperacusis has been described following abolition of the stapedial reflex as, e.g., seen in facial nerve palsy. Hyperacusis can also occur with Ménière's disease. Many people with hyperacusis have "normal" audiograms, thereby excluding hyperacute thresholds as well as hearing impairment (Anari et al., 1999). Recent work in animals has shown that noise exposure resulting in specific ribbon synapse damage can cause permanent degeneration of the cochlear nerve despite complete recovery from TTS (Kujawa and Liberman, 2009) and potentially hyperacusis (Hickox and Liberman, 2014). Hyperacusis is accompanied by increased amplitude of distortion product otoacoustic emissions in tinnitus patients without HL (Sztuka et al., 2010). This could point to involvement of the efferent system (see chapter: Hearing Basics).

4.4.1.2 Central Mechanisms

One mechanism that may maintain or enhance the neural response to a sound in the face of HL is homeostatic plasticity (Turrigiano, 1999), which enhances synaptic efficacy to maintain a relatively constant level of neural activity when input from the periphery in reduced (Schaette and Kempter 2006). However, there is a number of unanswered questions, including at what station(s) of the central pathway the plastic changes occur, in which populations of neurons, and how the changes interact with peripheral changes to produce enhanced central responses. Most studies have used evoked LFP recordings (e.g., Salvi et al., 1990; Wang et al., 2002) that did not allow identification of the neuronal sources of changes.

Hyperacusis and loudness recruitment may interact in the effects on firing rate and LFP amplitudes at elevated SPL. Following acoustic trauma, chopper neuron responses in the anteroventral cochlear nucleus (Cai et al., 2008) show sharply elevated amplitudes and discharge rates at high sound levels (loudness recruitment; see chapter: Types of Hearing Loss). This loudness recruitment may coincide with hyperacusis and is then called over-recruitment, in which intense sounds become too loud for comfortable listening (Sherlock and Formby, 2005; Nelson and Chen, 2004). In reality, hyperacusis patients often cannot stand relatively soft sounds either and complain that they are too loud. We thus have to look for a potential neural correlate that shows strongly increased neural responses starting just above threshold.

Clinical conditions may co-occur with hyperacusis as extensively described in two review papers (Marriage and Barnes, 1995; Katzenell and Segal, 2001). Here, I only mention a few of these conditions (more in Eggermont, 2012). Hyperacusis occurs in migraine, where the prevalence is reported to be between 70% and 83% during attacks and 76% between attacks. Hyperacusis co-occurs with unusual hypersensitivity or discomfort induced by exposure to light (Phillips and Hunter, 1982). Hyperacusis is also found in depression and posttraumatic stress disorder, and in benzodiazepine dependence. In each of these conditions (and in many others not mentioned) there is a disturbance in serotonin (5-HT) function (Marriage and Barnes, 1995; Simpson and Davies, 2000; Al-Mana et al., 2008). The serotonergic system modulates neuronal habituation responses to repetitive stimulation and may provide a "gain-control" of excitatory and inhibitory mechanisms (Hegerl and Juckel, 1993). Wutzler et al. (2008) found that the sound level dependence of cortical auditory evoked potentials and loudness of sound is inversely related to serotonergic neuronal activity. Serotonin already has a strong excitatory effect on fusiform cells in the dorsal CN (Tang and Trussell, 2015).

4.4.2 Tinnitus

Tinnitus is the conscious perception of sound heard in the absence of physical sound sources external or internal to the body (Eggermont, 2012). Sound perceived from physical sound sources inside the body such as blood flow and middle ear muscle twitching is generally called "objective tinnitus." Here, I will not deal with objective tinnitus, and refer to subjective tinnitus as "tinnitus." About 10–15% of adults experience tinnitus (see chapter: Epidemiology and Genetics of Hearing Loss and Tinnitus). Tinnitus is often a result of HL caused by noise exposure, but most chronic tinnitus is of central origin; that is, it is in the brain and not generated in the ear. The localization of tinnitus to one or both ears is thus likely attributable to a phantom sensation (Jastreboff et al., 1988). As in a true phantom sensation, the brain substitutes the sound of the missing frequencies in one ear, both ears, or inside the head. In case of low-frequency HL the tinnitus is low pitched ("roaring"), but in high-frequency NIHL the tinnitus has a high-pitched ringing or hissing sound. Describing how tinnitus sounds appears to be very personal and is typically referred to with known external sounds. As an illustration, MacNaughton-Jones (1890) who studied 260 cases of tinnitus described the sounds of tinnitus as follows:

> *The following were the noises I have recorded as complained of by patients. The sound resembling buzzing; sea roaring; trees agitated; singing of kettle; bellows; bee humming; noise of shell; horse out of breath, puffing; thumping noise; continual beating; crackling sounds in the head; train; vibration of a metal; whistle of an engine; steam engine puffing; furnace blowing; constant hammering; rushing water; sea waves; drumming; rain falling; booming; railway whistling; distant thunder; chirping of birds; kettle boiling; waterfall; mill wheel; music; bells.*

The following parts on tinnitus are based on the description in the work of Eggermont (2014), expanded by more recent studies.

4.4.2.1 Tinnitus Pitch

The pitch of tinnitus corresponds, when there is a HL, to the frequency region of that HL (Sereda et al., 2015). In case of low-frequency HL the tinnitus is low pitched ("roaring"), but in high-frequency NIHL the tinnitus has a high-pitched ringing or hissing sound. The tinnitus percept can often be synthesized by combining pure tones into a tinnitus spectrum (Langers et al., 2012; Noreña et al., 2002; Penner, 1993) (Fig. 4.6). Pitch-matching reliability varies widely across patients. Pitch matches can also vary from day to day or within a day. This may represent subtle shifts in the dominant frequencies in the tinnitus spectrum. In summary, the tinnitus percept is usually complex in quality. Although tinnitus can sometimes be matched by adjusting the frequency of a pure tone, the matches are often unreliable across sessions. The matching frequencies tend to fall in regions where the HL is greatest. In cases where the tinnitus is described as tonal, and for people with sloping audiograms, the frequency that matches the tinnitus may correspond to an edge in the audiogram, where the HL increases relatively abruptly. Again, more research is needed to confirm this finding. For temporary tinnitus produced by exposure to intense sounds, the frequency that matches the tinnitus may correspond to the upper edge of the region over which maximum TTS occurs.

4.4.2.2 Tinnitus Loudness

Tinnitus loudness is usually measured by a rating procedure or by matching it to the level of external sounds (Ward and Baumann, 2009). The loudness of tinnitus is typically matched to sound levels that are only a few dB above the hearing threshold at the tinnitus frequency. However, because of the recruitment type of HL that frequently underlies tinnitus these few dBs could still represent a fairly loud sound (Moore, 2012). The loudness level of

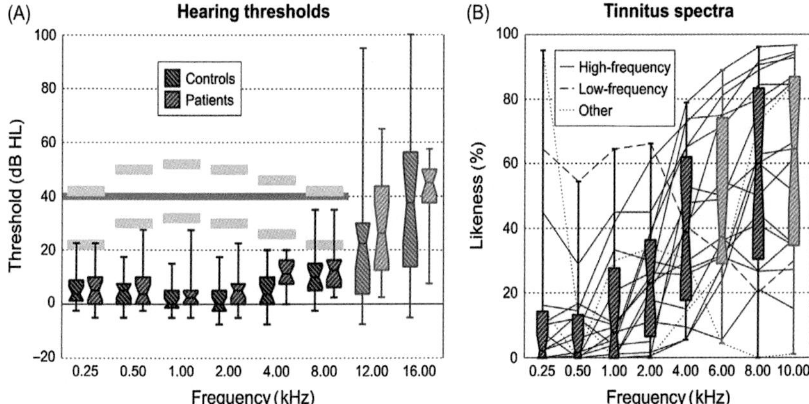

FIGURE 4.6 Similarity between the shape of the audiogram and the tinnitus spectrum. (A) HTs were measured at frequencies from 0.25 to 16.00 kHz. Results were averaged over both ears and shown by means of boxplots (showing interquartile ranges). Stimuli were presented at all octave frequencies from 0.25 to 8.00 kHz at two different intensity levels that differed by 20 dB. The light gray bars indicate the approximate presentation levels. In the analysis, the sound-evoked activation levels were interpolated to a uniform intensity level of 40 dB HL, indicated by the dark gray line. (B) Patients performed a tinnitus spectrum test in which they indicated the subjective "likeness" to their tinnitus percept of a range of sound stimuli with varying center frequencies. The majority of subjects showed high-frequency tinnitus (solid; likeness increasing with frequency); one subject showed a low-frequency tinnitus (dashed; likeness decreasing with frequency); two subjects showed a spectrum that could not be classified as high- or low-frequency (dotted; with a peak or a dip at intermediate frequencies). *From Langers, D.R., de Kleine, E., van Dijk, P., 2012. Tinnitus does not require macroscopic tonotopic map reorganization. Front. Syst. Neurosci. 6, 2, with permission from the authors.*

tinnitus fluctuates and this could be due to test–retest variability, actual fluctuation of the tinnitus loudness, and changes in tinnitus pitch or loudness produced by the measurement stimulus if presented to the tinnitus ear. Presenting a matching stimulus to the contralateral ear might reduce potential interference with tinnitus loudness, but because of central interactions might not completely eliminate them. One way to avoid this sound–tinnitus interaction is to use cross-modal loudness matching or using constraint psychophysical scaling (Ward and Baumann, 2009).

4.4.2.3 Tinnitus Masking and Residual Inhibition

Masking is based on two mechanisms: (1) a so-called "line-busy" effect where the masking sound activates the neurons and prevents them from firing to a probe sound (e.g., tinnitus) and (2) a suppression effect where the masker interferes with the mechanical activity pattern of the probe sound in the cochlea (Delgutte, 1990). Although pure tones can mask tinnitus completely in the majority of patients, masking of tinnitus does not follow the standard effects that a masker has on an external probe sound. It appears that the cochlear suppression mechanism is impaired in tinnitus patients likely because of the HL (based on a comparison of simultaneous masking or forward masking measurements of psychoacoustic frequency tuning). If the changes induced by the masker, and the generation site of tinnitus, were at the cochlear level, the masking of an external pure tone would be similar to the masking of tinnitus. This finding again points to central mechanisms of tinnitus.

Residual inhibition is a postmasking effect that, because of its long duration (usually seconds, but can last for minutes to hours), is a central effect. The residual inhibition is generally largest when using masking sounds in the HL range and that resembled the tinnitus spectrum (Roberts et al., 2008). The results suggest that cortical map reorganization induced by NIHL, which results in an overrepresentation of the edge frequency in the audiogram, is not the principal source of the tinnitus sensation. Because in that case one would expect the tinnitus pitch to match the edge frequency and that edge-frequency sounds would result in the largest residual inhibition. The duration of the residual inhibition is likely related to recovery from the habituation induced by the masker (see chapter: Brain Plasticity and Perceptual Learning).

4.4.2.4 The Role of Neural Synchrony in Tinnitus

Electrophysiological and functional imaging measurements in humans and animals suggest that elevated neural synchrony and increased spontaneous activity in the auditory system are potential neural correlates of tinnitus in humans (Eggermont and Roberts, 2015). Tinnitus could result from maladaptive plasticity of the CNS (Eggermont and Tass, 2015; Wu et al., 2015). The CNS aims to restore its evoked neural activity levels, reduced by the HL, by increasing the efficacy (or gain) of its synapses. But this gain also affects the SFRs and these will then generally increase. This increased SFR is interpreted as sound and called tinnitus (Schaette and Kempter, 2006). In these cases, tinnitus is a consequence of the central gain changes that also underlie some forms of hyperacusis. A puzzling aspect is that only 30% of people with HL experience tinnitus (Nondahl et al., 2011), so there must be other purely CNS aspects, such as attention, that promote or allow the perception of tinnitus (Eggermont, 2012; Roberts et al., 2013; Eggermont and Roberts, 2015).

According to Weisz and Obleser (2014) tinnitus needs increased local excitation, or excitation–inhibition imbalance, as well as synchronization. They further stated that: "enhanced synchronization could manifest itself in the gamma frequency range, increasing the postsynaptic impact of the respective neural population. Evidence exists for an increase of gamma activity in tinnitus as well as for its relationship to basic psychoacoustic features of this condition (van der Loo et al., 2009; Weisz et al., 2007)." Weisz and Obleser (2014) also assumed that reductions of alpha would go along with a greater integration of respective regions in a distributed tinnitus network (for review see, e.g., de Ridder et al., 2014). In this respect it must be noted that

Britvina and Eggermont (2008) found that in ketamine-anesthetized cat primary auditory cortex, moderate-level multifrequency tonal stimulation suppressed spontaneous alpha-frequency oscillations, as shown by the decrease of spectral power within the alpha-frequency range during stimulation as compared with the previous and following silent periods (Fig. 4.7). So, just as occipital alpha oscillations decrease when eyes are open (visual stimulation) compared to when they are closed, the transition from silence to sound decreases temporal alpha.

4.4.2.5 Brain Areas Involved in Tinnitus

Two studies by Melcher et al. set the tone for a positive identification of the auditory brain areas involved in generating tinnitus. Gu et al. (2010) reported physiological correlates of tinnitus and hyperacusis that very frequently co-occur: Patients with and without tinnitus, all with clinically normal HTs (i.e., <25 dB for frequencies ≤ 8 kHz) underwent both behavioral testing for the presence or absence of hyperacusis and fMRI to measure sound-evoked activation of central auditory centers IC, medial geniculate body (MGB), and A1. Subjects with hyperacusis showed elevated evoked activity in the auditory midbrain, thalamus, and primary auditory cortex compared with subjects with normal sound tolerance. This reflects the increased gain for processing external auditory stimuli. Primary auditory cortex, but not IC and MGB, showed elevated activation specifically related to tinnitus, i.e., in the absence of hyperacusis. The results directly link both hyperacusis and tinnitus to hyperactivity within the central auditory system. The authors hypothesized that the tinnitus-related elevations in cortical activation could reflect undue attention drawn to the auditory domain. This is consistent with the lack of tinnitus-related effects subcortically where activation is typically less modulated by attentional state. Melcher et al. (2013) then tested for differences in brain structure between tinnitus and control subjects. Voxel-based morphometry (VBM) was used to compare structural MRIs of tinnitus subjects and non-tinnitus controls, all with normal or near-normal thresholds at standard clinical frequencies (≤ 8 kHz). Mean HTs through 14 kHz, age, sex, and handedness were extremely well matched between groups. There were no significant differences in gray matter (GM) volume and concentration between tinnitus and non-tinnitus subjects.

Also using VBM, Boyen et al. (2013) showed that both hearing-impaired people without and with tinnitus, relative to the controls, had GM increases in the superior and middle temporal gyri and decreases in the superior frontal gyrus, occipital lobe, and hypothalamus. In agreement with Melcher et al. (2013) no significant GM differences were found between both patient groups. Subsequent region-of-interest analyses of all cortical areas, the cerebellum, and the subcortical auditory nuclei showed a GM increase in the left primary auditory cortex of the tinnitus patients compared to the

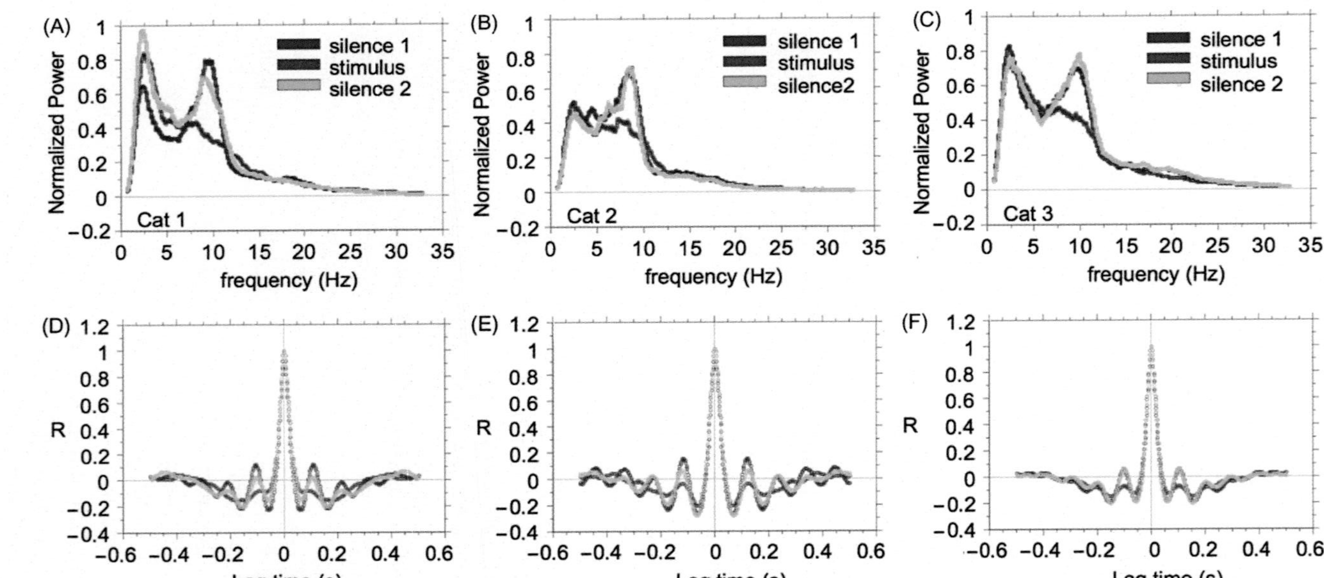

FIGURE 4.7 Suppression of spindle wave power by multifrequency sound. The average LFP power spectra are shown for cat 1 (A), cat 2 (B), and cat 3 (C). The power during the first period of silence, the subsequent period of multifrequency stimulation, and the second period of silence are represented by blue, red, and green lines, respectively. Parts D, E, and F show the corresponding autocorrelation functions. *Reprinted from Britvina, T., Eggermont, J.J., 2008. Multi-frequency stimulation disrupts spindling activity in anesthetized animals. Neuroscience 151, 888–900, with permission from Elsevier.*

hearing-impaired people without tinnitus and control groups. These results suggest a specific role of the left primary auditory cortex and the additional involvement of various nonauditory brain structures in tinnitus.

4.5 NEUROLOGICAL DISORDERS WITH HEARING PROBLEMS

This section is based on the study of Eggermont (2015). Schizophrenia and epilepsy are examples of neurological disorders that show auditory temporal processing deficits that are associated with abnormal neural synchronization. Synchronization of oscillatory responses in the EEG beta- and gamma-band is involved in a variety of cognitive functions, such as perceptual binding, routing of signals across distributed cortical networks, working memory, and perceptual awareness (Uhlhaas and Singer, 2006). Abnormalities in neuronal synchronization are correlated with cognitive dysfunctions, emphasizing the importance of temporal coordination. In general, the cortical distance over which synchronization is observed is related to the frequency of the synchronized oscillations. Short-distance synchronization tends to occur at higher EEG frequencies (gamma-band; 30–90 Hz) than long-distance synchronization, which often manifests itself in the beta (12–24 Hz) but also in the alpha (8–12 Hz) and theta (4–8 Hz) frequency range.

4.5.1 Schizophrenia

Individuals with schizophrenia showed impaired binaural temporal processing ability on an ITD discrimination task compared to age-, sex-, and verbal IQ-matched controls. In addition, patients had reduced ability to use binaural temporal cues to extract signal from noise in a masking level difference paradigm and to separate the location of a source sound in the presence of an echo in the precedence effect paradigm. Thus, individuals with schizophrenia have impairments in the accuracy with which simple binaural temporal information is encoded in the auditory system (Matthews et al., 2013).

Dale et al. (2010) used MEG to measure the time course of stimulus-locked activity over bilateral auditory cortices during discrimination of syllable pairs that differed either in voice-onset time (VOT) or place of articulation, in the presence or absence of noise. They also examined the association of these early neural response patterns to higher-order cognitive functions. The N_1 response from auditory cortex showed less attenuation to the second syllable in patients with schizophrenia than healthy controls during VOT-based discrimination in noise. Across participants, this lack of N_1 attenuation to the second syllable during VOT-based discrimination in noise was associated with poorer task accuracy and lower scores on measures of Verbal Learning and Memory and Global Cognition. Neural differences between healthy comparison subjects and patients occurred only in the presence of noise.

4.5.2 Epilepsy

Epilepsy is a common and diverse set of chronic neurological disorders characterized by seizures. A potential link between schizophrenia and epilepsy was suggested by Cascella et al. (2009); genes implicated in neurodevelopment may play a common role in both conditions and recently identified causative genes for partial complex seizures with auditory features might help explain the pathophysiology of schizophrenia. Seizures are not only a consequence of heightened neuronal excitability such as results from an imbalance between excitatory and inhibitory mechanisms (Uhlhaas and Singer, 2006). Alterations of the mechanisms that support the oscillatory patterning and the synchronization of neuronal activity appear to be equally important. Synchronization enhances the coupling among distributed neuronal populations (Fries, 2005). Both reduced synchronization, preceding some forms of epileptic activity, and enhanced synchronization associated with seizures proper, go along with the disturbance of cognitive functions.

In temporal lobe epilepsy (TLE) there may be functional abnormalities or structural lesions that can manifest as an auditory processing disorder (APD). Han et al. (2011) performed frequency pattern, duration pattern, and dichotic tests in 28 patients with TLE and with normal hearing on pure tone audiometry. The scores on the frequency pattern, duration pattern, and dichotic tests were abnormal in 78.6, 57.1, and 20.6% of patients, respectively, with no significant differences with the laterality of TLE. Patients with hippocampal sclerosis performed significantly worse than patients with normal MRI findings in frequency tests. Longer duration of epilepsy was negatively correlated with both the duration pattern and dichotic tests. This suggests that patients with TLE may be at risk for impairment of central auditory processing, which is increased in patients with hippocampal sclerosis or a longer duration of epilepsy.

4.6 HEARING DISORDERS WITHOUT HEARING SENSITIVITY LOSS

This section is based on the work of Eggermont (2015). APD is a common, heterogeneous, and poorly understood listening impairment that is increasingly diagnosed, especially in children and the elderly (where it could result from age-related damage to ribbon synapses in the IHC; Sergeyenko et al., 2013). The primary symptom in humans is poor speech perception despite clinically normal pure tone audiometry (Moore, 2006). Note that this primary symptom could also correspond to auditory neuropathy, where it is further defined with normal otoacoustic emissions and abnormal or absent ABR.

Dawes et al. (2009) examined the hypothesis that the underlying cause of APD is a modality-specific deficit in auditory temporal processing and also considered how far the auditory impairments in APD differ from those in children with dyslexia. Performance of children diagnosed with APD ($N = 22$) was compared with that of a normative group ($N = 98$) as well as with children with dyslexia ($N = 19$) on a battery of temporal auditory tasks; 2 Hz frequency modulation (FM), 40 Hz FM, and iterated rippled noise detection as well as a control task (240 Hz FM), which is thought to draw on peripheral spectral mechanisms. Visual tasks were coherent-form and coherent-motion detection. On average, the APD group performed more poorly than the normative group on the 40 Hz FM, 240 Hz FM, and iterated rippled noise tasks. There were no significant differences between the APD and dyslexia group's performance. A higher proportion of children in the APD group performed poorly (>1 SD below the mean) on the visual tasks than those in the normative group. The research did not support a modality-specific impairment of temporal auditory processing as being the underlying cause of APD. In both the APD and dyslexia groups, a similar proportion displayed poor auditory performance, and this does not seem entirely accounted for by attention or performance IQ. The authors found the significance of these auditory difficulties uncertain. They suggested that auditory temporal processing deficits are only part of a multifactorial description of learning problems.

Moore et al. (2010) tested whether APD is related to a sensory processing deficit. Randomly chosen, 6- to 11-year-old children with normal hearing ($N = 1469$) were tested in schools in four regional centers across the United Kingdom. Caregivers completed questionnaires regarding their participating children's listening and communication skills. Children completed a battery of audiometric, auditory frequency discrimination, speech-in-noise, cognitive (IQ, memory, language, and literacy), and auditory and visual attention tests. Auditory processing (AP) measures separated the sensory and nonsensory contributions to spectral and temporal perception. AP improved with age. Poor-for-age AP was significantly related to poor cognitive, communication, and speech-in-noise performance, which again may reflect poor temporal fine-structure discrimination. However, correlations between auditory perception and cognitive scores were generally low. Moore et al. (2010) suggested that APD is primarily an attention problem and that clinical diagnosis and management, as well as further research, should be based on that premise. As Moore (2012) rephrased it: "Auditory perception necessarily involves the integration in the brain of bottom-up, auditory 'sensory' information with top-down, multimodal 'cognitive' information [Chapter 3]. Bottom-up development of sensory function occurs very early—in humans mostly before the time of birth. However, top-down processes contributing to AP may continue to mature into adulthood. … Poor cognitive skills necessary for hearing can easily be confused with impaired sensory function. Recent data suggest that this confusion may lie at the heart of the notion, and diagnosis, of auditory processing disorder."

Recently (Eggermont, 2015) I phrased the description of APD as: "Currently, there are two major hypotheses concerning the nature of developmental APD. One is that APD results from impaired 'bottom-up' processing of sounds (the central gain hypothesis). While this view suggests that the problem with APD lies in the central nervous system, Moore and Hunter (2013) proposed that the problem may lie more peripherally, possibly in the ear, and could be the consequence of chronic otitis media with effusion affecting the inner ear via diffusion through the round window membrane. However, this would also trigger homeostatic plasticity. Alternatively, processing problems expressed in the ear, brainstem or auditory cortex could reflect the second hypothesis, 'top-down' influences from cortical cognitive processing centers that may exert their influence on the auditory cortex, and be conveyed to lower targets via efferent pathways. A variant of this hypothesis is that the problem may be entirely cognitive, primarily affecting language processing, attention or memory and exerting a non-specific (e.g. supramodal) effect on perception. Moore and Hunter (2013) proposed that the very high co-occurrence of APD with a wide variety of other auditory- and/or language-based learning disorders may reflect a more general 'neurodevelopmental syndrome', potentially resulting from (temporally) reduced peripheral input to the auditory brain. They conceptualized this as a supramodal communication disorder that captures a broad range of currently separate markers (e.g. language, literacy, attention and behavior problems) across age." Consequently, APD is likely the result of a central auditory dysfunction.

Long-term exposure to nondamaging sounds (Kujala et al., 2004; Pienkowski and Eggermont, 2009) may also lead to central changes that compromise speech understanding and mimic APD. An extensive discussion about (diagnostic) distinctions between dyslexia, specific language impairment, and APD is presented in chapter 14 of "Temporal Auditory Processing and its Related Disorders" (Eggermont, 2015) and in a broader context in the work of Musiek and Chermak (2014).

4.7 NONAUDITORY EFFECTS OF HEARING LOSS

4.7.1 Balance Problems

Hearing loss and balance problems tend to co-occur. Viljanen et al. (2009) measured HL as a part of the Finnish Twin Study on Aging in 103 monozygotic and 114 dizygotic female twin pairs aged 63–76 years. Twin analyses showed that the association between hearing acuity and postural balance was not explained by genetic factors in common for these traits. People with poor hearing acuity were found to have a higher risk for falls, which is partially explained by their poorer postural control. It is thus possible that hair cell loss is not restricted to the cochlea but that the hair cells in the utricle and saccule may also be involved. An alternative is a reduced efficacy of the central parts of the vestibular system.

More recently, Koh et al. (2015) tested 46 elderly individuals over 65 years of age, who were attending senior welfare centers, participated in this study. The speech frequency pure-tone average in the better ear was checked. HL was defined by a hearing level value of 25 dB or more in the better ear. Cognition ability was evaluated using the Korean mini-mental state examination. Dynamic balance ability was evaluated by the timed up and go test, and static balance ability was tested using a one-leg stance test. They found that as elderly individuals get older, their hearing and cognition, as well as their balance abilities deteriorated. The results indicated a significant correlation between HL and postural balance. In another study, de Souza Melo et al. (2015) observed that students (aged 7–18 years) showed greater instability in the postural control compared to normal hearing students of the same gender and age. Rumalla et al. (2015) found that wearing hearing aids may offer a significant public health benefit for avoiding falls in a group of bilateral hearing aid users aged over 65 years.

4.7.2 Effects on Quality of Life

Self-reported difficulties in judging the location and movement of sources of sound are among the auditory disabilities that are most strongly associated with restricted social participation and reduced emotional well-being (Noble and Gatehouse, 2004; Strawbridge et al., 2000).

Dalton et al. (2003) had already demonstrated that presbycusis may have a negative effect on quality of life (QoL) and psychological well-being. Social isolation, depression, anxiety, and even cognitive decline have been reported in affected persons. Ciorba et al. (2012) also found that only 39% of the population with HL perceive that they have an excellent QoL level or very good physical health, compared to 68% of those without HL. Nearly one-third of the population with HL reported being in fair or poor health, compared to only 9% of the population without HL; people with HL are less satisfied with their "life as a whole" than people without HL (The National Council on the Aging, 1999). There may be a confound with age in this conclusion. People with HL perceive their social skills as poor, and thus, they also may experience reduced self-esteem if a combination of hearing impairment and a poor coping strategy contributes to problems in their life (Ciorba et al., 2012).

Gopinath et al. (2012) examined 829 Blue Mountains Hearing Study participants (≥55 years) between 1997–1999 and 2007–2009. The QoL was assessed using the 36-Item Short-Form Survey (SF-36) where higher scores reflect better QoL. They compared hearing impairment at baseline with no impairment, and this was associated with significantly lower mean SF-36 mental composite score 10 years later. Physical composite score and mean scores for seven of the eight SF-36 domains after 10-year follow-up were also significantly lower among participants who self-reported hearing handicap at baseline. Gopinath et al. (2012) found that hearing aid users compared to nonusers showed a significant increase in SF-36 mental composite score of 1.82 point at baseline and a 3.32 point at the 10-year follow-up. Carlsson et al. (2014) conducted a retrospective study of data on 2319 patients with severe to profound hearing impairment in the Swedish Quality Register of Otorhinolaryngology, followed by a mailed questionnaire including the Hospital Anxiety and Depression Scale. Their results indicated greater levels of anxiety and depression among patients with severe or profound hearing impairment than in the general population, and annoying tinnitus and vertigo had strong negative effects on QoL.

4.7.3 A Greater Risk for Dementia

The Baltimore Longitudinal Study of Aging has provided massive data relating cognitive functions in the elderly. Lin et al. (2011a) observed that the risk of incident all-cause dementia increased with a factor 1.27 per 10 dB loss (95% confidence interval, 1.06–1.50). They found that compared with normal hearing, the hazard ratio (95% confidence interval) for incident all-cause dementia was 1.89 (1.00–3.58) for mild HL, 3.00 (1.43–6.30) for moderate HL, and 4.94 (1.09–22.40) for severe HL. Lin et al. (2011b) also analyzed a cross-sectional cohort of 347 participants at least 55 years in the Baltimore Longitudinal Study of Aging without mild cognitive impairment or dementia had audiometric and cognitive testing performed in 1990–94. Here, the reduction in cognitive performance associated with a 25 dB HL (pure-tone average 0.5–4 kHz) was equivalent to the reduction associated with an age difference of 6.8 years. Demented elderly tolerate hearing aids, however, hearing aid use did not improve their cognitive function or reduced behavioral or psychiatric symptoms (Allen et al., 2003). This suggests that the problem has to be addressed before dementia sets in, i.e., by timely hearing aid fitting.

Gurgel et al. (2014) followed 4463 subjects who were without dementia and 836 of whom had HL at baseline. Of those with HL, 16.3% developed dementia, compared with 12.1% of those without HL ($p < 0.001$). Mean time to dementia was 10.3 years in the HL group versus 11.9 years for non-HL (log rank test $p < 0.001$). Thus, HL is independently associated with lower scores on tests of memory and executive function. These cognitive changes in the elderly may be related to changes in brain volume. Lin et al. (2014) found that individuals with hearing impairment ($N = 51$) compared to those with normal hearing ($N = 75$) had accelerated volume declines in whole brain and regional volumes in the right temporal lobe. These findings demonstrated to Lin et al. (2014) that peripheral hearing impairment is independently associated with accelerated brain atrophy in whole brain and regional volumes concentrated in the right hemisphere. This corroborates previous MRI findings by Melcher et al. (2013), who found that high-frequency ($\geq$8 kHz) HL cause atrophy of auditory cortex.

4.7.4 Psychological Effects in Hearing-Impaired Children and Adolescents

This section discusses how hearing impairment and treatments by cochlear implants impact on self-esteem, social competence, and empathy. In addition, the occurrence of psychopathological symptoms is reviewed. All the reports reviewed originated from or were carried out in collaboration with a center for cochlear implantation (see chapter: Cochlear Implants).

Theunissen et al. (2014a) compared hearing-impaired children's self-esteem with that of normal hearing children and investigated the influence of communication, type of education, and audiological characteristics. They found that hearing-impaired children experienced lower levels of self-esteem compared to peers and parents than normal hearing controls. After correcting for their language development and intelligence, children who attended schools with special education for the deaf were still at risk. Theunissen et al. (2014b) investigated the level of psychopathological symptoms in hearing-impaired children and adolescents as compared with normally hearing peers using a systematic literature search. This suggested that hearing-impaired children and adolescents were more likely to develop depression, aggression, oppositional defiant disorder, conductive disorder, and psychopathy compared to their normal hearing peers. Theunissen et al. (2015) further analyzed the symptoms of psychopathology in children with cochlear implants or hearing aids and in normally hearing peers. They used a large sample (mean age = 11.8 years) included 57 with cochlear implants, 75 with conventional hearing aids, and 129 children, who were normally hearing that had comparable age, gender, socioeconomic status, and nonverbal intelligence. Theunissen et al. (2015) found that children with cochlear implants have lower levels of psychopathological symptoms than children who have hearing aids. They attributed this difference to the intensity of the rehabilitation program.

4.8 SUMMARY

In this chapter, we identified some effects of noise exposure on the CNS. The current standards for occupational noise exposure are apparently intended to just prevent PTSs, though it has long been suspected that problems such as poor speech intelligibility in noise, as well as tinnitus, could arise after long-term exposure at lower intensities. Likely consequences following TTSs in humans are difficulties understanding speech particularly in background noise, because of the loss of the high-threshold, low-SFR fiber activity, which is difficult to saturate by noise. Accumulation of afferent synapse damage over time with recurrent noise exposures could be a major contributor to age-related hearing impairment. In addition, subjects with cochlear HL are generally less able than normal to take advantage of spatial separation of speech and interfering noise. Combined with a poor sound localization, communication problems at cocktail parties are very likely and may lead to social isolation due to HL. I also reviewed more neurologically based hearing problems. In addition, we presented some annoying side effects of HL: the primary one being tinnitus, but an even more disturbing one is an increased probability of dementia. Finally, the psychological effects of HL or deafness in children and the effects of hearing aids or cochlear implants are reviewed.

REFERENCES

Allen, N.H., Burns, A., Newton, V., Hickson, F., Ramsden, R., Rogers, J., et al., 2003. The effects of improving hearing in dementia. Age Ageing 32, 189–193.

Al-Mana, D., Ceranic, B., Djahanbakhch, O., Luxon, L.M., 2008. Hormones and the auditory system: a review of physiology and pathophysiology. Neuroscience 153, 881–900.

Anari, M., Axelsson, A., Eliasson, A., Magnusson, L., 1999. Hypersensitivity to sound—questionnaire data, audiometry and classification. Scand. Audiol. 28, 219–230.

Baguley, D.M., 2003. Hyperacusis. J. R. Soc. Med. 96, 582–585.

Bee, M.A., Micheyl, C., 2008. The cocktail party problem: what is it? How can it be solved? And why should animal behaviorists study it? J. Comp. Psychol. 122, 235–251.

Bidet-Caulet, A., Fischer, C., Bauchet, F., Aguera, P.E., Bertrand, O., 2008. Neural substrate of concurrent sound perception: direct electrophysiological recordings from human auditory cortex. Front. Hum. Neurosci. 1, 5.

Boyen, K., Langers, D.R., de Kleine, E., van Dijk, P., 2013. Gray matter in the brain: differences associated with tinnitus and hearing loss. Hear. Res. 295, 67–78.

Brattico, E., Kujala, T., Tervaniemi, M., Alku, P., Ambrosi, L., Monitillo, V., 2005. Long-term exposure to occupational noise alters the cortical organization of sound processing. Clin. Neurophysiol. 116, 190–203.

Bregman, A.S., 1990. Auditory Scene Analysis. MIT Press, Cambridge, MA.

Briley, P.M., Summerfield, A.Q., 2014. Age-related deterioration of the representation of space in human auditory cortex. Neurobiol. Aging 35, 633–644.

Britvina, T., Eggermont, J.J., 2008. Multi-frequency stimulation disrupts spindling activity in anesthetized animals. Neuroscience 151, 888–900.

Bronkhorst, A.W., Plomp, R., 1989. Binaural speech intelligibility in noise for hearing-impaired listeners. J. Acoust. Soc. Am. 86, 1374–1383.

Cai, S., Ma, W.L., Young, E.D., 2008. Encoding Intensity in ventral cochlear nucleus following acoustic trauma: implications for loudness recruitment. J. Assoc. Res. Otolaryngol. 10, 5–22.

Carlsson, P.-I., Hjaldahl, J., Magnuson, A., Ternevall, E., Edén, M., Skagerstrand, Å., et al., 2014. Severe to profound hearing impairment: quality of life, psychosocial consequences and audiological rehabilitation. Disabil. Rehabil. 37, 1849–1856.

Cascella, N.G., Schretlen, D.J., Sawa, A., 2009. Schizophrenia and epilepsy: is there a shared susceptibility? Neurosci. Res. 63, 227–235.

Cherry, E.C., 1953. Some experiments on the recognition of speech, with one and two ears. J. Acoust. Soc. Am. 25, 975–979.

Ciorba, A., Bianchini, C., Pelucchi, S., Pastore, A., 2012. The impact of hearing loss on the quality of life of elderly adults. Clin. Interv. Aging 7, 159–163.

Cranston, C.J., Brazile, W.J., Sandfort, D.R., Gotshall, R.W., 2013. Occupational and recreational noise exposure from indoor arena hockey games. J. Occup. Environ. Hyg. 10, 11–16.

Dale, C.L., Findlay, A.M., Adcock, R.A., et al., 2010. Timing is everything: neural response dynamics during syllable processing and its relation to higher-order cognition in schizophrenia and healthy comparison subjects. Int. J. Psychophysiol. 75, 183–193.

Dalton, D.S., Cruickshanks, K.J., Klein, B.E., Klein, R., Wiley, T.L., Nondahl, D.M., 2003. The impact of hearing loss on quality of life in older adults. Gerontologist 43, 661–668.

Dawes, P., Sirimanna, T., Burton, M., Vanniasegaram, I., Tweedy, F., Bishop, D.V., 2009. Temporal auditory and visual motion processing of children diagnosed with auditory processing disorder and dyslexia. Ear Hear. 30, 675–686.

Delgutte, B., 1990. Physiological mechanisms of psychophysical masking: observations from auditory-nerve fibers. J. Acoust. Soc. Am. 87, 791–809.

De Ridder, D., Vanneste, S., Weisz, N., Londero, A., Schlee, W., Elgoyhen, A.B., et al., 2014. An integrative model of auditory phantom perception: tinnitus as a unified percept of interacting separable subnetworks. Neurosci. Biobehav. Rev. 44, 16–32.

de Souza Melo, R., Lemos, A., da Silva Toscano Macky, C.F., Falcão Raposo, M.C., Ferraz, C.M., 2015. Postural control assessment in students with normal hearing and sensorineural hearing loss. Braz. J. Otorhinolaryngol. 81, 431–438.

Dingle, R.N., Hall, S.E., Phillips, D.P., 2010. A midline azimuthal channel in human spatial hearing. Hear. Res. 268, 67–74.

Dingle, R.N., Hall, S.E., Phillips, D.P., 2012. The three-channel model of sound localization mechanisms: interaural level differences. J. Acoust. Soc. Am. 131, 4023–4029.

Dingle, R.N., Hall, S.E., Phillips, D.P., 2013. The three-channel model of sound localization mechanisms: interaural time differences. J. Acoust. Soc. Am. 133, 417–424.

Dobreva, M.S., O'Neill, W.E., Paige, G.D., 2011. Influence of aging on human sound localization. J. Neurophysiol. 105, 2471–2486.

Edmonds, B.A., Krumbholz, K., 2014. Are interaural time and level differences represented by independent or integrated codes in the human auditory cortex? J. Assoc. Res. Otolaryngol. 15, 103–114.

Eggermont, J.J., 1992. Neural interaction in cat primary auditory cortex. Dependence on recording depth, electrode separation and age. J. Neurophysiol. 68, 1216–1228.

Eggermont, J.J., 2006. Properties of correlated neural activity clusters in cat auditory cortex resemble those of neural assemblies. J. Neurophysiol. 96, 746–764.

Eggermont, J.J., 2012. The Neuroscience of Tinnitus. Oxford University Press, Oxford, UK.

Eggermont, J.J., 2014. Noise and the Brain. Experience Dependent Developmental and Adult Plasticity. Academic Press, London.

Eggermont, J.J., 2015. Auditory Temporal Processing and its Disorders. Oxford University Press, Oxford, UK.

Eggermont, J.J., Roberts, L.E., 2015. Tinnitus: animal models and findings in humans. Cell Tissue Res. 361, 311–336.

Eggermont, J.J., Tass, P., 2015. Maladaptive neural synchrony in tinnitus: origin and restoration. Front. Neurol. 6 (29).

Fries, P., 2005. A mechanism for cognitive dynamics: neuronal communication through neuronal coherence. Trends Cogn. Sci. 9, 474–480.

Glyde, H., Cameron, S., Dillon, H., Hickson, L., Seeto, M., 2013. The effects of hearing impairment and aging on spatial processing. Ear Hear. 34, 15–28.

Gopinath, B., Schneider, J., Hickson, L., McMahon, C.M., Burlutsky, G., Leeder, S.R., et al., 2012. Hearing handicap, rather than measured hearing impairment, predicts poorer quality of life over 10 years in older adults. Maturitas 72, 146–151.

Gourévitch, B., Edeline, J.-M., Occelli, F., Eggermont, J.J., 2014. Is the din really harmless? Experience-related neural plasticity for non-traumatic noise levels. Nat. Rev. Neurosci. 15, 483–491.

Gu, J.W., Halpin, C.F., Nam, E.C., Levine, R.A., Melcher, J.R., 2010. Tinnitus, diminished sound-level tolerance, and elevated auditory activity in humans with clinically normal hearing sensitivity. J. Neurophysiol. 104, 3361–3370.

Gurgel, R.K., Ward, P.D., Schwartz, S., Norton, M.C., Foster, N.L., Tschanz, J.T., 2014. Relationship of hearing loss and dementia: a prospective, population-based study. Otol. Neurotol. 35, 775–781.

Han, M.W., Ahn, J.H., Kang, J.K., et al., 2011. Central auditory processing impairment in patients with temporal lobe epilepsy. Epilepsy Behav. 20, 370–374.

Hegerl, U., Juckel, G., 1993. Intensity dependence of auditory evoked potentials as an indicator of central serotonergic neurotransmission: a new hypothesis. Biol. Psychiatry 33, 173–187.

Hickox, A.E., Liberman, M.C., 2014. Is noise-induced cochlear neuropathy key to the generation of hyperacusis or tinnitus? J. Neurophysiol. 111, 552–564.

Hudspeth, A.J., 2013. Snapshot: auditory transduction. Neuron 80, 536.

Jastreboff, P.J., Brennan, J.F., Coleman, J.K., Sasaki, C.T., 1988. Phantom auditory sensation in rats: an animal model for tinnitus. Behav. Neurosci. 102, 811–822.

Katzenell, U., Segal, S., 2001. Hyperacusis: review and clinical guidelines. Otol. Neurotol. 22, 321–326.

Kim, J., Morest, D.K., Bohne, B.A., 1997. Degeneration of axons in the brainstem of the chinchilla after auditory overstimulation. Hear. Res. 103, 169–191.

Kim, J.J., Gross, J., Morest, D.K., Potashner, S.J., 2004a. Quantitative study of degeneration and new growth of axons and synaptic endings in the chinchilla cochlear nucleus after acoustic overstimulation. J. Neurosci. Res. 77, 829–842.

Kim, J.J., Gross, J., Potashner, S.J., Morest, D.K., 2004b. Fine structure of long-term changes in the cochlear nucleus after acoustic overstimulation: chronic degeneration and new growth of synaptic endings. J. Neurosci. Res. 77, 817–828.

Kim, J.J., Gross, J., Potashner, S.J., Morest, D.K., 2004c. Fine structure of degeneration in the cochlear nucleus of the chinchilla after acoustic overstimulation. J. Neurosci. Res. 77, 798–816.

Kjellberg, A., 1990. Subjective, behavioral and psychophysiological effects of noise. Scand. J. Work Environ. Health 16, 29–38.

Koh, D.H., Lee, J.D., Lee, H.J., 2015. Relationships among hearing loss, cognition and balance ability in community-dwelling older adults. J. Phys. Ther. Sci. 27, 1539–1542.

Kryter, K.D., Ward, W.D., Miller, J.D., Eldredge, D.H., 1966. Hazardous exposure to intermittent and steady-state noise. J. Acoust. Soc. Am. 139, 451–464.

Kujala, T., Shtyrov, Y., Winkler, I., et al., 2004. Long-term exposure to noise impairs cortical sound processing and attention control. Psychophysiology 41, 875–881.

Kujawa, S.G., Liberman, M.C., 2009. Adding insult to injury: cochlear nerve degeneration after "temporary" noise-induced hearing loss. J. Neurosci. 29, 14077–14085.

Kujawa, S.G., Liberman, M.C., 2015. Synaptopathy in the noise-exposed and aging cochlea: primary neural degeneration in acquired sensorineural hearing loss. Hear. Res. 330, 191–199.

Langers, D.R., de Kleine, E., van Dijk, P., 2012. Tinnitus does not require macroscopic tonotopic map reorganization. Front. Syst. Neurosci. 6, 2.

Liberman, M.C., Beil, D.G., 1979. Hair cell condition and auditory nerve response in normal and noise-damaged cochleas. Acta Otolaryngol. 88, 161–176.

Liberman, M.C., Kiang, N.Y., 1978. Acoustic trauma in cats. Cochlear pathology and auditory-nerve activity. Acta Otolaryngol. Suppl. 358, 1–63.

Lim, D.J., 1986. Effects of noise and ototoxic drugs at the cellular level in the cochlea. Am. J. Otolaryngol. 7, 73–99.

Lin, F.R., Metter, E.J., O'Brien, R.J., Resnick, S.M., Zonderman, A.B., Ferrucci, L., 2011a. Hearing loss and incident dementia. Arch. Neurol. 68, 214–220.

Lin, F.R., Ferrucci, L., Metter, E.J., An, Y., Zonderman, A.B., Resnick, S.M., 2011b. Hearing loss and cognition in the Baltimore Longitudinal Study of Aging. Neuropsychology 25, 763–767.

Lin, F.R., Ferrucci, L., An, Y., Gohg, J.O., Doshi, J., Metter, E.J., et al., 2014. Association of hearing impairment with brain volume changes in older adults. Neuroimage 90, 84–92.

MacNaughton-Jones, H., 1890. The etiology of tinnitus aurium. Br. Med. J. 2, 667–672.

Magezi, D.A., Krumbholz, K., 2010. Evidence for opponent-channel coding of interaural time differences in human auditory cortex. J. Neurophysiol. 104, 1997–2007.

Makous, J.C., Middlebrooks, J.C., 1990. Two-dimensional sound localization by human listeners. J. Acoust. Soc. Am. 87, 2188–2200.

Marriage, J., Barnes, N.M., 1995. Is central hyperacusis a symptom of 5-hydroxytryptamine (5-HT) dysfunction?. J. Laryngol. Otol. 109, 915–921.

Matthews, N., Todd, J., Mannion, D.J., Finnigan, S., Catts, S., Michie, P.T., 2013. Impaired processing of binaural temporal cues to auditory scene analysis in schizophrenia. Schizophr. Res. 146, 344–348.

May, J.J., 2000. Occupational hearing loss. Am. J. Ind. Med. 37, 112–120.

McDermott, J.H., 2009. The cocktail party problem. Curr. Biol. 19, R1024–1027.

Melcher, J.R., Knudson, I.M., Levine, R.A., 2013. Subcallosal brain structure: correlation with hearing threshold at supra-clinical frequencies (>8 kHz), but not with tinnitus. Hear. Res. 295, 79–86.

Moore, B.C.J., 1996. Perceptual consequences of cochlear hearing loss and their implications for the design of hearing aids. Ear Hear. 17, 133–161.

Moore, B.C.J., 2012. The psychophysics of tinnitus. In: Eggermont, J.J., et al., (Eds.), Tinnitus, Springer Handbook of Auditory Research, 47. Springer Science + Business Media, New York, pp. 187–216.

Moore, D.R., 2006. Auditory processing disorder (APD)—potential contribution of mouse research. Brain Res. 1091, 200–206.

Moore, D.R., 2012. Listening difficulties in children: bottom-up and top-down contributions. J. Commun. Disord. 45, 411–418.

Moore, D.R., Hunter, L.L., 2013. Auditory processing disorder (APD) in children: a marker of neurodevelopmental syndrome. Hear. Balance Commun. 11, 160–167.

Moore, D.R., Ferguson, M.A., Edmondson-Jones, A.M., Ratib, S., Riley, A., 2010. Nature of auditory processing disorder in children. Pediatrics 126, e382–390.

Morest, D.K., Kim, J., Bohne, B.A., 1997. Neuronal and transneuronal degeneration of auditory axons in the brainstem after cochlear lesions in the chinchilla: cochleotopic and non-cochleotopic patterns. Hear. Res. 103, 151–168.

Morest, D.K., Kim, J., Potashner, S.J., Bohne, B.A., 1998. Long-term degeneration in the cochlear nerve and cochlear nucleus of the adult chinchilla following acoustic overstimulation. Microsc. Res. Tech. 41, 205–216.

Muly, S.M., Gross, J.S., Morest, D.K., Potashner, S.J., 2002. Synaptophysin in the cochlear nucleus following acoustic trauma. Exp. Neurol. 177, 2002–2221.

Munguia, R., Pienkowski, M., Eggermont, J.J., 2013. Spontaneous firing rate changes in cat primary auditory cortex following long-term exposure to non-traumatic noise. Tinnitus without hearing loss?. Neurosci. Lett. 546, 46–50.

Musiek, F.E., Chermak, G.D., 2014. second ed. Handbook of Central Auditory Processing. Auditory Neuroscience and Diagnosis, vol. 1. Plural Publishing, San Diego, CA.

Nelson, J.J., Chen, K., 2004. The relationship of tinnitus, hyperacusis, and hearing loss. Ear Nose Throat J. 83, 472–476.

Noble, W., Gatehouse, S., 2004. Interaural asymmetry of hearing loss, Speech, Spatial and Qualities of Hearing Scale (SSQ) disabilities, and handicap. Int. J. Audiol. 43, 100–114.

Nondahl, D.M., Cruickshanks, K.J., Huang, G.-H., Klein, B.E.K., Klein, R., Nieto, F.J., et al., 2011. Tinnitus and its risk factors in the Beaver Dam Offspring Study. Int. J. Audiol. 50, 313–320.

Noreña, A., Micheyl, C., Chery-Croze, S., Collet, L., 2002. Psychoacoustic characterization of the tinnitus spectrum: implications for the underlying mechanisms of tinnitus. Audiol. Neurootol. 7, 358–369.

Noreña, A.J., Eggermont, J.J., 2003. Changes in spontaneous neural activity immediately after an acoustic trauma: implications for neural correlates of tinnitus. Hear. Res. 183, 137–153.

Noreña, A.J., Eggermont, J.J., 2006. Enriched acoustic environment after noise trauma abolishes neural signs of tinnitus. Neuroreport 17, 559–563.

Noreña, A.J., Gourévitch, B., Aizawa, N., Eggermont, J.J., 2006. Enriched acoustic environment disrupts frequency representation in cat auditory cortex. Nat. Neurosci. 9, 932–939.

Nouvian, R., Eybalin, M., Puel, J.-L., 2015. Cochlear efferents in developing adult and pathological conditions. Cell Tissue Res. 361, 301–309.

Otte, R.J., Agterberg, M.J., Van Wanrooij, M.M., Snik, A.F., Van Opstal, A.J., 2013. Age-related hearing loss and ear morphology affect vertical but not horizontal sound-localization performance. J. Assoc. Res. Otolaryngol. 14, 261–273.

Penner, M.J., 1993. Synthesizing tinnitus from sine waves. J. Speech Hear. Res. 36, 1300–1305.

Phillips, H.C., Hunter, M., 1982. A laboratory technique for the assessment of pain behaviour. J. Behav. Med. 5, 283–294.

Pienkowski, M., Eggermont, J.J., 2009. Long-term, partially-reversible reorganization of frequency tuning in mature cat primary auditory cortex can be induced by passive exposure to moderate-level sounds. Hear. Res. 257, 24–40.

Pienkowski, M., Eggermont, J.J., 2010. Intermittent exposure with moderate-level sound impairs central auditory function of mature animals without concomitant hearing loss. Hear. Res. 261, 30–35.

Pienkowski, M., Eggermont, J.J., 2011. Cortical tonotopic map plasticity and behavior. Neurosci. Biobehav. Rev. 35, 2117–2128.

Puel, J.-L., D'aldin, C., Ruel, J., Ladrech, S., Pujol, R., 1997. Synaptic repair mechanisms responsible for functional recovery in various cochlear pathologies. Acta Otolaryngol. 117, 214–218.

Roberts, L.E., Moffat, G., Baumann, M., Ward, L.M., Bosnyak, D.J., 2008. Residual inhibition functions overlap tinnitus spectra and the region of auditory threshold shift. J. Assoc. Res. Otolaryngol. 9, 417–435.

Roberts, L.E., Husain, F.T., Eggermont, J.J., 2013. Role of attention in the generation and modulation of tinnitus. Neurosci. Biobehav. Rev. 37, 1754–1773.

Robertson, D., 1982. Effects of acoustic trauma on stereocilia structure and spiral ganglion cell tuning properties in the guinea pig cochlea. Hear. Res. 7, 55–74.

Robertson, D., Johnstone, B.M., 1980. Acoustic trauma in the guinea pig cochlea: early changes in ultrastructure and neural threshold. Hear. Res. 3, 167–179.

Rumalla, K., Karim, A.M., Hullar, T.E., 2015. The effect of hearing aids on postural stability. Laryngoscope 125, 720–723.

Salminen, N.H., May, P.J.C., Alku, P., Tiitinen, H., 2009. A population rate code of auditory space in the human cortex. PLoS One 4, e7600.

Salvi, R.J., Hamernik, R.P., Henderson, D., Ahroon, W.A., 1983. Neural correlates of sensori-neural hearing loss. Ear Hear. 4, 115–129.

Salvi, R.J., Saunders, S.S., Gratton, M.A., Arehole, S., Powers, N., 1990. Enhanced evoked response amplitudes in the inferior colliculus of the chinchilla following acoustic trauma. Hear. Res. 50, 245–257.

Saunders, J.C., Dear, S.P., Schneider, M.E., 1985. The anatomical consequences of acoustic injury: a review and tutorial. J. Acoust. Soc. Am. 78, 833–860.

Schaette, R., Kempter, R., 2006. Development of tinnitus-related neuronal hyperactivity through homeostatic plasticity after hearing loss: a computational model. Eur. J. Neurosci. 23, 3124–3138.

Sereda, M., Edmondson-Jones, M., Hall, D.A., 2015. Relationship between tinnitus pitch and edge of hearing loss in individuals with a narrow tinnitus bandwidth. Int. J. Audiol. 54, 249–256.

Sergeyenko, Y., Lall, K., Liberman, M.C., Kujawa, S.G., 2013. Age-related cochlear synaptopathy: an early-onset contributor to auditory functional decline. J. Neurosci. 33, 13686–13694.

Sheldrake, J., Diehl, P.U., Schaette, R., 2015. Audiometric characteristics of hyperacusis patients. Front. Neurol. 6, 105.

Sherlock, L.P., Formby, C., 2005. Estimates of loudness, loudness discomfort, and the auditory dynamic range: normative estimates, comparison of procedures, and test-retest reliability. J. Am. Acad. Audiol. 16, 85–100.

Simon, H.J., 2005. Bilateral amplification and sound localization: then and now. J. Rehabil. Res. Dev. 42 (Suppl. 2), 117–132.

Simpson, J.J., Davies, W.E., 2000. A review of evidence in support of a role for 5-HT in the perception of tinnitus. Hear. Res. 145, 1–7.

Slepecky, N., 1986. Overview of mechanical damage to the inner ear: noise as a tool to probe cochlear function. Hear. Res. 22, 307–321.

Spoendlin, H., 1976. Organisation of the auditory receptor. Rev. Laryngol. Otol. Rhinol. (Bord) 97 (Suppl), 453–462.

Strawbridge, W.J., Wallhagen, M.I., Shema, S.J., Kaplan, G.A., 2000. Negative consequences of hearing impairment in old age: a longitudinal analysis. Gerontology 40, 320–326.

Sztuka, A., Pospiech, L., Gawron, W., Dudek, K., 2010. DPOAE in estimation of the function of the cochlea in tinnitus patients with normal hearing. Auris Nasus Larynx 37, 55–60.

Tang, Z.-Q., Trussell, L.O., 2015. Serotonergic regulation of excitability of principal cells of the dorsal cochlear nucleus. J. Neurosci. 35, 4540–4551.

The National Council on the Aging, 1999. The Consequences of Untreated Hearing Loss in Older Persons. The National Council on the Aging, Washington, DC, <http://www.hearingoffice.com/download/UntreatedHearingLossReport.pdf>.

Theunissen, S.C., Netten, A.P., Rieffe, C., Briare, J.J., Soede, W., Kouwenberg, M., et al., 2014a. Self-esteem in hearing-impaired children: the influence of communication, education, and audiological characteristics. PLoS One 9, e94521.

Theunissen, S.C., Rieffe, C., Netten, A.P., Briare, J.J., Soede, W., Schoones, J.W., et al., 2014b. Psychopathology and its risk and protective factors in hearing-impaired children and adolescents: a systematic review. JAMA Pediatr. 168, 170–177.

Theunissen, S.C., Rieffe, C., Soede, W., Briare, J.J., Ketelaar, L., Kouwenberg, M., et al., 2015. Symptoms of psychopathology in hearing-impaired children. Ear Hear. 36, e190–e198.

Turrigiano, G., 1999. Homeostatic plasticity in neuronal networks: the more things change, the more they stay the same. Trends Neurosci. 22, 221–227.

Uhlhaas, P.J., Singer, W., 2006. Neural synchrony in brain disorders: relevance for cognitive dysfunctions and pathophysiology. Neuron 52, 155–168.

van der Loo, E., Gais, S., Congedo, M., Vanneste, S., Plazier, M., Menovsky, T., et al., 2009. Tinnitus intensity dependent gamma oscillations of the contralateral auditory cortex. PLoS One 4, e7396.

Viljanen, A., Kaprio, J., Pyykkö, I., Sorri, M., Pajala, S., Kauppinen, M., et al., 2009. Hearing as a predictor of falls and postural balance in older female twins. J. Gerontol. A Biol. Sci. Med. Sci. 64A, 312–317.

Wang, J., Ding, D., Salvi, R.J., 2002. Functional reorganization in chinchilla inferior colliculus associated with chronic and acute cochlear damage. Hear. Res. 168, 238–249.

Wang, Y., Ren, C., 2012. Effects of repeated "benign" noise exposures in young CBA mice: shedding light on age-related hearing loss. J. Assoc. Res. Otolaryngol. 13, 505–515.

Ward, L.M., Baumann, M., 2009. Measuring tinnitus loudness using constrained psychophysical scaling. Am. J. Audiol. 18, 119–128.

Weisz, N., Obleser, J., 2014. Synchronisation signatures in the listening brain: a perspective from non-invasive neuroelectrophysiology. Hear. Res. 307, 16–28.

Weisz, N., Müller, S., Schlee, W., Dohrmann, K., Hartmann, T., Elbert, T., 2007. The neural code of auditory phantomperception. J. Neurosci. 27, 1479–1484.

Wu, C., Stefanescu, R.A., Martel, D.T., Shore, S.E., 2015. Tinnitus: maladaptive auditory-somatosensory plasticity. Hear. Res. Available from: http://dx.doi.org/10.1016/j.heares.2015.06.005.

Wutzler, A., Winter, C., Kitzrow, W., Uhl, I., Wolf, R.J., Heinz, A., et al., 2008. Loudness dependence of auditory evoked potentials as indicator of central serotonergic neurotransmission: simultaneous electrophysiological recordings and in vivo microdialysis in the rat primary auditory cortex. Neuropsychopharmacology 33, 3176–3181.

Chapter 5

Types of Hearing Loss

Peripheral hearing loss comes in two broad types, conductive and sensorineural. Conductive hearing loss results from deficits in the sound-conducting apparatus of the outer and middle ear. Problems such as fluid in the middle ear and immobility of the middle ear bones are the main cause of a conductive hearing loss. Sensorineural hearing losses (SNHLs) that result from damage to the hair cells and the stria vascularis can be considered "sensory" hearing loss. Neural loss occurs when the auditory nerve fibers (ANFs) are involved, as might occur in a tumor of the eighth nerve (a vestibular schwannoma) or in a recently recognized type of hearing loss known (most commonly) as auditory neuropathy (ANP). Types of hearing loss that cause dominantly "central changes" have been described in Chapter 4, Hearing Problems.

5.1 SITE OF LESION TESTING

5.1.1 Air/Bone Conduction Audiograms

Conductive hearing losses are defined by an audiometric difference for air- and bone-conducted sound, the air–bone gap. Bone conducted sound, produced by a vibrator on the scalp (see chapter: Implantable Hearing Aids), bypasses the external and middle ear in stimulation of the cochlea, but is less effective than air conduction in normal hearing people.

5.1.2 Speech Discrimination Testing

Speech audiometry is a fundamental tool in hearing loss assessment. Together with pure-tone audiometry, it can aid in determining the degree and type of hearing loss. Speech audiometry provides information on word recognition and about discomfort or tolerance to speech stimuli. Speech audiometry outcomes help also in setting the gain and maximum output of hearing aids for patients with moderate to severe hearing losses. An adaptation of speech audiometry is the Hearing-in-Noise Test, in which the stimuli are presented to both ears together and the patient is required to repeat sentences

Hearing Loss. DOI: http://dx.doi.org/10.1016/B978-0-12-805398-0.00005-0

both in a quiet environment and with competing noise being presented from different directions.

5.1.3 Acoustic Immittance

Acoustic immittance testing evaluates the patency of the eardrum and the middle ear space behind the eardrum, as well as a middle ear muscle reflex (MEMR). The primary purpose of impedance audiometry is to determine the status of the tympanic membrane (TM) and middle ear via tympanometry. The secondary purpose of this test is to evaluate the MEMR pathways, which include the facial nerve, the audio-vestibular nerve, and the auditory brainstem. This test cannot be used to directly assess auditory sensitivity, although results are interpreted in conjunction with other threshold measures.

5.1.3.1 Tympanometry

Tympanometry is an examination used to test the condition of the middle ear, the mobility of the eardrum, and the conduction bones by creating variations of air pressure in the ear canal. A tone of 226 Hz is generated by the tympanometer into the ear canal, where the sound strikes the TM. Some of this sound is reflected back and picked up by the instrument. Middle ear problems often result in stiffening of the middle ear, which causes more of the sound to be reflected back. Tympanometry is an objective test of middle ear function. In evaluating hearing loss, tympanometry assists in diagnosing between sensorineural and conductive hearing loss.

5.1.3.2 Middle Ear Muscle Reflex

A standard clinical immittance test battery also includes the MEMR. The MEMR is the contraction of the stapedius muscle in response to high-level acoustic stimulation. The MEMR is a bilateral response, which means that presenting the loud tone or noise to one ear will elicit the response in both ears. MEMR measurements are used as a cross-check with the behavioral audiogram and as a way to separate cochlear from retrocochlear pathologies (Schairer et al., 2013).

5.1.4 Oto-Acoustic Emission Testing

Otoacoustic emissions (OAEs) can be measured with a sensitive microphone in the ear canal and provide a noninvasive measure of cochlear amplification (see chapter: Hearing Basics). There are two main types of OAEs in clinical use. Transient-evoked OAEs (TEOAEs) are evoked using a click stimulus. The evoked response from a click covers the frequency range up to around 4 kHz. Distortion product OAEs (DPOAEs) are evoked using a pair of primary tones f1 and f2 ($f1 < f2$) and with a frequency ratio $f2/f1 < 1.4$.

The most commonly measured DPOAE is at the frequency 2f1−f2. Recording of OAEs has become the main method for newborn and infant hearing screening (see chapter: Early Diagnosis and Prevention of Hearing Loss).

5.1.5 Electrocochleography

Electrocochleography (ECochG) is the recording of stimulus-related potentials generated in the human cochlea, including the first-order neurons forming the auditory nerve (see chapter: Hearing Basics; Appendix). These potentials are the cochlear microphonic (CM), the summating potential (SP), and the compound action potential (CAP). CM, SP, and CAP can be recorded from the promontory of the human cochlea by inserting (under local anesthesia) an electrode through the eardrum or from the eardrum itself. CM and SP are generated by the hair cells, CAP by the auditory nerve and is also represented by wave I of the auditory brainstem response (ABR). SP and CAP comparisons are used in the diagnosis of Ménière's disease. The presence of the CM and the absence of a CAP is a defining characteristic of ANP.

5.1.6 Auditory Brainstem Response Testing

5.1.6.1 The Auditory Brainstem Response

The ABR is an auditory evoked potential obtained by signal averaging from ongoing electrical activity in the brain and recorded via electrodes placed on the scalp. The resulting recording is a series of vertex positive waves of which I through V are evaluated. These waves occur in the first 10 ms after onset of an auditory stimulus. Wave I is generated in the proximal part of the auditory nerve (as is the CAP) and wave III in the lower brainstem. Wave IV and V are generated in the upper brainstem (Eggermont, 2007). The ABR represents initiated activity beginning at the base of the cochlea and moving toward the apex over a 4 ms period of time. For click stimuli, the peaks largely reflect activity from the most basal regions on the cochlea because the disturbance hits the basal end first and by the time it gets to the apex, a significant amount of phase cancellation occurs. Amplitude and latency of waves I, III, and V are the basic measures for quantifying the ABR. Amplitude is dependent on the number of neurons firing and above all their synchrony, latency depends on hearing loss and again neural synchrony, interpeak latency (the time between peaks) depends on conduction velocity along the brainstem, and interaural latency (the difference in wave V latency between ears) is sometimes used in acoustic neuroma diagnosis. The ABR is used for newborn hearing screening, auditory threshold estimation using tone-pip stimuli or high-pass mashed clicks, intraoperative monitoring, determining hearing loss type and degree, and auditory nerve and brainstem lesion detection.

5.1.6.2 The Stacked ABR

The stacked ABR is the sum of the synchronous neural activity generated from five frequency regions across the cochlea in response to click stimulation and high-pass noise masking. The technique is an application of high-pass noise masking of the click-evoked ABR resulting in derived ABRs (Don and Eggermont, 1978).

The derived waveforms representing activity from more apical regions along the basilar membrane have wave V latencies that are prolonged because of the nature of the traveling wave (see chapter: Hearing Basics). In order to compensate for these latency shifts, the wave V component for each derived waveform is stacked (aligned), added together, and then the resulting amplitude is measured. Don et al. (1997) showed that in a normal ear, the sum of the stacked derived ABRs has the same amplitude as the unmasked click-evoked ABR. However, the presence of even a small vestibular schwannoma results in a reduction in the amplitude of the stacked ABR in comparison with the unmasked click ABR.

5.1.6.3 The Cochlear Hydrops Analysis Masking Procedure

The Cochlear Hydrops Analysis Masking Procedure (CHAMP) test is a basically a high-pass noise masking ABR test that is used to screen for the presence of cochlear hydrops. It is hypothesized that cochlear hydrops alters the response properties of the basilar membrane which results in reduced masking effectiveness of high-pass noise on the ABR to click stimuli. It is the degree of this undermasking that is used as the diagnostic.

5.1.7 The Auditory Steady-State Response

The auditory steady-state response is an auditory evoked potential, elicited with modulated tones that is used to predict hearing sensitivity in patients of all ages. It is a cortical EEG or MEG response to rapid (or amplitude modulated) auditory stimuli and creates a statistically validated audiogram.

5.1.8 Tone Decay

The tone decay test is used in audiology to detect and measure auditory "fatigue" or abnormal adaptation. In people with normal hearing, a tone whose intensity is only slightly above their hearing threshold can be heard continuously for 60 s. The tone decay test produces a measure of the "decibels of decay," i.e., the number of decibels above the patient's absolute threshold of hearing that are required for the tone to be heard for 60 s. A decay of between 15 and 20 dB is indicative of cochlear hearing loss. A decay of more than 25 dB is indicative of damage to the auditory nerve.

The outcome of the tone decay test is used as a diagnostic criterion for retrocochlear pathology (damage to the auditory nerve).

5.2 CONDUCTIVE HEARING LOSS

Here I will first describe some forms of conductive hearing loss caused by the following conditions: ossicular chain interruption with intact TM, missing TM and ossicles, otosclerosis, TM collapse, and TM perforations. I base this overview on the work of Merchant et al. (1997).

5.2.1 Ossicular Interruption With Intact Tympanometry

Interruption of the ossicular chain in the presence of an intact TM produces an air–bone gap of 50–60 dB. When the ossicles are interrupted, there is no ossicle-coupled sound, and acoustic coupling provides the sole sound input to the cochlea (see chapter: Hearing Basics, Fig. 1.2). Peake et al. (1992) demonstrated that 50–60 dB air–bone gaps could be explained by the remaining acoustic coupling with the stapes.

5.2.2 Loss of Tympanometry, Malleus, and Incus

When the TM, malleus, and incus are missing, the resulting air–bone gap may be explained in terms of remaining acoustic coupling (Peake et al., 1992). With the TM and ossicles missing, ossicular coupling is abolished and acoustic coupling is approximately 10–20 dB larger than in the normal ear. Therefore, the air–bone gap for these conditions is 40–50 dB. Similar gaps should also occur when there is a large perforation of the TM in conjunction with ossicular disruption (Merchant et al., 1997).

5.2.3 Otosclerosis

Otosclerosis is a disease of the bony labyrinth (see chapter: Epidemiology and Genetics of Hearing Loss and Tinnitus) showing a progressive conductive hearing loss, a mixed-type hearing loss, or an SNHL (Rudic et al., 2015). The onset of the hearing loss caused by otosclerosis is generally at an age between 15 and 40 years. Otosclerosis is particularly widespread among Caucasians and very rare among African-Americans, Asians, and Native Americans (Cureoglu et al., 2006). Merchant et al. (1997) noted that the resulting stapes fixation leads to a decrease in stapes volume velocity and, therefore, a conductive hearing loss between 5 and 50 dB starting at low frequencies, and depending on the degree of fixation. The conductive hearing loss associated with malleus fixation is 15–25 dB, which is less than that observed with stapes fixation (up to 50 dB).

5.2.4 Collapse of the Tympanometry into the Middle Ear (Atelectasis)

Collapse of the TM into the middle ear (occurring without a TM perforation) in the presence of intact, mobile ossicles can result in a wide range of conductive hearing loss ranging from 6 to 50 dB. When the middle ear is aerated, any reduction in ossicular coupling would be the result of the TM abnormality. Aeration would also suggest normal acoustic coupling and stapes–cochlear impedance. Under these conditions, a 50% decrease in ossicular coupling will produce a 6 dB air–bone gap. As long as the middle ear and round window are aerated, the maximal loss will not exceed the amount of middle ear pressure gain for that particular ear. In TM collapse, air–bone gaps of 40–50 dB could occur if there is a loss of ossicular coupling. An example would be TM collapse such that there is no middle ear air space and the TM invaginates into the oval window and round window niches. Sound pressures at the oval and round windows would then be nearly the same as the ear canal pressure, and stapes volume velocity would be determined by acoustic coupling. As previously noted (see chapter: Hearing Basics), acoustic coupling is about 40–50 dB smaller than ossicular coupling, and this would predict an air–bone gap of approximately 40–50 dB (Merchant et al., 1997).

5.2.5 Perforations of the Tympanometry

The air–bone gaps associated with TM perforations can range from 0 to 40 dB and are generally greater for low frequencies (<1–2 kHz). Perforations might cause a conductive hearing loss by a combination of two mechanisms. First, a reduction in ossicular coupling expected to be proportional to the size of the perforation. Second, the middle ear pressure becomes similar to ear canal pressure, which is greatest at low frequencies and will also reduce ossicular coupling (Merchant et al., 1997).

5.3 USE OF TYMPANOMETRY IN DETECTING CONDUCTIVE HEARING LOSS

Different acoustic measurements of the mobility of the TM—typically known as tympanometry—including those of impedance, admittance, reflectance, and absorbance, are all examples of immittance measures (Rosowski et al., 2013). Wideband acoustic immittance (WAI) measurements often focus on absorbance and power reflectance of sound. Wideband stimuli allow rapid computation of the immittance or reflectance over a wide frequency range (e.g., 0.2–6 kHz) with a relatively fine frequency resolution (e.g., in steps of 100 Hz) (Rosowski et al., 2013).

The WAI response can be considered as an ordered series of single-frequency tympanograms. Vice versa, a single-frequency tympanogram can be extracted from the wideband absorbance plot. Here one assumes that neighboring frequencies do not interact (linearity assumption). Fig. 5.1A shows wideband absorbance plotted as a joint function of frequency and pressure, Fig. 5.1B shows absorbance plotted as a function of pressure, and in Fig. 5.1C absorbance is plotted as a function of frequency. Thus, WAI measurements might provide a view of the acoustic response properties of the middle ear over a broad range of frequencies, and ear canal pressures and appear to be good indicators of middle ear status, conductive hearing loss, and the effects of middle ear maturation (Sanford et al., 2013).

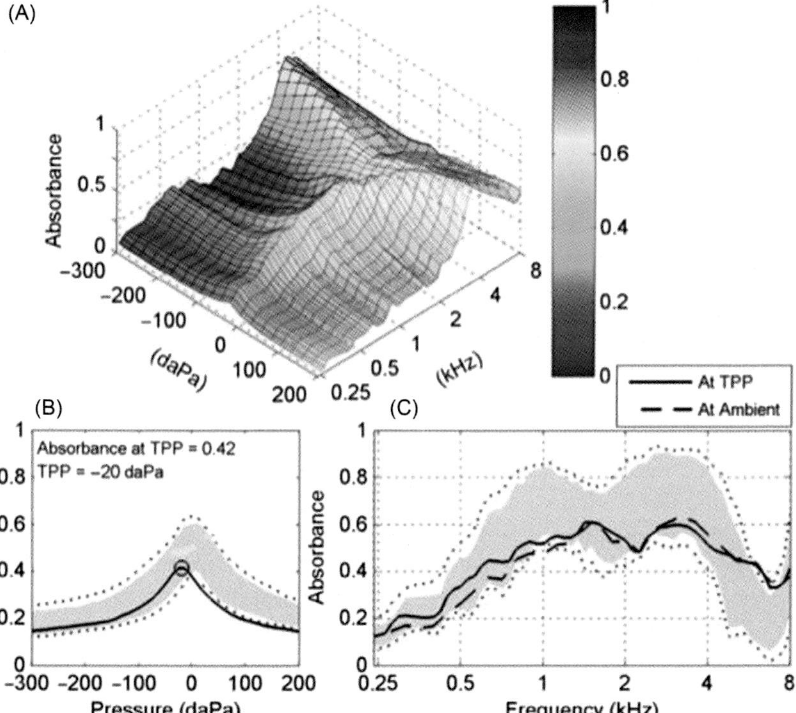

FIGURE 5.1 Wideband absorbance tympanometry data from one adult subject with normal middle ear function. (A) Absorbance as a joint function of frequency and pressure; (B) absorbance, which has been averaged across frequency (0.38–2.0 kHz), plotted as a function of pressure; (C) Absorbance plotted as a function of frequency. Absorbance in (C) is plotted for wideband responses at ambient ear canal pressure (dashed line) and at wideband tympanometric peak pressure (solid line). The shaded regions and fine dashed lines in (B) and (C) represent the 10th to 90th and 5th to 95th percentiles, respectively, of absorbance for adults with normal middle ear function. TPP, tympanometric peak pressure. *From Sanford, C., Hunter, L., Feeney, M., Nakajima, H., 2013. Wideband acoustic immittance: tympanometric measures. Ear Hear. 34, 65S–71S, with permission from Wolters Kluwer Health.*

TABLE 5.1 Expected Combinations of WAI, MEMR, and OAE Outcomes Compared With Diagnostic Threshold Air and Bone Audiometry or ABR Outcomes

Diagnosis	Normal	Conductive	Cochlear	Neural
WAI	Normal	*Flattened*	Normal	Normal
MEMR	Normal	*Elevated threshold*	Normal	*Absent*
OAE	Normal	*Absent*	*Absent*	Normal
ABR audiometry	Normal	*Elevated air* normal bone	*Elevated air and bone*	*Absent or abnormal*

Abnormal measures are italicized.
Source: After Hunter, L.L., Prieve, B.A., Kei, J., Sanford, C.A., 2013. Pediatric applications of wideband acoustic immittance measures. Ear Hear. 34, 36S–42S.

A diagnostic framework (Hunter et al., 2013) shown in Table 5.1, which includes WAI within the pediatric audiology cross-check framework, is helpful to consider how WAI and acoustic reflexes can be combined with OAEs and ABR tests to diagnose type of hearing loss when behavioral audiometry is not possible due to developmental level of the infant or child. Within this framework, ABR may be replaced or validated with behavioral audiometry whenever the child is able to provide reliable behavioral responses.

5.4 SENSORINEURAL HEARING LOSS

SNHL often results from noise exposure, either traumatic or resulting from accumulation of a life-long exposure to lower levels. If the loss is preneural it is called sensory loss, if it is neural it is called neural loss. A problem is that sensory loss may lead to neural loss and differentiating the two becomes near impossible, hence sensorineural is often appropriate. I will start with a brief description of the effects noise trauma has on the auditory periphery. This has been described in similar terms in chapter 3 of "Noise and the Brain" (Eggermont, 2014) but is presented here in an abbreviated form to make this chapter self-contained.

5.4.1 Noise-Induced Temporary Threshold Shifts

Davis et al. (1946, 1950), in a study financed by the Office of Scientific Research and Development of the United States Government, were the first to explore the limits of tolerance of the human ear to sustained sound. The group members took turns exposing one ear at a time and measuring the recovery of sensitivity from their temporary threshold shift (TTS). If recovery was complete (within 5 dB) within 24 h the exposure was considered "safe" in the military context. These criteria of tolerance were very different from those now invoked for control of noise pollution, and even from those for protection of hearing in industry. The output level of their "bullhorn" speaker (formerly used in a public address system on board of an aircraft carrier) was raised a full 20 dB above the prescribed settings. They discovered that when ears were exposed to a pure tone from the bullhorn the TTS was always greatest at a frequency about half an octave above that of the exposure tone (Fig. 5.2). Sometimes the threshold at the exposure tone frequency was shifted no more than 10 dB, while half an octave higher the shift 2 min after exposure might be 50 dB. Prolonged exposure to an intense 500 Hz tone or to noise of wide frequency spectrum caused severe speech discrimination loss at a low (40 dB SPL) presentation level but only moderate loss at a high (100 dB SPL) level.

Kujawa and Liberman (2009) showed that acoustic overexposure for 2 h with an 8–16 kHz band of noise at 100 dB SPL in the mouse caused a moderate, but completely reversible, threshold elevation as measured by ABR. The absence of permanent changes in the OAEs indicated that the exposure left outer hair cells (OHCs) intact. Using confocal imaging of the inner ear, they found that despite the normal appearance of the cochlea and normal hearing thresholds there was an acute loss of the ribbon synapses located on the medial side of the inner hair cell (IHC) connected to high-threshold, low-SFR ANFs followed by a delayed progressive diffuse degeneration of the

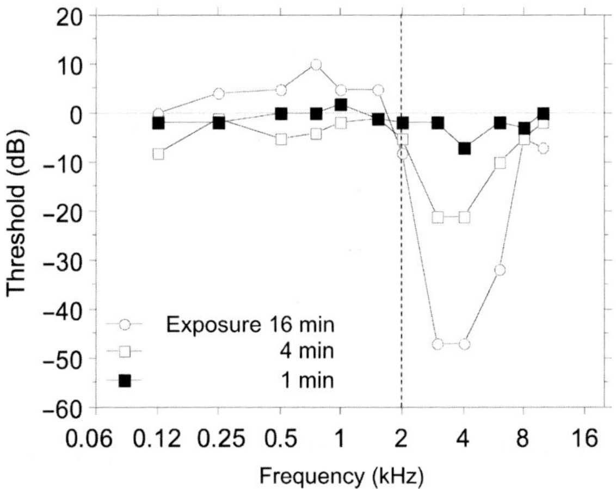

FIGURE 5.2 Audiograms after 1-, 4-, and 16-min exposure to a 2 kHz tone (dashed line) presented at 120 dB. *From Eggermont, J.J., 2014. Noise and the Brain. Experience Dependent Developmental and Adult Plasticity, Academic Press, London. Based on data from Davis, H., Morgan, C.T., Hawkins, J.E., Galambos, R., Smith, F.W., 1950. Temporary deafness following exposure to loud tones and noise. Acta Otolaryngol. Suppl. 88, 1–56.*

cochlear nerve. After more extensive studies Kujawa and Liberman (2015) concluded: "If our new view of hidden hearing loss is correct, then neither the audiogram nor otoacoustic emissions is the appropriate test to reveal the first signs of accumulating noise-induced hearing loss. What is required is a non-invasive test for cochlear synaptopathy that is applicable to human populations."

Partially supporting the Kujawa and Liberman (2009, 2015) studies in mice, Liu et al. (2012) found in guinea pigs exposed to broadband noise at 105 dB SPL for 2 h that "(1) contrary to reports in mice, the initial loss of ribbons largely recovered within a month after the noise exposure, although a significant amount of residual damage existed; (2) while the response threshold fully recovered in a month, the temporal processing continued to be deteriorated during this period." A follow-up study by Shi et al. (2013) found that "during the repair process, ribbons were temporally separated from the post-synaptic densities (PSDs). A plastic interaction between ribbons and postsynaptic terminals may be involved in the reestablishment of synaptic contact between ribbons and PSDs, as shown by location changes in both structures. Synapse repair was associated with a breakdown in temporal processing, as reflected by poorer responses in the compound action potential (CAP) of auditory nerves to time-stress signals." The results of Shi et al. (2013) were duplicated in the study of Song et al. (2016) who also used a 2 h exposure with broadband noise at 105 dB SPL and investigated the dynamic changes of ribbon synapses and the coding function of ANF single units in 1 month after a brief noise exposure that caused a massive damage of ribbon synapses but no permanent threshold shift (PTS). The synapse count and functional response measures indicated that a large portion of the disrupted synapses were reconnected at 1 month postexposure. This was consistent with the quick recovery of the change of the SFR distribution due to the initial loss of low spontaneous rate units. However, after reestablishment of the synapses ANF coding deficits were found in both intensity and temporal processing, revealing the nature of synaptopathy in hidden hearing loss. It seems that there may be a species difference in the ability to regenerate synapses after noise-induced damage. However, two recent studies using C57 mice (Shi et al., 2015a,b) reported that the loss of ribbon synapses induced by non-PTS-inducing noise is also largely reversible.

This raises the question whether such neural losses following ribbon synapse damage do occur in humans. An initial study that tends to support this (Viana et al., 2015) reported an analysis of five "normal" ears, aged 54–89 years, without any history of otologic disease, which suggested that cochlear synaptopathy and the degeneration of cochlear nerve peripheral axons, despite a near-normal hair cell population, may be an important component of human presbycusis. Stamper and Johnson (2015), in a study of normal hearing human ears that had been exposed with a range of recreational noise levels, between 67 and 83 dB(A), over 1 year, found a trend of smaller ABR wave I amplitudes in response to supra-threshold clicks and 4 kHz tone bursts for the larger exposures. In contrast, supra-threshold DPOAE level was not significantly related to noise exposure. They considered this a putative correlate of hidden hearing loss, as defined by Kujawa and Liberman (2009). However, Stamper and Johnson (2015) noted that an alternate explanation for the contrasting findings of wave I and wave V amplitude could be found in the study by Don and Eggermont (1978). There it was shown that for click-evoked ABR, using high-pass noise masking, wave I is dominantly generated by high-frequency ANFs (>3 kHz), whereas wave V is dominantly generated by activity from the cochlea generated in the less than 2 kHz region. Thus a high-frequency hearing loss, especially above 8 kHz, will reduce inputs to wave I and does not affect wave V. This would, however, have somewhat less of an effect for the 4 kHz tone bursts that were also used by Stamper and Johnson (2015) and showed similar results to the clicks. Based on their mouse model for cochlear neuropathy, Valero et al. (2016) suggested the MEMR as a diagnostic tool for ANP in humans.

5.5 LOUDNESS RECRUITMENT

In chapter 3 of "The Neuroscience of Tinnitus" (Eggermont, 2012), I wrote: "Loudness recruitment is an abnormal relationship between sound intensity and perceived loudness that may accompany NIHL. In this condition, the threshold of hearing is elevated, but a given increase in level above threshold causes significantly greater than normal increase in loudness. The rate of loudness growth in the impaired ear is often so steep that, at high intensities (60–100 dB SPL), the loudness in the impaired ear matches that in the normal one. It was commonly thought that the loudness of a sound reflects some aspects of the overall activation pattern of peripheral structures such as the auditory nerve, e.g., the total number of action potentials (e.g., Eggermont, 1976; Moore and Glasberg, 2004). From this point of view enhanced output from the cochlea would be expected based on the loss of the cochlear compressive nonlinearity consequent to damage to OHCs (Robles and Ruggero, 2001). In agreement, psychophysical results are consistent with an explanation of recruitment in terms of a loss of compression of the amplitude of basilar membrane vibration (Oxenham and Bacon, 2003). Steeper than normal rate-level functions (RLFs) of ANFs have also been observed in sensorineural hearing loss following ototoxic damage to OHCs in cats (e.g., Harrison, 1981)."

5.5.1 Compound Action Potentials and Recruitment

Using ECochG (see chapter: Hearing Basics; Appendix), I recorded CAPs from the cochlea promontory in humans with normal hearing and in patients with SNHL and recruitment (Eggermont, 1977a). In Fig. 5.3 results are shown

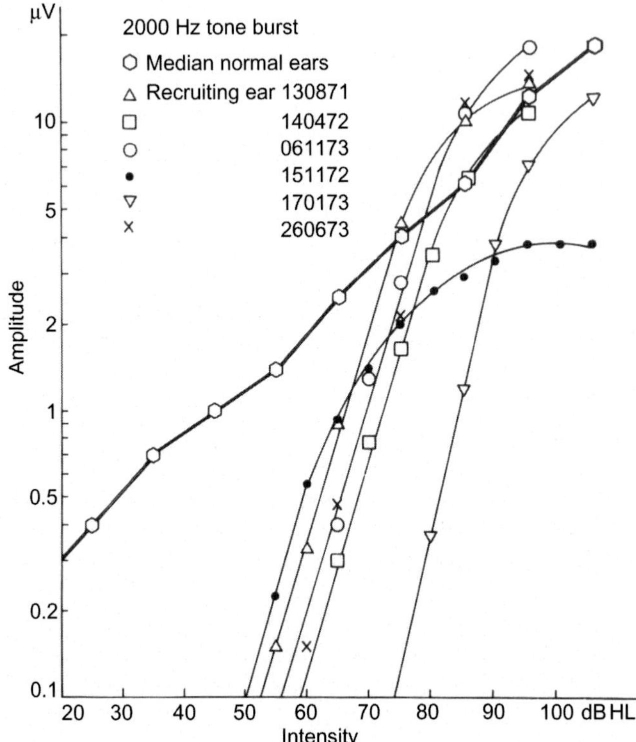

FIGURE 5.3 CAP amplitude-level functions for a group of 20 normal hearing people (median shown with fat line) and for 5 people with recruitment and SNHL. Data were recorded from the promontorium of the cochlea by inserting a thin needle through the TM under local anesthesia. *From Eggermont, J.J., 1977. Electrocochleography and recruitment. Ann. Otol. Rhinol. Laryngol. 86, 138–149, with permission from the Annals Publishing Company.*

for 2 kHz tone pips presented at a rate of 7/s. The median CAP amplitude-level function (fat line) for 20 normal ears typically continues down to levels of 0 dB HL (by definition, but not shown). The individual patient amplitude-level functions shown here reflect their 2 kHz hearing losses of 50–75 dB. Note that at high sound level these functions approach or exceed the median curve for normal hearing people. If loudness can be equated with CAP amplitude it would indicate a peripheral marker for the loudness recruitment that these patients showed behaviorally. This would be consistent with Ruggero and Rich (1991)'s finding that the velocity–intensity relationship of the basilar membrane becomes steeper after the loss of OHC amplification.

However, the overall story appears to be more complex. First of all, the examples for recruiting ears with Ménière's disease shown in Fig. 5.3 are not representative for recruiting ears in general. Fig. 5.4 shows for 20 normal hearing people, 23 people with hearing loss but no loudness recruitment

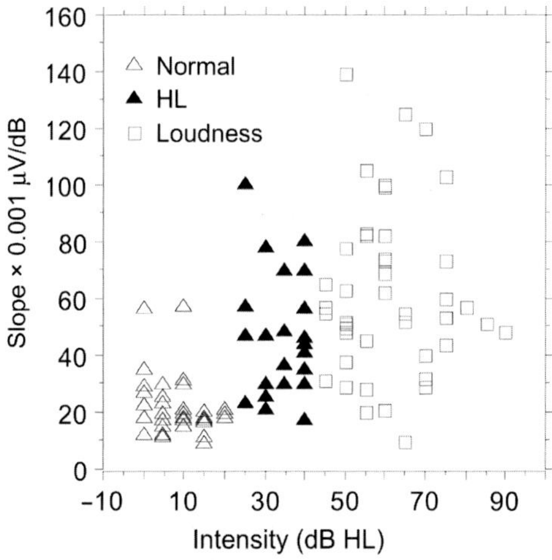

FIGURE 5.4 Steepness of input–output curves in normal and recruiting ears for 2000 Hz tone burst stimulation. For about 20 normal ears (thresholds <20 dB HL) and 80 hearing loss ears of which 23 without recruitment (filled triangles) and 60 with recruitment (open squares), the numerical values of the slope of the input–output curves below the 1.5 μV level are plotted as a function of threshold intensity. It appears that on the average the slope of the input–output curve increases with increasing threshold value. It also appears that no absolute boundary for recruiting and nonrecruiting ears exists. The diagnosis of loudness recruitment solely based on the slope of the input–output curve remains a question of probability. *Based on data from Eggermont, J.J., 1977. Electrocochleography and recruitment. Ann. Otol. Rhinol. Laryngol. 86, 138–149, with permission from the Annals Publishing Company.*

behaviorally, and 60 people with recruitment that the slope of the CAP amplitude–intensity function at 2 kHz is not an indicator of recruitment. However, loudness recruitment does (on average) depend on the amount of hearing loss and requires a hearing loss of 45 dB or more. Furthermore a detailed analysis of the CAP amplitude increments over 10 dB level increments as a function of intensity again indicated no difference between putative OHC loss ears, Ménière's ears, and normal hearing people (Eggermont, 1977a).

5.5.2 Single Auditory Nerve Fiber Responses and Recruitment

Following acoustic trauma, OHC loss is often accompanied by IHC damage (Liberman and Dodds, 1984), which compromises cochlear transduction and *lowers* the firing rates of ANFs (Liberman and Kiang, 1984). As a result, the slopes of their RLFs become *shallower* on average (Salvi et al., 1983; Heinz and Young, 2004). In cats, following acoustic trauma, the summed RLFs of ANFs over either a narrow or wide range of CFs surrounding the stimulus frequency did not show abnormally steep slopes (Heinz et al., 2005). Applying this to the CAP data of Fig. 5.3 we have to consider that the CAPs may reflect OHC damage and loss of compression (these changes signaled threshold shifts), but IHC damage cannot be excluded. In addition, increased neural synchrony of ANF firings after noise trauma will lead to larger CAPs and steeper amplitude-level functions.

5.5.3 Central Nervous System and Recruitment

Hyperactivity is seen in central auditory neurons after SNHL and is characterized by an abnormally rapid increase of response with stimulus intensity or by a heightened maximum response, or both (Salvi et al., 2000). This central hyperactivity is a potential contributor to amplitude growth of the local field potentials (LFPs; Fig. 5.5). Central hyperactivity must affect patterns of neural activity that code for perceived loudness. Coincidence-detecting cells in the ventral cochlear nucleus (VCN), which receive ANF inputs with a narrow region of CFs, can potentially decode the spatiotemporal cue for sound level (the synchronous firing of a neural population). Such spatiotemporal

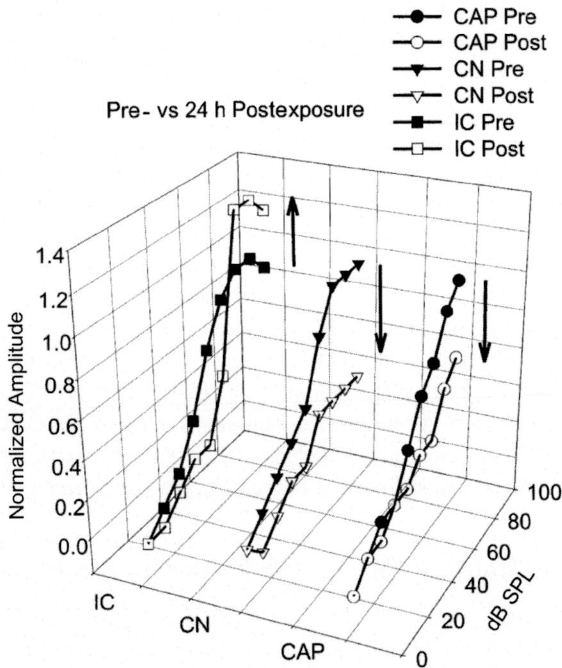

FIGURE 5.5 Schematic illustrating typical evoked response amplitude-level functions recorded at the round window (CAP), the cochlear nucleus (CN), and inferior colliculus (IC) at 1000 Hz pre- and 24 h postexposure. Traumatizing exposure: 105 dB SPL, 2.8 kHz, 2 h. Amplitude normalized to maximum preexposure response before the exposures. Arrows indicate the direction of amplitude change at moderate to high intensities. *Reprinted from Wang, J., Ding, D., Salvi, R.J., 2002. Functional reorganization in chinchilla inferior colliculus associated with chronic and acute cochlear damage. Hear. Res. 168, 238–249, with permission from Elsevier.*

processing could be performed at any level of the auditory brainstem where cells receive converging inputs from lower levels. For example, neurons recorded in VCN are typically classified based on temporal discharge patterns in response to tones. There is physiological evidence that cells in the anteroventral cochlear nucleus that receive convergent ANF inputs are sensitive to the temporal response pattern across the inputs (Carney, 1990). In particular, onset chopper response types (T-stellate cells) in the VCN, which are known to receive convergent inputs from ANFs with a wide range of CFs and to have wide dynamic ranges (Smith and Rhode, 1989), may act as coincidence detectors coding for sound intensity. As seen in Chapter 1, Hearing Basics synapses on IHCs span a range of different thresholds such that synchronous activity detected from a subset of them could give rise to a population response related to sound intensity.

Cai et al. (2008) analyzed RLFs summed over various populations of VCN neurons following acoustic trauma and reported that the responses of nonprimary-like types of neurons, especially chopper neurons, showed changes in rate responses consistent with loudness recruitment (as suggested by Carney, 1990), whereas primary-like neurons did not, except in cases where spread of excitation in the population was included. Cai et al. (2008) suggested that the VCN is the most peripheral stage of the auditory pathway at which over-excitability of neurons appears following acoustic trauma. Moreover, they again point to the synaptic processing in the nonprimary-like, especially the stellate, cells as the initial locus of central compensation for reduced rate responses in the ANFs. The LFP data shown in Fig. 5.5 are not corroborating this suggestion and put the source of recruitment in the midbrain. Neural hyperexcitability at the LFP level is commonly reported in cortex and IC (Qiu et al., 2000), but the results in CN have been mixed, with both positive (Saunders et al., 1972) and negative (Salvi et al., 1990) findings.

Summarizing, in considering recruitment, one assumes that changes in the loudness function are mirrored by similar changes in neural response growth functions, steeper neural growth functions corresponding to steeper loudness growth, for example. This behavior has been reported in the human auditory system through functional imaging at the level of the auditory cortex (Langers et al., 2007). These neural data match the expectation of recruitment qualitatively in that the neural responses of populations are stronger after cochlear damage.

5.6 AUDITORY NEUROPATHY

From the very beginning there has been confusion about the meaning of auditory neuropathy. This was best expressed by Rapin and Gravel (2003):

> *The term 'auditory neuropathy' is being used in the audiology/otolaryngology literature for a variety of individuals (mostly children) who fulfill the following*

criteria: (1) understanding of speech worse than predicted from the degree of hearing loss on their behavioral audiograms; (2) recordable otoacoustic emissions and/or cochlear microphonic; together with (3) absent or atypical auditory brain stem responses. To neurologists, the term neuropathy has a more precise connotation: it refers to pathology of peripheral nerve fibers rather than pathology in their neuronal cell bodies of origin.

With respect to potential confusing uses of terms, the one that exemplifies that best is the denotation "auditory neuropathy spectrum disorder" (ANSD). This definition is all encompassing but usually does not require more than the presence of a superficial phenomenology consisting of recordable OAEs and absent or very poorly defined ABRs, together with mild to moderate audiometric hearing loss but problems with speech understanding. We will see that there is quite a bit more with respect to underlying mechanisms. This has led to use of a new term "synaptopathy," which puts the main mechanism in the IHC ribbon synapses. Another umbrella term is "dys-synchrony," which can describe anything from the nonsynchronous transmitter substance release from the ribbon synapses, resulting in onset desynchrony in the ANF firings, to slowing down of action potentials along the auditory nerve which also results in a large spread of spike latencies and hence poorly shaped ABRs.

5.6.1 Identification

This section represents an abbreviated and updated version of chapter 13 of my book "Auditory Temporal Processing and its Disorders" (Eggermont, 2015). In the first paper to describe ANP, Starr et al. (1996) presented 10 children and young adults with hearing impairments that, by behavioral and physiological testing, were compatible with a disorder of the auditory nerve. Evidence of normal OHC function was provided by preservation of OAEs and CM (see chapter: Hearing Basics) in all of the patients. ABRs were absent in nine patients and severely distorted in one patient (Fig. 5.6). Auditory brainstem reflexes (stapedius reflex and contralateral suppression of OAEs) were absent in all of the tested patients. Behavioral audiometric testing showed a mild to moderate elevation of pure-tone threshold. The shape of the pure-tone audiogram varied, showing predominantly a low frequency loss in five patients, a flat loss across all frequencies in three patients and predominantly a high frequency loss in two patients. Speech intelligibility was tested in eight patients, and in six thereof was much worse than expected if the pure-tone loss were of cochlear origin. Starr et al. (1996) suggested that this type of hearing impairment is due to a disorder of auditory nerve function and may have a neuropathy as one of its causes. We note that in the same year, Kaga et al. (1996) identified the same disorder but called it auditory nerve disease.

Yasunaga et al. (1999) studied four families with nonsyndromic recessive ANP. They identified a novel human gene, otoferlin (*OTOF*), underlying an autosomal recessive, nonsyndromic prelingual deafness, DFNB9. In the mouse

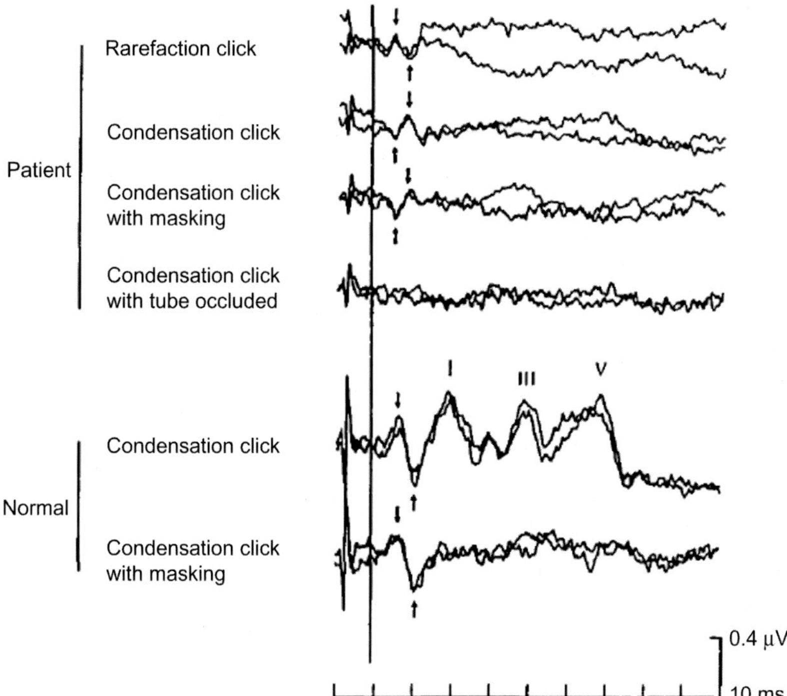

FIGURE 5.6 CM and ABRs in auditory neuropathy. The stimulus was a 75 dB nHL click presented at a rate of $10\ s^{-1}$. The stimuli were presented using insert earphones with a 0.9 ms delay from the microphone to the stimulus arriving at the TM. Recordings were obtained between the vertex and an electrode in the external auditory meatus. Each tracing represents the average of 2000 recordings for the normal subject (lower part) and 8000 recordings for the patient (upper part). The filter band pass was 20–10,000 kHz. Only the results for the right ear are shown. The recordings from the patient show no recognizable ABRs. The arrows point to the CM that reverses in polarity when the click is changed from rarefaction to condensation. When sufficient masking noise is added to prevent perception of the clicks, the CM remains, whereas any ABR is canceled. Closing off the tube from the microphone to the ear insert (trace four from top) shows that the recorded deflections are not delayed electrical artifacts from the microphone. *Reprinted from Starr, A., Picton, T.W., Sininger, Y., Hood, L.J., Berlin, C.I, 1996. Auditory neuropathy. Brain 119, 741–753, by permission of Oxford University Press.*

inner ear the expression of this gene was mainly in the sensory hair cells, indicates that its role could apply to synaptic vesicles. Subjects with this type of ANP are not helped by hearing aids but are helped by cochlear implants (Budenz et al., 2013), suggesting that the defect is peripheral to the ANFs.

5.6.2 Presynaptic Aspects of ANP

Hearing relies on faithful synaptic transmission at the ribbon synapse of IHCs (see chapter: Hearing Basics). Khimich et al. (2005) showed that anchoring of IHC ribbons is impaired in mouse mutants for the presynaptic scaffolding protein Bassoon. The lack of active-zone-anchored synaptic ribbons reduced the presynaptic readily releasable vesicle pool, and impaired synchronous auditory signaling as revealed by recordings of exocytic IHC capacitance changes and sound-evoked activation of spiral ganglion neurons. Both exocytosis of the hair cell releasable vesicle pool and the number of synchronously activated spiral ganglion neurons covaried with the number of anchored ribbons during development. Interestingly, ribbon-deficient IHCs were still capable of sustained exocytosis with normal Ca^{2+} dependence. It was concluded that ribbon-dependent synchronous release of multiple vesicles at the hair cell afferent synapse is essential for normal hearing. Ribbon synapse dysfunctions, termed auditory synaptopathies, impair the audibility of sounds to varying degrees but commonly affect neural encoding of acoustic temporal cues essential for speech comprehension (Moser et al., 2013). Clinical features of auditory synaptopathies are similar to those accompanying ANP. Genetic auditory synaptopathies include alterations of glutamate loading of synaptic vesicles, synaptic Ca^{2+} influx, or synaptic vesicle turnover. Acquired synaptopathies include noise-induced temporary hearing loss because of excitotoxic synaptic damage and subsequent gradual neural degeneration (Kujawa and Liberman, 2009). Alterations of ribbon synapses may also contribute to age-related hearing loss.

5.6.3 Postsynaptic Mechanisms of ANP

This section is based on the extensive review by Rance and Starr (2015) emphasizing that auditory nerve function disorders can occur at multiple sites along the auditory nerve. These are (1) dendrites within the cochlea, (2) myelinated dendrites and axons coursing centrally, and (3) myelinated auditory ganglion cells.

5.6.3.1 Dendritic Nerve Terminals

Pathology affecting the dendritic nerve terminals results in a pattern of objective measures similar to those outlined for ribbon synapse disorders; i.e., normal SPs reflecting normal IHC functions and absent CAPs (Rance and Starr, 2015). In an example of dendritic terminal abnormality (due to *OPA1* gene mutation), Santarelli et al. (2015) describe a broad low-amplitude negative potential that adapts at high stimulation rates, which is consistent with a neural origin. ABRs are absent or abnormal. Dendritic damage may also result from noise trauma that results in withdrawal of dendrites from synaptic connection with IHCs due to excessive neurotransmitter release (Kujawa and Liberman, 2009).

5.6.3.2 Axonal Neuropathies

Axonal neuropathies reduce neural activity in the auditory nerve and brainstem without affecting cochlear hair cells. Patients with pathology restricted to the nerve should have normal SPs (cf. chapter: Hearing Basics) and reduced amplitude or absent CAP/ABR depending on the degree of deafferentation (Rance and Starr, 2015). This could lead to increased SP/CAP amplitude ratios just as in Ménière's disease (see Section 5.8).

5.6.3.3 Auditory Ganglion Cell Disorders

The viability of auditory neural ganglion cells depends on a number of adverse metabolic factors including hyperbilirubinemia (Rance and Starr, 2015). Objective measures of auditory function in jaundiced patients with ANP (Santarelli and Arslan, 2001) showed normal SPs consistent with normal hair cell function, absent CAPs replaced by low-amplitude sustained negativities characteristic of reduced neural dendritic responsiveness, and absent ABRs.

5.6.3.4 Myelin Disorders

Some forms of ANP may reflect the attenuation of synchronous neural discharges due to demyelination. Demyelination disorders can accompany axonal damage. Wynne et al. (2013) evaluated patients with a range of disorders consistent with nerve fiber abnormality and found ABRs were abnormal with both reduced amplitude and delayed I–V conduction times. When stimuli were repeated trains of clicks, ABRs to the initial click were normal, but responses became attenuated and delayed to subsequent stimuli suggesting the development of a "conduction block" during repetitive stimulation (Rance and Starr, 2015).

5.6.3.5 Auditory Nerve Conduction Disorders

Pontine angle tumors such as vestibular neuromas and meningiomas frequently present with a hearing disorder resembling ANP due to compression of proximal auditory nerve. ABR results vary from complete absence to preserved waveforms with prolonged wave I–V conduction times. Multiple sclerosis is associated with demyelination of central auditory brainstem fibers. ABR wave I is typically unaffected whereas central components (waves III–V) can be absent or delayed in latency (Chiappa, 1997). As such, multiple sclerosis is a brainstem disease that has many features similar to auditory nerve disorders (Rance and Starr, 2015).

5.6.4 Electrocochleography Outcomes

Santarelli et al. (2009) recorded abnormal click-evoked cochlear potentials with transtympanic ECochG (Eggermont et al., 1974; Appendix) from four children with *OTOF* mutations to evaluate physiological effects in humans resulting from abnormal neurotransmitter release by IHCs. The subjects were profoundly deaf with absent ABRs and preserved OAEs consistent with ANP. Cochlear potentials evoked by clicks from 60 to 120 dB peak equivalent SPL were compared to recordings obtained from 16 normally hearing children. The CM was recorded with normal amplitudes from all but one ear, consistent with the preserved OAEs. After canceling CM, the remaining cochlear potentials were of negative polarity with reduced amplitude and prolonged duration compared to controls. These cochlear potentials were recorded as low as 50–90 dB below behavioral thresholds in contrast to the close correlation in normal hearing controls between cochlear potentials and behavioral threshold. SPs were identified in five out of eight ears with normal latency whereas CAPs were either absent or of low amplitude. Stimulation at high rates reduced amplitude and duration of the prolonged potentials, consistent with their neural generation site. This study suggests that mechano-electrical transduction and cochlear amplification are normal in patients with *OTOF* mutations. The low-amplitude prolonged negative potentials are consistent with sustained exocytosis and decreased phasic neurotransmitter release (Khimich et al., 2005) resulting in abnormal dendritic activation and impairment of auditory nerve firing (Fig. 5.7A).

Santarelli et al. (2011, 2013) then compared acoustically and electrically evoked potentials of the auditory nerve in patients with postsynaptic or presynaptic ANP with underlying mutations in the *OPA1* or *OTOF* gene, respectively. Among nonisolated ANP disorders, mutations in the *OPA1* gene are believed to cause disruption of auditory nerve discharge by affecting the unmyelinated portions of human ANFs. Transtympanic ECochG was used to record click-evoked responses from two adult patients carrying the R445H *OPA1* mutation and from five children with mutations in the *OTOF* gene. The CM amplitude was normal in all subjects. Prolonged negative responses were recorded as low as 50–90 dB below behavioral threshold in subjects with *OTOF* mutations (Fig. 5.7A), whereas in the *OPA1* disorder the prolonged potentials were correlated with hearing threshold (Fig. 5.7B). A CAP was superimposed on the prolonged activity at high stimulation intensity in two children with mutations in the *OTOF* gene while CAPs were absent in the *OPA1* disorder. Electrically evoked compound action potentials (eCAPs; see chapter: Cochlear Implants) could be recorded from subjects with OTOF mutations but not from *OPA1* mutations following cochlear implantation. The findings are consistent with abnormal function of distal portions of ANFs in patients carrying the *OPA1* mutation, whereas the low-threshold prolonged potentials recorded from children with mutations in the *OTOF* gene are consistent with abnormal neurotransmitter release resulting in reduced dendritic activation and impairment of spike initiation. Interestingly, ribbon-deficient IHCs are still capable of sustained exocytosis with normal Ca^{2+} dependence (Khimich et al., 2005), probably underlying the long-duration neural responses recorded from *OTOF* patients.

Santarelli et al. (2015) further characterized the hearing dysfunction in *OPA1*-linked disorders. Nine of 11 patients carrying *OPA1* mutations inducing haplo-insufficiency had normal hearing function. Eight patients carrying *OPA1* missense variants underwent cochlear implantation. The use of cochlear

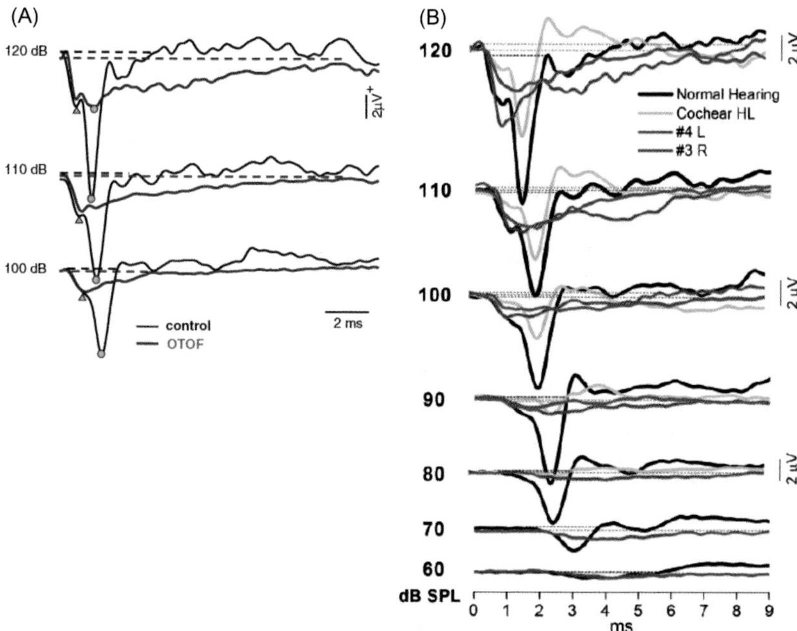

FIGURE 5.7 (A) Comparison between the SP–CAPs recorded from one subject with *OTOF* mutations and one control. The curves for the *OTOF* subject are superimposed on the recordings obtained from one control at intensities from 120 to 100 dB p.e. SPL to highlight the similarities of the SP component between controls and patients with *OTOF* mutations. Open circles and triangles refer to the CAP and SP peaks, respectively. (B) ECochG waveforms obtained after CM cancellation from two representative *OPA1*-M patients are superimposed on the corresponding responses recorded from one control with normal hearing and from one hearing-impaired child with cochlear hearing loss (cochlear HL) at decreasing stimulus intensity. *From (A) Santarelli, R., Del Castillo, I., Rodríguez-Ballesteros, M., et al., 2009. Abnormal cochlear potentials from deaf patients with mutations in the otoferlin gene. J. Assoc. Res. Otolaryngol. 10, 545–556, with permission of Springer; (B) Santarelli, R., Rossi, R., Scimemi, P., Cama, E., Valentino, M.L., La Morgia, C., et al., 2015. OPA1-related auditory neuropathy: site of lesion and outcome of cochlear implantation. Brain 138, 563–576, by permission of Oxford University Press.*

implant improved speech perception in all but one patient. ABRs were recorded in response to electrical stimulation in five of six subjects, whereas no eCAP was evoked from the auditory nerve through the cochlear implant. These findings corroborate that the impaired mechanism in patients carrying *OPA1* missense variants is desynchronized ANF firings resulting from neural degeneration affecting the terminal dendrites (Santarelli et al., 2015).

5.6.5 Evoked Potentials Following Cochlear Implantation

The various underlying defects leading to the same phenomenology of ANP has led to a more general term ANSD, but it remains to be seen if this is helpful. Jeon et al. (2013) studied 11 patients with ANSD and 9 control subjects with SNHL who did not have neural pathology. Patients and control subjects each received a cochlear implant and underwent electrically evoked ABR (eABR; chapter: Cochlear Implants) testing. eABR threshold, wave eV latency, and amplitude were measured as eABR parameters. All controls responded to eABR, whereas 6 of the 11 ANSD patients did not respond to eABR. After CI, two of the five patients in the eABR response group could understand conversation without visual cues, and one patient understood speech with lip reading. The remaining two were too young to be tested. In contrast, the nonresponse group demonstrated variable outcomes, although all of them still benefited from CI.

Cardon and Sharma (2013) investigated a cross-sectional (24 children) and a longitudinal (11 children) group with ANSD that were fitted with CIs. They compared P_1 cortical auditory evoked potential (CAEP) latency with the scores on the Infant Toddler Meaningful Auditory Integration Scale (IT-MAIS). P_1 responses were present in all children after implantation. P_1 CAEP latency was significantly correlated with participants' IT-MAIS scores. Furthermore, more children implanted before age 2 years showed normal P_1 latencies, while those implanted later mainly showed delayed latencies. Longitudinal analysis revealed that most children showed normal or improved cortical maturation after implantation. He et al. (2013) also investigated the feasibility of recording the electrically evoked auditory $P_1-N_1-P_2$ complex (cf. Fig. 8.1) and the electrically evoked auditory change complex (eACC) in response to temporal gaps, in cochlear implant children with ANSD. The acoustic change complex is a scalp-recorded negative–positive voltage transition evoked by a change in an otherwise steady-state sound (Martin and Boothroyd, 2000). The eACC is evoked by changing stimulation between two electrodes (Brown et al., 2008). The eERPs were recorded from all subjects with ANSD who participated in this study. He et al. (2013) found that in general, the eACC showed less mature morphological characteristics than the onset $P_1-N_1-P_2$ response recorded from the same subject. There was a robust correlation between word recognition scores and the eACC thresholds for gap detection. Subjects with poorer speech-perception performance showed larger eACC thresholds in this study.

5.6.6 Psychoacoustics

In chapter 13 of Eggermont (2015) I wrote: "Starr and colleagues (Zeng et al., 2005) studied the perceptual consequences of disrupted auditory nerve activity in 21 subjects who had been clinically diagnosed with auditory neuropathy. Psychophysical measures showed that the disrupted neural activity had minimal effects on intensity-related perception, such as loudness discrimination, pitch discrimination at high frequencies, and sound localization using inter-aural sound level differences. In contrast, the disrupted neural activity significantly impaired timing-related perception, such as pitch discrimination at low frequencies, temporal integration, gap detection, temporal modulation detection, backward and forward masking, signal detection in noise, binaural beats, and sound localization using inter-aural time differences. These perceptual consequences are the opposite of what is typically observed in subjects with cochlear hearing loss who have impaired intensity perception but relatively normal temporal processing after taking their impaired intensity perception into account."

Narne (2013) confirmed that listeners with ANP performed significantly poorer than normal hearing listeners in both amplitude modulation and frequency modulation detection. They saw significant correlation between measures of temporal resolution and speech perception in noise. Thus, an impaired ability to process speech envelope and fine structure may be causing poor speech perception by listeners with ANP.

Abnormal auditory loudness adaptation measurement (also known as tone decay) is a standard clinical tool for diagnosing auditory nerve disorders due to acoustic neuromas. Wynne et al. (2013) investigated this in ANP patients with either disordered function of IHC ribbon synapses (temperature-sensitive ANP) or ANFs (non-temperature-sensitive ANP). Non-feverish subjects were tested for (1) psychophysical loudness adaptation to comfortable loud sustained tones and (2) physiological adaptation of ABRs to clicks as a function of their position in brief 20-click stimulus trains. Results were compared with normal hearing listeners and listeners with other forms of hearing impairment. Subjects with ribbon synapse disorder had abnormally increased magnitude of loudness adaptation to both low (250 Hz) and high (8000 Hz) frequency tones (Fig. 5.8), but not as much as patients with vestibular schwannoma (neuroma) for 250 Hz. Subjects with ANF disorders had normal loudness adaptation to low-frequency tones; all but one had abnormal adaptation to high-frequency tones. Adaptation was both more rapid and of greater magnitude in ribbon synapse than in ANF disorders. ABR measures of adaptation in ribbon synapse disorder showed wave V to the first click in the train to be abnormal both in latency and amplitude, and these abnormalities increased in magnitude with click number. Wave V could also be absent for subsequent clicks. In contrast, ABRs in four of the five subjects with neural disorders were absent to every click in the train. The fifth subject had normal latency and abnormally reduced amplitude of wave V to the first click and abnormal or absent responses to subsequent clicks. Thus, dysfunction of both synaptic transmission and auditory neural function can be associated with abnormal loudness adaptation and the magnitude of the adaptation is significantly greater with ribbon synapse than neural disorders.

Ouabain application to the round window can selectively destroy type-I spiral ganglion cells, producing an animal model of ANP. To assess the long-term effects of this deafferentation on synaptic organization in the organ of Corti and CN, and to ask whether surviving cochlear neurons show

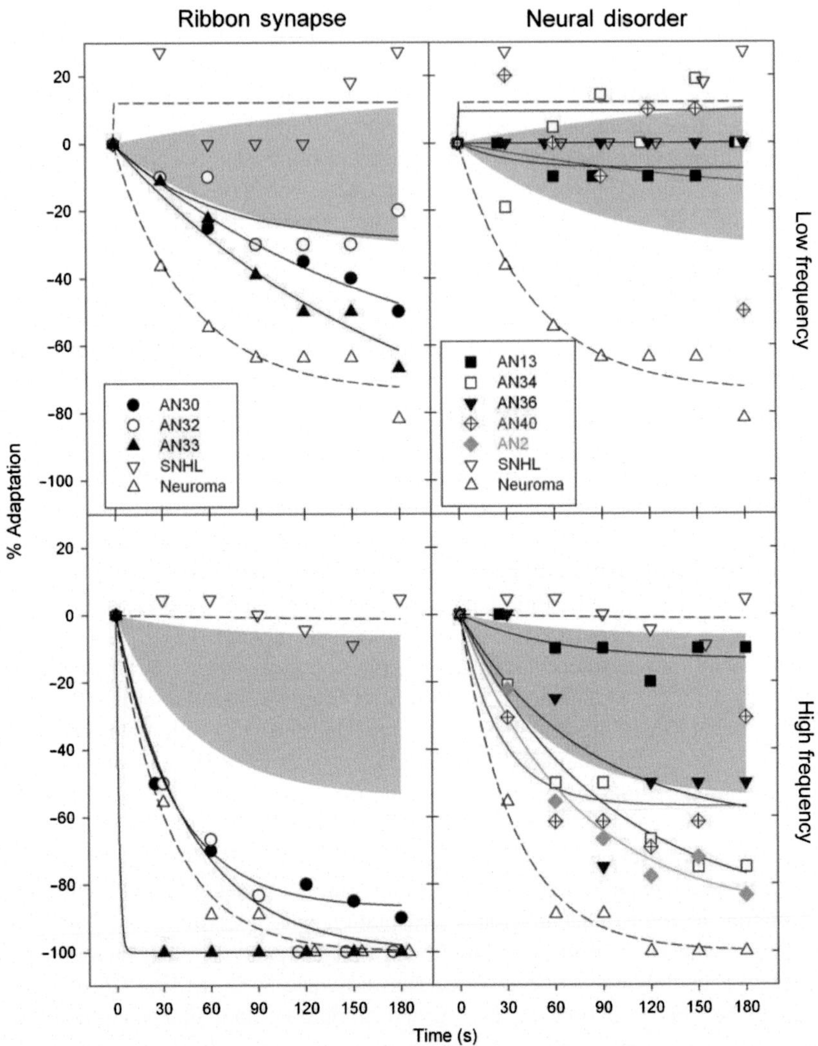

FIGURE 5.8 Individual adaptation profiles for subjects with ANP with ribbon synapse disorder (left) and neural disorder (right) for 250 Hz (top) and 8000 Hz (bottom) tones. The gray shaded area is the range of adaptation found in control subjects with normal hearing. Adaptation profiles are also shown for a subject with (1) SNHL (downward-pointing open triangles) and (2) another subject with acoustic neuroma (upward-pointing open triangles). The magnitude of adaptation in ribbon synapse disorders is similar to that found in the control subject with acoustic neuroma. *From Wynne, D.P., Zeng, F-G., Bhatt, S., Michalewski, H.J., Dimitrijevic, A., Starr, A, 2013. Loudness adaptation accompanying ribbon synapse and auditory nerve disorders. Brain, 136, 1626–1638, by permission of Oxford University Press.*

any postinjury plasticity in the adult, Yuan et al. (2014) in a mouse model of ANP quantified the peripheral and central synapses of type-I neurons at post-treatment times ranging from 1 to 3 months. Measures of normal DPOAEs and greatly reduced ABRs confirmed the neuropathy phenotype (Fig. 5.9). Surprisingly, at high SPLs, the ABR wave 1, classically considered to represent the summed activity of ANFs, may also include a robust contribution from receptor potentials in the hair cells (indicated as wave 1A in panel D) that is difficult to exclude from the "threshold" analysis based on latency alone. This is likely an SP.

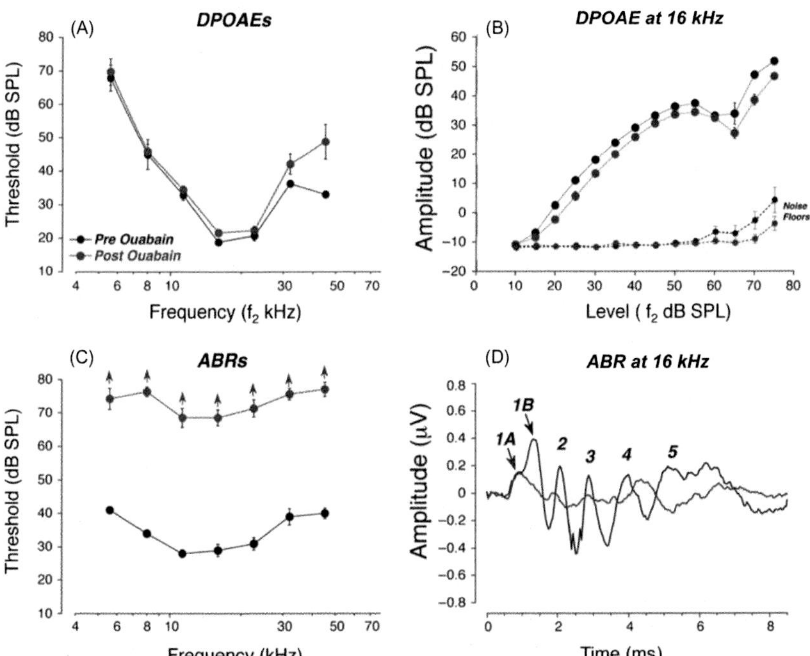

FIGURE 5.9 Ouabain treatment can elevate ABR thresholds without significant changes in DPOAEs. (A) Mean DPOAE thresholds (±SEMs) for control ears versus ears tested 1 week after ouabain application. (B) Mean amplitude versus level functions for $f_2 = 16$ kHz for the same animals shown in panel (A), with mean noise floors shown by dashed lines. (C) Mean ABR thresholds for the same animals shown in panel (A). Up arrows on post-ouabain data indicate that thresholds are underestimated because in two ears, no response was detected at the highest level presented (80 dB SPL). (D) Mean ABR waveforms in response to 16 kHz tone pips at 80 dB SPL for the same animals shown in (A) and (B), waves 1–5 are indicated. Key in (A) applies to all panels. *From Yuan, Y., Shi, F., Yin, Y., Tong, M., Lang, H., Polley, D.B., et al., 2014. Ouabain-induced cochlear nerve degeneration: synaptic loss and plasticity in a mouse model of auditory neuropathy. J. Assoc. Res. Otolaryngol. 15, 31–43, Figure 1, copyright 2012, with permission of Springer.*

5.7 VESTIBULAR SCHWANNOMA

A vestibular schwannoma is also called *acoustic neuroma*, *neurinoma*, or *neurilemmoma*. A vestibular schwannoma is a rare, usually slow-growing tumor of the eighth cranial nerve. Vestibular schwannomas are believed to develop from an overproduction of Schwann cells that press on the ANFs. Schwann cells are cells that normally wrap around and support peripheral nerve fibers, such as the ANFs. This type of brain tumor usually develops in adults between the ages of 30 and 60. People with the genetic condition neurofibromatosis 2 (NF-2) often develop acoustic neuromas in both ears: http://www.hopkinsmedicine.org/neurology_neurosurgery/centers_clinics/brain_tumor/center/acoustic-neuroma/vestibular-schwannoma.html (last accessed October 20, 2015).

5.7.1 Detection Using ABR

Vestibular schwannomas may often present auditory problems with normal DPOAEs and abnormal ABRs that mimic a form of ANP. The ABR test (Fig. 5.10) that used wave I–V latency differences was, until the advent of magnetic resonance imaging (MRI), an important component of the clinical test battery for vestibular schwannomas (Eggermont et al., 1980). Early studies claimed detection rates ranging as high as 95–98%, but the detected tumors were typically fairly large. Eggermont et al. (1980) showed that combining two standard ABR measures, the interaural wave V delay and the interpeak wave I to wave V delay, detected medium and large tumors very well but often missed small (<1 cm) tumors (Fig. 5.11). With the advent of MRI, studies over the past 10–15 years have repeatedly confirmed this conclusion and demonstrated that standard ABR tests missed 30–50% of these small tumors (e.g., Telian et al., 1999).

Tumor size can be correlated with only certain ABR parameters, notably the I–V delay. Eggermont et al. (1980) found no correlation of tumor size with the amount of hearing loss or the speech discrimination. ABR yielded high detection scores: up to 90%. However, it should be emphasized that in about 30% of the cases in this study, no wave I was detected in their recordings. For these cases, Eggermont et al. (1980) substituted the latency of the CAP recorded by transtympanic ECochG. Thus, the 95% score obtained on the basis of ABR depends on the ability to identify the latency of wave I.

5.7.2 Using the Stacked ABR

Let us first describe the derived ABR responses in some detail (Don and Eggermont, 1978). Fig. 5.12A shows a typical series of brainstem responses to a 60 dB SL click in combination with the various high-pass noise masking situations, which were in this case changed in cutoff frequency by 0.5-octave

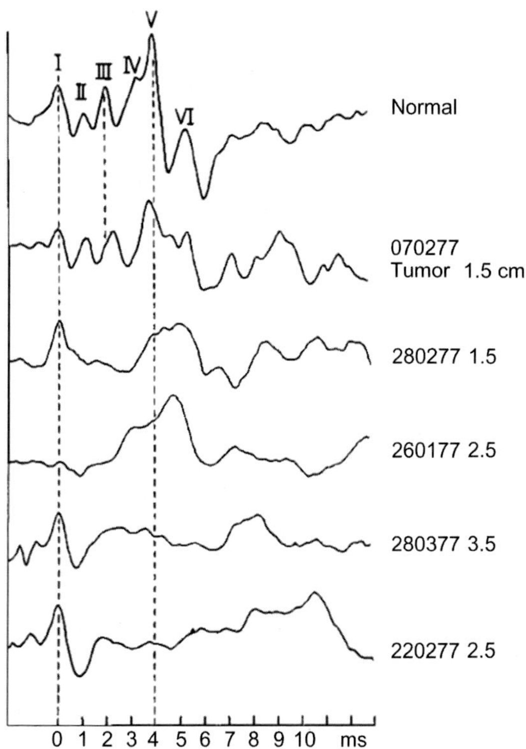

FIGURE 5.10 Waveforms of brainstem electric responses in vestibular schwannoma cases. The normal morphology indicated in the upper trace is only rarely observed in tumor cases. However, the second trace is an example of prolonged I–III and 1–V delays. In most cases, identification of the waves, except I, becomes difficult. The dashed lines bracket the normal I–V delay. The second trace from the top shows a false-negative case. *From Eggermont, J.J., Don, M., Brackmann, D.E., 1980. Electrocochleography and auditory brainstem electric response in patients with pontine angle tumors. Ann. Otol. Rhinol. Laryngol. 89 (Suppl. 75), 1–19, with permission from the Annals Publishing Company.*

steps. In clinical settings one typically uses 1-octave steps, which saves time. In the unmasked response, the normal ABR pattern is seen with all waves readily visible. This normal response pattern is preserved in the series of high-pass noise masking down to a cutoff frequency of about 4 kHz. With 4 kHz high-pass noise, wave IV begins to lose its distinctiveness. As the high-pass frequency cutoff is successively lowered, the earlier peaks begin to disappear. However, even at a cutoff frequency as low as 0.75 kHz, the broad wave V still has an amplitude that is not drastically reduced. This series suggests that, at sufficiently high click level, nearly the whole cochlear partition can contribute to the brainstem response. The contribution to the ABR of various regions along the cochlear partition can be seen in the

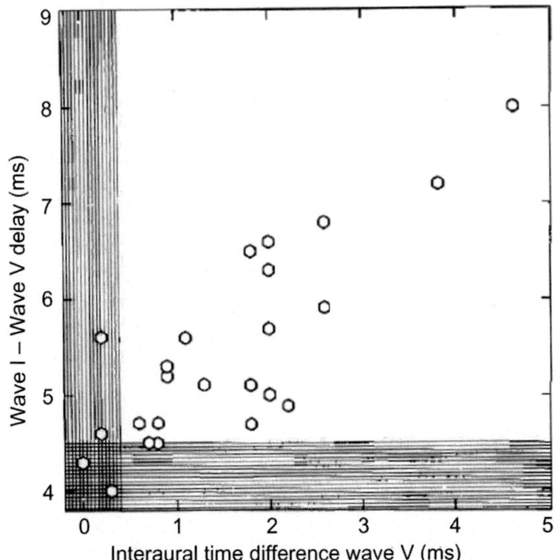

FIGURE 5.11 The relation between the wave I–V delay and the interaural latency difference between waves V is shown with the criteria for normal. The longer the I–V delay, the longer the interaural time difference; however, because audiograms differed for each ear, this relation is not simple. The responses may be normal for one criterion but abnormal for the other. For instance based on the I–V delay there are four missed tumors, including two borderline cases. Based on the interaural wave V delay there are also four missed cases. The combination of both criteria augments the detection score, in this case leading to only two missed cases. *From Eggermont, J.J., Don, M., Brackmann, D.E., 1980. Electrocochleography and auditory brainstem electric response in patients with pontine angle tumors. Ann. Otol. Rhinol. Laryngol. 89 (Suppl. 75), 1–19, with permission from the Annals Publishing Company.*

"narrow-band" waveforms derived by subtracting consecutive more masked waveforms (Fig. 5.12B). The characteristic frequency (CF) of the contributing cochlear region is indicated to the right of each derived waveform. High-frequency cochlear regions generate a normal "all-waves" response pattern. As the CF becomes lower, the derived response lack cochlear contributions to the earlier peaks, but clear and large contributions to wave V. The latency of the peaks, clearly seen in wave V, increases substantially for lower CFs.

A derived-ABR based method that circumvents the need to record wave I for the diagnosis of vestibular schwannoma is the stacked ABR (Don et al., 1997). The stacked ABR is obtained by taking the derived-band ABRs as shown in Fig. 5.13A (here obtained for octave band differences in high-pass noise) and shifting the waveforms to align the wave V peaks as shown in Fig. 5.13B. Shifting and aligning the waveforms minimize the phase cancellation of the substantial activity from lower frequency regions. The aligned (stacked) waveforms are then summed to form the stacked ABR shown as the fat bottom trace in Fig. 5.13B (Don, 2002). If contributions to wave V

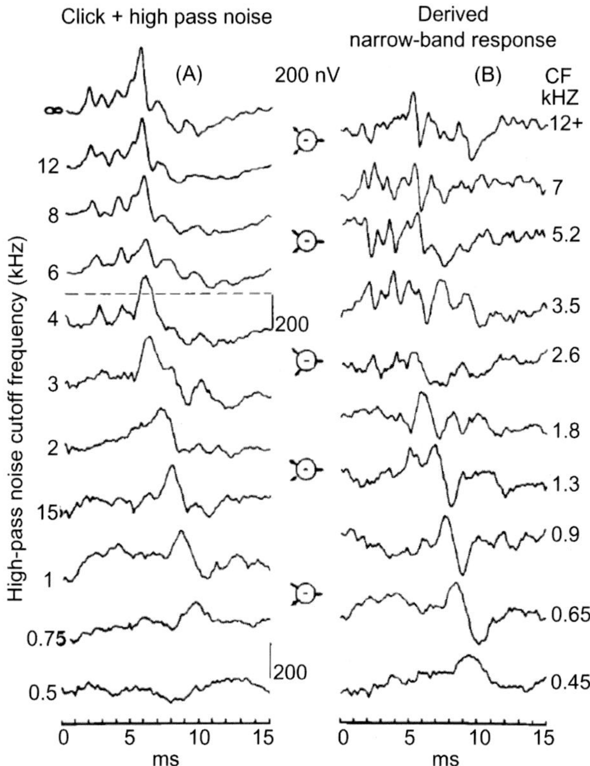

FIGURE 5.12 Waveforms for the high-pass noise masked ABR and the narrow-band ABR. (A) The waveforms of the ABR in the presence of high-pass noise of decreasing cutoff frequency, i.e., with increase in the masked area along the cochlear partition. One observes a general increase in latency with decreasing cutoff frequency, which is especially clear for the dominant peak: wave V. The horizontal dotted line indicates that the tracings which begin above this line are plotted at an amplitude scale different from those which begin below this line. The amplitude scale is denoted by the vertical bars to the right of the tracings. (B) The narrow-band ABRs obtained by subtracting a high-pass masked ABR from the one shown immediately above it, i.e., having a 0.5-octave-higher cutoff frequency. The CF assigned to each narrow band is always somewhat below the lowest cutoff frequency of both high-pass noise bands involved (due to the finite slope of the pattern used, i.e., 96 dB/oct). For instance, in the case of 3 kHz high-pass noise minus 2 kHz high-pass noise the CF is 1.8 kHz. For the higher CFs all waves are generally seen and appear nearly equally sharp. For CFs below 1.8 KHz generally wave V becomes dominant. *Reprinted from Don, M., Eggermont, J.J.,1978. Analysis of the click-evoked brainstem potentials in man using high-pass noise masking. J. Acoust. Soc. Am. 63, 1084–1092, with permission from Acoustic Society of America.*

from certain frequency bands are missing or are decreased in amplitude—as caused by the tumor—the stacked wave V becomes much smaller. In this way small (<1 cm) tumors that were missed by ABR measures based on I–V delay and interaural wave V differences (Eggermont et al., 1980) were detected (Don et al., 1997).

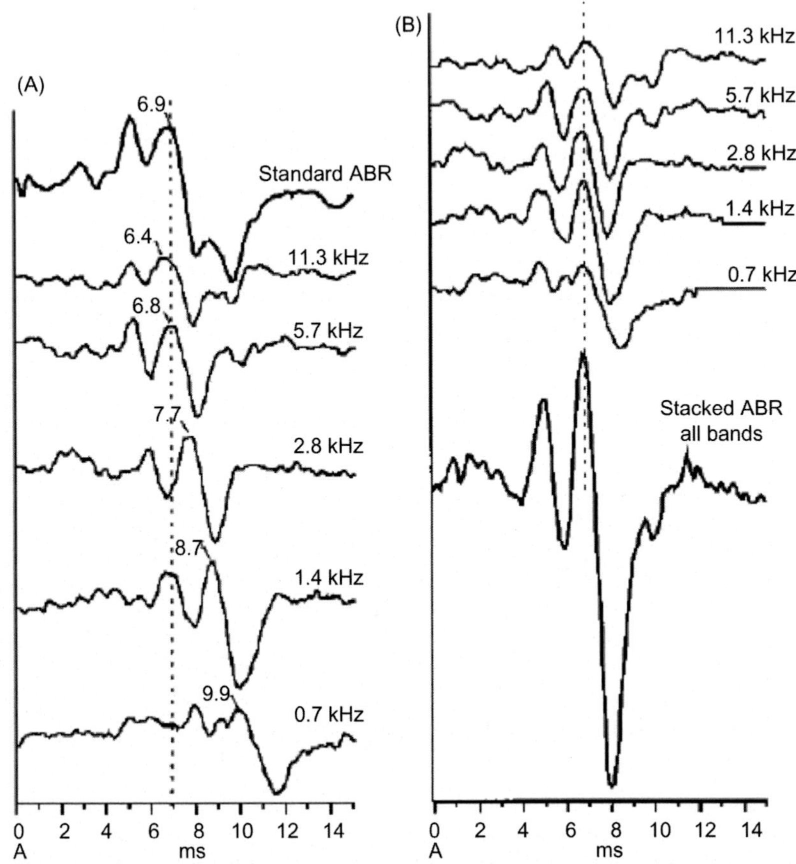

FIGURE 5.13 (A) Composition of wave V of the standard ABR, its dependence on neural activity from high frequencies, and the insensitivity of its latency and amplitude to loss of neural elements representing activity from mid- and low-frequency regions of the cochlea (A). Top (fat) trace is standard ABR to 60-dB normalized hearing level clicks from a normal hearing female subject. The succeeding five traces are derived-band ABRs representing ABRs initiated from octave wide regions around the labeled center frequencies in kilohertz. The standard ABR response is essentially composed of the sum of the five derived bands. (B) Formation of the stacked ABR and demonstration of the sensitivity of its amplitude to loss of neural elements representing activity from all frequency regions of the cochlea. Five derived-band ABR waveforms shown in part A are shifted to align the wave V peaks. The aligned waveforms are summed to form the stacked ABR, shown as the fat trace at the bottom. The stacked ABR wave V amplitude is nearly three times as large as the standard ABR and is a rough reflection of the total neural activity elicited by click stimulation. *From Don, M., 2002. Auditory brainstem response testing in acoustic neuroma diagnosis. Curr. Opin. Otolaryngol. Head Neck Surg. 10, 376–381, with permission from Wolters Kluwer Health.*

5.8 MÉNIÈRE'S DISEASE

5.8.1 Phenomenology and Pathology

Ménière's disease is characterized by intermittent spontaneous attacks of vertigo, fluctuating SNHL, aural fullness or pressure, and roaring (low frequency) tinnitus. This pathologic process involves distortion of the membranous labyrinth with the formation of an endolymphatic hydrops (Minor et al., 2004; Sajjadi and Paparella, 2008). Endolymphatic hydrops as a mechanism underlying Ménière's disease has to be understood against the fact that many possible factors may lead to hydrops, and that hydrops in turn generates the symptoms of Ménière's disease. However, there is still no conclusive proof for this (Wu et al., 2016). Merchant et al. (2005) found 28 cases with classical symptoms of Ménière's disease that showed hydrops in at least one ear. However, the reverse was not true. They found 9 cases with idiopathic hydrops and 10 cases with secondary hydrops, but none of the patients showed the classic symptoms of Ménière's disease. Merchant et al. (2005) concluded that endolymphatic hydrops should be considered as a histologic marker for Ménière's disease but not as a mechanism for its symptoms.

There are surprisingly few studies on tinnitus characteristics in Ménière's disease, although it is one of the four defining characteristics. Here I will present only two recent ones. Havia et al. (2002) studied 243 patients with Ménière's disease. The tinnitus was mild in 38%, moderate in 32%, and severe in 30% of patients. The intensity of tinnitus correlated with the occurrence of drop attacks, vertigo provoked by head positioning, by physical activity, or by atmospheric pressure changes. In 102 patients with Ménière's disease, Herraiz et al. (2006) found a statistical association between tinnitus intensity and worse hearing loss or hyperacusis, but it was not influenced by number of vertigo spells. Havia et al. (2002) reported that tinnitus in Meniere's disease was often severe and had most often a low-frequency pitch. Herraiz et al. (2006) also found that tinnitus pitch was more commonly identified in low and medium frequencies; in 28% of the cases it was matched to 125–250 Hz. In 44% of the cases Ménière's patients in this study described their tinnitus as a buzzing sound, 37% compared it to the sound from seawater waves, and only 11% reported a whistling sound. Tinnitus loudness was matched to 12.7 dB SL (range 0–35 dB), which combined with 50–60 dB HL thresholds and recruitment corresponds to a subjective loudness of approximately 70 dB. In an MRI study, Wu et al. (2016) assessed the presence of an endolymphatic hydrops in definite cases of unilateral Ménière's disease. They ruled out that an endolymphatic hydrops is the primary cause of tinnitus and aural fullness.

5.8.2 Natural History of Ménière's Disease

Eggermont and Schmidt (1985) followed the time course of hearing loss in 20 patients suffering from Ménière's disease or Lermoyez syndrome. In

Lermoyez's syndrome: Tinnitus and loss of hearing occur prior to an attack of vertigo, after which hearing improves, whereas in Ménière's disease they occur at the same time. Careful history taking as well as repeated audiometry over long time spans (5–20 years) resulted in a longitudinal characterization of hearing loss at the standard audiometric frequencies. Correlations between changes in hearing thresholds in both ears were more obvious in bilateral Ménière's disease than in unilateral disease, although there were still hearing threshold variations in the non-Ménière's ear. Correlations between changes in hearing loss at different frequencies in the same ear were more pronounced in Ménière's affected ears than in non-Ménière's ears (Eggermont and Schmidt, 1985).

From this longitudinal study Eggermont and Schmidt (1985) surmised that Ménière's disease starts either with vestibular complaints or with hearing loss, and there may be time lags before the other symptoms manifest themselves or both major symptoms are simultaneously present from the disease onset. When the vestibular as well as the auditory system is involved and both show fluctuations in their excitability, one recognizes the typically active Ménière's disease patient having vertiginous attacks combined with loss of hearing; the patient represented in Fig. 5.14 is an example. Note that in this patient, the hearing loss episodes are not occurring at the same time in both ears. From this state, remissions to the beginning condition (no hearing loss, labyrinths equally excitable) may occur but generally are rare. What

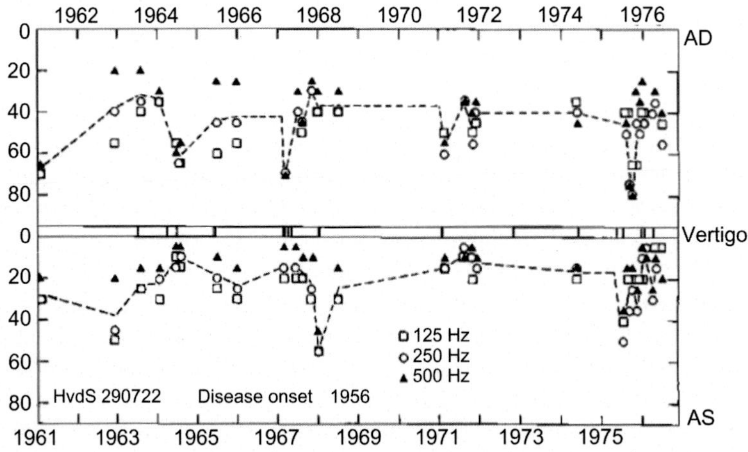

FIGURE 5.14 Time course of a bilaterally fluctuating case of Ménière's disease. Disease onset was in 1956. Note that the disease is very active even 20 years after onset. Vertiginous attack moments are indicated in the middle bar. Periods of hearing loss were not completely synchronous in both ears. The amount of change during attack is about 40 dB in both ears despite fact that they occur from 40 to 80 dB in right ear and from 20 to 60 dB in left ear. *From Eggermont, J.J., Schmidt, P.H., 1985. Ménière's disease: a long term follow-up of hearing loss. Ann. Otol. Rhinol. Laryngol. 94, 1–9, with permission from the Annals Publishing Company.*

one encounters is a reduced excitability of the labyrinth with some hearing loss remaining. However, the possibility that the threshold reaches a ceiling level, from which no increases are detectable with audiometry, becomes larger every time an attack occurs. When also the vestibular system becomes nearly unexcitable and sufficient central compensation occurs, the semifinal stage is reached. Patients do not suffer anymore from vertiginous attacks but the hearing loss and tinnitus remain; from this point on the vestibular as well as the auditory system show a slow progressive degeneration. Independent of the stage of the ear in question, the other ear may be affected too, giving rise to bilateral Ménière's disease. Eggermont and Schmidt (1985) found that recovery of hearing loss after an attack was nearly identical each time it occurred in the same ear, but differed between ears and patients. This deterministic finding in the otherwise random course of the hearing loss as a function of time may be important in deciding what mechanisms cause fluctuating hearing loss in Ménière's disease (Fig. 5.15).

Schmidt and Schoonhoven (1989) then performed clinical studies in 12 patients with Lermoyez syndrome during a close follow-up of 2–31 years. In their study material, the incidence of Lermoyez's syndrome, as compared with Ménière's disease, was almost 18%. In 6 out of 12 patients typical Ménière's attacks were found besides their Lermoyez attacks (Fig. 5.16). Almost every one of the 12 patients showed, besides the Lermoyez attacks, also hearing fluctuations without vertigo, and vertigo without hearing fluctuations. The type of vertigo was usually rotating, often with unsteadiness and in three patients drop attacks were sometimes observed. The typical Lermoyez attack lasted

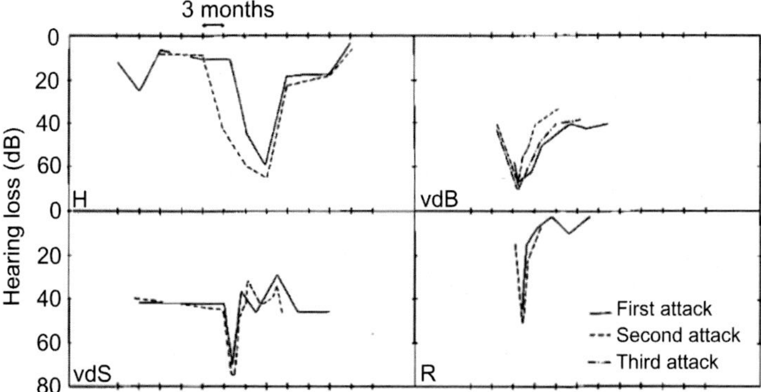

FIGURE 5.15 Recovery from hearing loss in one patient with Lermoyez disease and three patients with Ménière's disease. Time scale is enlarged with respect to other figures in that one division now indicates 3 months. Note the remarkable similarity of the recovery phases in each case. It appears that time course of hearing loss and recovery is much slower for the Lermoyez case (top left) as compared to the three ears with Ménière's disease (other panels). *From Eggermont, J.J., Schmidt, P.H., 1985. Ménière's disease: a long term follow-up of hearing loss. Ann. Otol. Rhinol. Laryngol. 94, 1–9, with permission from the Annals Publishing Company.*

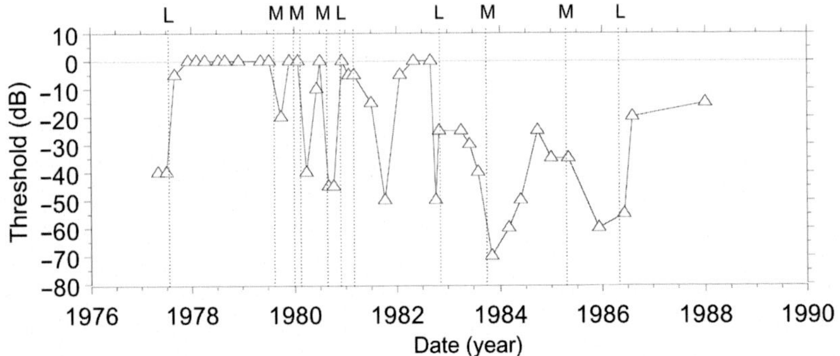

FIGURE 5.16 Vertigo and hearing fluctuations (means of 500, 1000, and 2000 Hz) during an 8-year period (patient HMR). Two types of event can be recognized: The occurrence of vertigo attacks of the Lermoyer type (L) and Ménière's type (M) is indicated by dotted lines. *Based on data from Schmidt, P.H., Schoonhoven, R., 1989. Lermoyez's syndrome. A follow-up study in 12 patients. Acta Otolaryngol. 107, 467–473.*

several hours. The hearing loss preceding the attack could occur for days to months earlier (Fig. 5.16). The hearing recovered after the attacks and remained stable for days to months. Improved hearing was noticed within some hours after the vertiginous attack in 9 out of 12 cases. Three patients noticed the hearing improvement already during the attack. In 8 out of 12 cases the disease became bilateral (Schmidt and Schoonhoven, 1989).

Schmidt and Schoonhoven (1989) believed that Ménière's disease and Lermoyez's syndrome are two clinical manifestations of the same underlying pathology. Their arguments for this view were the following: (1) The symptoms are similar. (2) Both the Ménière's as well as the Lermoyez type of attack can be found in the same ear (6 out of 12 patients; Fig. 5.16). (3) The Lermoyez data obtained with ECochG (cf. Section 5.8.3) fall entirely within those of the Ménière's group (Schmidt et al., 1974).

Some recent studies extend these two natural history reports: Huppert et al. (2010) analyzed the data of 46 mostly retrospective studies (with a total of 7852 patients). They found that "the frequency of vertigo attacks diminished within 5–10 years. Hearing loss (of about 50–60 dB) and vestibular function decrement (of about 35–50%) took place mainly in the first 5–10 years of disease. Drop attacks occurred early or late in the course of the disease, and remission was spontaneous in most cases. Bilaterality of the condition increased with increasing duration of the disease (up to 35% within 10 years, up to 47% within 20 years)." Belinchon et al. (2011) investigated the magnitude and configuration of hearing loss over the course of Ménière's disease, correcting the data according to patient age. This included a longitudinal study of pure-tone audiometries of 237 patients, who had been diagnosed with definitive Ménière's disease. They found that "in patients with unilateral disease, the mean hearing loss was characteristically low frequency, even in very advanced stages of the disease. Hearing loss was accentuated at 5 and 15 years from onset. In bilateral cases, hearing loss was slightly more severe and the average loss produced a flatter audiometric curve than in unilateral cases."

5.8.3 Electrochleography

Schmidt et al. (1974) were the first to examine patients ($N = 22$) suffering from Ménière's disease using transtympanic ECochG and pointing out the diagnostic value of the SP (Fig. 5.17). For the standard audiometric

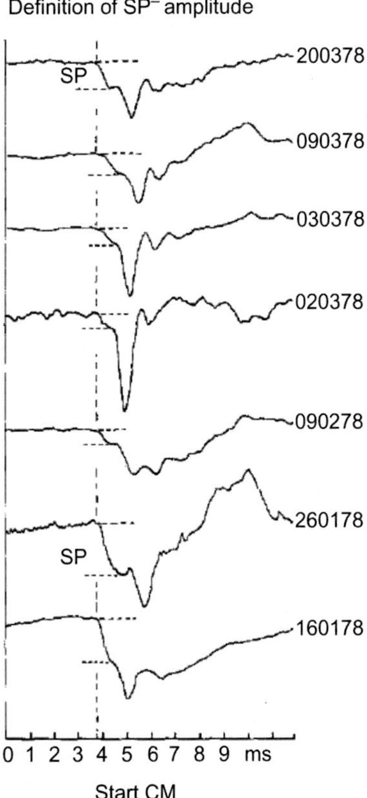

FIGURE 5.17 Identification of the SP in the compound SP–CAP recording. Ideally, the SP waveform equals the envelope of the stimulus used. For all recordings shown, the stimulus was a 4 kHz tone burst with 0.5 ms rise and fall times and 0.4 ms plateau. This is readily identified in the uppermost trace, and indication of the SP amplitude is easy. When the shape of the CAP is different, the clear SP waveform is lost. Its onset is, however, preserved since this is not modified by the CAP because of the zero-onset latency of the SP. In the subsequent traces the way in which the SP amplitude is determined is indicated. *From Eggermont, J.J., 1979. Summating potentials in Meniere's disease. Arch. Otorhinolaryngol. 222, 63–74, with permission of Springer.*

frequencies of 500 Hz and higher, the electrocochleographic and psychoacoustic data were compared with reference to the threshold values and loudness recruitment. A steep slope of the CAP input–output curves was considered indicative for positive recruitment in accordance with behavioral loudness tests (but see Section 5.4.1). The SP value, although very pronounced with respect to CAP amplitude, was almost equal to the values found in normal hearing, however, distinctly larger than the SP amplitude observed in high-frequency SNHL. This was discussed with respect to the differentiation of these two types of hearing loss, both with recruitment.

Subsequently, Eggermont (1979) studied the SPs in response to tone bursts of 2, 4, and 8 kHz in a group of 112 patients diagnosed with Ménière's disease and compared the results with electrocochleographic recordings from 27 normal human ears. Subdivision of the Ménière's ears into a low-threshold (<50 dB HL) and a high-threshold group (>50 dB HL) proved to be functional in relation to the behavior of the SP (Fig. 5.18). The data of the low-threshold Ménière's group did not differ significantly from the normal data, whereas the high-threshold Ménière's group showed significantly smaller SPs. Up to hearing thresholds of 50 dB, the detection threshold for the SP was independent of the hearing threshold; for greater hearing losses the increase in SP threshold equaled the increase in hearing loss. Analysis showed that the correlation found between the changes in SP and

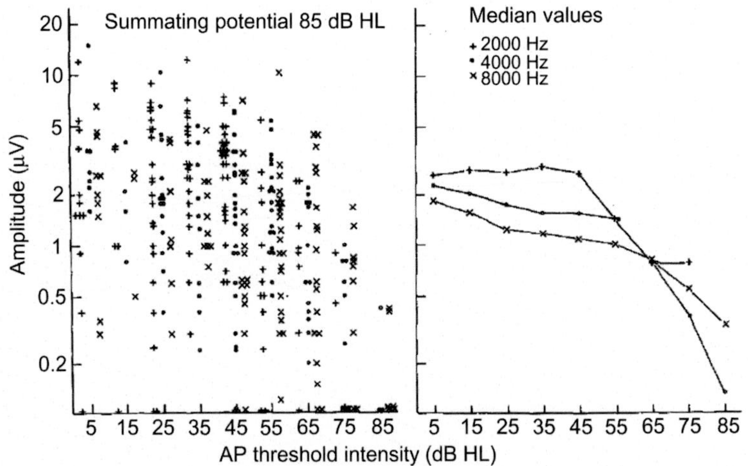

FIGURE 5.18 SP amplitudes at 85 dB HL plotted as a function of the AP threshold intensity for 2, 4, and 8 kHz. On the left the amplitudes are plotted in microvolts on a logarithmic scale for all of the individual Ménière's ears studied, the median values are connected and shown on the right. For 2 kHz stimulation, especially the SP amplitude drops considerably when the AP threshold is higher than 50 dB HL. For 8 kHz there is a more gradual decrease in SP amplitude with the amount of hearing loss. *From Eggermont, J.J., 1979. Summating potentials in Meniere's disease. Arch. Otorhinolaryngol. 222, 63–74, with permission of Springer.*

hearing threshold in individual patients did not differ significantly from zero for threshold values below 50 dB.

Eggermont (1979) suggested that since the SP amplitude below 50 dB HL is independent of the hearing loss, it is unlikely that the fluctuations in the hearing loss in the low- and middle-frequency regions are due to either functional changes in the hair cells or changes in the degree of hydrops. On the other hand, when the hearing loss exceeds this 50 dB level (on average; the value may differ from patient to patient) the SP amplitude gradually decreases (Fig. 5.18), which is an indication of hair cell loss or dysfunction. In this more stabilized situation, fluctuations in the hearing loss are less frequent. At the higher frequencies, where the SP amplitude decreases gradually with the amount of hearing loss, fluctuations become fewer the smaller the SP, also for threshold values below 50 dB. Association of the hearing loss in Ménière's disease with loudness recruitment is an obligatory diagnostic criterion. CAP-based frequency-tuning curves in Ménière's ears are broad just as for other types of cochlear loss such as resulting from noise trauma (Eggermont, 1977b).

Coats (1981) investigated the relationship between Ménière's disease and the putatively enlarged SP in a series of patients and normal subjects. CM, SP, and CAP in response to rectangular-pulse clicks delivered at 115 dB peak-equivalent SPL were recorded from ear canal skin. Coats (1981) performed regression for SP amplitudes as a function of CAP amplitudes in normal ears and obtained the 95% confidence interval for normal scatter around the fitted curve. Coats (1981) found that 68% of SP amplitudes from ears with Ménière's disease exceeded the upper confidence limit in normal hearing subjects. In contrast, only 7% of the cochlear ears and none of the retrocochlear ears had SP amplitudes above this normal upper limit. CM potentials and CAPs from the ears with Ménière's disease also tended to be larger than normal. The click-based SPs recorded in this study seem to give different results compared to the tone-burst evoked SPs in the work of Eggermont (1979). This difference may result from using a different recording site, transtympanic in the Eggermont (1979) study and from within the ear canal in the Coats (1981) study.

Mori et al. (1982) compared extratympanic and transtympanic ECochG recordings of SP. Mori et al. (1981) had found that in the transtympanic method a change in position of the electrode has a greater influence on amplitude and polarity of SP than in the extratympanic method. For these reasons, they considered the extratympanic method more suitable for a routine use in clinics. Goin et al. (1982), also recording from the external meatus, found that the SP/CAP amplitude ratio was a useful diagnostic measure. In their study, 62% of the Ménière's patients demonstrating abnormal ratios compared to 4% of the normal controls and 17% of patients with SNHL.

Gibson (1991) measured and compared click SP/CAP ratios and the absolute values of the SP for 1 kHz 90 dB nHL tone burst in 42 normal ears,

48 sensorineural loss ears, and 80 Ménière's ears. They found that using 1 kHz tone bursts increased the accuracy of diagnosis especially in ears with a hearing loss of less than 40 dB nHL. In a large study, Ge and Shea (2002) rediscovered that ECochG using tone bursts had the advantage of frequency selectivity (as decades earlier documented by Eggermont et al., 1974). They observed an increased CAP latency in 62.2% of ears with Ménière's disease. This increased CAP latency was significantly correlated with an enlarged SP/CAP amplitude ratio. An enlarged SP/CAP amplitude ratio was found 71% in Stage 1 (PTA ≤25 dB) of Ménière's disease, 82% in Stage 2 (26–40 dB), 85% in Stage 3 (41–70 dB), and 90% in Stage 4 (>70 dB). Enlarged SP/CAP amplitude ratios were also associated with the duration of the disease.

Note that an enlarged SP/CAP amplitude ratio also occurs in ANP, and there reflects desynchronized vesicle release form ribbon synapses and/or desynchronization of firings of ANFs (Section 5.6). This opens the possibility of a synaptopathy underlying Ménière's disease.

5.8.4 Diagnosis Using the Stacked ABR (CHAMP)

Don et al. (2005) devised an objective test for "active" Ménière's disease based on ABRs to clicks presented ipsilaterally with masking noise that was high-pass filtered at various frequencies. This was a variant of the "Stacked ABR" used in acoustic neuroma detection (Section 5.7). They compared 38 normal hearing subjects; and 23 patients who, at the time of testing, continued to have at least three of the four hallmark symptoms (i.e., tinnitus, vertigo, fluctuating hearing loss, and fullness) used in the diagnosis of Ménière's disease. Don et al. (2005) found that in Ménière's patients, the normal high-pass masking noise level was insufficient so that wave V was still present at a latency similar to that of wave V in the response to the unmasked clicks. In the control non-Ménière's normal hearing subjects, this undermasked component was either absent or significantly delayed because of the masking noise. The difference in the delays between these populations was diagnostically significant with 100% specificity and 100% sensitivity (Don et al., 2005). It is tempting to attribute this undermasking to a loss of mechanical lateral suppression in the basilar membrane response in Ménière's patients. This would imply a dysfunctional cochlear amplifier, and thus changes in OHC activity.

5.9 AGE-RELATED HEARING IMPAIRMENT (PRESBYCUSIS)

Age-related hearing impairment (ARHI) constitutes one of the most frequent sensory problems in the elderly. It presents itself as a bilateral SNHL that is most pronounced in the high frequencies. ARHI is a complex disorder, with both environmental and genetic factors contributing to the disease (Fransen et al., 2003; Eggermont, 2014).

Early on, Schuknecht (1955, 1964) distinguished four types of presbycusis on basis of audiogram shape and histopathology: (1) Sensory presbycusis with atrophy of the organ of Corti and auditory nerve in the basal end of the cochlea and characterized by a steep-sloping high-frequency hearing loss. (2) Neural presbycusis, with a loss of neural population in the auditory nerve and characterized by problems with speech discrimination. (3) Metabolic presbycusis potentially resulting from atrophy of the stria vascularis (Schuknecht et al., 1974) and characterized by flat audiograms. (4) Mechanical presbycusis believed to result from stiffening of the basilar membrane (also called inner ear conductive hearing loss). The first three types are still considered a viable framework for further research into the effects of genetic defects and environmental conditions on age-related hearing loss (Ohlemiller, 2004).

5.9.1 Changes in the Cochlea and Auditory Nerve

ARHI is the result of genetic predispositions combined with various environmental factors that affect the inner ear and accumulate during a lifetime. The environmental factors contributing to presbycusis are very heterogeneous and may include, for example, smoking, exposure to loud noise, and the use of ototoxic drugs (Gates and Mills, 2005). ARHI does not follow a single pattern and may potentially be accompanied by either one or more of the following: an age-dependent loss of sensory hair cells and/or ANFs, and degeneration of the stria vascularis (Gates and Mills, 2005).

Presbycusis may be the consequence of accumulating auditory stresses during life, superimposed upon the natural aging process. Subclinical damage accrued during employment may place the ear at higher risk for hearing impairment (see also section: Necrosis and Apoptosis in Noise-Induced Hearing Loss). Sergeyenko et al. (2013) described age-related cochlear synaptic and neural degeneration in CBA/CaJ mice never exposed to high-level noise. Cochlear hair cell and neuronal function was assessed via DPOAEs and ABRs, respectively. Immunostained cochlear sections were studied by confocal and conventional light microscopy to quantify hair cells, cochlear neurons, and presynaptic ribbons and postsynaptic glutamate receptors. Sergeyenko et al. (2013) found that ribbon synapse loss was seen throughout the cochlea long before age-related changes in thresholds or hair cell counts. Cochlear nerve loss paralleled the synaptic loss, after a delay of several months. ABRs appear to parallel the synaptic and neural fiber loss. Sergeyenko et al. (2013) labeled the changes as auditory synaptopathy/neuropathy, potentially corresponding to Schuknecht's "neural presbycusis" type.

5.9.2 Changes in Auditory Cortex

Ouda et al. (2015) described that aging may be accompanied by atrophy of the gray as well as white matter, and hence an enlargement of the cerebrospinal fluid space. There may also be changes in the content of some metabolites in the aged brain, as shown by magnetic resonance spectroscopy (MRS). MRS is a test associated with MRI for measuring biochemical changes in the brain. In addition to this, functional MRI (fMRI) has shown differences between activation of the central auditory system in the young and old brain (Profant et al., 2015). Altogether, the information reviewed by Ouda et al. (2015) suggested specific age-related changes in the central auditory system that occur mostly independently of the changes in the inner ear and that form the basis of the central presbycusis. In this respect, Humes et al. (2012) had disputed the existence of central presbycusis as an isolated entity. On basis of a literature review they found more evidence for a multifactorial condition that involves age- and/or disease-related changes in the auditory system and in the brain.

In an fMRI study of the central auditory system, Profant et al. (2015) compared two groups of elderly subjects (>65 years) with a group of young subjects (≤30 years). One elderly group had expressed presbycusis and the other elderly group had mild presbycusis. These two groups showed differences by pure-tone audiometry, in the presence of, and amplitudes of TEOAEs and DPOAEs, as well as in speech understanding in noisy conditions. The fMRI showed only minimal activation in response to the 8 kHz stimulation despite the fact that all subjects heard the stimulus. No statistically significant differences in activation of the auditory cortex were found between the mild and expressed presbycusis groups.

5.10 SUMMARY

Peripheral hearing loss comes in two broad types, conductive and sensorineural. We describe a diagnostic framework, which includes WAI and MEMR combined with OAEs and ABRs. This may diagnose the type of hearing loss when behavioral audiometry is not possible. Acoustic trauma causes retrograde degeneration of ANFs. It could be secondary when it follows IHC loss or primary as a result of damage of specific ribbon synapses in the IHCs. This primary degeneration can lead to "hidden hearing loss," characterized by normal audiograms and DPOAEs but great difficulty in understanding speech, especially in background noise. SNHL is accompanied by loudness recruitment, which is likely a central phenomenon. It often coexists with hyperacusis, also a central phenomenon, and so can lead to "over recruitment." ANP, only identified two decades ago, links to auditory temporal processing disorders and has a strong genetic component. Cochlear implantation very often solves this problem. The differences between ANP and the effects of a vestibular schwannoma are reviewed. Ménière's disease characterized by intermittent spontaneous attacks of vertigo, fluctuating SNHL, aural fullness or pressure, and roaring tinnitus still remains a mystery. We describe its natural history and the use of ECochG in detail. We conclude with ARHI and describe the epidemiology and its peripheral and cortical substrates.

REFERENCES

Belinchon, A., Perez-Garrigues, H., Tenias, J.M., Lopez, A., 2011. Hearing assessment in Ménière's disease. Laryngoscope 121, 622–626.

Brown, C.J., Etler, C., He, S., O'Brien, S., Erenberg, S., Kim, J.R., et al., 2008. The electrically evoked auditory change complex: preliminary results from nucleus cochlear implant users. Ear Hear. 29, 704–717.

Budenz, C.L., Telian, S.A., Arnedt, C., et al., 2013. Outcomes of cochlear implantation in children with isolated auditory neuropathy versus cochlear hearing loss. Otol. Neurotol. 34, 477–483.

Cai, S., Ma, W.L., Young, E.D., 2008. Encoding intensity in ventral cochlear nucleus following acoustic trauma: implications for loudness recruitment. J. Assoc. Res. Otolaryngol. 10, 5–22.

Cardon, G., Sharma, A., 2013. Central auditory maturation and behavioral outcome in children with auditory neuropathy spectrum disorder who use cochlear implants. Int. J. Audiol. 52, 577–586.

Carney, L., 1990. Sensitivities of cells in anteroventral cochlear nucleus of cat to spatiotemporal discharge patterns across primary afferents. J. Neurophysiol. 64, 437–456.

Chiappa, K., 1997. Principles of evoked potentials. In: Chiappa, K.H. (Ed.), Evoked Potentials in Clinical Medicine. Lippincott-Raven Publishers, Philadelphia, PA, pp. 1–30.

Coats, A.C., 1981. The summating potential and Meniere's disease. I. Summating potential amplitude in Meniere and non-Meniere ears. Arch. Otolaryngol. 107, 199–208.

Cruickshanks, K.J., Wiley, T.L., Tweed, T.S., et al., 1998. Prevalence of hearing loss in older adults in Beaver Dam, WI: the epidemiology of hearing loss study. Am. J. Epidemiol. 148, 879–886.

Cruickshanks, K.J., Nondahl, D.M., Tweed, T.S., et al., 2010. Education, occupation, noise exposure history and the 10-yr cumulative incidence of hearing impairment in older adults. Hear. Res. 264, 3–9.

Cureoglu, S., Schachern, P.A., Ferlito, A., Rinaldo, A., Tsuprun, V., Paparella, M.M., 2006. Otosclerosis: etiopathogenesis and histopathology. Am. J. Otolaryngol. 27, 334–340.

Davis, H., Morgan, C.T., Hawkins, J.E., Galambos, R., Smith, F.W., 1946. Temporary deafness following exposure to loud tones and noise. Laryngoscope 56, 19–21.

Davis, H., Morgan, C.T., Hawkins, J.E., Galambos, R., Smith, F.W., 1950. Temporary deafness following exposure to loud tones and noise. Acta Otolaryngol. 88 (Suppl), 1–56.

Don, M., 2002. Auditory brainstem response testing in acoustic neuroma diagnosis. Curr. Opin. Otolaryngol. Head Neck Surg. 10, 376–381.

Don, M., Eggermont, J.J., 1978. Analysis of the click-evoked brainstem potentials in man using high-pass noise masking. J. Acoust. Soc. Am. 63, 1084–1092.

Don, M., Kwong, B., Tanaka, C., 2005. A diagnostic test for Ménière's disease and cochlear hydrops: impaired high-pass noise masking of auditory brainstem responses. Otol. Neurotol. 26, 711–722.

Don, M., Masuda, A., Nelson, R., Brackmann, D., 1997. Successful detection of small acoustic tumors using the stacked derived-band auditory brain stem response amplitude. The American Journal of Otology 18, 608–621.

Eggermont, J.J., 1976. Physiological foundations for the interpretation of compound action potentials in electrocochleography. In: Hoke, M., von Bally, G. (Eds.), Symposium on Electrocochleography and Holography in Medicine. Münster University Press, Münster, Germany, pp. 115–140.

Eggermont, J.J., 1977a. Electrocochleography and recruitment. Ann. Otol. Rhinol. Laryngol. 86, 138–149.

Eggermont, J.J., 1977b. Compound action potential tuning curves in normal and pathological human ears. J. Acoust. Soc. Am. 62, 1247–1251.

Eggermont, J.J., 1979. Summating potentials in Meniere's disease. Arch. Otorhinolaryngol. 222, 63–74.

Eggermont, J.J., 2007. Electric and magnetic fields of synchronous neural activity propagated to the surface of the head: peripheral and central origins of AEPs. In: Burkard, R.R., Don, M., Eggermont, J.J. (Eds.), Auditory Evoked Potentials. Lippincott Williams & Wilkins, Baltimore, MD, pp. 2–21.

Eggermont, J.J., 2012. The Neuroscience of Tinnitus. Oxford University Press, Oxford, UK.

Eggermont, J.J., 2014. Noise and the Brain. Experience Dependent Developmental and Adult Plasticity. Academic Press, London.

Eggermont, J.J., 2015. Auditory Temporal Processing and its Disorders. Oxford University Press, Oxford, UK.

Eggermont, J.J., Schmidt, P.H., 1985. Ménière's disease: a long term follow-up of hearing loss. Ann. Otol. Rhinol. Laryngol. 94, 1–9.

Eggermont, J.J., Odenthal, D.W., Schmidt, P.H., Spoor, A., 1974. Electrocochleography. Basic principles and clinical application. Acta Otolaryngol. 316 (Suppl), 1–84.

Eggermont, J.J., Don, M., Brackmann, D.E., 1980. Electrocochleography and auditory brainstem electric response in patients with pontine angle tumors. Ann. Otol. Rhinol. Laryngol. 89 (Suppl. 75), 1–19.

Fransen, E., Lemkens, N., Van Laer, L., Van Camp, G., 2003. Age-related hearing impairment (ARHI): environmental risk factors and genetic prospects. Exp. Gerontol. 38, 353–359.

Gates, G.A., Mills, J.H., 2005. Presbycusis. Lancet 366, 1111–1120.

Ge, X., Shea Jr., J.J., 2002. Transtympanic electrocochleography: a 10-year experience. Otol. Neurotol. 23, 799–805.

Gibson, W.P.R., 1991. The use of electrocochleography in the diagnosis of Meniere's disease. Acta Otolaryngol. Suppl. 485, 46–52.

Goin, D.W., Staller, S.J., Asher, D.L., Mischke, R.E., 1982. Summating potential in Meniere's disease. Laryngoscope 92, 1383–1389.

Harrison, R.V., 1981. Rate-versus-intensity functions and related AP responses in normal and pathological guinea pig and human cochleas. J. Acoust. Soc. Am. 70, 1036–1044.

Havia, M., Kentala, E., Pyykko, I., 2002. Hearing loss and tinnitus in Meniere's disease. Auris Nasus Larynx 29, 115–119.

He, S., Grose, J.H., Teagle, H.F., Woodard, J., Park, L.R., Hatch, D.R., et al., 2013. Gap detection measured with electrically evoked auditory event-related potentials and speech-perception abilities in children with auditory neuropathy spectrum disorder. Ear Hear. 34, 733–744.

He, S., Grose, J.H., Teagle, H.F., Buchman, C.A., 2014. Objective measures of electrode discrimination with electrically evoked auditory change complex and speech-perception abilities in children with auditory neuropathy spectrum disorder. Ear Hear. 35, e63–e74.

Heinz, M.G., Young, E.D., 2004. Response growth with sound level in auditory-nerve fibers after noise-induced hearing loss. J. Neurophysiol. 91, 784–795.

Heinz, M.G., Issa, J.B., Young, E.D., 2005. Auditory-nerve rate responses are inconsistent with common hypotheses for the neural correlates of loudness recruitment. J. Assoc. Res. Otolaryngol. 26, 91–105.

Herraiz, C., Tapia, M.C., Plaza, G., 2006. Tinnitus and Ménière's disease: characteristics and prognosis in a tinnitus clinic sample. Eur Arch Otorhinolaryngol 263, 504–509.

Humes, L.E., Dubno, J.R., Gordon-Salant, S., Lister, J.J., Cacace, A.T., Cruickshanks, K.J., et al., 2012. Central presbycusis: a review and evaluation of the evidence. J. Am. Acad. Audiol. 23 (8), 635–666.

Hunter, L.L., Prieve, B.A., Kei, J., Sanford, C.A., 2013. Pediatric applications of wideband acoustic immittance measures. Ear Hear. 34, 36S–42S.

Huppert, D., Strupp, M., Brandt, T., 2010. Long-term course of Menière's disease revisited. Acta Otolaryngol. 130, 644–651.

Jeon, J.H., Bae, M.R., Song, M.H., Noh, S.H., Choi, K.H., Choi, J.Y., 2013. Relationship between electrically evoked auditory brainstem response and auditory performance after cochlear implant in patients with auditory neuropathy spectrum disorder. Otol. Neurotol. 34, 1261–1266.

Kaga, K., Nakamura, M., Shinogami, M., Tsuzuku, T., Yamada, K., Shindo, M., 1996. Auditory nerve disease of both ears revealed by auditory brainstem response, electrocochleography and otoacoustic emissions. Scand. Audiol. 25, 233–238.

Khimich, D., Nouvian, R., Pujol, R., Dieck, S.T., Egner, A., Gundelfinger, E.D., et al., 2005. Hair cell synaptic ribbons are essential for synchronous auditory signalling. Nature 434, 889–894.

Kujawa, S.G., Liberman, M.C., 2009. Adding insult to injury: cochlear nerve degeneration after "temporary" noise-induced hearing loss. J. Neurosci. 29, 14077–14085.

Kujawa, S.G., Liberman, M.C., 2015. Synaptopathy in the noise-exposed and aging cochlea: Primary neural degeneration in acquired sensorineural hearing loss. Hear Res. 330, 191–199.

Langers, D.R., van Dijk, P., Schoenmaker, E.S., Backes, W.H., 2007. fMRI activation in relation to sound intensity and loudness. Neuroimage 35, 709–718.

Lee, F.-S., Matthews, L.J., Dubno, J.R., Mills, J.H., 2005. Longitudinal study of pure-tone thresholds in older persons. Ear Hear. 26, 1–11.

Liberman, M.C., Dodds, L.W., 1984. Single-neuron labeling and chronic cochlear pathology. III. Stereocilia damage and alterations of threshold tuning curves. Hear. Res. 16, 55–74.

Liberman, M.C., Kiang, N.Y., 1984. Single-neuron labeling and chronic cochlear pathology. IV. Stereocilia damage and alterations in rate- and phase-level functions. Hear. Res. 116, 75–90.

Liu, L., Wang, H., Shi, L., Almuklass, A., He, T., et al., 2012. Silent damage of noise on cochlear afferent innervation in guinea pigs and the impact on temporal processing. PLoS One 7, e49550.

Martin, B.A., Boothroyd, A., 2000. Cortical, auditory evoked potentials in response to changes of spectrum and amplitude. J. Acoust. Soc. Am. 107, 2155–2161.

Merchant, S.N., Ravicz, M.E., Puria, S., Voss, S.E., Whittemore, K.R. Jr, Peake, W.T., et al., 1997. Analysis of middle ear mechanics and application to diseased and reconstructed ears. Am. J. Otol. 18, 139–154.

Merchant, S., Adams, J.C., Nadol, J.B., 2005. Pathophysiology of Ménière's syndrome: are symptoms caused by endolymphatic hydrops? Otol. Neurotol. 26, 74–81.

Minor, L.B., Schessel, D.A., Carey, J.P., 2004. Ménière's disease. Curr. Opin. Neurol. 17, 9–16.

Moore, B.C., Glasberg, B.R., 2004. A revised model of loudness perception applied to cochlear hearing loss. Hear. Res. 188, 70–88.

Mori, N., Matsunage, T., Asdai, H., 1981. Intertest reliability in non-invasive electrocochleography. Audiology 20, 290–299.

Mori, N., Saeki, K., Matsunaga, T., Asai, H., 1982. Comparison between AP and SP parameters in trans- and extratympanic electrocochleography. Audiology 21, 228–241.

Moser, T., Predoehl, F., Starr, A., 2013. Review of hair cell synapse defects in sensorineural hearing impairment. Otol. Neurotol. 34, 995–1004.

Narne, V.K., 2013. Temporal processing and speech perception in noise by listeners with auditory neuropathy. PLoS One 8, e55995.

Ohlemiller, K.K., 2004. Age-related hearing loss: the status of Schuknecht's typology. Curr. Opin. Otolaryngol. Head Neck Surg. 12, 439–443.

Ouda, L., Profant, O., Syla, J., 2015. Age-related changes in the central auditory system. Cell Tissue Res. 361, 337–358.

Oxenham, A.J., Bacon, S.P., 2003. Cochlear compression: perceptual measures and implications for normal and impaired hearing. Ear Hear. 24, 352–366.

Peake, W.T., Rosowski, J.J., Lynch, T.J., 1992. Middle-ear transmission: acoustic versus ossicular coupling in cat and human. Hear. Res. 57, 245–268.

Profant, O., Tintěra, J., Balogová, Z., Ibrahim, I., Jilek, M., Syka, J., 2015. Functional changes in the human auditory cortex in ageing. PLoS One 10 (3), e0116692.

Qiu, C., Salvi, R., Ding, D., Burkard, R., 2000. Inner hair cell loss leads to enhanced response amplitudes in auditory cortex of unanesthetized chinchillas: evidence for increased system gain. Hear. Res. 139, 153–171.

Rance, G., Starr, A., 2015. Pathophysiological mechanisms and functional hearing consequences of auditory neuropathy. Brain 138, 3141–3158.

Rapin, I., Gravel, J., 2003. "Auditory neuropathy": physiologic and pathologic evidence calls for more diagnostic specificity. Int. J. Pediatr. Otorhinolaryngol. 67, 707–728.

Robles, L., Ruggero, M.A., 2001. Mechanics of the mammalian cochlea. Physiol. Rev. 81, 1305–1352.

Rosowski, J.J., Stenfelt, S., Lilly, D., 2013. An overview of wideband immittance measurements techniques and terminology: you say absorbance, I say reflectance. Ear Hear. 34, 9S–16S.

Rudic, M., Keogh, I., Wagner, R., Wilkinson, E., Kiros, N., Ferrary, E., et al., 2015. The pathophysiology of otosclerosis: review of current research. Hear. Res. 330, 51–56.

Ruggero, M.A., Rich, N.C., 1991. Furosemide alters organ of Corti mechanics: evidence for feedback of outer hair cells upon the basilar membrane. J. Neurosci. 11, 1057–1067.

Sajjadi, H., Paparella, M.M., 2008. Meniere's disease. Lancet 372, 406–414.

Salvi, R.J., Henderson, D., Hamernik, R., Ahroon, W.A., 1983. Neural Correlates of Sensorineural Hearing Loss. Ear and Hearing 4, 115–129.

Salvi, R.J., Saunders, S.S., Gratton, M.A., Arehole, S., Powers, N., 1990. Enhanced evoked response amplitudes in the inferior colliculus of the chinchilla following acoustic trauma. Hear. Res. 50, 245–257.

Salvi, R.J., Wang, J., Ding, D., 2000. Auditory plasticity and hyperactivity following cochlear damage. Hear. Res. 147, 261–274.

Sanford, C.A., Hunter, L.L., Feeney, M.P., Nakajima, H.H., 2013. Wideband acoustic immittance: tympanometric measures. Ear Hear. 34, 65S–71S.

Santarelli, R., Arslan, E., 2001. Electrocochleography in auditory neuropathy. Hear. Res. 170, 32–47.

Santarelli, R., Del Castillo, I., Rodríguez-Ballesteros, M., et al., 2009. Abnormal cochlear potentials from deaf patients with mutations in the otoferlin gene. J. Assoc. Res. Otolaryngol. 10, 545–556.

Santarelli, R., Starr, A., Del Castillo, I., et al., 2011. Presynaptic and postsynaptic mechanisms underlying auditory neuropathy in patients with mutations in the *OTOF* or *OPA*1 gene. Audiol. Med. 9, 59–66.

Santarelli, R., Del Castillo, I., Starr, A., 2013. Auditory neuropathies and electrocochleography. Hear. Balance Commun. 11, 130–137.

Santarelli, R., Rossi, R., Scimemi, P., Cama, E., Valentino, M.L., La Morgia, C., et al., 2015. OPA1-related auditory neuropathy: site of lesion and outcome of cochlear implantation. Brain 138, 563–576.

Saunders, J.C., Bock, G.R., James, R., Chen, C.S., 1972. Effects of priming for audiogenic seizure on auditory evoked responses in the cochlear nucleus and inferior colliculus of BALB-c mice. Exp. Neurol. 37, 388–394.

Schairer, K.S., Feeney, M.P., Sanford, C.A., 2013. Acoustic reflex measurement. Ear Hear.43S–47S.

Schmidt, P.H., Schoonhoven, R., 1989. Lermoyez's syndrome. A follow-up study in 12 patients. Acta Otolaryngol. 107, 467–473.

Schmidt, P.H., Eggermont, J.J., Odenthal, D.W., 1974. Study of Meniere's disease by electrocochleography. Acta Otolaryngol. Suppl. 316, 75–84.

Schuknecht, H.F., 1955. Presbycusis. Laryngoscope 65, 402–419.

Schuknecht, H.F., 1964. Further observations on the pathology of presbycusis. Arch. Otolaryngol. 80, 369–382.

Schuknecht, H.F., Watanuki, K., Takahashi, T., Belal Jr, A.A., Kimura, R.S., Jones, D.D., et al., 1974. Atrophy of the stria vascularis, a common cause for hearing loss. Laryngoscope 84, 1777–1821.

Sergeyenko, Y., Lall, K., Liberman, M.C., Kujawa, S.G., 2013. Age-related cochlear synaptopathy: an early onset contributor to auditory function decline. J. Neurosci. 33, 13686–23694.

Shi, L., Liu, L., He, T., Guo, X., Yu, Z., Yin, S., et al., 2013. Ribbon synapse plasticity in the cochleae of guinea pigs after noise-induced silent damage. PLoS One 8, e81566.

Shi, L., Guo, X., Shen, P., Liu, L., Tao, S., Li, X., et al., 2015a. Noise-induced damage to ribbon synapses without permanent threshold shifts in neonatal mice. Neuroscience 304, 368–377.

Shi, L., Liu, K., Wang, H., Zhang, Y., Hong, Z., Wang, M., et al., 2015b. Noise induced reversible changes of cochlear ribbon synapses contribute to temporary hearing loss in mice. Acta Otolaryngol. 135, 1093–1102.

Smith, P.H., Rhode, W.S., 1989. Structural and functional properties distinguish two types of multipolar cells in the ventral cochlear nucleus. J. Comp. Neurol. 282, 595–616.

Song, Q., Shen, P., Li, X., Shi, L., Liu, L., Wang, J., et al., 2016. Coding deficits in hidden hearing loss induced by noise: the nature and impacts. Sci. Rep. 6, 25200.

Stamper, G.C., Johnson, T.A., 2015. Auditory function in normal-hearing, noise-exposed human ears. Ear Hear. 36, 172–184.

Starr, A., Picton, T.W., Sininger, Y., Hood, L.J., Berlin, C.I., 1996. Auditory neuropathy. Brain 119, 741–753.

Telian, S.A., Kileny, P.R., Niparko, J.K., Kemink, J.L., Graham, M.D., 1989. Normal auditory brainstem response in patients with acoustic neuroma. Laryngoscope 99, 10–14.

Valero, M.D., Hancock, K.E., Liberman, M.C., 2016. The middle ear muscle reflex in the diagnosis of cochlear neuropathy. Hear. Res. 332, 29–38.

Viana, L.M., O'Malley, J.T., Burgess, B.J., Jones, D.D., Oliveira, C.A., Santos, F., et al., 2015. Cochlear neuropathy in human presbycusis: confocal analysis of hidden hearing loss in postmortem tissue. Hear. Res. 327, 78–88.

Wang, J., Ding, D., Salvi, R.J., 2002. Functional reorganization in chinchilla inferior colliculus associated with chronic and acute cochlear damage. Hear. Res. 168, 238–249.

Wynne, D.P., Zeng, F.G., Bhatt, S., Michalewski, H.J., Dimitrijevic, A., Starr, A., 2013. Loudness adaptation accompanying ribbon synapse and auditory nerve disorders. Brain 136, 1626–1638.

Wu, Q., Dai, C., Zhao, M., Sha, Y., 2016. The correlation between symptoms of definite Meniere's disease and endolymphatic hydrops visualized by magnetic resonance imaging. Laryngoscope 126, 974–979.

Yasunaga, S., Grati, M., Cohen-Salmon, M., et al., 1999. A mutation in OTOF, encoding otoferlin, a FER-1-like protein, causes DFNB9, a nonsyndromic form of deafness. Nat. Gen. 21, 363–369.

Yuan, Y., Shi, F., Yin, Y., Tong, M., Lang, H., Polley, D.B., et al., 2014. Ouabain-induced cochlear nerve degeneration: synaptic loss and plasticity in a mouse model of auditory neuropathy. J. Assoc. Res. Otolaryngol. 15, 31–43.

Zeng, F.G., Kong, Y.Y., Michalewski, H.J., Starr, A., 2005. Perceptual consequences of disrupted auditory nerve activity. J. Neurophysiol. 93, 3050–3063.

Part III

The Causes

Chapter 6

Causes of Acquired Hearing Loss

Noise-induced hearing loss (NIHL) is most likely the dominant cause of acquired hearing loss. In turn, the fastest growing causes for NIHL is the use of personal listening devices (PLDs) and exposure to occupational noise including music performances. Extensive information has recently been published in the work of Eggermont (2014), and I use information from that book in this chapter to make it self-contained. Ototoxiciy, long-lasting effects of conductive hearing loss (CHL) in infants and children, and systemic disorders such as diabetes, bacterial and viral infections will also be covered. Genetic causes will be reviewed in Chapter 7, Epidemiology and Genetics of Hearing Loss and Tinnitus whereas auditory neuropathy, vestibular schwannoma (VS), Ménière's disease, and aging are covered as types of hearing loss in Chapter 5, Types of Hearing Loss.

6.1 OCCUPATIONAL NOISE EXPOSURE IN GENERAL

Excessive environmental noise levels are known to contribute to NIHL. Regulations of the Occupational Safety and Health Administration, or equivalents thereof, limit the level of daily noise exposure in the workplace. Unfortunately, regulations on admissible sound levels do not generally apply to recreational areas such as sports and concert venues, and if they do apply then typically it is only to what is audible outside the venue. Often sport games are attended in excess of the 8-h allowable 85 dB(A) (or 80 dB(A) in Europe) level of workplace noise exposure duration (cf. Fig. 4.1). Cumulative effects of repeated exposure to temporary threshold shift (TTS)-producing stimuli have been known to contribute to more permanent shifts, as recently also shown in animal studies (Wang and Ren, 2012; see chapter: Hearing Problems).

What levels of environmental noise are safe? Effective quiet is defined as the maximum noise level that does not interfere with the slow recovery from TTS. The upper limit of effective quiet has been suggested to be as low as 55 dB(A) (Kryter, 1985) and as high as 65–70 dB(A) (Ward et al., 1976) with lower limits of effective quiet required for exposures producing greater

Hearing Loss. DOI: http://dx.doi.org/10.1016/B978-0-12-805398-0.00006-2

TTS. The auditory injury threshold is the lowest level capable of producing any permanent threshold shift, regardless of exposure time. Based on measurements of the greatest TTS over extended exposure durations (i.e., the asymptotic threshold shift), the auditory injury threshold can be expected between approximately 75 and 78 dB(A) (Mills et al., 1981; Nixon et al., 1977). However, a recent study (Noreña et al., 2006) showed that an at least 4-month exposure of adult cats to a 4–20 kHz band-pass sound presented at 76 dB(A) (80 dB p.e. SPL) did not result in auditory brainstem response (ABR) threshold changes compared to controls. The study did show changes in the tonotopic map in primary auditory cortex, resulting from strongly reduced sensitivity of cortical neurons to frequencies between 4 and 20 kHz, and enhanced responsiveness at frequencies below and above that frequency range (Eggermont, 2014).

6.2 RECREATIONAL NOISE AND MUSIC

Attending music concerts and frequenting discos and bars is a major source of excessive music exposure. The sound level at a pop/rock concert is around 100–115 dB(A) (Meyer-Bisch, 1996). Several studies have reported that a high proportion of rock concert attendees had temporary hearing loss. For example, five out of six volunteers whose thresholds were measured before and after attending a rock concert had a TTS of more than 50 dB HL (Clark and Bohne, 1999). Schink et al. (2014) compared the incidence of hearing loss in professional musicians ($N = 2227$) among a general population aged between 19 and 68 years ($N \approx 3{,}000{,}000$) that was registered for occupation-related insurance purposes. During a 4-year period, 283,697 cases of hearing loss were seen, 238 of them among professional musicians (0.08%). The adjusted hazard ratio (for a definition see chapter: Epidemiology and Genetics of Hearing Loss and Tinnitus) of musicians was 1.45 (95% confidence interval (CI) = 1.28–1.65) for hearing loss (including conductive) and 3.61 (95% CI = 1.81–7.20) for NIHL. Thus, professional musicians have a high risk of NIHL (Eggermont, 2014).

6.2.1 Professional Musicians' Exposure in Symphony Orchestras

Symphony orchestral music is generally louder than allowed by occupational noise legislation. Classical musicians individually are often exposed to sound levels greater than 85 dB(A) for long periods of time, both during practice and performance. In one of the first studies, Westmore and Eversden (1981) measured the pure-tone hearing thresholds in 34 orchestral musicians. The audiometric results showed changes consistent with NIHL in 23 out of 68 ears (34%). The only other early large-scale study by Axelsson and Lindgren (1981) measured the hearing thresholds from 139 classical musicians. Following their criteria for hearing loss, defined as 20 dB or worse in one ear and at one frequency between 3 and 6 kHz, 80 (58%) musicians were identified as having hearing loss. After considering the age factor, they still found hearing loss in 51 of these cases (37%) being partially or wholly due to music exposure. Royster et al. (1991) described that classical musicians of the Chicago Symphony Orchestra over a standard working day (8 h) were exposed at 85.5 dB(A), which is only 0.5 dB above the recommended safe threshold in North American industrial settings. However, the maximal sound peaks reached much higher levels. McBride et al. (1992) measured sound pressure levels at five rehearsals and two concerts by the City of Birmingham Symphony Orchestra and found over 85 dB(A) during half of the rehearsal time. The maximal sound peaks were measured at over 110 dB(A) in front of the trumpet, piccolo, and bassoon positions. Jansen et al. (2009) tested 241 professional musicians of symphony orchestras, aged 23–64 years. Most musicians could be categorized as normal hearing, but their audiograms showed notches at 6 kHz (Fig. 6.1). Musicians had more NIHL than could be expected on the basis of age and gender, but only at 6 kHz. However, these musicians scored very well on the speech-in-noise test, compensating for the loss of hearing sensitivity. This is a known phenomenon in active musicians (chapter 9 in Eggermont, 2014).

Toppila et al. (2011) studied 63 musicians from four classical orchestras in Helsinki, Finland, and found that hearing loss in musicians was generally similar to that of the general population. However, highly exposed musicians had greater hearing loss at frequencies more than 3 kHz than less exposed

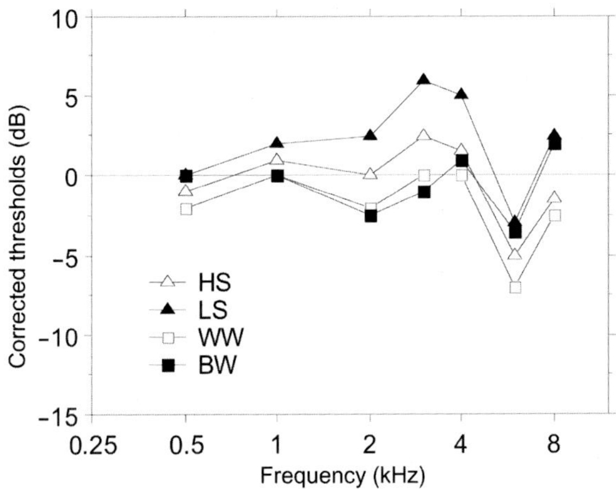

FIGURE 6.1 Average relative (i.e., corrected for age and gender) audiograms for instrument types. The instruments played by the musicians were classified into six groups: high strings (HS): violin and viola; low strings (LS): cello and double bass; woodwind (WW): oboe, clarinet, bassoon, flute; brass wind (BW). *Based on data from Jansen, E.J.M., Helleman, H.W., Dreschler, W.A., de Laat, J.A., 2009. Noise induced hearing loss and other hearing complaints among musicians of symphony orchestras. Int. Arch. Occup. Environ. Health 82, 153–164.*

ones. In a study on the hearing status of 182 classical symphony orchestra musicians, Schmidt et al. (2014) found that ears with the highest exposure (29 of 363 tested) had an additional threshold shift of 6.3 dB compared with the 238 ears with lowest exposure. However, the observed hearing loss of this group of musicians was less than the noise-induced permanent threshold shift expected from ISO1999. ISO1999 provides the basis for calculating hearing disability according to commonly measured audiometric frequencies, or combinations of such frequencies, which exceed a certain value. The measure of exposure to noise for a population at risk is the noise exposure level normalized to a nominal 8 h working day, Leq 8 h, for a given number of years of exposure. ISO1999 applies to noise at frequencies less than approximately 10 kHz which is steady, intermittent, fluctuating, or irregular (http://www.iso.org/iso/catalogue_detail.htm?csnumber=45103, accessed January 29, 2016). Whereas most of the musicians in the Schmidt et al. (2014) study had better hearing at 3, 4, and 6 kHz for age than expected, 29 ears with the highest exposure above 90.4 dB(A) with a mean exposure time of 41.7 years had significantly elevated hearing thresholds. Trumpet players and the left ear of first violinists showed significantly increased thresholds compared with the ears of other musicians (Eggermont, 2014).

Active musicians also practice outside their concert performances. O'Brien et al. (2013) estimated sound exposure during solitary practice of 35 professional orchestral musicians, representing players of most orchestral instruments. Sound levels were recorded between 60 and 107 dB(A) with (C-weighted) peak levels between 101 and 130 dB(C). For comparison between A- and C-weighting see Fig. 4.2. Note that the C-weighting is in fact an unweighted average. For average reported practice durations (2.1 h/day, 5 days a week) 53% would exceed accepted permissible daily noise exposure in solitary practice, in addition to sound exposure during orchestral rehearsals and performances. O'Brien et al. (2013) noted significant interaural differences in violin, viola, flute/piccolo, horn, trombone, and tuba. Only 40% used hearing protection while practicing. These findings indicate orchestral musicians that are already at risk of NIHL in ensemble performances are further at risk during solitary practice.

All in all, more than 30 years of evidence about the risk of hearing loss in professional classical musicians have not resulted in sufficient change in practice and prevention.

6.2.2 Active Musicians' Exposure at Pop/Rock Concerts

Axelsson and Lindgren (1977) reported that the prevalence of hearing loss among rock/pop musicians was 46% (38/83)—defining hearing loss the same as in their study in classical musicians (Section 6.2.1). In a follow-up study of the same pop/rock musicians 16 years later, Axelsson et al. (1995) found a significant deterioration in hearing thresholds at 4 and 8 kHz in the left ear, and at 4 kHz in the right among those who had shown hearing loss at these frequencies in the original study. In addition, 22% of participants showed a deterioration of their hearing greater than 15 dB HL at one or more frequencies in one or both ears. Kähäri et al. (2003) reported a similar prevalence of hearing loss in 139 rock/jazz musicians. According to their definition of hearing loss, which was the hearing threshold more than 25 dB at two frequencies or thresholds more than 30 dB at one frequency in one or both ears, 68/139 (49%) showed a hearing loss compared with the ISO7029 standard for matching age and gender. Twenty-three percentage of male musicians had hearing thresholds beyond the 90th percentile, whereas hearing thresholds obtained from female musicians were distributed at or just below the ISO7029 median according to their age. ISO7029 describes the normal changes in pure-tone thresholds with age (Stenklev and Laukli, 2004).

Schmuziger et al. (2006) assessed pure-tone audiometry in the conventional and extended high-frequency range in 42 nonprofessional pop/rock musicians, and in a control group of 20 normal hearing young adults with no history of long-term noise exposure. Relative to ISO7029, the mean hearing threshold in the frequency range of 3–8 kHz was 6 dB in the musicians and significantly higher than the 1.5 dB in the control group. Musicians, using regular hearing protection, had less hearing loss (average 3–8 kHz thresholds = 2.4 dB) compared to musicians who never used hearing protection (average 3–8 kHz thresholds = 8.2 dB). Samelli et al. (2012), in a prospective study, compared the effects of music exposure between professional pop/rock musicians ($N = 16$) and nonmusicians ($N = 16$). The musicians showed worse hearing thresholds in both conventional and high-frequency audiometry when compared to the nonmusicians. In addition, transient-evoked otoacoustic emissions (TEOAEs) were found to be smaller in the musicians group (Eggermont, 2014).

Halevi-Katz et al. (2015) examined the relationship between the years of professional experience of pop/rock/jazz musicians and their hearing loss. Forty-four pop/rock/jazz musicians were interviewed about symptoms of tinnitus and hyperacusis. Audiograms were measured for 1–8 kHz. All participants had a minimum of 4 years of experience playing their instrument with an average of 22.7 ± 10.4 years. The average weekly exposure of participants to pop/rock/jazz was 23.6 ± 17.7 h. Greater musical experience was positively and significantly linked to hearing loss in the frequency range of 3–6 kHz and to the presence of tinnitus. The number of hours/week playing had a greater effect on hearing loss in comparison to the number of years playing. The pure-tone average (1–8 kHz, age and gender corrected) was not significantly different for the right ear (2.8 ± 8.8 dB) and for the left ear (5.4 ± 10.4 dB).

6.2.3 Passive Exposure at Concerts and Discos

Serra et al. (2005) examined the effects of recreational noise exposure in adolescents during 4 years until 2001 when they turned 17. In the first year, there were 102 boys and 71 girls, and in the last year, 63 boys and 43 girls. They noted a tendency of the mean hearing threshold level to increase in both genders, especially at very high frequencies around 14–16 kHz. Boys had a higher mean hearing threshold level than girls and were more exposed to high sound levels than girls. Serra et al. (2014) subsequently reported on the exposure levels of the 14- to 15-year-old adolescents ($N = 172$) from this group. They found that the sound levels measured in their favorite discos were 107.8–112.2 dB(A) and for their PLDs 82.9–104.6 dB(A). The same group (Biassoni et al., 2014) compared the hearing of these adolescents at ages 14–15 (test, $N = 172$) and 17–18 (retest, $N = 59$) and found a significant increase in hearing thresholds (increasing from 2.7 dB at 250 Hz, to 7.0 dB at 8 kHz, and then decreasing to 4.5 dB at 12 kHz, and again increasing to 7.0 dB at 16 kHz). This was accompanied by a significant decrease in the amplitude of TEOAEs in the moderate and high exposure groups. The decrease was 5.0 dB at 1 kHz and decreasing with frequency to 3.4 dB at 4 kHz. This clearly illustrates the onset and progression of significant NIHL in these adolescents.

Hearing loss in disc jockeys may also be related to their exposure to music and length of time in the profession. Potier et al. (2009) surveyed a group with average age of approximately 26 years (SD = 6 years) who were on average 6.6 years in that profession and were on average exposed for approximately 22 (SD = 13) hours weekly. Their audiograms showed the expected hearing loss at 6 kHz, but also low frequency losses at 125–500 Hz (Fig. 6.2).

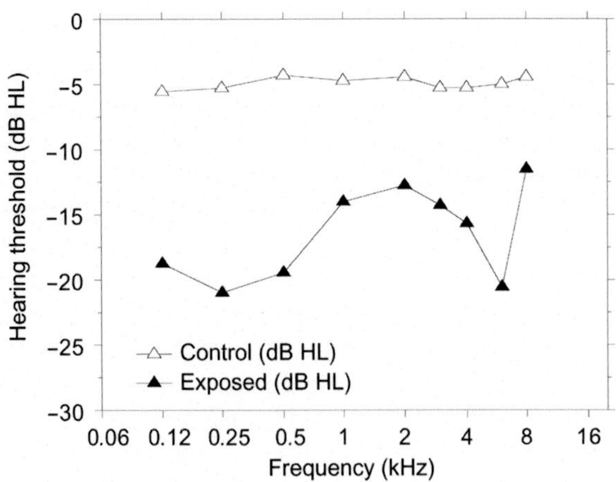

FIGURE 6.2 Audiogram in disc jockeys and control subjects averaged over both ears. Note the hearing losses at 6 kHz and in the low frequencies. *Based on data from Potier, M., Hoquet, C., Lloyd, R., Nicolas-Puel, C., Uziel, A., Puel, J.-L., 2009. The risks of amplified music for disc-jockeys working in nightclubs. Ear Hear. 30, 291–293.*

6.2.4 Personal Listening Devices

As has been clearly summarized by Ivory et al. (2014): "Concern is mounting as the use of personal listening devices (PLDs) has become de rigueur for today's teenagers. There is still debate and uncertainty as to the exact extent and contribution of PLD usage to hearing loss. However, there is an increasing body of recent literature supporting the connection between extended and/or elevated usage of PLDs and documented hearing loss, with a clear correlation between such usage patterns and hearing loss in the extended high frequency range of hearing (8–16 kHz)."

In 33 volunteers, Le Prell et al. (2012) measured effects of PLDs use on hearing. Subjects selected either rock or pop music, which was then presented at 93–95 ($N = 10$), 98–100 ($N = 11$), or 100–102 ($N = 12$) dB(A) in-ear exposure level for a period of 4 h. Audiograms and distortion product otoacoustic emissions (DPOAEs) were measured before and after music exposure. Post-music tests were done 15 min, 1 h 15 min, 2 h 15 min, and 3 h 15 min after the exposure ended. Additional tests were conducted the following day and 1 week later. They found that TTS was reliably detected after higher levels of sound exposure. This was reflected in audiometric thresholds with a "notch" configuration, with the largest changes at 4 kHz (mean ~6 ± 4 dB; range = 0–14 dB). Le Prell et al. (2012) found that threshold recovery was largely complete within the first 4 h postexposure, and all subjects showed complete recovery of both thresholds and DPOAE measures when tested 1 week postexposure. The study by Wang and Ren (2012) reviewed in Chapter 4, Hearing Problems which showed that repeated TTS exposures can result in permanent hearing loss should be kept in mind here.

Sulaiman et al. (2014) evaluated early hearing effects related to PLD usage in 35 young adult PLD users (listening for >1 h/day, at >50% of the maximum volume setting of their devices) and their age- and sex-matched controls using a combination of conventional and extended high-frequency audiometry as well as TEOAE and DPOAE measurements. The mean listening duration of the PLD users was 2.7 ± 1.0 h/day while their estimated average listening volume was 81.3 ± 9.0 dB(A) (free-field corrected). Sulaiman et al. (2014) did not detect typical signs of NIHL in the audiogram of PLD users and their audiometric thresholds at 0.25–8 kHz were comparable with those obtained from controls. As expected, mean hearing thresholds of PLD users at many of the extended high frequencies (9–16 kHz) were significantly higher than in controls. In addition, TEOAE and DPOAE amplitudes in users were significantly smaller than in controls. These results indicate the presence of an early stage of hearing damage in the PLD user group.

Despite the potential harmful music listening habits of the last decades, Hoffman et al. (2010) found that "for men and women of a specific age, high-frequency hearing thresholds were lower (better) in 1999–2004 than in 1959–1962. The prevalences of hearing impairment were also lower in the recent survey." To quantify this, we are talking here about a significant 5 dB difference on average for 6 kHz (the highest frequency measured in the 1959–62 survey) in the age groups above 45 years. In contrast for the youngest age groups (25–34 and 35–44 years) the difference was even larger at approximately 10 dB.

6.3 ANIMAL RESEARCH INTO EFFECTS OF NOISE EXPOSURE ON THE BRAIN

NIHL has its peripheral substrate in hair cell damage, diffuse or localized auditory nerve fiber (ANF) degeneration, and frequency-specific reduced auditory nerve output to the central auditory system. Because mostly high frequencies are affected, NIHL results in an imbalance of spontaneous as well as sound-driven firing rate in ANFs across frequency. This imbalance drives plastic changes in the efficacy of excitatory and inhibitory synapses in the brainstem, midbrain, and the thalamo-cortical auditory system. These changes cause a potential detrimental hearing impairment that comes on top of the loss in hearing sensitivity.

6.3.1 Necrosis and Apoptosis in Noise-Induced Hearing Loss

Cochlear damage following noise exposure occurs through two major routes. The first one is direct mechanical damage, which leads to both hair cell loss through mechanical disruption of the stereocilia and direct damage to supporting and sensory cells (Slepecky, 1986). The other route involves biochemical pathways leading to cell death through either apoptosis or necrosis. Apoptosis is an active, energy-requiring process that is initiated by specific pathways in the cell, while necrosis is a passive one requiring no metabolic energy and results in the rupture of the cell body. During necrosis, the cellular content is spilled onto adjacent cells, thereby possibly triggering inflammatory responses. Necrosis and apoptosis are easily distinguishable through differentially activated biochemical processes. The first studies evaluating the type of cochlear cell death following intense noise exposure date back to the mid-1980s. Swollen outer hair cells (OHCs) were observed in cochleae of animals subjected to loud noise (~120 dB SPL). As this is a hallmark of necrosis, it was assumed that necrosis was the major cause of cell death (Saunders et al., 1985).

Besides necrosis, apoptosis is a key mediator of NIHL. Several biochemical apoptotic markers, such as the caspase cascade, are activated in OHCs after noise trauma (Han et al., 2006). Two important factors seem to determine which cell death pathway is activated following intense noise exposure. The first is sound intensity level. Noises of approximately 115 dB SPL seem to favor necrosis, while only marginally louder noises (~120 dB) seem to favor apoptosis (Hu et al., 2000). In this study, two major types of morphological changes of OHC nuclei were noted in the noise-exposed cochleae. One was characterized by formation of chromatin fragments and by shrinkage of nuclei. Another was swelling of OHC nuclei. The finding of nuclear swelling and condensation in the noise-damaged cochleae suggested that two types of nuclear pathologies originated from two distinct biological processes or from a single biological process with two phases of the change. First, in the animals exposed to 110 or 115 dB noise, there was only swelling of nuclei. Second, in this study formation of chromatin fragments and shrinkage of nuclei predominately appeared 3 h after the noise exposure, whereas swelling of nuclei occurred in all the exposed cochleae, particularly in the cochleae obtained 3 and 14 days after the noise exposure. Finally, although both nuclear swelling and condensation coexisted in the animals exposed to 120 dB noise, their distribution along the organ of Corti was different. Considering these differences, Hu et al. (2000) concluded that nuclear swelling and nuclear condensation originated from two distinct biological processes leading to cell death. The typical changes of formation of chromatin fragments and shrinkage of nuclei noted in the animals exposed to 120 dB noise are morphologically similar to those nuclear changes described in previous studies for apoptosis, suggesting that apoptotic processes may be involved in intense noise-induced hair cell death (chapter 3 in Eggermont, 2014).

6.3.2 Delayed Effects of TTS Noise Exposure and Aging

Plastic changes in the auditory system (see chapter: Brain Plasticity and Perceptual Learning) often result from the loss of cochlear hair cells, regardless whether the loss is induced by mechanical intervention, traumatic noise exposure, or by the application of ototoxic drugs. Damage of hair cells may also result as a consequence of aging. Presbycusis in humans refers to age-related auditory deficits that include a loss of hearing sensitivity and a decreased ability to understand speech, particularly in the presence of background noise. The hearing loss tends to increase with age, with high-frequency losses exceeding low-frequency losses at all ages (Syka, 2002). Data from large populations screened for noise exposure and otologic disease (Willott, 1991) show a progressive increase of hearing loss amounting to 20 dB at frequencies below 1 kHz and increasing to 60 dB difference at 8 kHz over the age span from 30 to 70 years. Accumulating auditory stresses during life may lead to presbycusis. The involvement of environmental factors is implied, for example, by the fact that hearing levels are generally poorer in industrialized than in more isolated societies (Gates and Mills, 2005). Prolonged exposure to loud occupational noise has long been recognized as a cause of hearing loss but long-lasting effects, after the noise exposure has stopped, are not clear (Gates et al., 2000; Lee et al., 2005). Subclinical damage accumulated during employment may place the ear at higher risk for age-related hearing impairment.

Bao and Ohlemiller (2010) wrote a landmark review that we follow here. Survival of ANFs during aging depends on genetic and environmental interactions. Loss of ANFs without associated loss of hair cells is common among mammals during aging and is called primary degeneration. Apparent primary and secondary degeneration (following loss of inner hair cells (IHCs)) of ANFs may occur in the same cochlea (Hequembourg and Liberman, 2001), suggesting that age-related ANF and hair cell loss result from independent mechanisms. Primary degeneration of ANFs has been observed in the cochlea of CBA/CaJ mice after moderate noise exposure at a young age (Kujawa and Liberman, 2006). In this study, CBA/CaJ mice were exposed to an 8–16 kHz noise band at 100 dB SPL for 2 h at ages from 4 to 124 weeks and held with unexposed cohorts for postexposure times from 2 to 96 weeks. When evaluated 2 weeks after exposure, maximum threshold shifts in young-exposed animals (4–8 weeks) were 40–50 dB. Animals exposed at at least 16 weeks of age showed essentially no shift at the same postexposure time. However, when held for long postexposure times, these animals showed substantial ongoing deterioration of cochlear neural responses and corresponding primary neural degeneration throughout the cochlea without changes in OHC responses (as measured with DPOAEs). Delayed ANF loss was observed in all noise-exposed animals held 96 weeks after exposure, even in those that showed no NIHL 2 weeks after exposure. Thus, even in the case of clear hair cell loss, true primary versus secondary neuronal loss may be impossible to separate at the early degeneration stage. At the later stages, certain independent mechanisms may contribute to the uncoupling of age-related loss of hair cells and ANFs (Bao and Ohlemiller, 2010).

6.3.3 Noise-Induced Permanent Hearing Loss in Animals

6.3.3.1 Subcortical Findings

Cats that were exposed for 1–4 h to narrow-band or broadband noise with levels of 100–117 dB SPL showed a permanent hearing loss, and firing activity of ANFs was recorded 15–305 days after the trauma (Liberman and Kiang, 1978). Of the ANFs that still showed sharp frequency tuning, two typical forms of abnormal tuning-curve shape were found: the V-shaped tuning curve for which both the low-frequency tail and tip (at characteristic frequency (CF)) were elevated in level, and the W-shaped tuning curve resulting from an elevated tip threshold and a low-frequency-tail threshold that became at least as low as (and could even be lower than) the tip threshold.

Hamsters exposed to 120–125 dB SPL, 10 kHz tones for 2–6 h showed stereocilia lesions but no visible loss of hair cells (Kaltenbach et al., 1992). The purpose was to determine if and how the tonotopic map of the dorsal cochlear nucleus (DCN) was readjusted after hearing loss. Neural population thresholds and tonotopic organization were mapped over the surface of the DCN in normal unexposed animals and those showing tone-induced lesions. In many cases the center of the lesion was represented in the DCN as a distinct gap in the tonotopic map in which responses were either extremely weak or absent. The map area at the center of the lesion was nearly always surrounded by an expanded region of near-constant CF, which was suggestive of map reorganization. However, these expanded map areas had much increased frequency-tuning curve (FTC) tip (or tail) thresholds and showed other features suggesting that their CFs had been shifted downward by distortion and deterioration of their original tips. Such changes in neural tuning were similar to those observed in ANFs following acoustic trauma (Liberman and Kiang, 1978), and thus would seem to have a peripheral origin and not reflect plastic changes in the DCN. Rajan and Irvine (1998) also examined whether topographic map plasticity could be found in the adult auditory brainstem. Following partial cochlear lesions, they found no plasticity of the frequency map in the DCN confirming conclusions by Kaltenbach et al. (1992). This suggests that the DCN does not exhibit the type of plasticity that has been found in the auditory cortex and midbrain (see below).

Irvine et al. (2003) examined the effects of unilateral mechanical cochlear lesions on the tonotopic map of the central nucleus of the inferior colliculus (ICC) in adult cats. These lesions typically resulted in a broad high-frequency hearing loss in the range of 15–22 kHz. After recovery periods of 2.5–18 months, the frequency organization of ICC contralateral to the lesioned cochlea was determined separately for the onset and late components of multiunit responses to tone-burst stimuli. Most of the observed changes were explicable as passive consequences of the lesion and showed limited evidence for plasticity in the ICC of adult cats.

6.3.3.2 Findings in Auditory Cortex and Thalamus

The tonotopic maps of the ventral nucleus of the medial geniculate body (MGBv) in the thalamus of cats following mechanically induced restricted unilateral cochlear lesions were assessed by Kamke et al. (2003). These animals had severe mid-to-high frequency hearing losses and were investigated 40–186 days after lesioning. The region of MGBv in which mid-to-high frequencies were normally represented now showed an "expanded representation" of lesion-edge frequencies. Neuron clusters within these enlarged representation had "new" CFs and displayed latency, and FTC bandwidths very similar to those in normal animals. Thresholds of these neurons were close to normal for their prelesion frequency range. The tonotopic reorganization observed in MGBv was similar to that seen in primary auditory cortex (see below) and was more extensive than the partial reorganization found in the ICC, suggesting that the auditory thalamus plays an important role in cortical representational plasticity.

Rajan et al. (1993) had examined this in the primary auditory cortex of adult cats. In confirmation with the earlier study in adult guinea pigs (Robertson and Irvine, 1989) they found that 2–11 months after the unilateral cochlear lesion the map of the lesioned cochlea in the contralateral A1 was altered so that the A1 region in which frequencies with lesion-induced elevations in cochlear neural sensitivity would have been represented was occupied by an enlarged representation of lesion-edge frequencies (i.e., frequencies adjacent to those with elevated cochlear neural sensitivity). There was no topographic order within this enlarged representation. The normal threshold sensitivity at the CF for units in the reorganized regions of the map assured that the changes reflected a plastic reorganization rather than simply the residue of prelesion input.

At that time it was not clear whether gradual progressive changes in hearing loss were sufficient for topographic map changes to occur, or if they required very sharp audiogram boundaries as in the Rajan et al. (1993) study. Eggermont and Komiya (2000) thus exposed juvenile cats in an anechoic room twice for 1 h to a 6 kHz tone of 126 dB SPL. During this exposure the animals were awake, confined in a small cage and facing the loudspeaker. The first exposure was at 5 weeks after birth and it was repeated 1 week later. Recordings were made from A1 at least 6 weeks after the exposure under ketamine anesthesia. The trauma caused a reorganization of the tonotopic map for frequencies above 6 kHz such that the original CFs were now replaced by CFs from the near normal low-frequency edge of the induced hearing loss (Fig. 6.3). In the noise-damaged cats the highest CFs were about 10 kHz in the exposed litter of three kittens and about 7 kHz in the litter of two kittens. As a result, the mean CF–distance curve started to deviate from the normal progression at those frequencies, resulting in a 2–3 mm extent of cortex that had essentially the same CF. In addition, tonotopic order was not preserved in the reorganized region.

In order to study the effects of the recovery environment after noise trauma, Noreña and Eggermont (2005) exposed 14 adult cats to a one-third octave band of noise centered at 5 kHz. Measured at the cat's head, the noise had a level of 120 dB SPL. Five age-matched nonexposed cats served as controls. After a 2-h exposure, seven cats recovered in a quiet free-range room ("Quiet"). The other seven cats were exposed for 4 h ("EAE") and recovered—also in a free-range room—in the presence of continuous stimulation with random frequency tone pips between 4 and 20 kHz and presented at a level of 80 dB SPL. We called this an enhanced acoustic environment (EAE). This level was high enough to effectively stimulate the auditory system (Noreña et al., 2003; Eggermont and Komiya, 2000) but not to further damage the auditory system.

Peripheral hearing loss in these cats was estimated by comparing ABR thresholds to those obtained in a large reference group of normal hearing

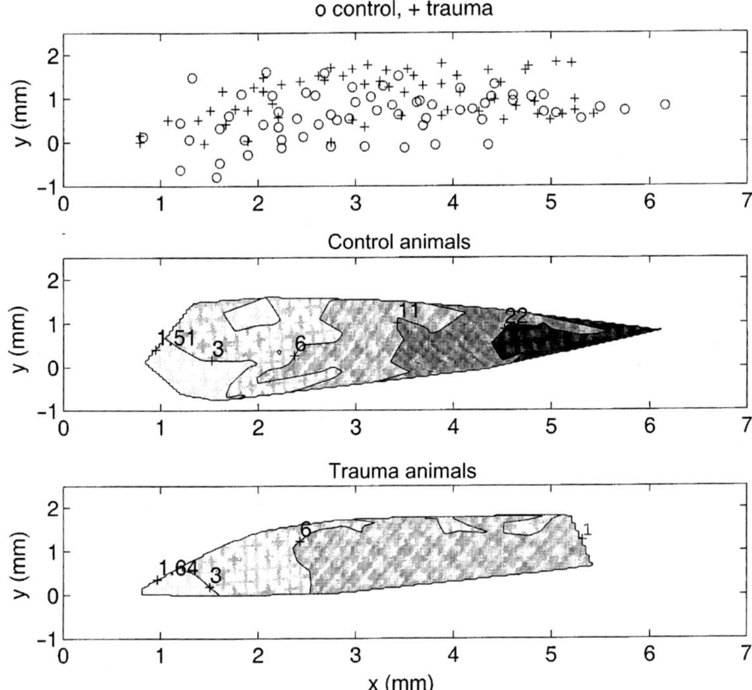

FIGURE 6.3 Topographic maps in cat primary auditory cortex. Top panel: Recording sites for control (o) and trauma (+) animals. Middle panel: CF map for control animals. This map is based on an interpolation of the recording site map. The contour lines are for 1, 3, 6, 11, and 22 kHz. The darker the shading the higher the CF. The lowest frequency was 1 kHz and the highest 40 kHz. Bottom panel: CF map for trauma animals. The contour lines are for 1.64, 3, and 6 kHz. The darker the shading the higher the CF. The lowest frequency was 1.64 kHz and the highest 10.4 kHz. *Reprinted from Eggermont, J.J., Komiya, H., 2000. Moderate noise trauma in juvenile cats results in profound cortical topographic map changes in adulthood. Hear. Res. 142(1–2), 89–101, with permission from Elsevier.*

cats (Fig. 6.4). A negative difference represents the amount of hearing loss. The control group did not show any significant threshold deviation from the reference cats except at the highest frequency tested. However, this hearing loss was small (11 dB, on average). All cats exposed to the traumatizing sound presented a hearing loss. Interestingly, the pattern of this hearing loss was very different for the two groups. "Quiet" cats (no additional sound stimulation after the trauma) presented a mild hearing loss in the middle-frequency range (~20 dB at 4 kHz) and a moderate hearing loss in the high-frequency range (~30–40 dB at 16–32 kHz). In contrast, "EAE" cats showed a significant hearing loss (note they were twice as long exposed as the "Quiet" group) in the middle-frequency range (~20 and 35 dB at 4 kHz

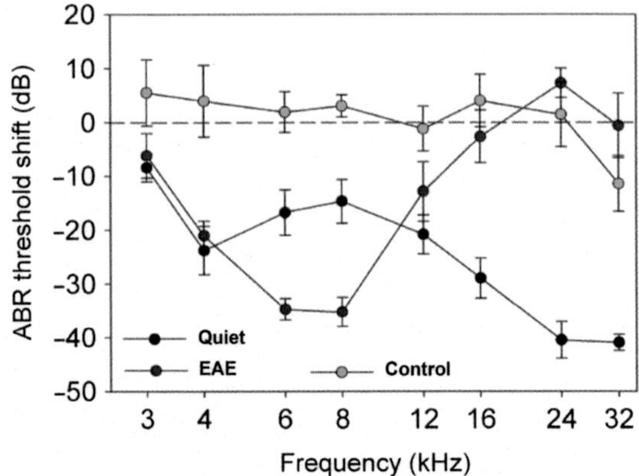

FIGURE 6.4 Averaged ABR threshold shifts across the three groups, compared to a large reference group of normal hearing cats. A negative shift is considered equal to the amount of hearing loss. "Quiet," recovery in quiet. "EAE," recovery in a high-frequency acoustic environment. Vertical bars indicate the SEM. *From Noreña, A.J., Eggermont, J.J., 2005. Enriched acoustic environment after noise trauma reduces hearing loss and prevents cortical map reorganization. J. Neurosci. 25, 699–705.*

and 6–8 kHz, respectively) however, ABR thresholds above 8 kHz were in the normal range. This reduction in high-frequency hearing loss in the "EAE" group compared with that in the "Quiet" group suggests that the high-frequency acoustic environment after the noise exposure limited the high-frequency hearing loss that was likely the result of glutamate neurotoxicity. We interpreted this recovery as the result of a reconnection of the ANF neurites to the IHCs, i.e., restoring the ribbon synapses, guided by the continued output of glutamate (in response to the EAE) by those still intact IHCs (Pujol and Puel, 1999; Nouvian et al., 2015).

The differing effects of the posttrauma sound environment were also evident in neural responses recorded in the middle layers (III–IV) of primary auditory cortex. The receptive fields of neurons were obtained simultaneously from up to 16 recording sites in A1. The composite maps of the CFs obtained at recording site across all cats in each group are shown in Fig. 6.5A–C. The reference coordinate (0,0) was taken as the tip of the posterior ectosylvian sulcus (PES). The horizontal axis is parallel to the posterior–caudal axis of the animal, and the vertical axis consequently runs ventromedial. Although I acknowledge that A1 does not consistently occupy the same region with reference to anatomical landmarks such as the PES, comparison of group data mapped onto a single anatomical map using the PES as a reference point will reveal any substantial changes in tonotopic maps.

The boundary between A1 and the anterior auditory field is usually found between approximately 5 and 7 mm from the tip (0,0) of the PES. The CF map obtained in control cats illustrates the tonotopic organization

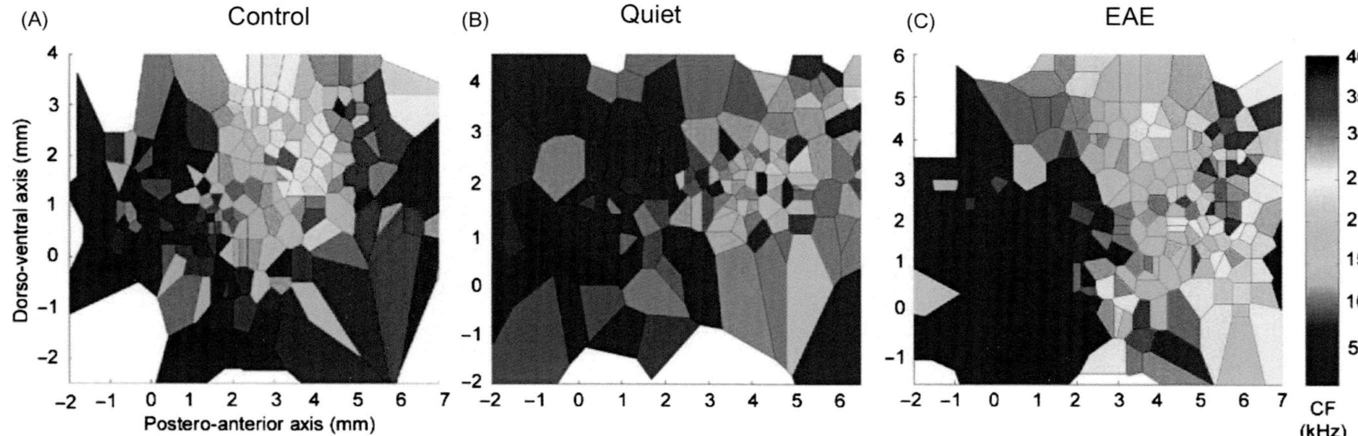

FIGURE 6.5 Compound CF maps in primary auditory cortex in control cats (A), group 1 cats (B), and group 2 cats (C). The center of each polygon, constructed using the tessellation method, corresponds to the coordinates of a recording site in auditory cortex along the anteroposterior axis (abscises) and the ventrodorsal axis (ordinates). The tip of the PES was taken as the (0,0) coordinate. The CF is represented by color; cold colors (blue-like) represent low CF, and hot colors (red-like) represent high CF as indicated by the color bar. *From Noreña, A.J., Eggermont, J.J., 2005. Enriched acoustic environment after noise trauma reduces hearing loss and prevents cortical map reorganization. J. Neurosci. 25, 699–705.*

of A1 (Fig. 6.5A). The CF smoothly increases along the anteroposterior axis up to the anterior auditory field from where it starts to decrease. In contrast, "Quiet" cats showed a modified tonotopic organization (Fig. 6.5B) (i.e., CFs above 30 kHz are no longer present, and those from 20–30 kHz are severely underrepresented). Instead, neurons in the anterior part of A1, those that were normally sensitive to high frequencies (Fig. 6.5A), are now mostly sensitive to a narrow frequency range corresponding to the low-frequency boundary of maximum hearing loss (i.e., ~15 kHz) (Fig. 6.4). EAE cats, despite having a hearing loss in the mid-frequency range, showed an apparent normal tonotopic organization (Fig. 6.5C). Similar to the control group, the CF smoothly increased along the anteroposterior axis. Noreña and Eggermont (2005) showed that this mapping was not related to the shape of the audiogram, but was the result of the posttrauma high-frequency sound exposure since the tonotopic map was vastly abnormal in a cat with the same audiogram that had recovered in quiet.

Summarizing, recovery in quiet from NIHL causes a reorganization of the cortical tonotopic map in primary auditory cortex. However, if acoustic stimulation with a spectrum corresponding to the frequency band of the hearing loss is provided after the trauma, the hearing loss in the high-frequency region is reduced or even disappeared, and the tonotopic map remains normal.

6.4 OTOTOXICITY

Ototoxicity is the functional impairment and cellular degeneration of the tissues of the inner ear caused by therapeutic agents. The result of exposure to these agents is a loss of hearing and/or vestibular function (Rybak and Ramkumar, 2007). Schacht et al. (2012) reviewed the range of ototoxic agents—aminoglycoside (AG) antibiotics, platinum-based chemotherapeutic agents (cisplatin (CP) and carboplatin), loop diuretics, salicylate, and antimalarial drugs (quinine)—and stated that: "Salicylate (aspirin) has long been associated with elevated hearing threshold and tinnitus (ringing in the ear), two effects that disappear with cessation of drug intake. Loop diuretics such as furosemide, ethacrynic acid, and bumetanide have transient side effects on the inner ear given singly, but create disastrous repercussions in combination with aminoglycoside antibiotics." Recently, a safety warning has been issued that administration of some phosphodiesterase inhibitors (such as Cialis, Viagra, Revatio, and Levitra) may be associated with sudden hearing loss and/or vestibular disturbances (Maddox et al., 2009).

6.4.1 Salicylate

Eggermont (2012) reviewed salicylate effects: Numerous biochemical processes underlying the effects of salicylate have been identified: inhibition of prostaglandin synthesis through the inhibition of cyclooxygenase, inhibition of numerous metabolic enzymes, inhibition of free radicals, insertion into membranes and interference with ion transport, uncoupling of oxidative phosphorylation, and activation of heat shock transcription factor. Among them, the inhibition of cyclooxygenase activity is the best known pharmacological effect of salicylate. With respect to causing hearing loss, a single dose of salicylate interferes with the cochlear amplifier by affecting prestin in the OHCs (see chapter: Hearing Basics). This is likely the mechanism that leads to transient hearing loss, whereas chronic salicylate causes a compensatory temporary enhancement in DPOAE amplitudes and upregulation of prestin mRNA and protein expression in animals (Sheppard et al., 2014) and can also impair ganglion cell neurons (Deng et al., 2013). Most likely, salicylate primarily influences electromotility and the nonlinear capacitance of the OHCs via a direct interaction with prestin (Greeson and Raphael, 2009). A number of animal studies have used salicylate as the preferred inducer of tinnitus because its reliable effects in doing so. One of the drawbacks of using salicylate is that it acts on both the peripheral and the central auditory nervous system. In the periphery, salicylate affects first of all the motility of the OHCs. In addition, as described above, salicylate interferes with the arachidonic acid (AA) cycle in the IHCs and so upregulates the action of N-methyl-D-aspartate (NMDA) receptors in the cochlea, which increases the channel opening probability of the NMDA receptor and potentially leads to increased spontaneous firing rate for high doses (Guitton et al., 2003). A different mechanism for salicylate-induced hearing loss was recently suggested by Wu et al. (2010). Instead of an action on prestin, which would require a very high dose in humans compared to animals in order to cause a hearing loss, salicylate could also act via blocking the KCNQ4 outward current $I_{K,n}$. This would subsequently cause depolarization of OHCs, resulting in a reduction of the driving force (by a ceiling effect) for the transduction current and electromotility (Eggermont, 2012). More studies are required to reveal the exact mechanisms of salicylate ototoxicity.

6.4.2 Platin Chemotherapy Drugs

One of the first electrophysiological studies on the effect of these drugs was carried out by Coupland et al. (1991), who investigated "early cisplatin ototoxicity using both the broadband click and derived ABR" (see chapter: Types of Hearing Loss) and monitored progressive hearing loss with repeated drug trials in 18 patients studied over a 2-year period. ABRs were obtained serially prior to and following intravenous administration of CP. For click ABRs, the cumulative dosage of CP at age of ABR examination was correlated with hearing loss in only those patients under 3 years of age. No significant correlation was found between cumulative CP dosage when tested and degree of hearing loss in those patients over 3 years of age. The need for longitudinal ABR testing cannot be overemphasized as

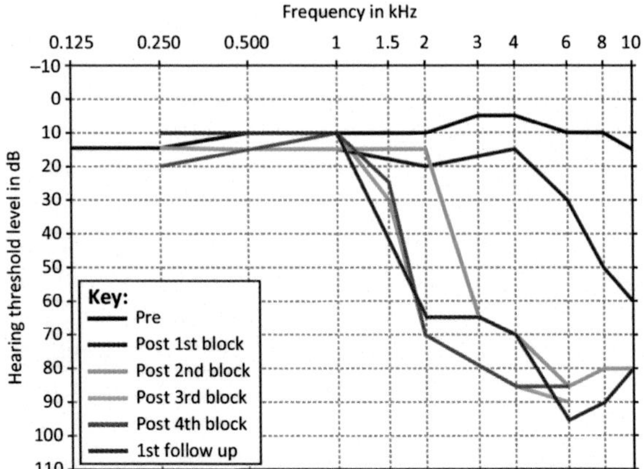

FIGURE 6.6 Progressive hearing loss during CP therapy. Shown are left ear audiograms of a patient, recorded at baseline, that is, before start of CP therapy (pre), after each of four CP cycles (post-first block, post-second block, post-third block, post-fourth block), and during follow-up (first follow up). *Reprinted from Langer, T., am Zhnhoff-Dinnese, A., Radtke, S., Meitert, J., Zolk, O., 2013. Understanding platinum-induced ototoxicity. Trends Pharmacol. Sci. 34(8), 458–469, with permission from Elsevier.*

Coupland et al. (1991) have observed changes in the derived ABR occurring some 12 weeks following chemotherapy treatment. It would appear that some of the effects of ototoxicity are not immediately detectable in the brainstem response or that the degree of neurotoxicity is cumulative with time.

Ding et al. (2012) noted that: "Cisplatin, carboplatin, nedaplatin, and oxaliplatin are widely used in contemporary oncology. For these compounds, a major factor in the damage is drug uptake from stria vascularis into the cochlear fluids. Disrupting the blood–labyrinth barrier with diuretics or noise exposure enhances drug uptake and significantly increases the amount of damage." Langer et al. (2013) found that platinum ototoxicity usually results in bilateral, symmetrical, high-frequency sensorineural hearing loss often accompanied by tinnitus and vertigo. Platinum-induced hearing loss initially affects higher frequencies (≥4 kHz) and can progress to involve speech frequencies (<4 kHz) (Fig. 6.6). Hearing loss may progress after completion of the therapy and is likely caused by retention of platinum in the body for up to 20 years after administration. It is clear that monitoring during chemotherapy treatments with particular attention to high-frequency behavioral threshold or otoacoustic emission changes is warranted (Langer et al., 2013).

6.4.3 Aminoglycosides

The first therapeutic drugs to show obvious ototoxic effects were in the AG group, starting with streptomycin and numerous newer related compounds including gentamicin, kanamycin, amikacin, etc. (Guthrie, 2008). Many studies have been devoted to the mechanisms by which the AGs cause inner ear damage. A few salient findings are noted here. After systemic injection of gentamycin conjugated with a fluorophore, Warchol (2010) noted that fluorescence is first observed in the apical portions of hair cell stereocilia. Over the next several hours, the fluorescent signal appears to migrate through the stereocilia bundle and into the hair cell body. This indicates that AGs enter hair cells, at least initially, through their transduction channels. Guthrie (2008) found no statistical difference between pre- and post-pure-tone standard audiometric. However, otoacoustic emissions revealed statistically significant changes between pre- and post-treatment. Because AG treatment induces cochlear damage, the effectiveness of OAE testing, which assesses the function of the OHCs, is obvious. However, extending conventional pure-tone audiometry with frequencies more than 8 kHz and one-sixth octave protocols significantly improved the detection of ototoxicity (Guthrie, 2008).

6.4.4 Mechanisms for Cisplatin and Aminoglycoside Ototoxicity

Rybak and Ramkumar (2007) in a comprehensive review described that AGs and CP target the OHCs in the basal turn of the cochlea to cause high-frequency sensorineural hearing loss in a substantial percentage of patients treated with these drugs. The mechanisms appear to involve the production of reactive oxygen species (ROS), which can trigger cell death (Fig. 6.7). The steps that appear to be involved in AG ototoxicity include (Rybak and Ramkumar, 2007): (1) AG entry into OHC through the mechano-electrical transducer channels; (2) formation of an AG–iron complex can react with electron donors, such as AA to form ROS, like superoxide, hydroxyl radical, and hydrogen peroxide; (3) ROS can then active JNK, which can then (4) translocate to the nucleus to activate genes in the cell death pathway; (5) these genes can then translocate to the mitochondria, causing (6) the release of cytochrome c (cyt c), which can trigger (7) apoptosis via caspases. Cell death may also result from caspase-independent mechanisms.

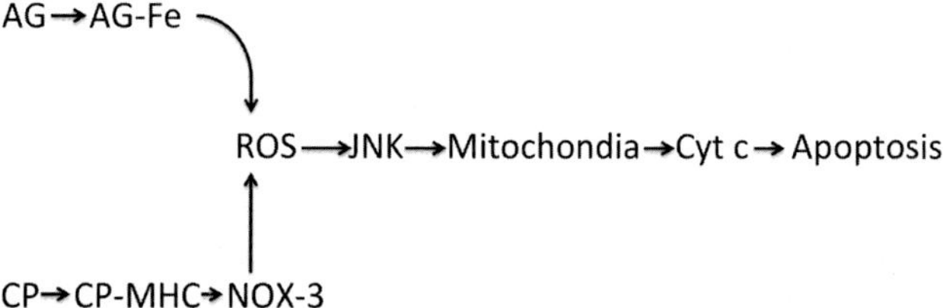

FIGURE 6.7 Mechanisms of AG and CP-induced OHC death. AG and CP entry into OHC results in cell death by either caspase-dependent or caspase-independent mechanisms. *After Rybak, L.P., Ramkumar, V., 2007. Ototoxicity. Kidney Int. 72, 931–935.*

CP targets three areas in the cochlea: the hair cells in the basal turn of organ of Corti, the spiral ganglion cells, and the lateral wall tissues (spiral ligament and stria vascularis). OHCs, cells in the stria vascularis, and spiral ligament undergo apoptosis and platinated DNA immunoreactivity has been localized to the nuclei of OHCs, and cells in the stria vascularis and spiral ligament. ROS mimic the effects of CP on OHCs in vitro, and CP reacts with cochlear tissue explants to generate ROS. This can lead to calcium influx within cochlear cells resulting in apoptosis (Rybak and Ramkumar, 2007). CP entry into OHC results in cell death, which appears to be primarily caspase-dependent (Fig. 6.7). The steps that may be involved include: (1) CP entry into the OHC through mechanotransducer channels; (2) CP within cells can be aquated to form the monohydrate complex (MHC), which is more highly reactive; (3) CP and/or MHC can activate NOX-3, resulting in ROS production; then it follows the same route to apoptosis as for the AG. "The generation of ROS involves the formation of an aminoglycoside-iron (AG-Fe) complex, which catalyzes their production from unsaturated fatty acids" (Rybak and Ramkumar, 2007).

6.4.5 Diuretics

It has long been known that loop diuretics (acting on the renal loop system) also effect ion pumping mechanisms in the stria vascularis to produce hearing threshold elevations. Acting alone, the hearing loss is typically temporary; normal hearing is restored as the diuretic is cleared from the blood. Importantly however the concomitant administration with other ototoxic drugs will potentiate their damaging effects.

6.4.5.1 Furosemide

Furosemide is an ototoxic diuretic. It has the following properties (Rybak, 1985): (1) Furosemide in ototoxic doses (>85 to 90 μg/mL) causes a reversible reduction of the endocochlear potential and endolymph potassium concentration. (2) Furosemide differentially reduces the amplitude of the eighth nerve action potential without much change in the cochlear microphonic potential (Evans and Klinke, 1982). (3) Combination of furosemide and kanamycin results in increased ototoxicity in animals. Ruggero and Rich (1991) found that furosemide reversibly altered the responses to tones and clicks of the chinchilla basilar membrane, causing response-magnitude reductions that were largest (up to 61 dB, averaging 25–30 dB) at low stimulus intensities at the CF and small or nonexistent at high intensities and at frequencies far removed from CF.

6.4.5.2 Ethacrynic Acid

Ethacrynic acid is a potent diuretic, exacerbates the action of other ototoxic drugs, and causes hearing loss. Ding et al. (2003) reported that concurrent administration of a high dose of gentamicin and ethacrynic acid results in rapid destruction of virtually all cochlear hair cells.

6.4.6 Bacterial and Viral Infections

6.4.6.1 Bacterial Infections

> Bacterial meningitis *is the most common etiology for acquired hearing loss. Five to 35% of patients with bacterial meningitis will develop permanent sensorineural hearing loss, and profound bilateral hearing loss will occur in up to 4% of patients. Streptococcus pneumonia (pneumococcus) and Neisseria meningitides (meningococcus) are by far the most common organisms responsible for bacterial meningitis. The exact mechanism of hearing loss in patients with bacterial meningitis is not well understood and is likely due to multiple factors that include direct labyrinth involvement, cochlear neuroepithelial damage, and vascular insult.*
>
> Kutz et al. (2006).

In a study by Richardson et al. (1997) 92 children (74%) had meningococcal and 18 (15%) had pneumococcal meningitis. Three children (2.4%, 95% CI 0.5–6.9%) had permanent sensorineural hearing loss. Thirteen children (10.5%) had reversible hearing loss of whom nine had an impairment that resolved within 48 h of diagnosis. Permanent deafness was rare but 10% of the patients had a rapidly reversible cochlear dysfunction. In the study of Kutz et al. (2006), 134 patients underwent audiologic testing during their initial hospitalization, and 41 (30.6%) were found to have at least a unilateral mild sensorineural hearing loss. The incidence of hearing loss was greater in patients with pneumococcal meningitis than in patients with meningococcal meningitis (35.9% and 23.9%, respectively). Jit (2010) in a systematic review found that in pneumococcal meningitis survivors the prevalence of hearing loss was 20.9% (range 17.1–24.7%).

Otitis media. Besides CHL during the infection, 5–20% of all cases of otitis media will result in sensorineural hearing loss; the majority of these cases are bilateral, high-frequency SNHL.

6.4.6.2 Virus Infections

Viral infections can cause generally sensorineural hearing loss, which can be congenital or acquired, unilateral or bilateral. Certain viral infections can directly damage inner ear structures, others can induce inflammatory responses, which then cause this damage, and still others can increase susceptibility or bacterial or fungal infection, leading to hearing loss (Cohen et al., 2014).

Viral meningitis is also called aseptic meningitis. Based on clinical symptoms, viral meningitis cannot be differentiated from bacterial meningitis. Both appear as headache, fever, and neck stiffness but viral meningitis has no evidence of bacteria present in CSF. Viral meningitis is less serious than bacterial meningitis. Most cases are caused by enteroviruses (common stomach viruses). Other viruses can also cause viral meningitis. For instance: West Nile virus (WNV), mumps, measles, herpes simplex virus (HSV), varicella zoster virus (VZV), and lymphocytic choriomeningitis virus (https://en.wikipedia.org/wiki/Viral_meningitis; accessed on January 31, 2016).

I review the most common cases of virus-induced hearing loss.

Cytomegalovirus (CMV) is a member of the herpes virus family, along with HSV and VZV. CMV is the leading nongenetic cause of childhood SNHL and causes bilateral progressive severe SNHL with an incidence of 6–23% if asymptomatic, and 22–65% if symptomatic. Antiviral therapy may lead to recovery (Cohen et al., 2014). CMV is typically acquired early in life and may even be acquired in utero. In the United States, up to 1% of newborns are infected (Fowler et al., 1997; Smith et al., 2005). CMV transmission to fetuses can occur during primary maternal infection (accounting for 40–50% of cases of congenital CMV) or reactivation during pregnancy (1% of cases of congenital CMV). Madden et al. (2005) described 21 patients who were identified with symptomatic congenital CMV infection at birth. The median initial PTA for the 21 subjects was 86 dB and the median final PTA was 100 dB. Delayed manifestations of congenital CMV infection, particularly SNHL, can manifest months or years after birth (Fowler et al., 1997). Initial hearing screening will miss the majority of cases of SNHL in CMV-infected children (Fowler et al., 1999; see chapter: Early Diagnosis and Prevention of Hearing Loss).

Herpes zoster virus (HZV) can cause unilateral or bilateral moderate to profound neural hearing loss and is relatively rare (but in congenital HZV the incidence is up to 33%). Hearing loss following neonatal infection can be bilateral or unilateral severe to profound SNHL (Westerberg et al., 2008).

VZV causes mild to moderate unilateral neural hearing loss with an incidence of 7–85%, and if treated with prednisone the hearing loss may improve. Reactivation of latent VZV within the geniculate ganglion causes Ramsay Hunt syndrome through the development of geniculate ganglionitis and inflammation of the facial nerve. In Ramsay Hunt syndrome, eighth nerve involvement results from transfer of the virus from the nearby geniculate ganglion or directly from the facial nerve within the internal auditory canal. Symptoms include SNHL (24% of affected patients), tinnitus (48%), and vertigo (30%) (Cohen et al., 2014).

Human immunodeficiency virus (HIV) can cause mild to moderate SNHL or mild to maximal CHL or mixed hearing loss. The incidence of hearing loss is 27.5–33.5%, recovery may occur depending on the type of hearing loss (Cohen et al., 2014). However, patients with HIV often have abnormal ABRs, suggesting involvement of the auditory nerve and/or brainstem (Rarey, 1990).

WNV may cause bilateral mild to profound SNHL but is very rare and often recovers spontaneously (Cohen et al., 2014).

Measles may cause profound bilateral SNHL with an incidence of 0.1–3.4%, from which there is no recovery. Prevention occurs through vaccination (Cohen et al., 2014).

Rubella, also known as German measles, leads from bilateral mild to severe flat SNHL with an incidence of 12–19%. It affects the stria vascularis and the organ of Corti, and there is no recovery. SNHL is the most common finding of congenital rubella infection (58%) and is most often seen when maternal rubella infection occurs within the first 16 weeks of pregnancy. Hearing loss typically manifests in the first 6–12 months of life, although it can be present at birth (Cohen et al., 2014).

Mumps may cause unilateral variable SNHL, as it attacks the stria vascularis and organ of Corti, with a widely diverging incidence of 0.005–4%. SNHL tends to occur suddenly 4–5 days after the onset of flu-like symptoms and parotitis. Typically, hearing loss is unilateral and reversible but can be severe and permanent (Cohen et al., 2014).

6.5 LONG-TERM EFFECTS OF CONDUCTIVE HEARING LOSS IN INFANCY

6.5.1 Effects in Humans

CHL in infancy can disrupt central auditory processing. After the hearing returns to normal, months or years may be required for a return to normal perception (Hogan et al., 1997; Wilmington et al., 1994). Clinical studies that target the subset of children with a history of otitis media, also accompanied by hearing loss, consistently show perceptual and physiological deficits that can last for years after peripheral hearing becomes audiometrically normal. These studies suggest that infants with otitis media severe enough to cause CHL are particularly at risk to develop lasting central auditory impairments (Whitton and Polley, 2011). In Chapters 4 and 8, Hearing Problems and Early Diagnosis and Prevention of Hearing Loss we describe the central consequences of hearing loss, conductive as well as sensorineural, in children and ways to prevent this.

6.5.2 Animal Studies

Xu et al. (2007) induced CHL in developing gerbils, reared the animals for 8–13 days, and subsequently assessed the responses of auditory cortex layer 2/3 pyramidal neurons in a thalamocortical brain slice preparation with whole-cell recordings. Periodic stimulation of the MGBv evoked robust short-term depression of the postsynaptic potentials (PSPs) in control neurons, and this depression increased monotonically at higher stimulation frequencies. In contrast, CHL neurons displayed a faster rate of synaptic depression and smaller asymptotic amplitude. Moreover, the latency of PSPs in MGBv was consistently longer in CHL neurons for all stimulus rates.

A separate assessment of spike frequency adaptation in response to trains of injected current pulses revealed that CHL neurons displayed less adaptation compared with controls, although there was an increase in temporal jitter. For each of these properties, nearly identical findings were observed for SNHL neurons. Hearing loss increased firing probability and reduced spike latency. Together, these data show that CHL significantly alters the temporal properties of neural activity in auditory cortex. This may contribute to processing deficits that attend mild to moderate hearing loss.

It has been commonly assumed that during conductive loss, there is still normal spontaneous activity in cochlear afferents, such that overall, despite the reduced cochlear activation, the auditory central nervous system is still excited at normal levels. Harrison and Negandhi (2012) showed that during an approximately 40 dB conductive loss induced in mice by blocking the ear canals for 3 days, activity levels in the cochlear nucleus and inferior colliculus, as shown by c-fos labeling, were reduced during this CHL. The implication is that with a lack of cochlear activation, transmitter synthesis and release at the IHC level is downregulated. This resembles the effects of sensorineural hearing loss and suggests that aspects of SNHL can accompany CHL.

To test whether developmental hearing loss resulted in comparable changes to perception and sensory coding, Rosen et al. (2012) examined behavioral and neural detection thresholds for sinusoidally amplitude modulated (SAM) stimuli. Behavioral SAM detection thresholds for slow (5 Hz) modulations were significantly worse for animals reared with bilateral CHL, as compared to controls. This difference could not be attributed to hearing thresholds, proficiency at the task, or attention.

6.6 VESTIBULAR SCHWANNOMA

VS, also called acoustic neuroma, results in typically one-sided slowly progressive mild to moderate sloping hearing loss accompanied by tinnitus (Lee et al., 2015), which does not subside after surgery (Overdevest et al., 2016). When the VS is intracanalicular, the hearing loss slowly increases up to the time of surgery regardless of VS growth (Pennings et al., 2011) In Chapter 5, Types of Hearing Loss we have extensively reviewed the diagnostic methods, here we point to the way VS causes a hearing loss. Frequently a VS causes significant degeneration in the cochlea (Roosli et al., 2012) as observed in comparing temporal bones of the VS side with the contralateral non-VS temporal bones. Roosli et al. (2012) found significantly more IHC and OHC loss, cochlear neuronal loss, and precipitate in endolymph and perilymph. As another indication of a cochlear origin of hearing loss in VS, Gouveris et al. (2007) observed that DPOAE amplitudes started to decrease at the early stages of hearing loss. On the other hand, findings that the hearing loss can have a sensory as well as a neural origin with about equal probability were reported as well (Ferri et al., 2009; Odabashi et al., 2002).

6.7 MÉNIÈRE'S DISEASE

Ménière's disease is characterized by fullness in the ear, vertigo, tinnitus, and hearing loss. Over time, the number of vertigo episodes decreases and the overall hearing loss increases. Typically, the loss is fluctuating at the onset of the disease, particularly at low frequencies, but tends to be irreversible and nonfluctuating at high frequencies. In later periods, the disease often stabilizes at a moderate to severe level, but generally does not reach profound levels (Eggermont and Schmidt, 1985; Goin et al., 1982).

We illustrate this with an example of a unilateral Ménière's patient from our longitudinal study (Eggermont and Schmidt, 1985). Nine audiograms covering this 10-year period are shown in Fig. 6.8. The first audiogram in a panel (full lines; see dates beside the audiograms) precedes the vertigo attack. Note the fluctuating loss at low frequencies and recovery to near normal (as shown in the first audiogram in the next series) and the progressively increasing hearing loss at high frequencies. Detailed descriptions of the time course of hearing loss in Ménière's disease together with electrocochleographic studies are provided in Chapter 5, Types of Hearing Loss.

6.8 DIABETES

Diabetes mellitus Type 2 (formerly called noninsulin-dependent diabetes mellitus or adult-onset diabetes) is a metabolic disorder that is characterized by hyperglycemia (high blood sugar) in the context of insulin resistance and relative lack of insulin. This is in contrast to diabetes mellitus Type 1, in which there is an absolute lack of insulin due to breakdown of islet cells in the pancreas (wikipedia.org/wiki/Diabetes_mellitus_type_2; accessed May 27, 2015). The epidemiology of hearing loss caused by diabetes is reviewed in Chapter 7, Epidemiology and Genetics of Hearing Loss and Tinnitus.

6.8.1 Hearing Loss in Diabetes

Pessin et al. (2008) found that the most frequent auditory symptoms in Type 1 diabetes were tinnitus and hearing loss. Sensorineural hearing loss was found in 10% of patients predominantly bilateral, symmetric, and affecting the high frequencies, coexisting with normal speech discrimination. These patients had a longer time since diabetes diagnosis and had poor glycemia control. An increase of ABR interpeak latency I–III was observed in approximately 11% of the ears. Gupta et al. (2015) also reported that in Type 2 patients the latency of waves III, V and interpeak latencies III–V, I–V showed a significant delay bilaterally in diabetic males. In a systematic review, Akinpelu et al. (2014) found that Type 2 patients had significantly higher incidence for at least the mild degree of HL when compared with controls. Mean PTA thresholds were greater in diabetics for all frequencies but

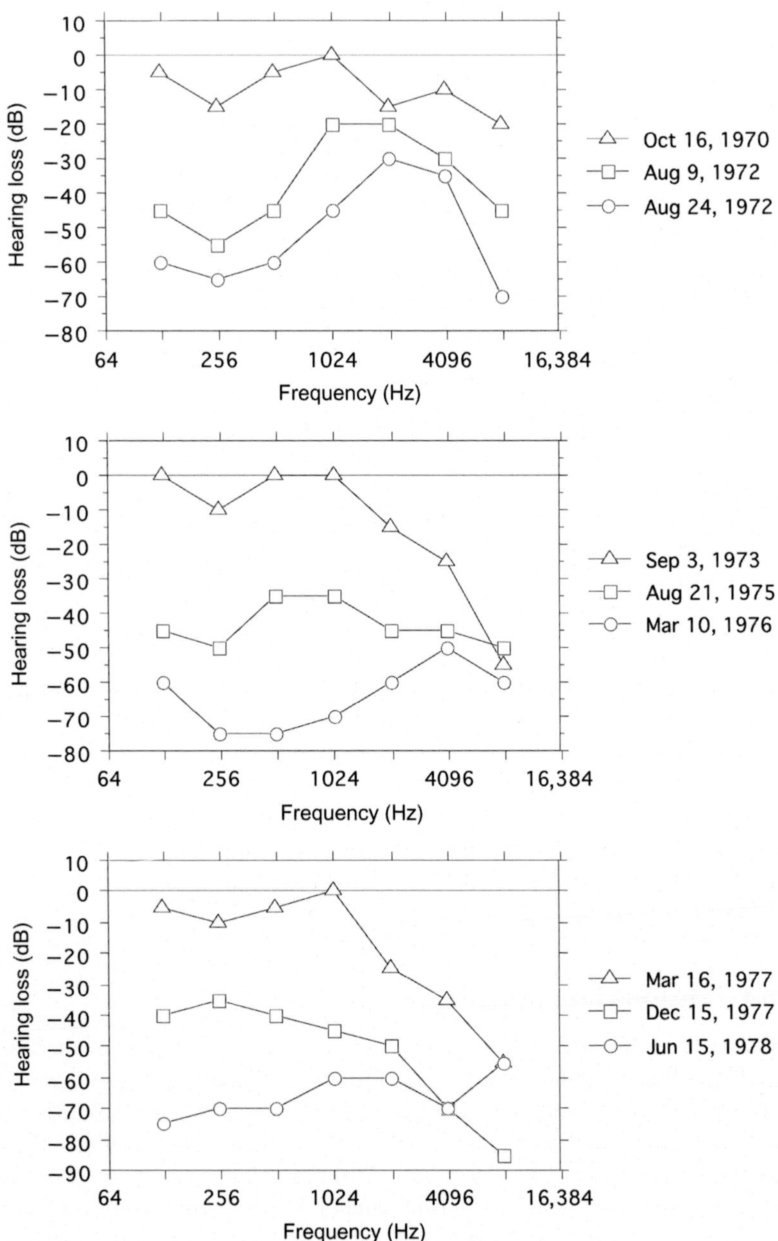

FIGURE 6.8 Selected audiograms for a Ménière's patient collected over an 8-year time span. Each set of three audiograms represents one period containing attacks. Note fluctuation in low frequencies and gradual increase in hearing loss for high frequencies. *Based on data from Eggermont, J.J., Schmidt, P.H., 1985. Ménière's disease: a long term follow-up of hearing loss. Ann. Otol. Rhinol. Laryngol. 94, 1–9.*

were more clinically relevant at 6000 and 8000 Hz. Prolonged ABR wave V latencies in the diabetic group suggest retrocochlear involvement. Botelho et al. (2014) studied 40 adolescents with Type I diabetes and 40 healthy subjects. They found sensorineural hearing loss at 6 and 8 kHz in 7.7% of Type I patients compared to 0% in controls. A higher prevalence of cochlear damage was detected by DPOAE responses, 32% for the Type I group, versus 3.7% in the controls. Absent TEOAE responses were observed in only three individuals (5.1%) from the diabetic group, and none in the controls. Hearing thresholds were significantly better in diabetic subjects with good control when compared to ones with regular or poor control.

6.8.2 Pathology

Hong et al. (2013) found that hearing loss in the context of Type 2 diabetes may result from microangiopathic processes that affect the ears, which are vulnerable to metabolic and circulatory stress, and also impact neurological function. There is increasing evidence to suggest a role for oxidative stress in the pathway leading to acoustic trauma causing hearing loss. Hong and Kang (2014) found that "hyperglycemia associated with Type 1 or Type 2 diabetes causes auditory nerve dysfunction, while hyperinsulinemia associated with Type 2 causes dysfunction to both the central auditory pathways and cochlear hair cells."

6.9 SUMMARY

The dominant causes of acquired hearing loss are occupational and recreational noise exposure and being treated with ototoxic drugs. Professional musicians are at risk, regardless if they play in a rock band or in a classical symphony or chamber orchestra. Passive listening to music either at public venues or using PLDs adds to the effects of occupational noise. Subclinical effects, i.e., with normal audiograms, of these exposures may be an important factor contributing to age-related hearing loss. Animal studies have highlighted the central nervous system changes of noise exposure that are additional to the peripheral induced hearing loss, but also are present in exposures that do not result in detectable peripheral threshold changes. These changes include reorganization of cortical tonotopic maps and increased spontaneous neural activity, and reduced temporal processing ability. Ototoxic drugs, such as AGs and platinum-based chemotherapy drugs, result in hearing loss, especially when combined with diuretics. We also stress that there may be long-term aftereffects on the central nervous system of CHL in infancy. We conclude with the auditory neuropathy mimicking effects of acoustic neuroma and the recently recognized effects of diabetes on the auditory system.

REFERENCES

Akinpelu, O.V., Mujca-Mota, M., Daniel, S.J., 2014. Is type 2 diabetes mellitus associated with alterations in hearing? A systematic review and meta-analysis. Laryngoscope 124, 767–776.

Axelsson, A., Lindgren, F., 1977. Does pop music cause hearing damage? Audiology 16, 432–437.

Axelsson, A., Lindgren, F., 1981. Hearing in classical musicians. Acta Otolaryngol. Suppl. 377, 3–74.

Axelsson, A., Eliasson, A., Israelsson, B., 1995. Hearing in pop/rock musicians: a follow up study. Ear Hear. 16, 245–253.

Bao, J., Ohlemiller, K.K., 2010. Age-related loss of spiral ganglion neurons. Hear. Res. 264, 93–97.

Biassoni, E.C., Serra, M.R., Hinalaf, M., Abraham, M., Pavlik, M., Villalobo, J.P., et al., 2014. Hearing and loud music exposure in a group of adolescents at the ages of 14–15 and retested at 17–18. Noise Health 16, 331–341.

Botelho, C.T., da Silva Carvalh, S., Silva, I.N., 2014. Increased prevalence of early cochlear damage in young patients with type 1 diabetes detected by distortion product otoacoustic emissions. Int. J. Audiol. 53, 402–408.

Clark, W.W., Bohne, B.A., 1999. Effects of noise on hearing. JAMA. 281, 1658–1659.

Cohen, B.E., Durstenfeld, A., Roehm, P.C., 2014. Viral causes of hearing loss: a review for hearing health professionals. Trends Hear. 18, 1–17.

Coupland, S.G., Ponton, C.W., Eggermont, J.J., Bowen, T., Grant, R.M., 1991. Assessment of cisplatin-induced ototoxicity using derived-band ABRs. Int. J. Pediatr. Otorhinolaryngol. 22, 237–248.

Deng, L., Ding, D., Su, J., Manohar, S., Salvi, R., 2013. Salicylate selectively kills cochlear spiral ganglion neurons by paradoxically up-regulating superoxide. Neurotox. Res. 24, 307–319.

Ding, D., McFadden, S.L., Browne, R.W., Salvi, R.J., 2003. Late dosing with ethacrynic acid can reduce gentamicin concentration in perilymph and protect cochlear hair cells. Hear. Res. 185, 90–96.

Ding, D., Allman, B.L., Salvi, R., 2012. Ototoxic characteristics of platinum antitumor drugs. Anat. Rec. (Hoboken) 295, 1851–1867.

Eggermont, J.J., 2012. The Neuroscience of Tinnitus. Oxford University Press, Oxford, UK.

Eggermont, J.J., 2014. Noise and the Brain. Experience Dependent Developmental and Adult Plasticity. Academic Press, London.

Eggermont, J.J., Komiya, H., 2000. Moderate noise trauma in juvenile cats results in profound cortical topographic map changes in adulthood. Hear. Res. 142, 89–101.

Eggermont, J.J., Schmidt, P.H., 1985. Ménière's disease: a long term follow-up of hearing loss. Ann. Otol. Rhinol. Laryngol. 94, 1–9.

Evans, E.F., Klinke, R., 1982. The effects of intracochlear and systemic furosemide on the properties of single cochlear nerve fibres in the cat. J. Physiol. 331, 409–427.

Ferri, G.G., Modugno, G.C., Calbucci, F., Ceroni, A.R., Antonio Pirodda, A., 2009. Hearing loss in vestibular schwannomas: analysis of cochlear function by means of distortion-product otoacoustic emissions. Auris Nasus Larynx 36, 644–648.

Fowler, K.B., McCollister, F.P., Dahle, A.J., Boppana, S., Britt, W.J., Pass, R.F., 1997. Progressive and fluctuating sensorineural hearing loss in children with asymptomatic congenital cytomegalovirus infection. J. Pediatr. 130, 624–630.

Fowler, K.B., Dahle, A.J., Boppana, S., Pass, R.F., 1999. Newborn hearing screening: will children with hearing loss caused by congenital cytomegalovirus infection be missed? J. Pediatr. 135, 60–64.

Gates, G.A., Mills, J.H., 2005. Presbycusis. Lancet 366, 1111–1120.

Gates, G.A., Schmid, P., Kujawa, S.G., Nam, B., D'Agostino, R., 2000. Longitudinal threshold changes in older men with audiometric notches. Hear. Res. 141, 220–228.

Goin, D.W., Staller, S.J., Asher, D.L., Mischke, R.E., 1982. Summating potential in Meniere's disease. Laryngoscope 92, 1383–1389.

Gouveris, H.T., Victor, A., Mann, W.J., 2007. Cochlear origin of early hearing loss in vestibular schwannoma. Laryngoscope 117, 680–683.

Greeson, J.N., Raphael, R.M., 2009. Amphipath-induced nanoscale changes in outer hair cell plasma membrane curvature. Biophys. J. 96, 510–520.

Guitton, M.J., Caston, J., Ruel, J., Johnson, R.M., Pujol, R., Puel, J.L., 2003. Salicylate induces tinnitus through activation of cochlear NMDA receptors. J. Neurosci. 23, 3944–3952.

Gupta, S., Bajewa, P., Mittal, S., Kumar, A., Singh, K.D., Sharma, R., 2015. Brainstem auditory evoked potential abnormalities in type 2 diabetes mellitus. North Am. J. Med. Sci. 5, 60–65.

Guthrie, O.W., 2008. Aminoglycoside induced ototoxicity. Toxicology 249, 91–96.

Halevi-Katz, D.N., Yaakobi, E., Putter-Katz, H., 2015. Exposure to music and noise-induced hearing loss (NIHL) among professional pop/rock/jazz musicians. Noise Health 17, 158–164.

Han, W., Shi, X., Nuttall, A.L., 2006. AIF and endoG translocation in noise exposure induced hair cell death. Hear. Res. 211, 85–95.

Harrison, R.V., Negandhi, J., 2012. Resting neural activity patterns in auditory brainstem and midbrain in conductive hearing loss. Acta Otolaryngol. 132, 409–414.

Hequembourg, S., Liberman, M.C., 2001. Spiral ligament pathology: a major aspect of age-related cochlear degeneration in C57BL/6 mice. J. Assoc. Res. Otolaryngol. 2, 118–129.

Hoffman, H.J., Dobie, R.A., Ko, C.-W., Themann, C.L., Murphy, W.J., 2010. Americans hear as well or better today compared with 40 years ago: hearing threshold levels in the unscreened adult population of the United States, 1959–1962 and 1999–2004. Ear. Hear. 31, 725–734.

Hogan, S.C., Stratford, K.I., Moore, D.R., 1997. Duration and recurrence of otitis media with effusion in children from birth to 3 years: prospective study using monthly otoscopy and tympanometry. Br. Med. J. 314, 350–353.

Hong, B.N., Kang, T.H., 2014. Distinction between auditory electrophysiological responses in type 1 and type 2 diabetic animal models. Neurosci. Lett. 566, 309–314.

Hong, O., Buss, J., Thomas, E., 2013. Type 2 diabetes and hearing loss. Dis. Mon. 59, 139–146.

Hu, B.H., Guo, W., Wang, P.Y., Henderson, D., Jiang, S.C., 2000. Intense noise-induced apoptosis in hair cells of guinea pig cochleae. Acta Otolaryngol. 120, 19–24.

Irvine, D.R., Rajan, R., Smith, S., 2003. Effects of restricted cochlear lesions in adult cats on the frequency organization of the inferior colliculus. J. Comp. Neurol. 467, 354–374.

Ivory, R., Kane, R., Diaz, R.C., 2014. Noise-induced hearing loss: a recreational noise perspective. Curr. Opin. Otolaryngol. Head Neck Surg. 22, 394–398.

Jansen, E.J.M., Helleman, H.W., Dreschler, W.A., de Laat, J.A., 2009. Noise induced hearing loss and other hearing complaints among musicians of symphony orchestras. Int. Arch. Occup. Environ. Health 82, 153–164.

Jit, M., 2010. The risk of sequelae due to pneumococcal meningitis in high-income countries: a systematic review and meta-analysis. J. Infect. 61, 114–124.

Kähäri, K., Zachau, G., Eklöf, M., Sandsjö, L., Möller, C., 2003. Assessment of hearing and hearing disorders in rock/jazz musicians. Int. J. Audiol. 42, 279–288.

Kaltenbach, J.A., Czaja, J.M., Kaplan, C.R., 1992. Changes in the tonotopic map of the dorsal cochlear nucleus following induction of cochlear lesions by exposure to intense sound. Hear. Res. 59, 213–223.

Kamke, M.R., Brown, M., Irvine, D.R., 2003. Plasticity in the tonotopic organization of the medial geniculate body in adult cats following restricted unilateral cochlear lesions. J. Comp. Neurol. 459, 355–367.

Kryter, K.D., 1985. The Effects of Noise on Man, 2nd ed. Academic Press, Orlando, FL.

Kujawa, S.G., Liberman, M.C., 2006. Acceleration of age-related hearing loss by early noise exposure: evidence of a misspent youth. J. Neurosci. 26, 2115–2123.

Kutz, J.W., Simon, L.M., Chennupati, S.K., Giannoni, C.M., Manolidis, S., 2006. Clinical predictors for hearing loss in children with bacterial meningitis. Arch. Otolaryngol. Head Neck Surg. 132, 941–945.

Langer, T., am Zehnhoff-Dinnesen, A., Radtke, S., Meitert, J., Zolk, O., 2013. Understanding platinum-induced ototoxicity. Trends Pharmacol. Sci. 34, 458–469.

Le Prell, C.G., Dell, S., Hensley, B., et al., 2012. Digital music exposure reliably induces temporary threshold shift in normal-hearing human subjects. Ear Hear. 33, e44–e58.

Lee, F.-S., Matthews, L.J., Dubno, J.R., Mills, J.H., 2005. Longitudinal study of pure-tone thresholds in older persons. Ear Hear. 26, 1–11.

Lee, S.H., Choi, S.K., Lim, Y.J., Chung, H.Y., Yeo, J.H., Na, S.Y., et al., 2015. Otologic manifestations of acoustic neuroma. Acta Otolaryngol. 2015, 140–146.

Liberman, M.C., Kiang, N.Y., 1978. Acoustic trauma in cats. Cochlear pathology and auditory-nerve activity. Acta Otolaryngol. Suppl. 358, 1–63.

Madden, C., Wiley, S., Schleiss, M., Benton, C., Meinzen-Derr, J., Greinwald, J., et al., 2005. Audiometric, clinical and educational outcomes in a pediatric symptomatic congenital cytomegalovirus (CMV) population with sensorineural hearing loss. Int. J. Pediatr. Otorhinolaryngol. 69, 1191–1198.

Maddox, P.T., Saunders, J., Chandrasekhar, S.S., 2009. Sudden hearing loss from PDE-5 inhibitors: a possible cellular stress etiology. Laryngoscope 119, 1586–1589.

McBride, D., Gill, F., Proops, D., Harrington, M., Gardiner, K., Attwell, C., 1992. Noise and the classical musician. BMJ 305, 1561–1563.

Meyer-Bisch, C., 1996. Epidemiological evaluation of hearing damage related to strongly amplified music (personal cassette players, discotheques, rock concerts)—high-definition audiometric survey on 1364 subjects. Audiology 35, 121–142.

Mills, J.H., Adkins, W.Y., Gilbert, R.M., 1981. Temporary threshold shifts produced by wideband noise. J. Acoust. Soc. Am. 70, 390–396.

Nixon, C.W., Johnson, D.L., Stephenson, M.R., 1977. Asymptotic behavior of temporary threshold shift and recovery from 24- and 48-hour exposures. Aviat. Space Environ. Med. 48, 311–315.

Noreña, A.J., Eggermont, J.J., 2005. Enriched acoustic environment after noise trauma reduces hearing loss and prevents cortical map reorganization. J. Neurosci. 25, 699–705.

Noreña, A.J., Tomita, M., Eggermont, J.J., 2003. Neural changes in cat auditory cortex after a transient pure-tone trauma. J. Neurophysiol. 90, 2387–2401.

Noreña, A.J., Gourévitch, B., Aizawa, N., Eggermont, J.J., 2006. Enriched acoustic environment disrupts frequency representation in cat auditory cortex. Nat. Neurosci. 9, 932–939.

Nouvian, R., Eybalin, M., Puel, J.L., 2015. Cochlear efferents in developing adult and pathological conditions. Cell Tissue Res. 361, 301–309.

O'Brien, I., Driscoll, T., Ackermann, B., 2013. Sound exposure of professional orchestral musicians during solitary practice. J. Acoust. Soc. Am. 134, 2748–2754.

Odabasi, A.O., Telischi, F.F., Gomez-Marin, O., Stagner, B., Martin, G., 2002. Effect of acoustic tumor extension into the internal auditory canal on distortion-product otoacoustic emissions. Ann. Otol. Rhinol. Laryngol. 111, 912–915.

Overdevest, J.B., Pross, S.E., Cheung, S.W., 2016. Tinnitus following treatment for sporadic acoustic neuroma. Laryngoscope 126 (7), 1639–1643.

Pennings, R.J.E., Morris, D.P., Clarke, L., Allen, S., Walling, S., Bance, M.L., 2011. Natural history of hearing deterioration in intracanalicular vestibular schwannoma. Neurosurgery 68, 68–77.

Pessin, A.B.B., Martins, R.H., de Paula Pimenta, W., Simöes, A.C.P., Marsiglia, A., Amaral, A. V., 2008. Auditory evaluation in patients with type 1 diabetes. Ann. Otol. Rhinol. Laryngol. 117, 366–370.

Potier, M., Hoquet, C., Lloyd, R., Nicolas-Puel, C., Uziel, A., Puel, J.-L., 2009. The risks of amplified music for disc-jockeys working in nightclubs. Ear Hear. 30, 291–293.

Pujol, R., Puel, J.L., 1999. Excitotoxicity, synaptic repair, and functional recovery in the mammalian cochlea: a review of recent findings. Ann. N. Y. Acad. Sci. 884, 249–254.

Rajan, R., Irvine, D.R., 1998. Absence of plasticity of the frequency map in dorsal cochlear nucleus of adult cats after unilateral partial cochlear lesions. J. Comp. Neurol. 399, 35–46.

Rajan, R., Irvine, D.R., Wise, L.Z., Heil, P., 1993. Effect of unilateral partial cochlear lesions in adult cats on the representation of lesioned and unlesioned cochleas in primary auditory cortex. J. Comp. Neurol. 338, 17–49.

Rarey, K.E., 1990. Otologic pathophysiology in patients with human immunodeficiency virus. Am. J. Otolaryngol. 11, 366–369.

Richardson, M.P., Reid, A., Tarlow, M.J., Rudd, P.T., 1997. Hearing loss during bacterial meningitis. Arch. Dis. Child. 76, 134–138.

Robertson, D., Irvine, D.R., 1989. Plasticity of frequency organization in auditory cortex of guinea pigs with partial unilateral deafness. J. Comp. Neurol. 282, 456–471.

Roosli, C., Linthicum Jr., F.H., Cureoglu, S., Merchant, S.M., 2012. Dysfunction of the cochlea contributing to hearing loss in acoustic neuromas: an underappreciated entity. Otol. Neurotol. 33, 473–480.

Rosen, M.J., Sarro, E.C., Kelly, J.B., Sanes, D.H., 2012. Diminished behavioral and neural sensitivity to sound modulation is associated with moderate developmental hearing loss. PLoS One 7, e41514.

Royster, J.D., Royster, L.H., Killion, M.C., 1991. Sound exposure and hearing thresholds of symphony orchestra musicians. J. Acoust. Soc. Am. 89, 2793–2803.

Ruggero, M.A., Rich, N.C., 1991. Furosemide alters organ of Corti mechanics: evidence for feedback of outer hair cells upon the basilar membrane. J. Neurosci. 11, 1057–1067.

Rybak, L.P., 1985. Furosemide ototoxicity: clinical and experimental aspects. Laryngoscope 95 (Suppl. 38), 1–14.

Rybak, L.P., Ramkumar, V., 2007. Ototoxicity. Kidney Int. 72, 931–935.

Samelli, A.G., Matas, C.G., Carvallo, R.M.M., Gomes, R.F., de Beija, C.S., Magliaro, F.C.L., et al., 2012. Audiological and electrophysiological assessment of professional pop/rock musicians. Noise Health 14, 6–12.

Saunders, J.C., Dear, S.P., Schneider, M.E., 1985. The anatomical consequences of acoustic injury: a review and tutorial. J. Acoust. Soc. Am. 78, 833–860.

Schacht, J., Talaska, A.E., Rybak, L.P., 2012. Cisplatin and aminoglycoside antibiotics: hearing loss and its prevention. Anat. Rec. (Hoboken) 295, 1837–1850.

Schink, T., Kreutz, G., Busch, V., Pigeot, I., Ahrens, W., 2014. Incidence and relative risk of hearing disorders in professional musicians. Occup. Environ. Med. 71, 472–476.

Schmidt, J.H., Pedersen, E.R., Paarup, H.M., Christense-Dalgaard, J., Andersen, T., Poulsen, T., et al., 2014. Hearing loss in relation to sound exposure of professional symphony orchestra musicians. Ear Hear. 35, 448–460.

Schmuziger, N., Patscheke, J., Probst, R., 2006. Hearing in nonprofessional pop/rock musicians. Ear Hear. 27, 321–330.

Serra, M.R., Biassoni, E.C., Richter, U., Minoldo, G., Franco, G., Abraham, S., et al., 2005. Recreational noise exposure and its effects on the hearing of adolescents. Part I: an interdisciplinary long-term study. Int. J. Audiol. 44, 65–73.

Serra, M.R., Biassoni, E.C., Hinalaf, M., Abraham, M., Pavlik, M., Villalobo, J.P., et al., 2014. Hearing and loud music exposure in 14–15 years old adolescents. Noise Health 16, 320–330.

Sheppard, A., Hayes, S.H., Chen, G.-D., Ralli, M., Salvi, R., 2014. Review of salicylate-induced hearing loss, neurotoxicity, tinnitus and neuropathophysiology. Acta Otorhinolaryngol. Ital. 34, 79–93.

Slepecky, N., 1986. Overview of mechanical damage to the inner ear: noise as a tool to probe cochlear function. Hear. Res. 22, 307–321.

Smith, R.J., Bale Jr., J.F., White, K.R., 2005. Sensorineural hearing loss in children. Lancet 365, 879–890.

Stenklev, N.C., Laukli, E., 2004. Presbyacusis–hearing thresholds and the ISO 7029. Int. J. Audiol. 43, 295–306.

Sulaiman, A.H., Husain, R., Seluakumaran, K., 2014. Evaluation of early hearing damage in personal listening device users using extended high-frequency audiometry and otoacoustic emissions. Eur. Arch. Otorhinolaryngol. 271, 1463–1470.

Syka, J., 2002. Plastic changes in the central auditory system after hearing loss, restoration of function, and during learning. Physiol. Rev. 82, 601–636.

Toppila, E., Koskinen, H., Pyykkö, I., 2011. Hearing loss among classical-orchestra musicians. Noise Health 13, 45–50.

Wang, Y., Ren, C., 2012. Effects of repeated "benign" noise exposures in young CBA mice: shedding light on age-related hearing loss. J. Assoc. Res. Otolaryngol. 13, 505–515.

Warchol, M.E., 2010. Cellular mechanisms of aminoglycoside ototoxicity. Curr. Opin. Otolaryngol. Head Neck Surg. 18, 454–458.

Ward, W.D., Cushing, E.M., Burns, E.M., 1976. Effective quiet and moderate TTS: implications for noise exposure standards. J. Acoust. Soc. Am. 59, 160–165.

Westerberg, B.D., Atashband, S., Kozak, F.K., 2008. A systematic review of the incidence of sensorineural hearing loss in neonates exposed to Herpes simplex virus (HSV). Int. J. Pediatr. Otorhinolaryngol. 72, 931–937.

Westmore, G.A., Eversden, I.D., 1981. Noise-induced hearing loss and orchestral musicians. Arch. Otolaryngol. 107, 761–764.

Whitton, J.P., Polley, D.B., 2011. Evaluating the perceptual and pathophysiological consequences of auditory deprivation in early postnatal life: a comparison of basic and clinical studies. J. Assoc. Res. Otolaryngol. 12, 535–547.

Willott, J.F., 1991. Aging and the Auditory System: Anatomy, Physiology, and Psychophysics. Singular Publishing Group, San Diego, CA.

Wilmington, D., Gray, L., Jahrsdoerfer, R., 1994. Binaural processing after corrected congenital unilateral conductive hearing loss. Hear. Res. 74, 99–114.

Wu, T., Lv, P., Kim, H.J., Yamoah, E.N., Nuttall, A.L., 2010. Effect of salicylate on KCNQ4 of the guinea pig outer hair cell. J. Neurophysiol. 103, 1969–1977.

Xu, H., Kotak, V.C., Sanes, D.H., 2007. Conductive hearing loss disrupts synaptic and spike adaptation in developing auditory cortex. J. Neurosci. 27, 9417–9426.

Chapter 7

Epidemiology and Genetics of Hearing Loss and Tinnitus

Hearing loss occurs in children and in the elderly, in war veterans and factory workers, and in classical musicians and disc jockeys. Epidemiology is the study of factors such as hearing loss that affect the health and illness of populations and serves as the foundation for interventions made in public health and preventive medicine. Two quantifiers are frequently used; *prevalence* is defined as the percentage of people with hearing loss in a certain age group (usually with a range of 10 years), and *incidence* is the percentage of people who did not have hearing loss at the onset of the study period but acquired it over a (typically) 5- to 10-year span. Both odds ratios (ORs) and hazard ratios (HRs) are used often interchangeably. Here are the definitions from a statistics Web site: "An odds ratio (OR) is a measure of association between an exposure and an outcome. The OR represents the odds that an outcome will occur given a particular exposure, compared to the odds of the outcome occurring in the absence of that exposure." Furthermore, "in survival analysis, the hazard ratio (HR) is the ratio of the hazard rates corresponding to the conditions described by two levels of an explanatory variable. For example, in a drug study, the treated population may die at twice the rate per unit time as the control population." A note on OR versus HR: "In logistic regression, an odds ratio of 2 means that the event is 2 time more probable given a one-unit increase in the predictor. In Cox regression, a hazard ratio of 2 means the event will occur twice as often at each time point given a one-unit increase in the predictor. They're almost the same thing when doubling the odds of the event is almost the same as doubling the hazard of the event. They're not automatically similar, but under some (fairly common) circumstances they may correspond very closely." Typically 95% confidence intervals (CIs) are used; when the CI does not bracket the "1" ratio, the findings are significant ($p < 0.05$). When the CI refers to another level than 95%, I will indicate that (http://stats.stackexchange.com; accessed May 20, 2015).

This chapter adds general surveys on the prevalence and incidence of hearing loss and tinnitus to complement the causes described in Chapter 6,

Hearing Loss. DOI: http://dx.doi.org/10.1016/B978-0-12-805398-0.00007-4

Causes of Acquired Hearing Loss. In addition it presents an introduction to the genetic causes of hearing loss. Some of the earlier data that I use here are also reviewed in my book "Noise and the Brain" (Eggermont, 2014).

7.1 EPIDEMIOLOGY OF SENSORINEURAL HEARING LOSS

Worldwide, 16% of disabling hearing loss in adults is attributed to occupational noise exposure, ranging from 7% to 21% in various regions of the world (Nelson et al., 2005). At the time of that study, in the United States an estimated 9.4 million workers were exposed at levels more than 80 dB(A), of which 3.4 million were exposed to levels more than 90 dB(A). In a UK sample ($N = 21{,}201$; age ≤ 65 years), about 2% of subjects reported severe hearing difficulties, i.e., wearing a hearing aid or having great difficulty hearing a conversation in a quiet room. In men, the prevalence of this outcome rose steeply with age, from below 1% in those aged 16–24 years to 8% in those aged 55–64. In the United Kingdom overall some 153,000 men and 26,000 women aged 35–64 years were estimated to have severe hearing difficulties attributable to noise exposure at work (Palmer et al., 2002). Hasson et al. (2010) analyzed the questionnaires answered by 9756 working people and 1685 nonworkers in a Swedish population. They found that 31% in the working population and 36% in the nonworking population reported either hearing loss or tinnitus or both. Nonoccupational (e.g., recreational) noise exposure was not taken into account, but the close numbers suggest that it likely accounts for an important part of the hearing loss (see chapter: Causes of Acquired Hearing Loss).

The prevalence of audiometric hearing loss among all individuals (age ≥ 12 years) in the United States was estimated using an extrapolation from a nationally representative data set (Lin et al., 2011). Pure-tone thresholds from people ($N = 7490$) aged from 12 years to well over 70 years from the 2001 through 2008 cycles of the National Health and Nutritional Examination Surveys (NHANES) were analyzed. A pure-tone average (PTA) of the hearing thresholds at 0.5, 1, 2, and 4 kHz of at least 25 dB HL (hearing level) in both ears as recommended by the World Health Organization was taken as an indication of hearing loss. From this sample, the authors extrapolated that 30.0 million (12.7%) of Americans 12 years and older had bilateral hearing loss from 2001 through 2008, and this estimate increased to 48.1 million (20.3%) when individuals with unilateral hearing loss were included. Overall, the prevalence of hearing loss increased with every age decade (Fig. 7.1). The prevalence of hearing loss was lower in women than in men and in African-American versus Caucasian individuals across nearly all age decades.

More than 3 million people in Germany with employment subject to social insurance contributions to three health-insurance providers were included in the cohort studied by Schink et al. (2014). During the study period (January 1,

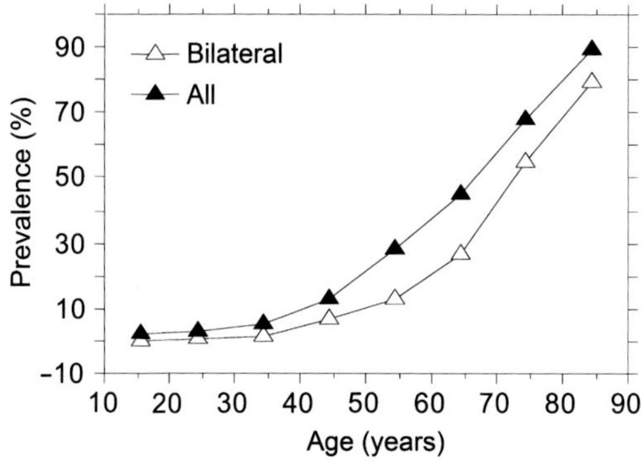

FIGURE 7.1 Prevalence of bilateral and all (unilateral + bilateral) hearing loss in the United States as a function of age (± 5 years). *Data from Lin, F.R., Niparko, J.K., Ferrucci, L., 2011. Hearing loss prevalence in the United States. Arch. Intern. Med. 171, 1851–1852.*

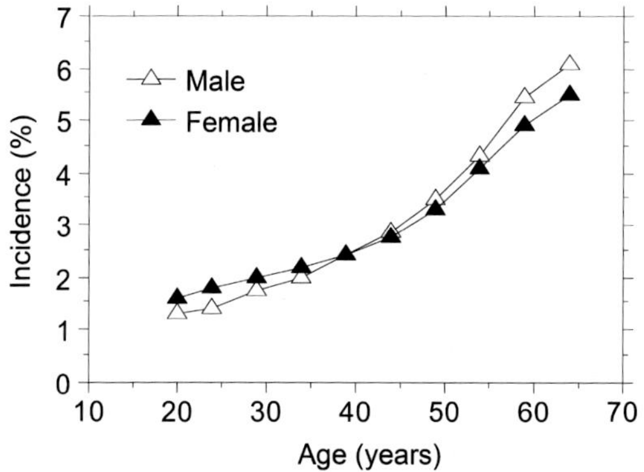

FIGURE 7.2 Incidence rate of hearing loss by sex and age group (± 2 years). *Data from Schink, T., Kreutz, G., Busch, V., Pigeot, I., Ahrens, W., 2014. Incidence and relative risk of hearing disorders in professional musicians. Occup. Environ. Med. 71, 472–476.*

2004–December 31, 2008), 283,697 cases of hearing loss were newly diagnosed. As expected, these incidence rates increased with age. Up to the age of 36 years the incidence of hearing loss in 4-year age groups was higher in women, but above that age it was higher for men (Fig. 7.2).

In the period from 1996 to 1998, Borchgrevink et al. (2005) collected audiometric data from 50,723 of 82,141 unscreened invited subjects in Nord-Trondelag, Norway (age range 20–101 years, mean = 50.2 years). The PTA of hearing thresholds at 0.5, 1, 2, and 4 kHz showed hearing impairment more than 25 dB in the worst ear in 32% of males and 23% of females. The overall prevalence of this hearing impairment was 18.8% for the better ear and 27.2% for the worse ear, respectively. In the same subjects, Tambs et al. (2006) compared the effects of occupational noise and firearms noise.

Reported noise exposure levels and observed threshold shifts were moderate among women. Threshold shifts averaged over both ears among subjects in the group of the highest 2% of exposure levels were 13 dB for 65-year-old men and generally largest at 3–4 kHz. The shifts induced by impulse noise were approximately 8 dB and strongest at 3–8 kHz among men aged 45–65 years. Comparable results for firearms noise were obtained from 3753 participants aged 48–92 years in the Beaver Dam study in Wisconsin (Nondahl et al., 2009). After age and other factors were adjusted, men ($N = 1538$) who had regularly participated in target shooting (OR = 1.57; CI = 1.12–2.19) or who had done so in the past year (OR = 2.00; CI = 1.15–3.46) were more likely to have a marked high-frequency hearing loss than those who had not. The risk of having a marked high-frequency hearing loss increased 7% for every 5 years the men had hunted (OR = 1.07; CI = 1.03–1.12). Thirty-eight percent of the target shooters and 95% of the hunters reported they did not use hearing protection.

For a representative sample of 705 subjects from a rural Danish population aged 31–50 years, Karlsmose et al. (2000) reported changes in hearing sensitivity over 5 years. Hearing deterioration was defined as an average at least 10 dB/5 years at 3–4 kHz in at least one ear and was present in 23.5% of the sample. The 41- to 50-year-olds had an OR = 1.32 (CI = 1.01–1.73) compared with the 31- to 40-year-olds. Males had an OR = 1.35 (CI = 1.03–1.76) compared with females. These example data suggest that a large percentage of the adult population has noise-induced hearing loss (NIHL), regardless being from a rural or urban environment.

7.2 EPIDEMIOLOGY OF AGE-RELATED HEARING LOSS

Cruickshanks et al. (2010) determined the 10-year cumulative incidence of hearing impairment and its associations with education, occupation, and noise exposure history in a population-based cohort study of 3753 adults who were 48–92 years of age at the baseline examinations during 1993–95 in Beaver Dam, WI. Hearing thresholds were measured at baseline and at 2.5-year, 5-year, and 10-year follow-up examinations. Hearing impairment was defined as a PTA more than 25 dB HL at 0.5, 1, 2, and 4 kHz. Demographic characteristics and occupational histories were obtained by questionnaire. The 10-year cumulative incidence of hearing impairment was 37.2%. Age (5 year; HR = 1.81), sex (men vs women; HR = 2.29), occupation based on longest held job (production/operations/farming vs others; HR = 1.34), marital status (unmarried vs married; HR = 1.29), and education (<16 vs 16+ years; HR = 1.40) were associated with the 10-year incidence, whereas a history of occupational noise was not. In this largely retired population, occupational noise exposure may have contributed to hearing impairments present at the baseline examination (Cruickshanks et al., 1998), but there was no evidence of any residual effect on long-term risk of declining hearing sensitivity among people with normal hearing at the baseline examination. Even among those exposed to occupational noise at the baseline examination, there was no evidence of an effect. These results are consistent with the study by Lee et al. (2005), which measured hearing repeatedly, and reported no difference in the rate of change between people with and without positive noise histories. Cruickshanks et al. (2010)'s study suggests that, on a population basis, there is little evidence that prior occupational noise exposure plays an important role in the onset or progression of hearing impairment in older adults followed for 10 years. Note that these audiometric data do not reflect "hidden hearing loss" resulting from damage to inner hair cell ribbon synapses or loss of high-threshold auditory nerve fibers (see chapters: Hearing Problems and Types of Hearing Loss).

7.3 EPIDEMIOLOGY OF TINNITUS

Tinnitus prevalence in the general population was extracted from three review papers, the original publications contributing to those overviews, and from more recent papers. One review provided an in-depth reanalysis of a few large epidemiology studies (Hoffman and Reed, 2004). The second study also covered some older epidemiology where different criteria for inclusion of tinnitus were used (Davis and El-Rafaie, 2000). The third study presented a more general (but without a prevalence by age group) overview of a larger number of epidemiology studies (Sanchez, 2004). All in all they covered 14 papers that illustrate an upward trend of tinnitus prevalence with age that is generally the same for all studies but where the absolute levels depend on the questions asked and the type of tinnitus included. This was previously reviewed more extensively in my book "The Neuroscience of Tinnitus" (Eggermont, 2012). The prevalence of significant tinnitus across the adult lifespan is illustrated in Fig. 7.3. Significant tinnitus has to be longer than 5 min in duration and not immediately (and transiently) following exposure to loud noise (Davis and El-Rafaie, 2000). Sometimes, more stringent definitions, such as that tinnitus has to be bothersome, are used. This typically lowers the prevalence a few percentage points. This bifurcation can be seen in Fig. 7.3. As these prevalence studies across the lifespan show, tinnitus is about twice as frequent in the elderly as in young adults.

Hearing loss, resulting from exposure to loud noise, is considered an important risk factor for developing tinnitus. Consequently, a history of recreational, occupational, or firearm noise exposure may be associated with increased likelihood of acquiring tinnitus. The relation between noise exposure and significant tinnitus, however, differs depending on the presence or absence of hearing impairment. It should be noted that clinical hearing impairment generally requires a loss of at least 25 dB over the audiometric frequencies (0.25–8 kHz). Especially, hearing loss for frequencies more than 8 kHz typically is accompanied by tinnitus (Langers et al., 2012;

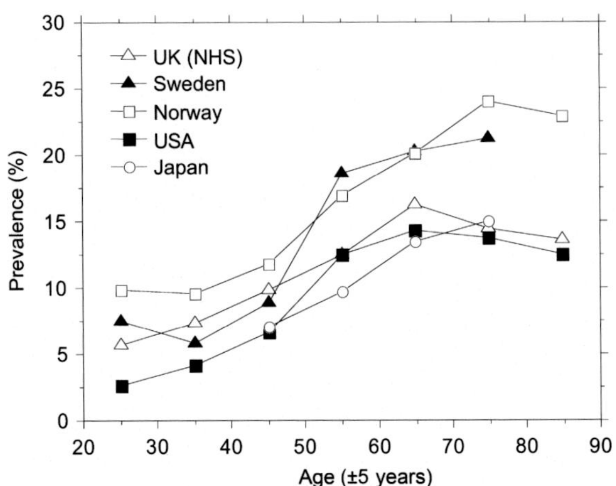

FIGURE 7.3 Prevalence of significant tinnitus for adults. The UK and Swedish data reflect significant tinnitus, whereas the other studies reflect bothersome tinnitus. The UK(NHS) data are from Davis and El-Rafaie (2000), the US data are from Shargorodsky et al. (2010), the Swedish data are from Axelsson and Ringdahl (1989), the Norway study was by Tambs et al. (2003), and the Japanese data are from Fujii et al. (2011). *Modified from Eggermont, J.J., 2012. The Neuroscience of Tinnitus. Oxford University Press, Oxford, UK.*

Melcher et al., 2013). Occupational noise exposure was more likely to correlate with significant tinnitus in participants with hearing impairment, while recreational noise exposure was more associated with increased occurrence of significant tinnitus in participants without (clinical) hearing impairment (Shargorodsky et al., 2010). Engdahl et al. (2012) confirmed that occupation had a marked effect on tinnitus prevalence. In men, age-adjusted prevalence ORs of tinnitus (in relation to a reference population of teachers) ranged from 1.5 (workshop mechanics) to 2.1 (crane and hoist operators) in the 10 occupations with the highest tinnitus prevalence. In women, the most important contribution to the tinnitus prevalence was from the large group of occupationally inactive persons, with a prevalence OR of 1.5.

Using data from the Epidemiology of Hearing Loss Study (1993–95, 1998–2000, 2003–05, and 2009–10) and the Beaver Dam Offspring Study (2005–08) in the United States, Nondahl et al. (2012) examined birth cohort patterns in the report of tinnitus for adults aged 45 years and older ($N = 12{,}689$ observations from 5764 participants). They found that tinnitus prevalence tended to increase in more recent birth cohorts compared to earlier birth cohorts. On average, participants in a given generation were significantly more likely to report tinnitus than participants from a generation 20 years earlier (OR = 1.78, CI = 1.44–2.21). This also may underlie the leveling off of tinnitus prevalence in Fig. 7.3 for the age group above 65 years and may thus refer back to the much lower prevalence in cohorts born before 1950.

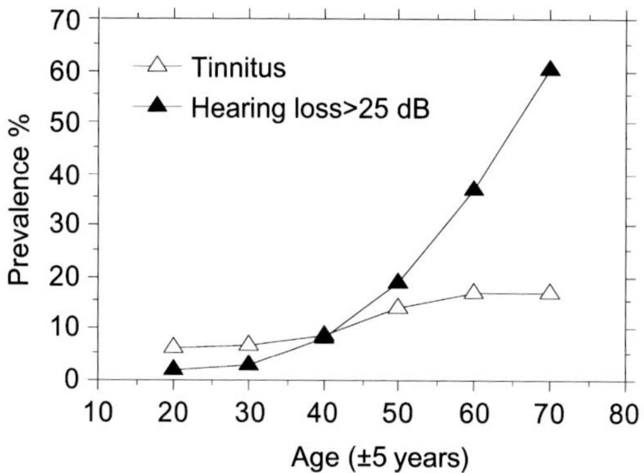

FIGURE 7.4 Mean prevalence of significant tinnitus and significant hearing loss. *Redrawn from Eggermont, J.J., 2014. Noise and the Brain. Experience Dependent Developmental and Adult Plasticity, Academic Press, London, UK.*

The average prevalence of significant tinnitus by age group in the two Scandinavian countries, the United States and the United Kingdom (from Fig. 7.3), is shown in Fig. 7.4. The most recent study covered 14,178 participants in the 1999–2004 National Health and Nutrition Examination Surveys (Shargorodsky et al., 2010). The overall prevalences of the tinnitus in the sample groups were: the United Kingdom 10.1%, Sweden 14.2%, the United States 8.4%, and Norway 15.1%. One observes a tendency for the prevalence of tinnitus to level off in the seventh decade of life. In contrast, the prevalence for significant hearing loss (>25 dB HL, from 0.5–4 kHz) continues to increase (Leensen et al., 2011). This suggests that the prevalence of clinical hearing loss in the standard audiometric frequency range is not related to tinnitus prevalence.

A comprehensive new study (Martinez et al., 2015) described the incidence rate of clinical significant tinnitus in the United Kingdom. They identified 14,303 incident cases of significant tinnitus among 26.5 million person-years of observation. They found an incidence rate of 5.4 (CI = 5.3–5.5) cases per 10,000 person-years. The incident rate peaked in the 60–69 years age group (Fig. 7.5) and was the same for males and females. The incident rate increased with approximately 0.21 per year over the time span 2002–11.

7.4 EPIDEMIOLOGY OF SMOKING AND ALCOHOL CONSUMPTION

Taken on their own, smoking and alcohol consumption have opposite effects on hearing loss. Nicotine affects the cochlea (Maffei and Miani, 1962; Abdel-Hafez et al., 2014) through its effects on antioxidant mechanisms and/or on the vasculature supplying the auditory system. Smoking is accompanied

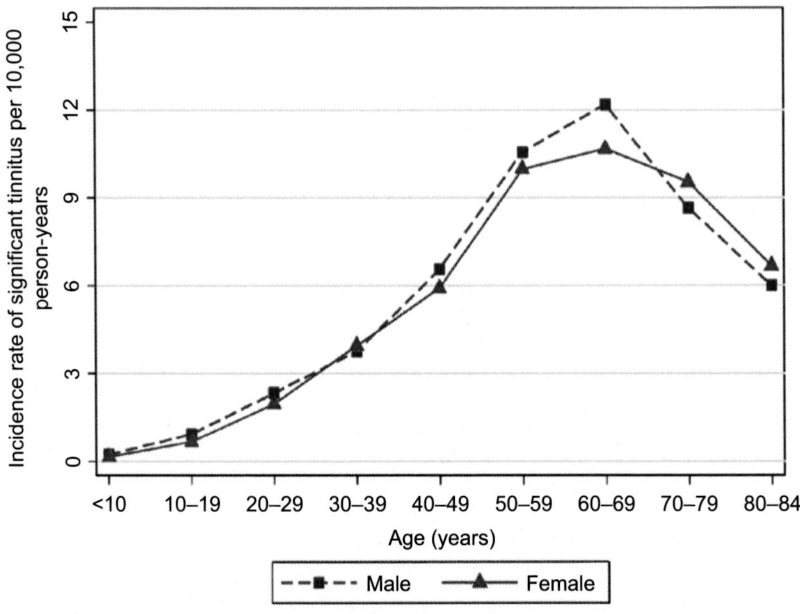

FIGURE 7.5 Gender- and age-specific incidence rates of clinically significant tinnitus. *From Martinez, C., Wallenhorst, C., McFerran, D., Hall, D.A., 2015. Incidence rates of clinically significant tinnitus: 10-year trend from a cohort study in England. Ear Hear. 36, e69–e75.*

by a higher incidence of high-frequency hearing loss. Moderate alcohol consumption appears to have a protective effect on hearing.

Cruickshanks et al. (1998) found that current smokers were 1.69 times as likely to have a hearing loss as nonsmokers (CI = 1.31–2.17). This relationship remained for those without a history of occupational noise exposure and in analyses excluding those with non-age-related hearing loss. Nonsmoking participants who lived with a smoker were more likely to have a hearing loss than those who were not exposed to a household member who smoked (OR = 1.94; CI = 1.01–3.74). Follow-up studies in the "Established Populations for Epidemiologic Studies of the Elderly" sample (Ferrucci et al., 1998) found among the 10,118 participants, 1406 (12.4%) that reported hearing problems at baseline. Of those with no baseline hearing problems and complete follow-up information (N = 8495), 1120 (13.2%) developed new hearing problems. Smoking was associated with higher prevalence and incidence rates of hearing impairment. In both cases the association was weak although statistically significant. Compared with participants without a history of smoking, those who had ever smoked were more likely to report hearing problems at baseline (OR = 1.2; CI = 1.0–1.3) and more likely to develop new hearing problems over the follow-up period (OR = 1.6; CI = 1.4–1.8).

Three studies measured the overall effect of both smoking and alcohol consumption on hearing loss in the same population. The first study (Fransen et al., 2008) collected audiometric data in 4083 subjects between 53 and 67 years from nine audiological centers across Europe. PTAs for 0.5, 1, and 2 kHz were adjusted for age and sex and tested for association with exposure to risk factors. Noise exposure was associated with a significant loss of hearing at frequencies of 1–8 kHz. Smoking significantly increased high-frequency hearing loss, and the effect was dose dependent. Moderate alcohol consumption was inversely correlated with hearing loss. Significant associations were found in the high as well as in the low frequencies.

In the Blue Mountain Hearing Study (Gopinath et al., 2010) of 2956 participants (aged >50 years) alcohol consumption and smoking status were measured by using an interviewer-administered questionnaire. Logistic regression was used to obtain ORs with 95% CIs that compared the chances of having hearing loss in participants, who did or did not smoke or consume alcohol, after adjusting for other factors previously reported to be associated with hearing loss. Cross-sectional analysis demonstrated a significant protective association between the moderate consumption of alcohol (> 1 but ≤ 2 drinks/day) and hearing function in older adults (compared with nondrinkers), OR = 0.75 (CI = 0.57–0.98). Smokers who were not exposed to occupational noise had a significantly higher likelihood of hearing loss after adjusting for multiple variables, OR = 1.63 (CI = 1.01–2.64). The interaction between smoking and noise exposure was not significant.

Dawes et al. (2014) evaluated the association between smoking, passive smoking, alcohol consumption, and hearing loss in 164,770 adults aged between 40 and 69 years who completed a speech-in-noise hearing test (the Digit Triplet Test). Current smokers were more likely to have a (speech-in-noise) hearing loss than nonsmokers (OR = 1.15, CI = 1.09–1.21). Nonsmokers who reported passive exposure to tobacco smoke were also more likely to have a hearing loss (OR = 1.28, CI = 1.21–1.35). Those who consumed alcohol were less likely to have a hearing loss than lifetime teetotalers. The association was similar across three levels of consumption by volume of alcohol (lightest 25%, OR = 0.61, CI = 0.57–0.65; middle 50%, OR = 0.62, CI = 0.58–0.66; heaviest 25%, OR = 0.65, CI = 0.61–0.70). They defined one unit of alcohol equal to 8 g, consequently one glass of wine and beer were rated as 2.5 units, and one shot of strong liquor at 1 unit. The lightest consumption level was less than 118.4 g/week (not including teetotalers; i.e., <6 glasses of wine or beer), middle level 118.4–196.8 g/week (6–10 glasses), and the highest consumption level more than 196.8 g/week (> 10 glasses). Regardless of the dose level, alcohol consumption was associated with a protective effect.

The incidence of tinnitus also depends on the smoking and alcohol consumption history, much in the same way as hearing loss does. Nondahl et al. (2010) described the 10-year cumulative incidence of tinnitus and its risk factors. Participants ($N = 2922$, aged 48–92 years) who did not report tinnitus at the baseline study (1993–95) were followed for up to 10 years. In addition to audiometric testing data on tinnitus, health, and other history were obtained via questionnaire. Potential risk factors were assessed with discrete-time proportional hazards models. They found that "the 10-year cumulative incidence of tinnitus was 12.7%. The risk of developing tinnitus was significantly associated with a history of ever smoking (HR = 1.40), and among women, hearing loss (HR = 2.59). Alcohol consumption (HR = 0.63 for ≥ 141 grams/week vs. <15 grams/week) was associated with decreased risk."

There have, so far, not been studies into the interaction of smoking and alcohol consumption, i.e., by separating smokers in a drinking and nondrinking group, on hearing loss or tinnitus in the same subjects.

7.5 EPIDEMIOLOGY OF DIABETES

As we have seen in Chapter 6, Causes of Acquired Hearing Loss, uncontrolled diabetes may cause hearing loss. The epidemiological studies reviewed here detail the risks. Austin et al. (2009) found greater amounts of hearing loss in less than 50 years old adult diabetes subjects compared to controls. Significant hearing differences were present at all frequencies for Type 2 diabetes subjects, but for Type 1 subjects (for definitions see chapter: Causes of Acquired Hearing Loss), differences were found ≤ 1 kHz and ≥ 10 kHz. Over age 50 years, there were significant associations between hearing at low frequencies for Type 1 only. Bainbridge et al. (2008) found that the association between diabetes and hearing impairment was independent of known risk factors for hearing impairment, such as noise exposure, ototoxic medication use, and smoking. They phrased this as: "The ORs, adjusted for low- or mid-frequency and high-frequency hearing impairment, were 1.82 (CI = 1.27–2.60) and 2.16 (CI = 1.47–3.18), respectively." Cheng et al. (2009) reported for adults aged 25–69 years, tested between 1971 and 2004, that "the adjusted prevalence ratios of hearing impairment for persons with diabetes vs. those without diabetes was 1.17 (CI = 0.87–1.57) for the NHANES 1971–1973 and 1.53 (95% CI, 1.28–1.83) for NHANES 1999–2004." This suggests a clear increase in hearing loss prevalence over the 25-year span between the two surveys.

Mitchell et al. (2009) reported the relationship between Type 2 diabetes and the prevalence, the 5-year incidence, and the progression of hearing impairment in a representative, older, Australian population (the Blue Mountain study). They found that age-related hearing loss was present in 50% of diabetic participants ($N = 210$) compared with 38.2% of nondiabetic participants ($N = 1648$), OR = 1.55 (CI = 1.11–2.17), after adjusting for multiple risk factors. Diabetes duration and hearing loss were positively correlated. After 5 years, incident hearing loss occurred in 18.7% of participants with, and 18.0% of those without diabetes, adjusted OR = 1.01 (CI = 0.54–1.91). Progression of existing hearing loss (> 5 dB HL) was significantly greater in participants with newly diagnosed diabetes (69.6%) than in those without diabetes (47.8%) over this period, adjusted OR = 2.71 (CI = 1.07–6.86). Type 2 diabetes was associated with the larger prevalent, but not incident hearing loss in this older population.

Cruickshanks et al. (2015) conducted a longitudinal population-based cohort study (1993–95 to 2009–10). Follow-up examinations were obtained from 87.2% ($N = 1678$; mean baseline age 61). The 15-year cumulative incidence of hearing impairment was 56.8%. Adjusting for age and sex, current smoking (HR = 1.31, $p = 0.048$) and poorly controlled diabetes mellitus (HR = 2.03, $p = 0.048$) were associated with greater risk of hearing impairment. People with better-controlled diabetes mellitus were not at greater risk.

7.6 EPIDEMIOLOGY OF OTITIS MEDIA

Conductive hearing loss resulting from acute otitis media or otitis media with effusion is mostly intermittent mild to moderate hearing loss in infants and young children. Aarhus et al. (2014) studied a population-based cohort of 32,786 participants who had their hearing tested by pure-tone audiometry in primary school and again at ages between 20 and 56 years. Hearing loss was diagnosed in 3066 children; the remaining sample had normal childhood hearing. Significantly reduced adult hearing thresholds in the whole frequency range were found in those diagnosed with childhood hearing loss caused by otitis media with effusion ($N = 1255$; 2 dB), chronic suppurative otitis media (CSOM; $N = 108$; 17–20 dB), or hearing loss after recurrent acute otitis media (rAOM; $N = 613$; 7–10 dB) compared those with normal childhood hearing. The effects of CSOM and hearing loss after rAOM on adult hearing thresholds were larger in participants tested in middle adulthood (ages 40–56 years) than in those tested in young adulthood (ages 20–40 years). Aarhus et al. (2014) concluded that CSOM as well as rAOM in childhood are associated with adult hearing loss.

7.7 EPIDEMIOLOGY OF AUDITORY NEUROPATHY SPECTRUM DISORDER

Auditory nerve myelinopathy and/or deficits in synchrony of neural discharges are the most probable underlying pathophysiological mechanisms of auditory neuropathy spectrum disorder (ANSD) (see chapter: Types of Hearing Loss). Korver et al. (2012) collected all available published evidence on the prevalence of auditory neuropathy in the "well-baby" population in the Netherlands. The population-based prevalence in children in population hearing screening was found to vary between 0.006% (SD = 0.006) and 0.03% (SD = 0.02). The false-negative rate, based on otoacoustic emission testing in newborn hearing screening programs, caused by missed children with auditory neuropathy, was estimated between 4% and 17%. Bielecki et al. (2012) investigated in the period from 2002 to 2011, 9419 infants whose hearing ability was uncertain or who had risk factors for hearing loss were investigated, and 352 were diagnosed with sensorineural hearing loss (SNHL). Of these 352 children, 18 (5.1%) were diagnosed with ANSD, suggesting that is not an extremely rare hearing disorder. Auditory neuropathy, as defined in the audiology/otolaryngology literature, occurs frequently and is responsible for approximately 8% of newly diagnosed cases of hearing loss in children per year (Vlastarakos et al., 2008).

7.8 GENETICS OF SENSORINEURAL HEARING LOSS

Genetic deafness that affects about 0.1% of individuals by severe or profound deafness at birth or during early childhood, i.e., in the prelingual period can be distinguished in syndromic and nonsyndromic deafness (Petit et al., 2001). Syndromic deafness is associated with other defects and contributes to about 30% of the cases during early childhood and may be conductive, sensorineural, or mixed. Nonsyndromic hereditary deafness is classified by the mode of inheritance; DFNX, DFNA, and DFNB refer to deafness forms inherited on the X chromosome-linked, autosomal (i.e., any chromosome other than a sex chromosome) dominant, and autosomal recessive modes of transmission, respectively. About 80% of the cases of prelingual nonsyndromic deafness are DFNB forms, whereas most of the late-onset forms are DFNA forms. Prelingual nonsyndromic deafness is almost exclusively sensorineural (Petit et al., 2001).

7.8.1 Syndromic Hearing Loss

Forms of dominant syndromic hearing loss include Waardenburg syndrome (deafness, structural defects derived from the neural crest, and pigmentation anomalies), branchio-oto-renal syndrome (branchial fistulas, renal anomalies, and abnormal development of the ear), Stickler syndrome (distinctive facial abnormalities, eye problems, hearing loss, and joint problems), and neurofibromatosis 2 (characterized by tumors, including acoustic neuromas) (Raviv et al., 2010). Recessive syndromic hearing loss includes the deafness–blindness of Usher syndrome, Pendred syndrome (inner ear and enlargement of the thyroid gland, known as goiter), Jervell and Lange-Nielsen syndrome (characterized by electrocardiographic changes as well as hearing loss), and Cockayne syndrome (described with dwarfism, retinal atrophy, and deafness). X-linked syndromic deafness includes Alport syndrome (also known as hereditary nephritis) and Norrie disease (congenital ocular symptoms with progressive hearing loss) (Raviv et al., 2010).

7.8.1.1 Usher Syndrome as an Example

Usher syndrome (USH) is an autosomal recessive disorder characterized by combined (syndromic) deafness–blindness. It affects about one child out of 25,000 and accounts for approximately 50% of all hereditary deafness–blindness cases. Three clinical subtypes, referred to as USH1, USH2, and USH3, are defined according to the severity of the hearing impairment, the presence/absence of vestibular dysfunction, and the age at onset of retinitis pigmentosa (Bonnet and Amraoui, 2012). Defects in myosin-VIIa (Fig. 7.6), the PDZ-domain-containing protein harmonin, cadherin-23 and protocadherin-15 (two cadherins with large extracellular regions), and the putative scaffolding protein Sans underlie five genetic forms of Usher syndrome type I (USH1). PDZ domains are protein-interaction domains that are found for instance in the post-synaptic density of neuronal excitatory synapses (Kim and Sheng, 2004). The most frequent cause of hereditary deafness–blindness in humans is the aggregate of the three subtypes (El-Amraoui and Petit, 2005). The human genome contains over 115 genes encoding cadherins and cadherin-like proteins. Defects in about 21 of these proteins have been linked to inherited disorders in

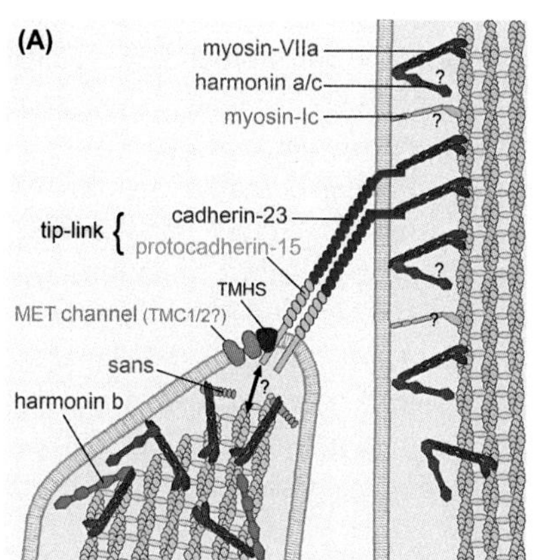

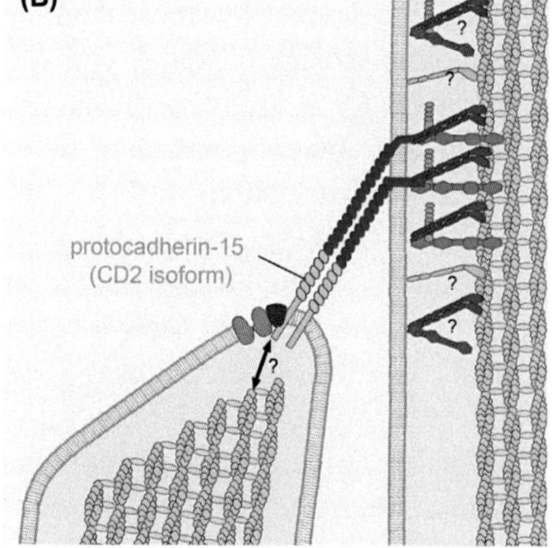

FIGURE 7.6 The mechano-electrical transmission (MET) machinery in cochlear hair cells. (A) In the developing hair bundle, the MET machinery comprises the MET channel(s) and TMHS at the lower tip-link insertion point. Sans and myosin-VIIa are also present, but the nature of their interaction with the MET complex is still unknown. The nature of the interaction between the MET complex and the actin cytoskeleton is also unknown at the lower tip-link insertion point. At the upper tip-link insertion point, myosin-VIIa and harmonin b interact with cadherin-23. The role of myosin-Ic remains unclear in the cochlea because its function has not yet been tested in mice mutant for this protein. In addition, the location of myosin-Ic cannot be investigated by immunohistochemistry due to the absence of the appropriate mutant mice to confirm the specificity of antibodies directed against this protein. (B) Mature MET machinery. Sans, myosin-VIIa, and harmonin b are located at the upper tip-link insertion point. *From Michalski, N., Petit, C. 2015. Genetics of auditory mechano-electrical transduction. Pflugers Arch. 467, 49–72.*

humans, including sensory defects associated with deafness and blindness (El-Amraoui and Petit, 2005).

One of the defects in Usher's syndrome harmonin localizes to the upper tip link of the stereocilia, so a lack of harmonin results in impaired mechano-transduction in sensory cochlear hair cells (Fig. 7.6). Harmonin seems also to play a functional role at synchronous exocytosis in IHC ribbon synapses. (Wichmann, 2015). Studies of USH1 demonstrated the crucial role of transient fibrous links formed by cadherin-23 and protocadherin-15 in the cohesion of the developing hair bundle, the mechano-receptive structure of the hair cells, as well as the involvement of these cadherins in the formation of the tip link (El-Amraoui and Petit, 2010).

7.8.2 Nonsyndromic Hearing Loss

The various nonsyndromic deafness forms so far studied appear as monogenic disorders (i.e., controlled by a single gene). They are fairly uncommon with the exception of one, caused by mutations in the gene (*GJB2*) encoding the gap junction protein connexin26, which accounts for between one-third and one-half of the cases of prelingual inherited deafness in Caucasian populations (Petit et al., 2001). *GJB2* is also the most common form of severe to profound age-related hearing impairment (ARHI). The most common causes of mild to moderate ARHI are *STRC* causative variants (*DFNB16*) (Sloan-Heggen et al., 2016).

Typically the cooperation of several genes underlies hereditary or progressive hearing loss. Nishio and Usami (2015), using massively parallel DNA sequencing of target genes, found that the number of *GJB2* mutations was exceptionally high, followed by those in *CDH23*, *SLC26A4*, *MYO15A*, *COL11A2*, *MYO7A*, and *OTOF*. Miyagawa et al. (2015) noted that in Japanese people, mutations in the myosin gene *MYO15A* are a notable cause of nonsyndromic hearing loss. Patients with *MYO15A* mutations present 1 of 2 types of hearing impairment phenotype: (1) prelingual onset and severe or profound hearing loss or (2) a milder phenotype with postlingual onset and progressive hearing loss. Myosins are a large family of actin-dependent molecular motors and play a role in the hydrolysis of ATP to generate the force required for the movement of actin filaments. These figures do not pertain to African-American or European-Americans where the prevalence numbers are *GJB2*, followed by *STRC*, *SLC26A4*, *TECTA*, *MYO15A*, *MYO7A*, and *CDH23* (Sloan-Heggen et al., 2016).

7.8.2.1 GJB2 *Mutations as an Example*

Mutations in the *GJB2* (*Connexin26*) gene account for up to 50% of cases of severe to profound ARHI (Sloan-Heggen et al., 2016). *GJB2* encodes a gap junction protein that is widely expressed in the inner ear. Cells in the cochlea communicate through gap junctions that regulate the fluid and ion balance.

Mutations in *GJB2* are thought to result in an altered function of gap junctions and a disturbance of potassium homeostasis, leading to hearing loss. *GJB2* may contribute to potassium homeostasis through recycling of potassium ions from hair cells back to the endolymph. The most common *GJB2* mutation in the European-American population is 35delG, a single guanine deletion at cDNA position 35, which accounts for 85% of the incidence of *GJB2*-induced hearing loss (Van Eyken et al., 2007). In the Jewish population the most common mutation is 167delT, and in the Japanese 235delC. 35delG can cause some reduction in distortion-product otoacoustic emissions (DPOAEs) at very high frequencies, suggesting alterations in the outer hair cells of the cochlea and potentially a higher risk for the development of ARHI (Engel-Yeger et al., 2002). In *GJB2* there is excellent phenotype–genotype correlation; two truncating mutations result in the worst hearing loss, one truncating and one missense mutation result in less hearing loss, and two missense mutations result in the least amount of hearing loss (Snoeckx et al., 2005). Van Eyken et al. (2007) and Carlsson et al. (2004) did not find a significant association between 35delG and NIHL in Polish and Swedish noise-exposed factory workers, respectively.

7.9 GENETICS OF OTOSCLEROSIS

Thys and Van Camp (2009) described otosclerosis as a pathologic increased bone turnover in the otic capsule, which in most cases leads to stapes fixation, resulting in a conductive hearing loss. Although environmental factors have been implicated, it is clear that genetic factors play a significant role. Thys and Van Camp (2009) considered otosclerosis as a complex disease with rare autosomal dominant forms caused by a single gene. Although, seven monogenic loci have been published, none of the genes involved have been identified. Data from both genetic association studies and gene expression analysis of otosclerotic bone showed that the transforming growth factor beta-1 (TGF-β1) pathway is most likely an important factor in the pathogenesis of otosclerosis. The TGFβ signaling pathway is involved in many cellular processes including cell growth. Recently, Ziff et al. (2016) used whole-exome sequencing in four families exhibiting dominantly inherited otosclerosis. Multiple mutations were found in the *SERPINF1* (Serpin Peptidase Inhibitor, Clade F) gene, which encodes pigment epithelium derived factor, a potent inhibitor of angiogenesis and known regulator of bone density.

7.10 GENETICS OF AUDITORY NEUROPATHY

It is not clear if ANSD is more than a convenient label based on normal otoacoustic emissions and abnormal auditory brainstem response (ABR). Genes underlying two forms of ANSD are *OTOF* resulting in synaptopathy and *OPA1* resulting in neuropathy of the auditory nerve fiber dendrites; both are characterized as ANSD, but the real phenomenological differentiation also requires (preferably transtympanic) electrocochleography (see chapter: Types of Hearing Loss).

7.10.1 *Otoferlin*

Wichmann (2015)'s review showed that because exocytosis was almost completely abolished in an otoferlin knock-out mouse model, otoferlin should have a role in a late step of exocytosis. Otoferlin appears to mediate the replenishment of the ready releasable vesicle pool and plays a role in the vesicle recruitment to the active zone membrane via filamentous tethers.

7.10.2 The *OPA1* gene

Huang et al. (2012) investigated cochlear abnormalities accompanying hearing loss and the effects of cochlear implantation. They sequenced the *OPA1* gene and recorded the cochlear potentials, cochlear microphonic, summating potential, and compound action potentials (CAPs), by electrocochleography before cochlear implantation. Genetic analysis identified a R445H mutation in the *OPA1* gene. Audiological studies showed preserved DPOAEs and absent or abnormally delayed ABRs. Trans-tympanic electrocochleography showed prolonged low-amplitude negative potentials without auditory nerve CAPs. After cochlear implantation, hearing thresholds, speech perception, and synchronous activity in auditory brainstem pathways were restored. This suggests that deafness accompanying this *OPA1* mutation is due to altered function of terminal unmyelinated portions of auditory nerve (see chapter: Types of Hearing Loss).

7.10.3 The *AIFM1* gene

Zong et al. (2015) found that variants in *AIFM1* gene are a common cause of familial and sporadic ANSD and provide insight into the expanded spectrum of *AIFM1*-associated diseases. "*AIFM1* encodes apoptosis-inducing factor 1, a flavoprotein located in the mitochondrial intermembrane space. AIFM1 has at least two functions. First, . . . it mediates caspase-independent programmed cell death when translocating from mitochondria to the nucleus upon apoptotic stimuli. And second, . . . it plays an important role in oxidative phosphorylation, redox control and respiratory chain activity in healthy cells."

7.10.4 The *PVJK* gene

Delmaghani et al. (2006) reported on *DFNB59*, a newly identified gene mutated in four families segregating autosomal recessive auditory neuropathy. They found that the primary lesion in DFNB59 hearing impairment is outside the cochlea. Their findings suggested a dysfunction in neurons along the auditory pathway, which is consistent with the observed distribution of pejvakin in the cell bodies of neurons in the spiral ganglion and the brainstem auditory nuclei. More recent findings (Delmaghani et al., 2015) described that: "A deficiency in pejvakin (*Pvjk*), a protein of unknown function, causes a strikingly heterogeneous form of human deafness. Noise exposure rapidly upregulates *Pvjk* cochlear transcription in wildtype mice and triggers peroxisome proliferation in hair cells and primary auditory neurons." Delmaghani et al. (2015) concluded: "In patients with hearing impairment, the amplification of sound by hearing aids or direct electrical stimulation of the auditory nerve by a cochlear implant delivers a stimulus with an energy level similar to that shown here to worsen the hearing impairment of $Pvjk^{-/-}$ mice within 1 min of sound exposure." This suggests that cochlear implantation or hearing aid prescription in this form of ANSD may not be without problems.

7.11 GENE NETWORKS

Petersen and Willems (2006) grouped the genes underlying hearing loss into a network of genes regulating potassium homeostasis in the cochlea and a network of genes forming stereocilia. In the cochlea network, the hair cells are repolarized when the potassium ions leave these cells via channels probably formed by the *KCNQ4* gene and enter the supporting Deiter cells. They then diffuse through gap junctions formed by connexins to the stria vascularis. These potassium ions are secreted back into the endolymph through potassium channels formed by the *KCNQ1* and *KCNE1* gene products, thereby resetting the mechano-electrical transduction system. The pendrin gene *SLC26A4*, different gap junction genes encoding connexins (*GJB2*, *GJB3*, and *GJB6*), and the tight junction protein gene *CLDN14* are also involved in ion homeostasis in the cochlea and implicated in deafness (Petersen and Willems, 2006).

The second network frequently implicated in deafness is located in the stereocilia (Fig. 7.6) and consists of the myosin-VIIa, myosin-XV, whirlin, harmonin, cadherin-23, and protocadherin-15 genes. These proteins are located in the stereocilia and/or tip links connecting stereocilia, which are the mechano-electrical transducers transducing sound vibrations into electrical signals (Petersen and Willems, 2006).

Nishio et al. (2015) in addition distinguished a third group of genes associated with vesicle transport, neuronal transmission, and calcium-binding functions. This group includes *CABP2* (calcium-binding protein 2), membrane traffic protein OTOF (otoferlin), *SLC17A8* (vesicular glutamate transporter 3), and *TBC1D24* (TBC domain-containing RAB-specific GTPase-activating protein 24).

7.12 HEREDITARY VERSUS ACQUIRED HEARING LOSS

In an authoritative review that I will follow here, Smith et al. (2005) wrote that the incidence of acquired SNHL in children living in more developed countries has fallen during the past three to four decades. This was largely the result of improved neonatal care and the widespread use of immunization. This resulted in a relative increase in the proportion of inherited forms of SNHL.

7.12.1 Neonates

More than half of neonates with SNHL have inherited hearing loss. Mutations in a gene called *GJB2* (Section 7.8.2.1) account for roughly half of hereditary cases of severe to profound SNHL. The most common form of syndromic hereditary SNHL is Pendred's syndrome. The worldwide prevalence of SNHL resulting from congenital rubella syndrome also remains high (Smith et al., 2005). However, today congenital cytomegalovirus (cCMV) is the most common cause of nonhereditary SNHL in childhood. The prevalence of cCMV in developed countries is approximately 0.6%. Among these newborns, approximately 13% will experience hearing loss. Among symptomatic children, the majority has bilateral loss; among asymptomatic children, unilateral loss predominates. In both groups the hearing loss is mainly severe to profound. Among hearing-impaired children, cCMV is the causative agent in 10–20% (Goderis et al., 2014).

7.12.2 Infants and School Age

Bacterial meningitis is most commonly the cause of acquired SNHL in infants and children and accounts for about 6% of all cases of SNHL in children (Smith et al., 2005). The prevalence is about 7/100,000 with 75% of affected children younger than 2 years, 15% are aged 2–5 years, and 10% are older than 5 years. Late-onset hearing loss that is at least moderate in severity is diagnosed in 1.2–3.3 per 10,000 school-aged children (Smith et al., 2005).

7.12.3 Genetic Susceptibility for Noise-Induced Hearing Loss

This section and the following one is based on, and an extension of, my previous review (Eggermont, 2014). The mechanisms of sensory hair cell degeneration in response to different ototoxic stimuli share a final common pathway: caspase activation. Inhibition of caspases prevents or delays hair cell death and may preserve hearing. Inhibition of mitogen-activated protein kinases protects against noise-induced hair cell death. Individual animals (and humans) show different susceptibility to noise damage even under very carefully controlled exposure conditions, likely due to genetic differences. Common experimental animals (rats, guinea pigs, chinchillas, cats) are typically outbred and their genomes contain a mixture of many genes. In contrast, many mouse strains have been inbred over many generations thereby reducing the individual variability and making them ideal candidates for studying the genetic modulation of individual susceptibility. The recessive adult hearing loss gene (*Ahl*), mapped to chromosome 10, has been identified in the C57BL/6J and DBA/2J inbred strains of mice and is the presumed cause of the progressive hearing loss of this species (Erway et al., 1993). Mice homozygous for the Ahl allele are more sensitive to the damaging effects of noise and also are probably damaged in a different manner by noise than mice containing the wild-type gene (Davis et al., 2001). This suggests that interactions between ARHI and NIHL are very likely. Di Palma et al. (2001) have shown that the wild-type *Ahl* gene codes for a hair cell specific cadherin *Cdh23* that may form the lateral links between stereocilia. Reduction of, or missing, *Cdh23* weakens the cell and may allow stereocilia to be more easily physically damaged by loud sounds and by aging. The *Ahl* gene product was determined to be cadherin-23 (Noben-Trauth et al., 2003).

Mouse mutations that promote both NIHL and apparent sensory ARHI, such as Cdh23^{Ahl}, suggest that there is often no useful distinction between the sensory and neural substrates of NIHL and ARHI. In these mice environmental noise levels, that would normally be harmless, may cause permanent hearing loss. A nonlinear interaction between ARHI and NIHL was observed in gerbils born and raised in a quiet environment (Mills et al., 1997). They were exposed monaurally at 18 months of age to a 3.5 kHz pure tone for 1 h at 113 dB SPL. Six weeks after the exposure permanent threshold shifts in the exposed ear were approximately 20 dB in the 4–8 kHz region. Thresholds in the nonexposed ear were unaffected. The nonexposed ear would then reflect the pure ARHI, whereas in the exposed ear it would be combined with NIHL. This of course assumes that there is no central interaction (via the olivo-cochlear bundle) between the activity from the exposed ear and the other ear. Animals were then allowed to age in quiet until 36 months of age when thresholds were assessed again. The effects of NIHL and ARHI were nonadditive—i.e., the resulting hearing loss in the exposed ear was larger than expected on the basis of the loss in the pure NIHL and pure ARHI groups. Thus, sensory ARHI may represent cumulative damage, and alleles that promote this condition may make affected individuals prone to damage from otherwise benign exposures. Noise exposure early in life may also trigger progressive neuronal loss, the hallmark of neural ARHI (Ohlemiller, 2006; Kujawa and Liberman, 2006).

Twin studies estimate heritability for NIHL of approximately 36%, and strain-specific variation in sensitivity has been demonstrated in mice. Lavinsky et al. (2015) conducted a genome-wide association study (GWAS) in mice, with an emphasis on a significant peak for susceptibility to NIHL on chromosome 17 within a haplotype block containing NADPH oxidase-3 (Nox3). This peak was detected after an 8 kHz tone burst stimulus. A haplotype is, in the simplest terms, a specific group of genes or alleles that progeny inherited from one parent. A specific meaning of the term haplotype: a set of single-nucleotide polymorphisms (SNPs) on one chromosome that tend to always occur together, i.e., are associated statistically. It is thought that identifying these statistical associations and few alleles of a specific haplotype sequence can facilitate identifying all other such polymorphic sites that are nearby on the chromosome. Knowledge of haplotype structure might make it possible to conduct GWASs at greatly reduced cost, since haplotype-tagging SNPs can be chosen to capture most of the genetic information in a region that has a block structure <wikipedia.org/wiki/Haplotype>.

The Nox3 mutants and heterozygotes demonstrated a greater susceptibility to NIHL specifically at 8 kHz both on measures of DPOAEs and on the ABR. This sensitivity resides within the synaptic ribbons of the cochlea in the mutant animals specifically at 8 kHz (Lavinsky et al., 2015).

7.12.4 Genetic Susceptibility for Age-Related Hearing Impairment

Age-related hearing impairment constitutes one of the most frequent sensory problems in the elderly. It presents itself as a bilateral SNHL that is most pronounced in the high frequencies. ARHI is a complex disorder, with both environmental and genetic factors contributing to the disease (Gates et al., 1999). Genetic association studies on a small number of candidate ARHI susceptibility genes showed significant associations for *NAT2* and *KCNQ4* (Van Eyken et al., 2006), but failed to link other genes to ARHI (Van Laer et al., 2002). The *NAT2* (N-acetyltransferase 2) gene encodes an enzyme that both activates and deactivates arylamine and hydrazine drugs and carcinogens. Konings et al. (2007) investigated whether variations SNPs in the catalase gene (*CAT*), one of the genes involved in oxidative stress, influence noise susceptibility. Audiometric data from 1261 Swedish and 4500 Polish noise-exposed workers were analyzed. DNA samples were collected from the 10% most susceptible and the 10% most resistant individuals. Twelve SNPs were selected and genotyped. Significant interactions were observed between noise exposure levels and genotypes of two SNPs for the Swedish population and of five SNPs for the Polish population. Two of these SNPs were significant in both populations. This study identified significant associations between catalase SNPs and haplotypes and susceptibility to development of NIHL. Konings et al. (2007) also showed that the effect of *CAT* polymorphisms on NIHL was only present for high noise exposure levels (>92 dB). This suggested that *CAT* polymorphisms have a larger effect when people are exposed to higher levels of noise.

TABLE 7.1 Gene Mutations in Congenital Deafness, NIHL, and ARHI

	Congenital	NIHL Susceptibility	ARHI
Potassium homeostasis network	*KCNQ1*	*KCNQ1*	*KCNQ1*
	KCNQ4	*KCNQ4*	*KCNQ4*
	KCNE1	*KCNE1*	*KCNE1*
	GJB2	*GJB2*	
	TJP2	*TJP2*	
	CLDN14		
	SLC26A4		
Stereocilia network	*Myosin*		
	Whirlin		
	Harmonin		
	CDH23	*CDH23*	*CDH23*
Calcium-binding network	*CABP2*		
	OTOF		
	SLC17A8	*SLC17A8*	*SLC17A8*
	TBC1D24		

Ruel et al. (2008) identified SLC17A8, which encodes the vesicular glutamate transporter-3 (VGLUT3), as the gene responsible for DFNA25, an autosomal-dominant form of progressive, high-frequency nonsyndromic deafness. Autosomal-dominant SNHL is genetically heterogeneous, with a phenotype closely resembling presbycusis, the most common sensory defect associated with aging in humans. Friedman et al. (2009) identified common alleles of *GRM7*, the gene encoding metabotropic glutamate receptor type 7, contribute to an individual's risk of developing ARHI, possibly through a mechanism of altered susceptibility to glutamate excitotoxicity. Newman et al. (2012) showed that *GRM7* alleles are associated primarily with peripheral measures of hearing loss, and particularly with speech detection in older adults. Bonneux et al. (2011) investigated the role of mitochondrial DNA variants in ARHI and found no association with ARHI.

Table 7.1 presents an overview of genes carrying variants associated with congenital deafness, NIHL, and ARHI. The genes are grouped according to the three networks discussed above. Some overlap is evident, the more so between susceptibility for NIHL and ARHI, as it is often unclear if the ARHI is only based on genetic makeup, on environmental exposure, or both.

7.13 SUMMARY

Worldwide, 16% of disabling hearing loss in adults is attributed to occupational noise exposure. Overall, the prevalence of hearing loss increases with every age decade from 5–10% at age 40 to about 80–90% at age 85. The incidence of new cases of hearing loss rises sharply from about 2% at age 40 to 6% at age 65. Tinnitus prevalence rises from about 5% at age 20 to 15–25% (depending on the criterion) at age 65, after which, in contrast to the hearing loss, it levels off. The prevalence of hearing loss in the standard audiometric frequency range is not related to tinnitus prevalence. Smoking is accompanied by a higher incidence of high-frequency hearing loss. Alcohol consumption, in moderation, appears to have a protective effect on hearing. Poorly controlled diabetes is now recognized as a risk factor of hearing loss. A genetic diagnosis of hearing loss can now be made in about 40% of all persons with symmetric loss if comprehensive genetic testing is done. Gene networks for potassium homeostasis, stereocilia functioning, and calcium binding have been identified and those responsible for increased susceptibility for NIHL and presbycusis are described. Several gene mutations are now identified in auditory neuropathy spectrum disorder, including those of *OTOF*, *OPA1*, *AIFM1*, and *PVJK*.

REFERENCES

Aarhus, L., Tambs, K., Kvestad, E., Engdahl, B., 2014. Childhood otitis media: a cohort study with 30-year follow-up of hearing (The HUNT Study). Ear Hear. 36, 302–308.

Abdel-Hafez, A.M.M., Elgayar, S.A.M., Husain, O.A., Thabet, H.A.S., 2014. Effect of nicotine on the structure of cochlea of guinea pigs. Anat. Cell Biol. 47, 162–170.

Austin, D.F., Konrad-Martin, D., Griest, S., McMillan, G.P., McDermott, D., Fausti, S., 2009. Diabetes-related changes in hearing. Laryngoscope 119, 1788–1796.

Axelsson, A., Ringdahl, A., 1989. Tinnitus—a study of its prevalence and characteristics. Br. J. Audiol. 23, 53–62.

Bainbridge, K.E., Hoffman, H.J., Cowie, C.C., 2008. Diabetes and hearing impairment in the United States: audiometric evidence from the National Health and Nutrition Examination Survey, 1999 to 2004. Ann. Intern. Med. 149, 1–10.

Bielecki, I., Horbulewicz, A., Wolan, T., 2012. Prevalence and risk factors for auditory neuropathy spectrum disorder in a screened newborn population at risk for hearing loss. Int. J. Pediatr. Otorhinolaryngol. 76, 1668–1670.

Bonnet, C., Amraoui, A., 2012. Usher syndrome (sensorineural deafness and retinitis pigmentosa): pathogenesis, molecular diagnosis and therapeutic approaches. Curr. Opin. Neurol. 25, 42–49.

Bonneux, S., Fransen, E., Van Eyken, E., Van Laer, L., Huyghe, J., Van de Heyning, P., et al., 2011. Inherited mitochondrial variants are not a major cause of age-related hearing impairment in the European population. Mitochondrion 11, 729–734.

Borchgrevink, H.M., Tambs, K., Hoffman, H.J., 2005. The Nord-Trøndelag Norway Audiometric Survey 1996–98: unscreened thresholds and prevalence of hearing impairment for adults >20 years. Noise Health 7, 1–15.

Carlsson, P.I., Borg, E., Grip, L., Dahl, N., Bondeson, M.L., 2004. Variability in noise susceptibility in a Swedish population: the role of 35delG mutation in the connexin 26 (GJB2) gene. Audiol. Med. 2, 123–130.

Cheng, Y.J., Gregg, E.W., Saaddine, J.B., Imperatore, G., Zhang, X., Albright, A.L., 2009. Three decade change in the prevalence of hearing impairment and its association with diabetes in the United States. Prev. Med. 49, 360–364.

Cruickshanks, K.J., Wiley, T.L., Tweed, T.S., Klein, B.E., Klein, R., Mares-Perlman, J.A., et al., 1998. Prevalence of hearing loss in older adults in Beaver Dam, Wisconsin. The Epidemiology of Hearing Loss Study. Am. J. Epidemiol. 148, 879–886.

Cruickshanks, K.J., Nondahl, D.M., Tweed, T.S., et al., 2010. Education, occupation, noise exposure history and the 10-yr cumulative incidence of hearing impairment in older adults. Hear. Res. 264, 3–9.

Cruickshanks, K.J., Nondahl, D.M., Dalton, D.S., Fischer, M.E., Klein, B.E.K., Nieto, F.X., et al., 2015. Smoking, central adiposity, and poor glycemic control increase risk of hearing impairment. J. Am. Geriatr. Soc. 63, 918–924.

Davis, A., El-Rafaie, A., 2000. Epidemiology of tinnitus. In: Tyler, R.S. (Ed.), Tinnitus Handbook. Singular Press, San Diego, CA, pp. 1–23.

Davis, R.R., Newlander, J.K., Ling, X.-B., Cortopassi, G., Krieg, E.F., Erway, L.C., 2001. Genetic basis for susceptibility to noise-induced hearing loss in mice. Hear. Res. 155, 82–90.

Dawes, P., Cruickshanks, K.J., Moore, D.R., Edmondson-Jones, M., Mccormack, A., Fortnum, H., et al., 2014. Cigarette smoking, passive smoking, alcohol consumption, and hearing loss. J. Assoc. Res. Otolaryngol. 15, 663–674.

Delmaghani, S., del Castillo, F.J., Michel, V., Leibovici, M., Aghaie, A., Ron, U., et al., 2006. Mutations in the gene encoding pejvakin, a newly identified protein of the afferent auditory pathway, cause DFNB59 auditory neuropathy. Nat. Genet. 38, 770–778.

Delmaghani, S., Defourny, J., Aghaie, A., Beurg, M., Dulon, D., Thelen, N., et al., 2015. Hypervulnerability to sound exposure through impaired adaptive proliferation of peroxisomes. Cell 163, 894–906.

Di Palma, F., Holme, R.H., Bryda, E.C., et al., 2001. Mutations in Cdh23, encoding a new type of cadherin, cause stereocilia disorganization in waltzer, the mouse model for Usher syndrome type 1D. Nat. Genet. 27, 103–107.

Eggermont, J.J., 2012. The Neuroscience of Tinnitus. Oxford University Press, Oxford, UK.

Eggermont, J.J., 2014. Noise and the Brain. Experience Dependent Developmental and Adult Plasticity. Academic Press, London, UK.

Eggermont, J.J., 2015. Auditory Temporal Processing and its Disorders. Oxford University Press, Oxford, UK.

El-Amraoui, A., Petit, C., 2005. Usher I syndrome: unravelling the mechanisms that underlie the cohesion of the growing hair bundle in inner ear sensory cells. J. Cell Sci. 118, 4593–4603.

El-Amraoui, A., Petit, C., 2010. Cadherins as targets for genetic diseases. Cold Spring Harb. Perspect. Biol. 2, a003095.

Engdahl, B., Krog, N.H., Kvestad, E., Hoffman, H.J., Tambs, K., 2012. Occupation and the risk of bothersome tinnitus: results from a prospective cohort study (HUNT). BMJ Open 2, e000512.

Engel-Yeger, B., Zaaroura, S., Zlotogora, J., Shalev, S., Hujeirat, Y., Carrasquillo, M., et al., 2002. The effects of a connexin 26 mutation—35delG—on oto-acoustic emissions and brainstem evoked potentials: homozygotes and carriers. Hear. Res. 163, 93–100.

Erway, L.C., Willott, J.F., Archer, J.R., Harrison, D.E., 1993. Genetics of age-related hearing loss in mice: I. Inbred and F1 hybrid strains. Hear. Res. 65, 125–132.

Ferrucci, L., Guralnik, J.M., Penninx, B.W.J.H., Leveille, S. (Eds.), 1998. Letter to the editor. JAMA 280, 963.

Fransen, E., Topsakal, V., Hendrickx, J.-J., et al., 2008. Occupational noise, smoking, and a high body mass index are risk factors for age-related hearing impairment and moderate alcohol consumption is protective: a European population-based multicenter study. J. Assoc. Res. Otolaryngol. 9, 264–276.

Friedman, R.A., Van Laer, L., Huentelman, M.J., et al., 2009. GRM7 variants confer susceptibility to age-related hearing impairment. Hum. Mol. Genet. 18 (4), 785–796.

Fujii, K., Nagata, C., Nakamura, K., Kawachi, T., Takatsuka, N., Oba, S., et al., 2011. Prevalence of tinnitus in community-dwelling Japanese adults. J. Epidemiol. 21, 299–304.

Gates, G.A., Couropmitree, N.N., Myers, R.H., 1999. Genetic associations in age-related hearing thresholds. Arch. Otolaryngol. Head Neck Surg. 125, 654–659.

Goderis, J., De Leenheer, E., Smets, K., Van Hoecke, H., Keymeulen, A., Dhooge, I., 2014. Hearing loss and congenital CMV infection: a systematic review. Pediatrics 134, 972–982.

Gopinath, B., Flood, V.M., McMahon, C.M., Burlutsky, G., Smith, W., Mitchell, P., 2010. The effects of smoking and alcohol consumption on age-related hearing loss: the blue mountains hearing study. Ear Hear. 31, 277–282.

Hasson, D., Theorell, T., Westerlund, H., Canlon, B., 2010. Prevalence and characteristics of hearing problems in a working and non-working Swedish population. J. Epidemiol. Community Health 64, 453–460.

Hoffman, H.J., Reed, G.W., 2004. Epidemiology of tinnitus. In: Snow, J.B. Jr (Ed.), Tinnitus: Theory and Management. BC Dekker, Hamilton, ON, pp. 16–41.

Huang, T., Santarelli, R., Starr, A., 2012. Mutation of *OPA1* gene causes deafness by affecting function of auditory nerve terminals. Brain Res. 1300, 97–104.

Karlsmose, B., Lauritzen, T., Engberg, M., Parving, A., 2000. A five-year longitudinal study of hearing in a Danish rural population aged 31–50 years. Br. J. Audiol. 34, 47–55.

Kim, E., Sheng, M., 2004. PDZ domain proteins of synapses. Nat. Rev. Neurosci. 5, 771–781.

Konings, A., Van Laer, L., Pawelczyk, M., et al., 2007. Association between variations in CAT and noise-induced hearing loss in two independent noise-exposed populations. Hum. Mol. Genet. 16, 1872–1883.

Korver, A.M., van Zanten, G.A., Meuwese-Jongejeugd, A., van Straaten, H.L., Oudesluys-Murphy, A.M., 2012. Auditory neuropathy in a low-risk population: a review of the literature. Int. J. Pediatr. Otorhinolaryngol. 76, 1708–1711.

Kujawa, S.G., Liberman, M.C., 2006. Acceleration of age-related hearing loss by early noise exposure: evidence of a misspent youth. J. Neurosci. 26, 2115–2123.

Langers, D.M., de Kleine, E., van Dijk, P., 2012. Tinnitus does not require macroscopic tonotopic map reorganization. Front. Syst. Neurosci. 6, 2.

Lavinsky, J., Crow, A.L., Pan, C., Wang, J., Aaron, K.A., Ho, M.K., et al., 2015. Genome-wide association study identifies nox3 as a critical gene for susceptibility to noise-induced hearing loss. PLoS Genet. 11 (4), e1005094. Available from: http://dx.doi.org/10.1371/journal.pgen.1005094.

Lee, F.-S., Matthews, L.J., Dubno, J.R., Mills, J.H., 2005. Longitudinal study of puretone thresholds in older persons. Ear Hear. 26, 1–11.

Leensen, M.C., van Duivenbooden, J.C., Dreschler, W.A., 2011. A retrospective analysis of noise-induced hearing loss in the Dutch construction industry. Int. Arch. Occup. Environ. Health 84, 577–590.

Lin, F.R., Niparko, J.K., Ferrucci, L., 2011. Hearing loss prevalence in the United States. Arch. Intern. Med. 171, 1851–1852.

Maffei, G., Miani, P., 1962. Experimental tobacco poisoning. Resultant structural modifications of the cochlea and tuba acustica. Arch. Otolaryngol. 75, 386–396.

Martinez, C., Wallenhorst, C., McFerran, D., Hall, D.A., 2015. Incidence rates of clinically significant tinnitus: 10-year trend from a cohort study in England. Ear Hear. 36, e69–e75.

Melcher, J.R., Knudson, I.M., Levine, R.A., 2013. Subcallosal brain structure: correlation with hearing threshold at supra-clinical frequencies (> 8 kHz), but not with tinnitus. Hear. Res. 295, 79–86.

Michalski, N., Petit, C., 2015. Genetics of auditory mechano-electrical transduction. Pflugers Arch. 467, 49–72.

Mills, J.H., Boettcher, F.A., Dubno, J.R., 1997. Interaction of noise-induced permanent threshold shift and age-related threshold shift. J. Acoust. Soc. Am. 101, 1081–1086.

Mitchell, P., Gopinath, B., McMahon, C.M., Rochtchina, E., Wang, J.J., Boyages, S.C., et al., 2009. Relationship of type 2 diabetes to the prevalence, incidence and progression of age-related hearing loss. Diabet. Med. 26, 483–488.

Miyagawa, M., Nishio, S.Y., Ichinose, A., Iwasaki, S., Murata, T., Kitajiri, S., et al., 2015. Mutational spectrum and clinical features of patients with ACTG1 mutations identified by massively parallel DNA sequencing. Ann. Otol. Rhinol. Laryngol. 124 (Suppl. 1), 84S–93S.

Nelson, D.I., Nelson, R.Y., Concha-Barrientos, M., Fingerhut, M., 2005. The global burden of occupational noise-induced hearing loss. Am. J. Ind. Med. 48, 446–458.

Newman, D.L., Fisher, L.M., Ohmen, J., Parody, R., Fong, C.-T., Frisina, S.T., et al., 2012. GRM7 variants associated with age-related hearing loss based on auditory perception. Hear. Res. 294, 125–132.

Nishio, S.Y., Usami, S., 2015. Deafness gene variations in a 1120 nonsyndromic hearing loss cohort: molecular epidemiology and deafness mutation spectrum of patients in Japan. Ann. Otol. Rhinol. Laryngol. 124 (Suppl. 1), 49S–60S.

Nishio, S.Y., Hattori, M., Moteki, H., Tsukada, K., Miyagawa, M., Naito, T., et al., 2015. Gene expression profiles of the cochlea and vestibular endorgans: localization and function of genes causing deafness. Ann. Otol. Rhinol. Laryngol. 124 (Suppl. 1), 6S–48S.

Noben-Trauth, K., Zheng, Q., Johnson, K.R., 2003. Association of cadherin 23 with polygenetic inheritance and genetic modification of sensorineural hearing loss. Nat. Genet. 35, 21–23.

Nondahl, D.M., Shi, X., Cruickshanks, K.J., Dalton, D.S., Tweed, T.S., Wiley, T.L., et al., 2009. Notched audiograms and noise exposure history in older adults. Ear Hear. 30, 696–703.

Nondahl, D.M., Cruickshanks, K.J., Wiley, T.L., Klein, B.E., Klein, R., Chappell, R., et al., 2010. The ten-year incidence of tinnitus among older adults. Int. J. Audiol. 49, 580–585.

Nondahl, D.M., Cruickshanks, K.J., Huang, G.H., Klein, B.E., Klein, R., Tweed, T.S., et al., 2012. Generational differences in the reporting of tinnitus. Ear Hear. 33, 640–644.

Ohlemiller, K.K., 2006. Contributions of mouse models to understanding of age-and noise-related hearing loss. Brain Res. 1091, 89–102.

Palmer, K.T., Griffin, M.J., Syddall, H.E., Davis, A., Pannett, B., Coggon, D., 2002. Occupational exposure to noise and the attributable burden of hearing difficulties in Great Britain. Occup. Environ. Med. 59, 634–639.

Petersen, M.B., Willems, P.J., 2006. Non-syndromic, autosomal-recessive deafness. Clin. Genet. 69, 371–392.

Petit, C., Levilliers, J., Hardelin, J.P., 2001. Molecular genetics of hearing loss. Annu. Rev. Genet. 35, 589–646.

Raviv, D., Dror, A.A., Avraham, K.B., 2010. Hearing loss: a common disorder caused by many rare alleles. Ann. N. Y. Acad. Sci. 1214, 168–179.

Ruel, J., Emery, S., Nouvian, R., Bersot, T., Amilhon, B., Van Rybroek, J.M., et al., 2008. Impairment of SLC17A8 encoding vesicular glutamate transporter-3, VGLUT3, underlies nonsyndromic deafness DFNA25 and inner hair cell dysfunction in null mice. Am. J. Hum. Genet. 83, 278–292.

Sanchez, L., 2004. The epidemiology of tinnitus. Audiol. Med. 2, 8–17.

Schink, T., Kreutz, G., Busch, V., Pigeot, I., Ahrens, W., 2014. Incidence and relative risk of hearing disorders in professional musicians. Occup. Environ. Med. 71, 472–476.

Shargorodsky, J., Curhan, G.C., Wildon, R., Farwell, W.R., 2010. Prevalence and characteristics of tinnitus among US adults. Am. J. Med. 123, 711–718.

Sloan-Heggen, C.M., Bierer, A.O., Shearer, A.E., Kolbe, D.L., Nishimura, C.J., Frees, K.L., et al., 2016. Comprehensive genetic testing in the clinical evaluation of 1119 patients with hearing loss. Hum. Genet. 135, 441–450.

Smith, R.J., Bale Jr., J.F., White, K.R., 2005. Sensorineural hearing loss in children. Lancet 365, 879–890.

Snoeckx, R.L., Huygen, P.L., Feldmann, D., et al., 2005. GJB2 mutations and degree of hearing loss: a multicenter study. Am. J. Hum. Genet. 77, 945–957.

Tambs, K., Hoffman, H.J., Borchgrevink, H.M., Holmen, J., Samuelsen, S.O., 2003. Hearing loss induced by noise, ear infections, and head injuries: results from the Nord-Trøndelag Hearing Loss Study. Int. J. Audiol. 42, 89–105.

Tambs, K., Hoffman, H.J., Borchgrevink, H.M., Holmen, J., Engdahl, B., 2006. Hearing loss induced by occupational and impulse noise: results on threshold shifts by frequencies, age and gender from the Nord-Trøndelag Hearing Loss Study. Int. J. Audiol. 45, 309–317.

Thys, M., Van Camp, G., 2009. Genetics of otosclerosis. Otol. Neurotol. 30, 1021–1032.

Van Eyken, E., Van Laer, L., Fransen, E., et al., 2006. KCNQ4: a gene for age-related hearing impairment?. Hum. Mutat. 27, 1007–1016.

Van Eyken, E., Van Laer, L., Fransen, E., Topsakal, V., Hendrickx, J.J., Demeester, K., et al., 2007. The contribution of GJB2 (connexin 26) 35delG to age-related hearing impairment and noise-induced hearing loss. Otol. Neurotol. 28, 970–975.

Van Laer, L., DeStefano, A.L., Myers, R.H., et al., 2002. Is DFNA5 a susceptibility gene for age-related hearing impairment?. Eur. J. Hum. Genet. 10, 883–886.

Vlastarakos, P.V., Nikolopoulos, T.P., Tavoulari, E., Papacharalambous, G., Korres, S., 2008. Auditory neuropathy: endocochlear lesion or temporal processing impairment? Implications for diagnosis and management. Int. J. Pediatr. Otorhinolaryngol. 72, 1135–1150.

Wichmann, C., 2015. Molecularly and structurally distinct synapses mediate reliable encoding and processing of auditory information. Hear. Res. 330, 178–190.

Ziff, J.L., Crompton, M., Powell, H.R.F., Lavy, J.A., Christopher, P., Aldren, C.P., et al., 2016. Mutations and altered expression of *SERPINF1* in patients with familial otosclerosis. Hum. Mol. Genet. Available from: http://dx.doi.org/10.1093/hmg/ddw106.

Zong, L., Guan, J., Ealy, M., et al., 2015. Mutations in apoptosis-inducing factor cause X-linked recessive auditory neuropathy spectrum disorder. J. Med. Genet. 52, 523–531.

Chapter 8

Early Diagnosis and Prevention of Hearing Loss

Humans hear long before they are born. The human cochlea is fully developed by 24 weeks of gestation. A blink-startle response can already be elicited (acoustically) at 24–25 weeks and is constantly present at 28 weeks. Hearing thresholds are 40 dB HL at 27–28 weeks and reach the adult threshold by 42 weeks of gestation, i.e., at term birth (Birnholz and Benacerraf, 1983). Still, the central auditory system needs adequate sound stimulation to develop in normal fashion. Studies in altricial animals—such as cats and mice that develop hearing about a week after birth—have demonstrated that sound deprivation during early development reduces the number of neuronal dendritic processes in the auditory cortex and distorts their normal geometry (Kral et al., 2000). In addition, it has been shown that reduced or absence of action potential firing in axons delays normal myelin development, and thereby the onset of auditory function. Thus, sound deprivation negatively impacts the maturation of both synaptic organization and axonal conduction in auditory cortex. These findings provide at least a partial explanation for the observed abnormalities in evoked potentials (Fig. 8.1) in deafness, as well as for the degree of recovery occurring after restoration of sound with a cochlear implant (see chapter: Cochlear Implants).

8.1 NORMAL HUMAN AUDITORY DEVELOPMENT

There are gradients of maturation in the auditory system. One gradient is characterized by early maturation of the brainstem and reticular activating system (RAS) pathways, followed by a later and very extended maturation of thalamocortical and intracortical connections. It is, however, clear that the specific lemniscal and extralemniscal auditory pathways mature at different and slower rates than the nonspecific RAS. The parallel processing in these three pathways may offer ultimately a top–down influence on processing of auditory information (Eggermont and Moore, 2012) and involve top–down mechanisms of perceptual learning (see chapter: Brain Plasticity and Perceptual Learning). Another gradient is that of the maturation of cells and

Hearing Loss. DOI: http://dx.doi.org/10.1016/B978-0-12-805398-0.00008-6

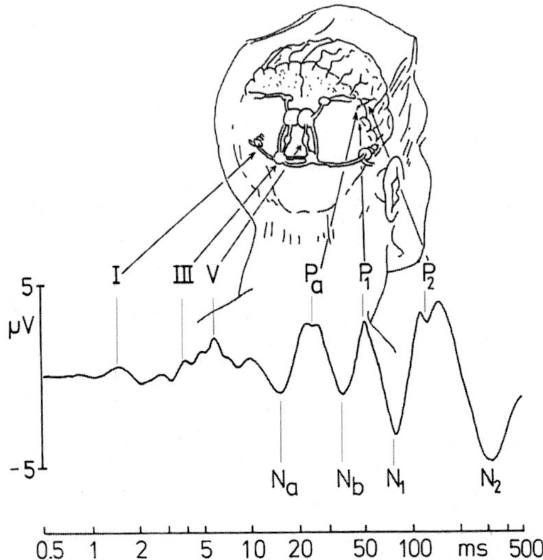

FIGURE 8.1 Auditory brainstem responses (ABRs) and obligatory cortical auditory evoked potentials (CAEPs) on a logarithmic timescale. The ABR components ("waves") are labeled I, III, and V. The middle latency components are indicated with N_a, P_a, N_b, P_b (P_1). The long-latency components are indicated with P_1, N_1, P_2, and N_2. Note that P_b typically overlaps with P_1. *Modified from Picton, T.W., Hillyard, S.A., Krausz, H.I., Galambos, R., 1974. Human auditory evoked potentials. I. Evaluation of components. Electroencephalogr. Clin. Neurophysiol. 36, 179–190, with permission from Elsevier.*

axons in cortical layers, initially in layer I followed by layers IV–VI and then upward to the superficial layers II–III (Moore and Guan, 2001).

The peripheral to cortical maturational processes suggest that the roughly two decades of human auditory maturation can be divided into several periods dictated by structural or functional temporal landmarks. One may divide anatomical development into a perinatal period (third trimester to 6 months postnatal), early childhood (6 months to 5 years), and late childhood (5–12 years) as done by Moore and Linthicum (2007). Although the division into these periods has its merits, Eggermont and Moore (2012) explored a separation of the maturational sequence as reflecting two major auditory processes: discrimination and perception of sound. The first developmental period is manifested by early auditory discrimination and is characterized by maturation in the brain stem, the RAS, and cortical layer I. This process is determined by increasing axonal conduction velocity and is largely complete at age 1 year, though fine-tuning occurs into the second year of life (Fig. 8.2).

Infants younger than the age of 6 months have the ability to discriminate phonemic speech contrasts in nearly all languages, a capability they later lose when raised in a one-language environment. In contrast, the histology of the brain in the first half-year of life indicates only a poor and very partial maturation of the auditory cortex. This discrepancy suggests either that

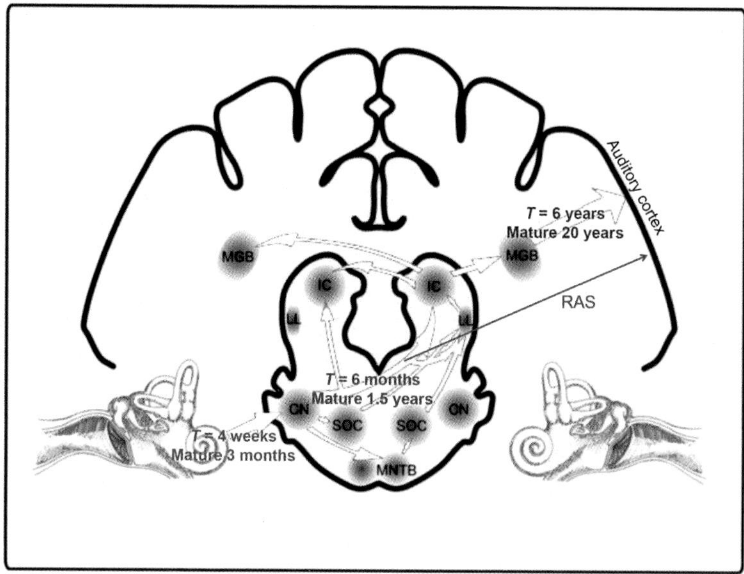

FIGURE 8.2 Functional maturation of the human auditory system. The fastest maturing system is formed by the cochlea and auditory nerve with a time constant $T = 4$ weeks and reaches maturation at approximately 3 months of age. The brain stem up to and including the fibers into the medial geniculate body matures with a time constant of approximately 6 months and reaches maturity at approximately 1.5 years of age. This also includes the maturation of the RAS pathway that innervates cortical layer I. The thalamus, the auditory radiation, and the cortex do not appear mature until approximately 20 years of age. CN, cochlear nucleus; IC, inferior colliculus; MGB, medial geniculate body; MNTB, medial nucleus of the trapezoid body; SOC, superior olivary complex. *Modified from Kral, A., Eggermont, J.J., 2007. What's to lose and what's to learn: development under auditory deprivation, cochlear implants and limits of cortical plasticity. Brain Res. Rev. 56, 259–269, with permission from Elsevier.*

infants rely largely on subcortical processing for this discrimination, or that the methods used in quantifying the structural and physiological properties of the auditory system are incomplete, or at least insensitive. It is likely that the cortical input in this period is mainly due to that provided by the early maturing RAS. A detailed argument is provided in the work of Eggermont and Moore (2012).

The second major maturational period reflects the development of auditory perception, the attribution of meaning to sound, with its neural substrate in cortical maturation. This process depends on synapse formation and increasing axonal conduction velocity and has a maturational onset between 6 months and 1 year. The age of 6 months is a behavioral turning point, with changes occurring in the infant's phoneme discrimination. This is more or less paralleled by regressive changes in the constituent makeup of layer I axons in the auditory cortex and the onset of maturation of input to the deep cortical layers. One could entertain the idea that, at about 6 months of age, the cortex starts to exert a modulating or gating influence on subcortical processing via efferents from the maturing layers IV–VI, resulting among others in the loss of discrimination of foreign language contrasts. The period between 2 and 5 years of age, the time of development of perceptual language, is characterized by a relatively stable level of cortical synaptic density that declines by 14 years of age (Huttenlocher and Dabholkar, 1997). In later childhood, a continued improvement of speech in reverberation and noise, and of sound localization, is noted. At the end of the maturational timeline, one usually considers the hearing of young adolescents as completely adult like. However, speech perception in noise and reverberant acoustic environments does not mature until around age 15. This should be considered in the design of class rooms and the management of sound levels during class.

The maturation of auditory anatomy and behavior is reflected in progressive changes in electrophysiological responses. At approximately 2 years of age, the electrophysiological measures of auditory function in the form of the auditory brain stem response, middle latency response, the late P_2 component of the cortical auditory-evoked potentials, and the mismatch negativity are fully mature. At about 6 years of age, the long-latency ($\sim$100 ms) N_1 component of the CAEP is typically not recordable with stimulus repetition rates above 1/3 s. Reliability improves over the next 5 years, and the N_1 is detectable in all 9- to 10-year-olds at stimulus rates of approximately 1/s. Age 12 and up is characterized by major transient changes in the cortical evoked potentials that are likely related to the onset of puberty (Ponton et al., 2000) and functional aspects of this perceptual process continue to change well into adulthood (Fig. 8.3). Although behavioral measures of auditory perception are mostly adult-like by age 15, the maturation of long-latency CAEPs continues for at least another 5 years thereafter. This may suggest the need for additional behavioral studies in adolescents and provides yet another example of the relative strengths and weaknesses of alternative methods in the evaluation and interpretation of human auditory maturation. However, not all detectable electrophysiological or structural changes need to be behaviorally relevant. Early diagnosis as provided by universal newborn hearing screening (UNHS) is the most relevant and will be discussed in Section 8.3.

8.2 EFFECTS OF EARLY HEARING LOSS ON SPEECH PRODUCTION

The value of early determination of hearing loss was reported by Yoshinaga-Itano et al. (1998). They found that early identification of hearing loss and early intervention resulted in significantly better language development. This was exemplified by children whose hearing losses were identified by 6 months of age that demonstrated significantly better language scores than children identified after 6 months of age. Downs and Yoshinaga-Itano (1999)

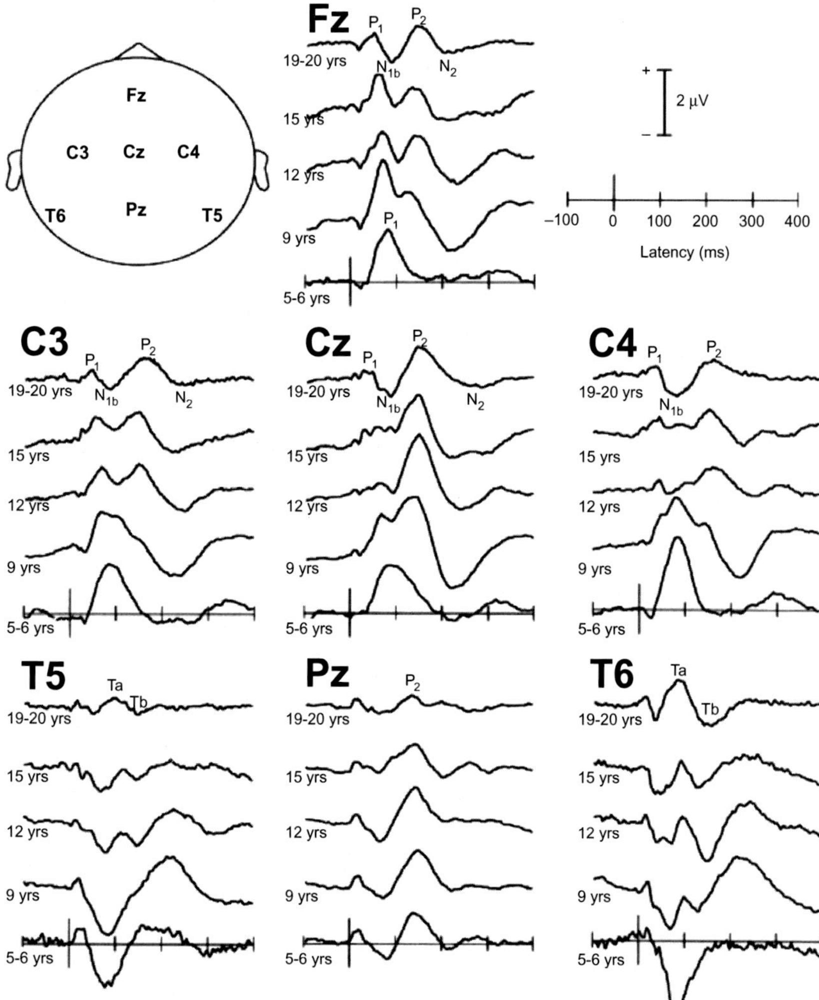

FIGURE 8.3 Age-dependent morphology of the AEPs for different recording sites. Note the late (~9 years) appearance of N_1 (here labeled N_{1b}) in the Fz and Cz recordings. Note the (incomplete) phase reversal for the temporal (T) and parietal (P) electrodes compared to the central (C) and frontal (F) ones. *Reprinted from Ponton, C.W., Eggermont, J.J., Kwong, B., Don, M., 2000. Maturation of human central auditory system activity: evidence from multichannel evoked potentials. Clin. Neurophysiol. 111, 220–236, with permission from Elsevier.*

drew the conclusion that "identification of hearing loss by 6 months of age, followed by appropriate intervention, is the most effective strategy for the normal development of language in infants and toddlers with hearing loss. Identification of hearing loss by 6 months can only be accomplished through universal newborn hearing screening." Moeller et al. (2007a) compared the vocalizations 21 infants with normal hearing and 12 early-identified infants with hearing loss over a period of 14 months (from 10 to 24 months of age). They found that children with hearing loss were delayed in the onset of consistent canonical babble relative to age-matched controls. This suggested to them that consonant development in infants with hearing loss was delayed but not qualitatively different from children with normal hearing. Moeller et al. (2007b) extended this to the transition from babble to words and found that both groups increased the purposeful use of voice between 16 and 24 months of age. However, the hearing loss group was much slower to develop expressive vocabulary and demonstrated larger individual differences than the normal hearing group. The delay had significant effects on expressive vocabulary development. Sininger et al. (2010) looked at the effects of the age of hearing aid or cochlear implant fitting. They measured potential predictors such as age at fitting of amplification (1 month to 6 years), degree of hearing loss in the better hearing ear (ranging from mild to profound), and cochlear implant use. Age at fitting of amplification showed the largest influence and was a significant factor in all outcome models. The degree of hearing loss and cochlear implant use were important factors in the modeling of speech production and spoken language outcomes. These studies strongly suggest the need for early identification of hearing loss.

8.3 EARLY DETECTION

Early studies into the feasibility of newborn hearing screening include Alberti et al. (1983) who evaluated the use of ABRs in 321 high-risk newborns. Of those, 234 received predischarge tests and 200 had follow-up tests. Screening ABR with 40 dB nHL clicks appeared appropriate in the nursery. Screening sensitivity was good, and only 8% of babies failed. Follow-up ABR after 3 months confirmed hearing loss in eight babies. Swigonski et al. (1987) screened an extremely high-risk group of 137 infants from a neonatal intensive care unit (NICU). Of the 137 infants tested, 82 passed the initial ABR, 22 conditionally passed, and 34 failed. Follow-up behavioral and audiometric testing was done in 82 infants. Four infants had severe sensorineural hearing loss (SNHL), each of whom had failed the initial ABR. None of the infants who initially passed or conditionally passed the ABR had SNHL on follow-up testing. An important precursor for UNHS was the study by White et al. (1994) who screened 1850 infants from the well-baby nursery and NICU using transient evoked otoacoustic emissions (TEOAEs) (see chapter: Hearing Basics) in a two-stage process. Infants referred from the first stage prior to being discharged from the hospital were rescreened 4–6 weeks later. Those who did not pass the second-stage TEOAE screening were referred for diagnostic ABR and/or behavioral audiological evaluation for confirmation of hearing loss, fitting with amplification, and enrollment in early intervention programs. They identified 11 infants with unilateral or bilateral SNHL more than 25 dB (a prevalence of 5.95 per 1000) and 37 with unilateral or bilateral recurrent conductive hearing loss more than 25 dB (a prevalence of 20.0 per 1000). Sininger et al. (1999) emphasized critical periods in neural development of hearing as an important incentive for UNHS. In another prequel to UNHS procedures, Brown et al. (2000) assessed 149 newborns who had normal automated ABRs and also had distortion product otoacoustic emissions (DPOAEs; chapter: Hearing Basics) measured. They found that the DPOAE pass rates were lower at low frequencies, likely due to lower signal-to-noise ratios, and that test results could be improved by eliminating frequencies below 2.0 kHz.

A Joint Committee on Infant Hearing statement (JCIH, 1995) had "endorsed the goal of universal detection of infants with hearing loss and encourages continuing research and development to improve techniques for detection of and intervention for hearing loss as early as possible."

8.3.1 Universal Newborn Hearing Screening: A Survey

An overview of a representative group of large-scale neonatal hearing screening programs is shown in Table 8.1. In the following we present some additional details, more or less in chronological order.

Mason and Hermann (1998) performed newborn hearing screening with automated ABR in 10,372 infants born during a 5-year period in Hawaii. The false-positive rate was 3.5% after the initial screening and 0.2% when a two-stage screening procedure was used. They found an incidence of congenital bilateral hearing loss in the well population of 1/1000 and in the NICU population of 5/1000.

Following the Joint Committee on Infant Hearing statement, a feasibility study sponsored by the National Institutes of Health in the United States was set up to determine the accuracy of three measures of peripheral auditory system status (TEOAEs, DPOAEs, and ABR thresholds) applied in the perinatal period. In this study, all graduates of involved NICUs and healthy babies with one or more risk factors for hearing loss were targeted for follow-up testing using visual reinforcement audiometry (VRA) at 8–12 months of age (Norton et al., 2000a). Automated ABR was implemented using a 30 dB nHL click stimulus, which appeared to be reliable for the rapid assessment of hearing in newborns. More than 99% of infants could complete the ABR protocol. More than 90% of NICUs and well-baby nursery infants "passed" given the strict criteria for response, whereas 86% of those with high-risk factors met the criterion for ABR detection (Sininger et al., 2000). DPOAE measurements in neonates and infants resulted in robust responses in the vast majority of ears for f2 frequencies for at least 2.0, 3.0, and 4.0 kHz (Gorga et al., 2000). All NICU infants and healthy babies with risk factors (including healthy babies who failed neonatal tests) were targeted for follow-up VRA evaluation once they had reached the 8 months corrected

TABLE 8.1 Newborn Hearing Screening

Country	Year of Report	*N* (% of Population)	TEOAE	AABR	Pass (%)	Clinical Follow-Up	Incidence Bilateral Hearing Loss
USA (Hawaii)	1998[1]	10,372		+	96%	ABR	1/1000
USA	2000[2–5]	7179	+	+		VRA	20/1000[a]
Italy	2006[6]	158,048 (79.5%)	+	+		ABR	0.72/1000
Poland	2008[7]	1,392,427 (96.3%)	+		95.6%	ABR	1.3/1000
Sweden	2007[8]	14,287 (99.1%)	+		97%		
Germany (Hessen)	2006[9]	17,439	+	+	97%	ABR	1.14/1000
Netherlands	2010[10]	335,560	+			At age 3–5	0.78/1000
Sweden	2011[11]	31,092	+	+	98.8%	ABR	1.8/1000
Belgium (Flanders)	2012[12]	103,835 (100%)		+	99.3%	ABR	0.87/1000
Belgium (French part)	2014[13]	263,508 (100%)	+		97.6%	ABR	1.41/1000
Taipei City	2013[14]	15,790 (99.1%)		+			1.4/1000
Israel (Zefat)	2013[15]	5212 (94.8%)	+	+	94.8%	ABR	1.5/1000
Turkey (Corlu)	2014[16]	11,575	+	+	94.9%	ABR	1.3/1000
Brazil (São Paulo)	2014[17]	929	+		95.9%	ABR	9.1/1000[b]

VRA, visual reinforcement audiometry; AABR, automated ABR.
[1]Mason and Hermann (1998), [2]Sininger et al. (2000), [3]Gorga et al. (2000), [4]Widen et al. (2000), [5]Norton et al. (2000a,b), [6]Bubbico et al. (2008), [7]Szyfter et al. (2008), [8]Hergils (2007), [9]Neumann et al. (2006), [10]Korver et al. (2010), [11]Berninger and Westling (2011), [12]Van Kerschaver et al. (2012), [13]Vos et al. (2014), [14]Huang et al. (2013), [15]Gilbey et al. (2013), [16]Ulusoy et al. (2014), and [17]Colella-Santos et al. (2014).
[a]*NICU plus high-risk group only.*
[b]*NICU only*

age. More than 95% of the infants could reliably be tested and 90% provided complete tests (Widen et al., 2000). Accuracy for the OAE measurements was best when the speech awareness threshold or the pure-tone average for 2.0 kHz and 4.0 kHz was used as the gold standard. ABR accuracy varied little as a function of the frequencies included in the gold standard (Norton et al., 2000b).

In Italy (Bubbico et al., 2008) UNHS coverage had undergone a steep increase from 29.3% in 2003 to 48.4% in 2006. The majority of UNHS programs were implemented in the northwest and northeast areas. The Polish Universal Neonatal Hearing Screening Program started in 2002 in all neonatal units in Poland (Szyfter et al., 2008). Between 2003 and 2006 a total number of 1,392,427 children were screened for hearing impairment. The first Swedish UNHS program included over 33,000 measurement files from 14,287 children at two maternity wards (Hergils, 2007). Test performance was clearly better when the children were tested day 2 after birth or later. In a cohort study the outcome of the UNHS program in the German state of Hessen in 2005 was analyzed (Neumann et al., 2006). Forty-nine hearing-impaired children out of 17,439 tested were diagnosed at a median age of 3.1 months and treated at a median age of 3.5 months.

Between 2002 and 2006, all 65 regions in the Netherlands replaced distraction hearing screening, conducted at 9 months of age, with newborn hearing screening (Korver et al., 2010). Consequently, the type of hearing screening offered was based on availability at the place and date of birth and was independent of developmental prognoses of individual children. All children born in the Netherlands between 2003 and 2005 were included. At the age of 3−5 years, all children with permanent childhood hearing impairment were identified. Evaluation ended in December 2009. During the study period, 335,560 children were born in a newborn hearing screening region and 234,826 children in a distraction hearing screening region (not shown in Table 8.1). At follow-up, 263 children in newborn hearing screening regions (0.78 per 1000 children) and 171 children in distraction hearing screening regions (0.73 per 1000 children) had been diagnosed with permanent childhood hearing impairment. Compared with distraction hearing screening, a newborn hearing screening program was associated with better developmental outcomes at age 3−5 years among children with permanent childhood hearing impairment.

Van Kerschaver et al. (2012) tested the entire population of term newborns in Flanders, Belgium, by a UNHS program. Follow-up diagnosis was done in specialized referral centers. Lammens et al. (2013) determined the etiology of hearing loss detected by this Flemish screening program. From 1997 to 2011, 569 neonates out of 103,835 were referred to the referral center after failed neonatal screening. A retrospective review was performed. In 35% of the subjects no obvious etiology could be determined. This series showed a genetic syndromic cause in 80% of the genetic bilateral hearing loss cases, whereas connexin26-positive diagnoses were underrepresented. Vos et al. (2014) presented data from the 2007 to 2012 screening period from the French speaking area in Belgium. Over the screening period, only 62.21% of the referred newborns had a follow-up; the follow-up rate was particularly low for the first year (44.91%) and then strongly increased (+19.52% in 2008) but never exceeded 70%. Berninger and Westling (2011) reported data from bedside UNHS programs at Karolinska University Hospital and at Södertälje Hospital in Sweden. The recorded multiple TEOAEs and found that this reduced the need for ABRs.

Examples of some smaller scale or exploratory screenings included 85% of the delivery units in Taipei City, which include 20 hospitals and 14 obstetrics clinics, were recruited into the screening program in two stages from September 2009 to December 2010 (Huang et al., 2013). Sixty-four percent (14/22) of babies with bilateral hearing loss completed the full diagnostic hearing tests within 3 months of birth. Gilbey et al. (2013) evaluated a newly established universal newborn hearing screening program at the Ziv Medical Center in Zefat, Israel. Screening results of all neonates born from the initiation of the program on March 15, 2010 until the end of 2011 were reviewed. Ulusoy et al. (2014) reviewed the newborns from Çorlu State Hospital in the west part of Turkey or referred from other Health Care Centers, between September 2009 and November 2012. Colella-Santos et al. (2014) described the outcome of screening using ABR and audiological diagnosis in 929 neonates from the NICU at the Woman's University Hospital (CAISM) at State University of Campinas (São Paulo, Brazil). See Table 8.1 for outcome details.

8.3.2 Potential Problems with UNHS and Follow-Up Studies

One of the first realizations that UNHS could face potential problems was voiced by Durieux-Smith et al. (1991). They compared the results of click ABR in infants of an NICU to those obtained on the same children with pure-tone audiometry at 3 years of age. They initially tested 600 infants and follow-up in 333. In 297 (89%) the click ABRs accurately predicted the hearing status at the age of 3 years. Twenty-nine (9%) of the discrepancies were related to conductive hearing losses. Six patients assessed as normal by click ABR had significant hearing losses at the age of 3 years. Five of these had normal hearing at one frequency between 1000 and 4000 Hz, so that the click response was still normal, the remaining one may have acquired SNHL.

In the United Kingdom, a 10-year cohort of 35,668 births was enrolled into a UNHS and was followed up until the children had completed the first year of primary school (Watkin and Baldwin, 2011). The cohort followed up was born from September 1992, when UNHS coverage had been optimized, until 2002 when the UNHS program was introduced UK wide. There were 3.65/1000 children with a permanent hearing impairment of any degree embarking on their education; 1.51/1000 had a moderate or worse bilateral deafness but only 0.9/1000 with this degree of deafness had been identified by UNHS. An additional postneonatal yield of 1.2/1000 had mild or unilateral impairments. When all degrees of impairment were considered, half of the children with a permanent hearing impairment had required identification by postneonatal care testing. Watkin and Baldwin (2011) concluded that UNHS is an insufficient diagnostic and that postneonatal testing remains essential.

Stevens et al. (2013) determined the accuracy with which tone pip ABR and click ABR, carried out in babies referred from UNHS, is able to predict the hearing outcome from follow-up tests. The cohort consisted of all babies referred for hearing assessment from the UNHS in Sheffield, UK for the period January 2002 to September 2007, who were found to have a significant hearing impairment. The results of hearing assessment following referral from the UNHS were compared with follow-up test results carried out up to an age when behavioral testing had established ear- and frequency-specific thresholds at 0.5, 1, 2, and 4 kHz. The study showed that tone pip ABR following referral from newborn hearing screening has a similar accuracy to that reported in older subjects and is a much better predictor compared to click ABR (as already suggested by Eggermont, 1982).

Dedhia et al. (2013) described the findings of children who passed their UNHS in the Children's Hospital in Pittsburgh (PA) and were subsequently found to have childhood hearing loss. They identified 923 children with hearing loss from 2001 to 2011 of which 78 were included in the follow-up. The suspicion of hearing loss in patients who passed the UNHS was most often from parental concerns and failed school hearing screens. Thirty-seven patients (47%) had severe or profound hearing loss. The etiology was unknown in 42 patients (54%); the remaining was attributed to genetics (17%), anatomic abnormality (14%), acquired perinatal (12%), and auditory neuropathy (4%). This study suggests that follow-up screening is needed in children who passed the UNHS but received subsequent concerns about their hearing.

Ching et al. (2013) addressed the question whether early detection and amplification improved outcomes of children with hearing impairment. They used regression analysis to predict global outcomes for 356 children including 15 predictor variables and found that this explained 40% of the total variance. They did not find evidence for age of amplification as a significant predictor of children's outcomes at 3 years of age. However, the age at which the first cochlear implant was switched on was associated with outcomes at 5% significance level. This implied to Ching et al. (2013) that UNHS and early auditory intervention are important for improving outcomes of children with severe or profound hearing loss, as early implantation would not have been possible without early detection.

Ohmori et al. (2015) studied the effect of the UNHS program in Okayama prefecture in Japan on language development. For that purpose,

105 7-year-old children with hearing impairment were tested. They found that the adjusted Picture Vocabulary Test score and number of productive words were significantly higher in the post-NHS group than the pre-NHS group. For receptive vocabulary the odds ratios were 2.63 (95% confidence interval (CI) = 1.17–5.89) and for productive vocabulary 4.17 (95% CI = 1.69–10.29).

Wood et al. (2015) assessed the performance of the UNHS in England in a retrospective analysis of 4,645,823 children born April 1, 2004 to March 31, 2013. They found that 97.5% of the eligible population completed screening by 4–5 weeks of age and 98.9% completed screening by 3 months of age. The rate of referral for the 2012–13 birth cohort was 2.6%. From the referred rate the percentage of babies followed up by 4 weeks of age was 82.5% and at 6 months of age was 95.8%. Consequently, the performance of the newborn hearing screening program has improved with time.

Wake et al. (2016) studied the efficacy of UNHS in comparison with two other cohorts that used "risk factor screening (RFC)" and "opportunistic detection" in three Australian states with comparable demographics and services. RFC involves systematically identifying and referring for audiologic testing all infants with risk factors for hearing loss, coupled with predischarge hearing screening of infants admitted to the NICU. UNHS was applied to 172,523 newborns, and 123,855 were subjected to RFC. In the two screening programs, 216 were detected with UNHS, and 171 with RFC, of which 69 and 65 in their early school years participated respectively. They found that children were diagnosed at a younger age with UNHS than with RFC (adjusted mean difference −8.0 months, 95% CI −12.3 to −3.7). They concluded that: "Population language outcomes by the early school years benefited incrementally on moving from opportunistic detection to systematic risk factor screening to universal newborn hearing screening."

Summarizing, newborn hearing screening is still not universal, but the implementation becomes more reliable, albeit not the last word in predicting hearing loss. Postneonatal testing remains essential.

Regardless the early detection of hearing loss and follow-ups, the children and adolescents who passed the UNHS are still likely to be exposed to potentially damaging sound levels, after school, at work or by long duration exposure to personal listening devices (PLDs). We will introduce the potential damaging conditions and continue with protection methods including education to prevent acquired hearing loss. We then continue with potential roles of drug protection against NIHL.

8.4 NOISE EXPOSURE DURING ADOLESCENCE AND YOUNG ADULTHOOD

Noise exposure during adolescence and young adulthood becomes more and more of a concern. This is not limited to occupational noise but also to the extensive use of the PLDs. We illustrate this with a few example studies (see also chapter: Causes of Acquired Hearing Loss and Eggermont, 2014). This sets the stage for hearing protection in Section 8.5.

Ivory et al. (2014) stressed that: "Noise exposure is often thought in terms of occupational hazard; however, there are many recreational activities that can expose individuals to hazardous levels of noise." Engard et al. (2010) collected personal noise exposure samples from five workers at a large-sized college football stadium and five workers at a medium-sized college football stadium in northern Colorado (USA) during three home football games, for a total of 30 personal noise exposures. None of the workers' noise doses were above the Occupational Safety and Health Administration (OSHA) permissible exposure limit of 90 dB(A). However, 11 of 28 (39%) workers' noise doses exceeded the OSHA action level of 85 dB(A) that would require enrollment in a hearing conservation program. One has to consider that nonathlete participants, such as sporting officials, are also at risk for noise-related hearing loss for instance from chronic whistle use (Flamme and Williams, 2013).

8.5 PHYSICAL HEARING PROTECTION

8.5.1 After Work Music

Reducing the risk of hearing damage from music venues is particularly challenging because the exposure is voluntary, occurs during leisure time, and varies widely between individuals, making blanket regulation difficult. Although workplace health and safety laws are in place to ensure a healthy environment for workers at such venues, no such regulations are in place to limit the noise level to which the patrons are exposed. Rather, the responsibility for taking the initiative to reduce the risk of hearing damage is left to the patrons themselves (Beach et al., 2016).

Beach et al. (2010) conducted structured telephone interviews with 20 regular nightclub patrons. Participants were asked about their experience of wearing earplugs and, in particular, what they perceive to be the advantages and disadvantages of earplugs. Foam earplugs were considered less satisfactory than more expensive earplugs, which are relatively discreet and comfortable, and facilitate communication with others. Beach et al. (2013) subsequently investigated 1000 18- to 35-year-old Australian adults participating in high-noise leisure activities. Annual noise exposure from the leisure activities ranged from 0 to 6.77 times the acceptable noise exposure and was correlated with early warning signs of hearing damage and perceived risk of damage.

Kelly et al. (2012) examined the compliance with new noise regulations in nine nightclubs in Ireland. Using two logging dosimeters and a fixed-position sound level meter they found that the average bar employee daily noise exposure was 92 dB(A), almost four times (12 dB) more than the accepted legal limit. Barlow and Castilla-Sanchez (2012) assessed four public music venues in the United Kingdom where live and/or recorded music is regularly played. Thirty staff members in different roles in the venues were monitored using noise dosimetry to determine noise exposure. The majority of staff (70%) in all venues exceeded the daily noise exposure limit value in their working shift. Use of hearing protection was rare (<30% of staff) and not enforced by most venues. The implication is that the music venue industry is failing to meet regulatory requirements. Johnson et al. (2014) investigated whether excessive nightclub sound levels, i.e., more than 85 dB(A) are a direct cause of auditory symptoms related to noise-induced hearing loss. A questionnaire was completed by 325 students and showed that "88.3 per cent of students experienced tinnitus after leaving a nightclub and 66.2 per cent suffered impaired hearing the following morning. In terms of behavior, 73.2 per cent of students said that the risk of hearing damage would not affect their nightclub attendance."

8.5.2 An Interlude About Earplugs

In a systematic review (Kraaijenga et al., 2016) found only one well-conducted randomized clinical trial indicating that wearing earplugs to concerts is effective in reducing postconcert threshold shifts. Bockstael et al. (2015) compared the listening experience and temporary effects on hearing with different types of earplugs after exposure to contemporary club music (study not included in the Kraaijenga et al., 2016 review). Five different types of commercially available and commonly used hearing protectors were worn. Four of them were premolded earplugs and the fifth type was a foam earplug frequently distributed for free at music events (Fig. 8.4A). During five different test sessions of 30 min each, participants not professionally involved in music wore one particular type of protector. Contemporary club music was played at sound pressure levels (SPLs) representative of concerts and bars. After each listening session, a questionnaire on sound quality and general appreciation was completed. OAEs were measured directly before and after music exposure. The reported appreciation also clearly differed per earplug type, whereas the reported appreciation mainly depended on comfort and looks. Differences in sound quality were less noticeable. The changes in OAE amplitude before and after noise exposure were small.

The assumed attenuation reported by the manufacturers for the five earplugs tested is shown in Fig. 8.4B. Earplugs 2, 3, and 4 showed a more or less flat spectrum. By contrast, for Earplug 1 the attenuation clearly increased with increasing frequency. As expected, standard Earplug 5 had the highest attenuation. Bockstael et al. (2015) found that Earplug 1 was significantly more likely to be bought than both the standard earplug and Earplug 2 (Fig. 8.4C). Earplug 4 was also chosen significantly over the

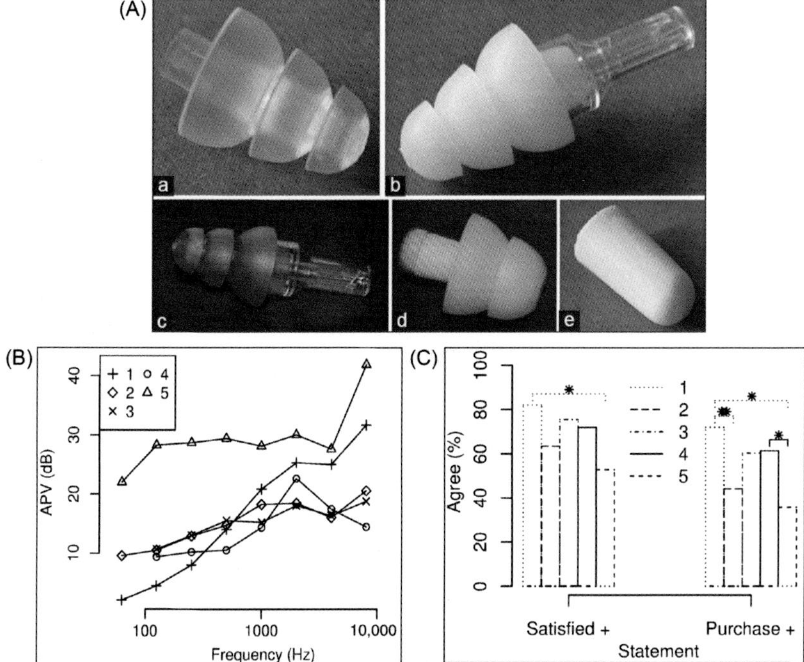

FIGURE 8.4 (A) Included earplugs in the test (1–5). Note that in the graphs a correspondence between a–e and 1–5 respectively is used. Between brackets, the approximate maximal diameter of the plug fitting the ear canal is given, followed by the length of the stem for the musician earplugs. (a) Earplug 1 (11 mm; 14 mm); (b) Earplug 2 (13 mm; 16 mm); (c) Earplug 3 (9 mm; 16 mm); (d) Earplug 4 (11 mm; 8 mm); (e) Earplug 5 (11 mm). (B) Assumed protection value (APV) reported by the manufacturer for the five hearing protectors tested. (C) Percentage of participants agreeing with the statement that they are generally satisfied (satisfied +) with their protector and that they would purchase the protector (purchase +) per hearing protector type (1–5). Significant pairwise contrasts are indicated by * for $p < 0.05$ and ** for $p < 0.01$. *From Bockstael, A., Keppler, H., Botteldooren, D., 2015. Musician earplugs: appreciation and protection. Noise Health 17, 198–208.*

standard earplug. The participants significantly preferred wearing Earplug 1 compared to the standard Earplug 5.

Ackermann et al. (2014) conducted a cross-sectional population (377 musicians (about 70%) from eight Australian symphony orchestras) survey. They found that noise exposure was high in private practice, although awareness of risk and earplug use in this environment was lower than in orchestral settings. O'Brien et al. (2014) tested active (electronic) earplugs, i.e., the MP-915 Musicians Electronic Earplug, produced by Etymotic Research using acoustic equipment and musician's ears. The earplug uses a 312 zinc-air battery and transmits sound to the ears via a processor and receiver, similar to hearing aid technology. The earplug may be fitted using generic "ear buds" of various sizes or by using sleeves custom-molded to the shape of the user's ear canal. O'Brien et al. (2014) concluded: "Although the active earplugs used in this study are yet to produce true hi-fidelity sound, they are more suited to the orchestral environment than previously available musician's earplugs and less likely to cause problems common with passive earplugs such as players hearing other musicians and the need to frequently remove and replace the devices as sound levels vary."

8.6 EDUCATION

8.6.1 Changing the Attitude About Noise Exposure

Are adolescents aware of the effects of noise exposure? Instilling awareness about noise levels and their impact on the auditory system may be the most important front-line method in prevention.

Vogel et al. (2010a) assessed adolescents' music-listening risk and protective behaviors in discotheques. The majority of the adolescents visited discotheques, of which (according to self-reports) about 25% were classified as frequent visitors, and a significant proportion of visitors reported to stand mostly within 2 m of the loudspeakers. Hardly any of them used hearing protection despite volume levels of about 105 dB(A). To arrive at more general prevention strategies and interventions, Vogel et al. (2010b) asked 1687 adolescents (12–19 years old) at Dutch high schools to complete questionnaires about their music-listening behaviors. Over 70% of participants reported to have visited discotheques; 24.6% of them were categorized as visitors at risk for hearing loss due to estimated exposure of 100 dB(A) for at least 1.25 h per week without the use of hearing protection. Vogel et al. (2010b) suggested that voluntary behavior change among adolescents might be difficult to achieve, because visiting discotheques seems to be strongly linked to current adolescent lifestyle. From the same questionnaires, it was estimated whether and to what extent a group of adolescents were at risk of developing permanent hearing loss as a result of voluntary exposure to high-volume music (Vogel et al., 2010c). About half of the adolescents exceeded safety standards for occupational noise exposure. About one-third of the respondents exceeded safety standards solely as a result of listening to MP3 players. Hearing symptoms that occurred after using an MP3 player or going to a discotheque were associated with exposure to high-volume music. Continuing with this same population, Vogel et al. (2011) investigated correlates of adolescents' risky MP3-player listening behavior. Of all participants, 90% reported listening to music through earphones on MP3 players; 28.6% were categorized as listeners at risk for hearing loss due to estimated exposure of 89 dB(A) for more than 1 h per day. Habit strength was the strongest correlate of risky listening behavior, again suggesting that voluntary behavior change among adolescents might be difficult to achieve and that a multiple strategy approach may be needed to prevent MP3-induced hearing loss. Vogel et al. (2014) then estimated the extent to which exposure to music through earphones or headphones with MP3 players or at discotheques and pop/rock concerts exceeded current occupational safety standards for noise exposure. For that purpose, 943 students in Dutch inner-city senior-secondary vocational schools completed questionnaires about their sociodemographics, music listening behaviors, and health. They found that about 60% exceeded safety standards for occupational noise exposure. About 10% of the participants experienced permanent hearing-related symptoms. Temporary hearing symptoms were also associated with exposure to high-volume music.

Compared to the study of Vogel et al. (2011) lower MP3 volume settings were found in by McNeill et al. (2010). Twenty-eight university students (12 males, 16 females; aged 17–23) completed a 49-item questionnaire assessing user listening habits and subjective measures of hearing health. Sound level measurements of participants' self-identified typical and "worst case" volume levels were taken in different classrooms with background sound levels between 43 and 52 dB(A). The median frequency and duration of use was 2 h per day, 6.5 days a week. The median sound levels at typical and "worst case" volume settings were 71 and 79 dB(A), respectively. When typical sound levels were considered with self-reported duration of daily use, none of the participants surpassed Leq(8) = 85 dB(A). On the questionnaire, 19 students reported experiencing at least one symptom of possible noise-induced hearing loss. Significant differences in MP3 user listening patterns were found between respondents who had experienced tinnitus and those who had not.

Dell and Holmes (2012) assessed the effectiveness of a hearing conservation program in changing knowledge and attitudes toward exposures to high-intensity sounds or noise among adolescents. The participants were grade six to eight students enrolled in physical education classes from the University of Florida's Developmental Research School. Sixty-four of these students between the ages of 12 and 14 received pre- and posttest measures and participated in the hearing conservation program. A statistically significant reduction in pro-noise attitudes among the adolescents was present after the intervention. This hearing conservation program positively changed noise attitudes among adolescents.

Portnuff et al. (2013) evaluated the use of PLDs and the relationships between self-report measures and long-term dosimetry measures of listening habits. Participants ($N = 52$) were 18- to 29-year-old men and women who completed surveys. A randomly assigned subset ($N = 24$) of participants had their listening monitored by dosimetry for 1 week. They found that "median weekly noise doses reported were low (9–93%), but 14.3% of participants reported exceeding a 100% noise dose weekly. Dosimetry indicated that 16.7% of participants exceeded a 100% noise dose weekly. Thus, among the participants in this study, a small but substantial percentage of PLD users incurred exposure from PLD use alone that increases their risk of music-induced hearing loss."

Beach et al. (2016) allocated 51 regular patrons of a nightclub to either a low-information (lo-info) or high-information (hi-info) group. The hi-info group was also provided with audiovisual and written information about the risks of excessive noise exposure. Both groups were provided with one-size-fits-all filtered music earplugs. After 4 weeks, and again after an additional 12 weeks, participants were asked about their recent earplug usage, intention to use earplugs in the future, and perceived risk of hearing damage. Beach et al. (2016) found that "the provided information was unnecessary to motivate behavioral change. Rather, the simple act of providing access to earplugs appears to have effectively encouraged young at-risk adults to increase their earplug use."

Martin et al. (2013) evaluated the effectiveness of four NIHL prevention interventions at improving knowledge, attitudes, and intended behaviors regarding sound exposure and appropriate use of hearing protective strategies in children. Questionnaires were completed prior to, immediately after, and 3 months after each intervention, which included: (1) a classroom presentation by older-peer educators, (2) a classroom presentation by health professionals, (3) exploration of a museum exhibition, and (4) exploration of an Internet-based virtual museum. A comparison group received no intervention. Fifty-three fourth grade classrooms (1120 students) participated in the study. Martin et al. (2013) conclude that: "all interventions produced significant improvements but the number of improvements decreased over time. In terms of effectiveness, the classroom programs were more effective than the internet-based virtual exhibit, which was more effective than the visit to the museum exhibition."

8.6.2 National Campaigns

Meinke and Morata (2012) described the rationale and creation of a national award to recognize and promote hearing loss prevention. In 2007, the National Institute for Occupational Safety and Health partnered with the National Hearing Conservation Association to create the Safe-in-Sound Excellence in Hearing Loss Prevention Award (www.safeinsound.us, accessed October 14, 2015). An expert committee developed specific and explicit award evaluation criteria of excellence in hearing loss prevention for organizations in different industrial sectors. The general approach toward this was to incorporate current "best practices" and familiar benchmarks of hearing loss prevention programs. In addition, mechanisms were identified to measure the impact of the award itself. Interest in the award was recorded through the monitoring of the visitor traffic registered by the award Web site and is increasing yearly. The Safe-in-Sound Award has obtained high-quality field data, identified practical solutions, disseminated successful strategies to minimize the risk of hearing loss, generated new partnerships, and shared practical solutions with others in the field. More recent information about advocacy programs and their implementation in some industries can be found in the study of Murphy (2016).

Gilles and Van de Heyning (2014) investigate whether a preventive campaign could alter attitudes toward noise in adolescents and whether this resulted in an increase of hearing protection use in this population. The governmental preventive campaign (PrevC) was released in the Flanders, the Dutch speaking part of Belgium, in order to prevent hearing damage caused by noise exposure. PrevC had the intention to make young people more aware of the risks of loud music and therefore increase the use of hearing protection in noisy situations and to effectuate a more controlled and responsible use of PLDs. The campaign was promoted via television and radio commercials, social network sites (Facebook/Twitter), posters, and a Web site (www.ietsminderisdemax.be). A cohort of 547 Flemish high school students, aged 14–18 years old, completed a questionnaire prior to and after a governmental campaign focusing on the harmful effects of recreational noise and the preventive use of hearing protection. Gilles and Van de Heyning (2014) found a more negative attitude toward noise and a more positive attitude toward hearing protection. The use of hearing protection increased significantly from 3.6% prior to 14.3% post campaign in students who were familiar with the campaign.

8.7 DRUG PROTECTION AGAINST NOISE-INDUCED HEARING LOSS

The following is partly based on a previous review by Eggermont (2014). In essence two major routes lead to cochlear hair cell loss: apoptosis and necrosis. Apoptosis is an active, energy-requiring process that is initiated by specific pathways in the cell, while necrosis is a passive one that results in the rupture of the cell body membrane. During necrosis, the cell bursts and its content is spilled onto adjacent cells, thereby possibly triggering inflammatory responses. Necrosis and apoptosis are distinguishable through differentially activated biochemical processes. Normally, a healthy cell maintains a balance between pro- and anti-apoptotic factors. Disturbance of this balance may result in damage. Apoptosis contributes to several acquired forms of hearing impairment. Noise-induced hearing loss is the result of prolonged exposure to excessive noise, which triggers apoptosis in cochlear hair cells. Moreover, hearing loss caused by the use of therapeutic, but ototoxic, drugs such as aminoglycoside antibiotics and platins potentially may also result in the activation of apoptosis in hair cells and thus leading to hearing loss. The outer hair cells are the primary target for cell death following excessive noise exposure. Finally, apoptosis is a key contributor to the development of presbycusis (Op de Beeck et al., 2011).

Pharmacological strategies in animals that have proved successful in protecting against acoustic trauma include calcium channel blockers (Maurer et al., 1999), glutamate antagonists (Duan et al., 2000), D-methionine (Campbell et al., 1996, 2007, 2011) and glutathione (Hight et al., 2003) reduced hearing threshold shifts after noise exposure and decreased the amount of apoptosis in hair cells in animals (Op de Beeck et al., 2011), and inhibitors of the generation of reactive oxygen species, i.e., antioxidants (Seidman et al., 1993). Recently, Campbell et al. (2016) described the first experiment to determine dose-dependent D-met otoprotection against kanamycin-induced hearing loss with sub- and supra-optimal dosing protection. They found that 300 mg/kg/day fractionated D-met dosages appear to provide optimal protection against kanamycin-induced ototoxicity in guinea pigs with 34–41 dB threshold shift reductions.

Some negative findings offset these potential optimistic results. Hamernik et al. (2008) exposed three groups of chinchillas to a continuous broadband noise at 105 dB(A), 8 h per day for 5 days. One group received only the noise. A second group was exposed to noise and was additionally given *N*-acetyl-L-cysteine (L-NAC; 325 mg/kg, i.p.). The third group was exposed to the noise and received saline injections on the same schedule as the L-NAC treated animals. Pure-tone thresholds were obtained from local field potential recordings from the IC, and hair cell loss was quantified. In all three groups, the permanent threshold shift exceeded 50 dB at frequencies more than 2.0 kHz accompanied by severe hair cell loss in the basal half of the cochlea. There was no statistically significant difference among the three groups in those measures of noise-induced trauma. Davis et al. (2010) confirmed this negative finding in C57BL/6J (B6) mice.

There are a few studies for *N*-acetylcysteine (NAC) in humans. Kramer et al. (2006) measured pure-tone thresholds and DPOAEs in 31 normal hearing humans to evaluate the protective effect of NAC against the effects of loud music exposure. They used a randomized, double-blind, placebo-controlled design, with administration of NAC before and after 2 h of live music in a nightclub. The average music level was 98.1 dB(A) (range 92.5–102.8 dB(A)). There were no statistically significant differences between participants who received NAC versus a placebo for any of the outcome measures. Note that the threshold shifts were very mild in all tested subjects, so finding significant differences would be difficult. Also testing the use of NAC in humans, Lindblad et al. (2011) explored the hearing loss before and after a shooting session in a bunker-like room. Twenty shots were fired in 2 min, with a mean sound level in the ear canal under the protector of 137 dB SPL. A control group of 23 military officers was exposed without NAC and another group of 11 officers received peroral administration of NAC, directly after the shooting. They were tested on tone thresholds, TEOAEs, with and without contralateral noise, as well as thresholds for brief tones in modulated noise. The effects from shooting on hearing thresholds in these military persons without NAC administration were small. Interestingly, the nonlinearity of the cochlea was strongly reduced in the group without NAC, whereas it was practically unchanged in the NAC group throughout the study. This suggests that early effects of noise trauma can be prevented. Kopke et al. (2015) conducted a prospective, randomized, double-blinded, placebo-controlled clinical trial investigating the safety profile and the efficacy of NAC to prevent hearing loss in a military population after weapons training. Of the 566 total study subjects, 277 received NAC while 289 were given placebo. No significant differences were found for the primary outcome, rate of threshold shifts, but the right ear threshold shift rate difference did approach significance ($p = 0.0562$).

8.8 SUMMARY

This chapter deals with hearing loss prevention, but starts with newborn hearing screening as an important tool to prevent problems resulting from hearing loss. I first review briefly normal hearing development in humans. After an overview of various UNHS studies, I focus on potential problems therewith and the need for follow-up and conclude that newborn hearing screening is still not universal, but the implementation becomes more reliable, albeit not the last word in predicting hearing loss. Postneonatal testing remains essential. The second part of the chapter describes the effects of hearing protection both through ear plugs—including active ones, changing attitudes toward recreational noise exposure, and the potential of using drugs against noise-induced hearing loss.

REFERENCES

Ackermann, B.J., Kenny, D.T., O'Brien, I., Driscoll, T.R., 2014. Sound practice—improving occupational health and safety for professional orchestral musicians in Australia. Front. Psychol. 5, 973.

Alberti, P.W., Hyde, M.L., Riko, K., Corbin, H., Abramovich, S., 1983. An evaluation of BERA for hearing screening in high-risk neonates. Laryngoscope 93, 1115–1121.

Barlow, C., Castilla-Sanchez, F., 2012. Occupational noise exposure and regulatory adherence in music venues in the United Kingdom. Noise Health 14, 86–90.

Beach, E., Williams, W., Gilliver, M., 2010. Hearing protection for clubbers is music to their ears. Health Promot. J. Austr. 21, 215–221.

Beach, E.F., Gilliver, M., Williams, W., 2013. Leisure noise exposure: participation trends, symptoms of hearing damage, and perception of risk. Int. J. Audiol. 52 (Suppl. 1), S20–S25.

Beach, E.F., Nielsen, L., Gilliver, M., 2016. Providing earplugs to young adults at risk encourages protective behaviour in music venues. Glob. Health Promot. 23 (2), 45–56.

Berninger, E., Westling, B., 2011. Outcome of a universal newborn hearing-screening programme based on multiple transient-evoked otoacoustic emissions and clinical brainstem response audiometry. Acta Otolaryngol. 131 (7), 728–739.

Birnholz, J.C., Benacerraf, B.R., 1983. The development of human fetal hearing. Science 222, 516–518.

Bockstael, A., Keppler, H., Botteldooren, D., 2015. Musician earplugs: appreciation and protection. Noise Health 17, 198–208.

Brown, D.K., Tobolski, C.J., Shaw, G.R., Dort, J.C., 2000. Towards determining distortion product otoacoustic emission protocols for newborn hearing screening. J. Speech Lang. Pathol. Audiol. 24, 68–73.

Bubbico, L., Tognola, G., Greco, A., Grandori, F., 2008. Universal newborn hearing screening programs in Italy: survey of year 2006. Acta Otolaryngol. 128, 1329–1336.

Campbell, K.C., Rybak, L.P., Meech, R.P., Hughes, L.F., 1996. D-methionine provides excellent protection from cisplatin ototoxicity in the rat. Hear. Res. 102, 90–98.

Campbell, K.C., Meech, R.P., Klemens, J.J., et al., 2007. Prevention of noise- and drug-induced hearing loss with D-methionine. Hear. Res. 226, 92–103.

Campbell, K., Claussen, A., Meech, R., Verhulst, S., Foz, D., Hughes, L., 2011. D-methionine (D-met) significantly rescues noise-induced hearing loss: timing studies. Hear. Res. 282, 138–144.

Campbell, K.C., Martin, S.M., Meech, R.P., Hargrove, T.L., Verhulst, S.J., Fox, D.J., 2016. D-methionine (D-met) significantly reduces kanamycin-induced ototoxicity in pigmented guinea pigs. Int. J. Audiol. 10, 1–6.

Ching, T.Y., Dillon, H., Marnane, V., Hou, S., Day, J., Seeto, M., et al., 2013. Outcomes of early- and late-identified children at 3 years of age: findings from a prospective population-based study. Ear Hear. 34 (5), 535–552.

Colella-Santos, M.F., Hein, T.A.D., de Souza, G.L., Do Amaral, M.I.R., Casali, R.L., 2014. Newborn hearing screening and early diagnostic in the NICU. BioMed Res. Int. doi: 10.1155/2014/845308.

Davis, R.R., Custer, D.A., Krieg, E., Alagramam, K., 2010. *N*-Acetyl-L-cysteine does not protect mouse ears from the effects of noise. J. Occup. Med. Toxicol. 5, 11.

Dedhia, K., Kitsko, D., Sabo, D., Chi, D.H., 2013. Children with sensorineural hearing loss after passing the newborn hearing screen. JAMA Otolaryngol. Head Neck Surg. 139, 119–123.

Dell, S.M., Holmes, A.E., 2012. The effect of a hearing conservation program on adolescents' attitudes towards noise. Noise Health 22, 39–44.

Downs, M.P., Yoshinaga-Itano, C., 1999. The efficacy of early identification and intervention for children with hearing impairment. Pediatr. Clin. North Am. 46 (1), 79–87.

Duan, M., Agerman, K., Ernfors, P., Canlon, B., 2000. Complementary roles of neurotrophin 3 and a *N*-methyl-D-aspartate antagonist in the protection of noise and aminoglycoside-induced ototoxicity. Proc. Natl. Acad. Sci. U.S.A 97, 7597–7602.

Durieux-Smith, A., Picton, T.W., Bernard, P., MacMurray, B., Goodman, J.T., 1991. Prognostic validity of brainstem electric response audiometry in infants of a neonatal intensive care unit. Audiology 30, 249–265.

Eggermont, J.J., 1982. The inadequacy of click-evoked auditory brainstem responses in audiological applications. Ann. N. Y. Acad. Sci. 388, 707–709.

Eggermont, J.J., 2014. Noise and the Brain. Experience Dependent Developmental and Adult Plasticity. Academic Press, London.

Eggermont, J.J., Moore, J.K., 2012. Morphological and functional development of the auditory nervous system. In: Werner, L.A., Popper, A.N., Fay, R.F. (Eds.), Human Auditory Development, Springer Handbook of Auditory Research, 42. Springer Science & Business Media, New York, pp. 61–105.

Engard, D.J., Sandfort, D.R., Gotshall, R.W., Brazile, W.J., 2010. Noise exposure, characterization, and comparison of three football stadiums. J. Occup. Environ. Hyg. 7, 616–621.

Flamme, G.A., Williams, N., 2013. Sports officials' hearing status: whistle use as a factor contributing to hearing trouble. J. Occup. Environ. Hyg. 10, 1–10.

Gilbey, P., Kraus, C., Ghanayim, R., Sharabi-Nov, A., Bretler, S., 2013. Universal newborn hearing screening in Zefat, Israel: the first two years. Int. J. Pediatr. Otorhinolaryngol. 77, 97–100.

Gilles, A., Van de Heyning, P., 2014. Effectiveness of a preventive campaign for noise-induced hearing damage in adolescents. Int. J. Pediatr. Otorhinolaryngol. 78, 604–609.

Gorga, M.P., Norton, S.J., Sininger, Y.S., et al., 2000. Identification of neonatal hearing impairment: distortion product otoacoustic emissions during the perinatal period. Ear Hear. 21, 400–424.

Hamernik, R.P., Qiu, W., Davis, B., 2008. The effectiveness of *N*-acetyl-L-cysteine (L-NAC) in the prevention of severe noise-induced hearing loss. Hear. Res. 239, 99–106.

Hergils, L., 2007. Analysis of measurements from the first Swedish universal neonatal hearing screening program. Int. J. Audiol. 46, 680–685.

Hight, N.G., McFadden, S.L., Henderson, D., Burkard, R.F., Nicotera, T., 2003. Noise-induced hearing loss in chinchillas pre-treated with glutathione monoethylester and R-PIA. Hear. Res. 179, 21–32.

Huang, H.-M., Chiang, S.-H., Shiau, Y.-S., Yeh, W.-Y., Ho, H.C., Wang, L., et al., 2013. The universal newborn hearing screening program of Taipei City. Int. J. Pediatr. Otorhinolaryngol. 77, 1734–1737.

Huttenlocher, P.R., Dabholkar, A.S., 1997. Regional differences in synaptogenesis in human cerebral cortex. J. Comp. Neurol. 387, 167–178.

Ivory, R., Kane, R., Diaz, R.C., 2014. Noise-induced hearing loss: a recreational noise perspective. Curr. Opin. Otolaryngol. Head Neck Surg. 22, 394–398.

JCIH, 1995. Joint Committee on Infant Hearing 1994 Position Statement. Int. J. Pediatr. Otorhinolaryngol. 32 (3), 265–374.

Johnson, O., Walker, B.A., Morgan, S., Aldren, A., 2014. British university students' attitudes towards noise-induced hearing loss caused by nightclub attendance. J. Laryngol. Otol. 128, 29–34.

Kelly, A.C., Boyd, S.M., Henehan, G.T.M., Chambers, G., 2012. Occupational noise exposure of nightclub bar employees in Ireland. Noise Health 14, 148–154.

Kopke, R., Slade, M.D., Jackson, R., Hammill, T., Fausti, S., Lonsbury-Martin, B., et al., 2015. Efficacy and safety of *N*-acetylcysteine in prevention of noise induced hearing loss: A randomized clinical trial. Hear. Res. 323, 40–50.

Korver, A.M., Konings, S., Dekker, F.W., et al., 2010. Newborn hearing screening vs. later hearing screening and developmental outcomes in children with permanent childhood hearing impairment. JAMA 304, 1701–1708.

Kraaijenga, V.J., Ramakers, G.G., Grolman, W., 2016. The effect of earplugs in preventing hearing loss from recreational noise exposure: a systematic review. JAMA Otolaryngol. Head Neck Surg. 142, 389–394.

Kral, A., Eggermont, J.J., 2007. What's to lose and what's to learn: development under auditory deprivation, cochlear implants and limits of cortical plasticity. Brain Res. Rev. 56, 259–269.

Kral, A., Hartmann, R., Tillein, J., Heid, S., Klinke, R., 2000. Congenital auditory deprivation reduces synaptic activity within the auditory cortex in a layer-specific manner. Cereb. Cortex 10, 714–726.

Kramer, S., Dreisbach, L., Lockwood, J., et al., 2006. Efficacy of the antioxidant *N*-acetylcysteine (NAC) in protecting ears exposed to loud music. J. Am. Acad. Audiol. 17, 265–278.

Lammens, F., Verhaert, N., Devriendt, K., Debruyne, F., Desloovere, C., 2013. Aetiology of congenital hearing loss: a cohort review of 569 subjects. Int. J. Pediatr. Otorhinolaryngol. 77, 1385–1391.

Lindblad, A., Rosenhall, U., Olofsson, Å., Hagerman, B., 2011. The efficacy of *N*-acetylcysteine to protect the human cochlea from subclinical hearing loss caused by impulse noise: a controlled trial. Noise Health 13, 392–401.

Martin, W.H., Griest, S.E., Sobel, J.L., Howarth, L.C., 2013. Randomized trial of four noise-induced hearing loss and tinnitus prevention interventions for children. Int. J. Audiol. 52 (Suppl. 1), S41–S49.

Mason, J.A., Hermann, K.R., 1998. Universal infant hearing screening by automated auditory brainstem response measurement. Pediatrics 101, 221–228.

Maurer, J., Heinrich, U.R., Hinni, M., Mann, W., 1999. Alteration of the calcium content in inner hair cells of the cochlea of the guinea pig after acute noise trauma with and without application of the organic calcium channel blocker diltiazem. J. Otorhinolaryngol. 61, 328–333.

McNeill, K., Keith, S.E., Feder, K., Konkle, A.T.M., Michaud, D.S., 2010. MP3 player listening habits of 17 to 23 year old university students. J. Acoust. Soc. Am. 128, 646–653.

Meinke, D.K., Morata, T.C., 2012. Awarding and promoting excellence in hearing loss prevention. Int. J. Audiol. 51 (Suppl. 1), S63–S70.

Moeller, M.P., Hoover, B., Putman, C., Arbataitis, K., Bohnenkamp, G., Peterson, B., et al., 2007a. Vocalizations of infants with hearing loss compared with infants with normal hearing: Part I—phonetic development. Ear Hear. 28, 605–627.

Moeller, M.P., Hoover, B., Putman, C., Arbataitis, K., Bohnenkamp, G., Peterson, B., et al., 2007b. Vocalizations of infants with hearing loss compared with infants with normal hearing: Part II—transition to words. Ear Hear. 28, 628–642.

Moore, J.K., Guan, Y.L., 2001. Cytoarchitectural and axonal maturation in human auditory cortex. J. Assoc. Res. Otolaryngol. 2, 297–311.

Moore, J.K., Linthicum Jr., F.H., 2007. The human auditory system: a timeline of development. Int. J. Audiol. 46, 460–478.

Murphy, W.J., 2016. Preventing occupational hearing loss—time for a paradigm shift. Acoust. Today 12 (1), 28–35.

Neumann, K., Gross, M., Bottcher, P., Euler, H.A., Spormann-Lagodzinski, M., Polzer, M., 2006. Effectiveness and efficiency of a universal newborn hearing screening in Germany. Folia Phoniatr. Logop. 58, 440–455.

Norton, S.J., Gorga, M.P., Widen, J.E., et al., 2000a. Identification of neonatal hearing impairment: a multicenter investigation. Ear Hear. 21, 348–356.

Norton, S.J., Gorga, M.P., Widen, J.E., et al., 2000b. Identification of neonatal hearing impairment: transient evoked otoacoustic emissions during the prenatal period. Ear Hear. 21, 425–442.

O'Brien, I., Driscoll, T., Williams, W., Ackermann, B., 2014. A clinical trial of active hearing protection for orchestral musicians. J. Occup. Environ. Hyg. 11, 450–459.

Ohmori, S., Sygaya, A., Toida, N., Suzuki, E., Izutsu, M., Tsutsui, T., et al., 2015. Does the introduction of newborn hearing screening improve vocabulary development in hearing-impaired children? A population-based study in Japan. Int. J. Pediatr. Otorhinolaryngol. 79, 196–201.

Op de Beeck, K., Schacht, L., Van Camp, G., 2011. Apoptosis in acquired and genetic hearing impairment: the programmed death of the hair cell. Hear. Res. 281, 18–27.

Picton, T.W., Hillyard, S.A., Krausz, H.I., Galambos, R., 1974. Human auditory evoked potentials. I. Evaluation of components. Electroencephalogr. Clin. Neurophysiol. 36, 179–190.

Ponton, C.W., Eggermont, J.J., Kwong, B., Don, M., 2000. Maturation of human central auditory system activity: evidence from multi-channel evoked potentials. Clin. Neurophysiol. 111, 220–236.

Portnuff, C.D., Fligor, B.J., Arehart, K.H., 2013. Self-report and long-term field measures of MP3 player use: how accurate is self-report?. Int. J. Audiol. 52 (Suppl. 1), S33–S40.

Seidman, M.D., Shivapuja, B.G., Quirk, W.S., 1993. The protective effects of allopurinol and superoxide dismutase on noise-induced cochlear damage. Otolaryngol. Head Neck Surg. 109, 1052–1056.

Sininger, Y.S., Doyle, K.J., Moore, J.K., 1999. The case for early identification of hearing loss in children. Auditory system development, experimental auditory deprivation, and development of speech perception and hearing. Pediatr. Clin. North Am. 46 (1), 1–14.

Sininger, Y.S., Cone-Wesson, B., Folsom, R.C., et al., 2000. Identification of neonatal hearing impairment: auditory brainstem responses in the perinatal period. Ear Hear. 21, 383–399.

Sininger, Y.S., Grimes, A., Christensen, E., 2010. Auditory development in early amplified children: factors influencing auditory-based communication outcomes in children with hearing loss. Ear Hear. 31, 166–185.

Stevens, J.C., Boul, A., Lear, S., Parker, G., Ashall-Kelly, K., Gratton, D., 2013. Predictive value of hearing assessment by the auditory brainstem response following universal newborn hearing screening. Int. J. Audiol. 52, 500–506.

Swigonski, N., Shallop, J., Bull, M.J., Lemons, J.A., 1987. Hearing screening of high risk newborns. Ear Hear. 8, 26–30.

Szyfter, W., Wróbel, M., Radziszewska-Konopka, M., Szyfter-Harris, J., Karlik, M., 2008. Polish universal neonatal hearing screening program-4-year experience (2003–2006). Int. J. Pediatr. Otorhinolaryngol. 72, 1783–1787.

Ulusoy, S., Ugras, H., Cingi, C., Yilmaz, H.B., Muluk, N.B., 2014. The results of national newborn hearing screening (NNHS) data of 11,575 newborns from west part of Turkey. Eur. Rev. Med. Pharmacol. Sci. 18, 2995–3003.

Van Kerschaver, E., Boudewyns, A.N., Declau, F., Van de Heyning, P.H., Wuyts, F.L., 2012. Socio-demographic determinants of hearing impairment studied in 103 835 term babies. Eur. J. Publ. Health 23, 55–60.

Vogel, I., Brug, J., Van der Ploeg, C.P.B., Raat, H., 2010a. Young people: taking few precautions against hearing loss in discotheques. J. Adolesc. Health 46, 499–502.

Vogel, I., Brug, J., Van der Ploeg, C.P.B., Raat, H., 2010b. Discotheques and the risk of hearing loss among youth: risky listening behavior and its psychosocial correlates. Health Educ. Res. 25, 737–747.

Vogel, I., Verschuure, H., Van der Ploeg, C.P.B., Brug, J., Raat, H., 2010c. Estimating adolescent risk for hearing loss based on data from a large school-based survey. Am. J. Public Health 100, 1095–1100.

Vogel, I., Brug, J., Van der Ploeg, C.P.B., Raat, H., 2011. Adolescents risky MP3-player listening and its psychosocial correlates. Health Educ. Res. 26, 254–264.

Vogel, I., van de Looij-Jansen, P.M., Mieloo, C.L., Burdorf, A., de Waart, F., 2014. Risky music listening, permanent tinnitus and depression, anxiety, thoughts about suicide and adverse general health. PLoS One 9, e98912.

Vos, B., Lagasse, T., Levêque, A., 2014. Main outcomes of a newborn hearing screening program in Belgium over six years. Int. J. Pediatr. Otorhinolaryngol. 78, 1496–1502.

Wake, M., Ching, T.Y., Wirth, K., Poulakis, Z., Mensah, F.K., Gold, L., et al., 2016. Population outcomes of three approaches to detection of congenital hearing loss. Pediatrics 137 (1), 1–10.

Watkin, P.M., Baldwin, M., 2011. Identifying deafness in early childhood: requirements after the newborn hearing screen. Arch. Dis. Child. 96, 62–66.

White, K.R., Vohr, B.R., Maxon, A.B., Behrens, T.R., McPherson, M.G., Mauk, G.W., 1994. Screening all newborns for hearing loss using transient evoked otoacoustic emissions. Int. J. Pediatr. Otorhinolaryngol. 29, 203–217.

Widen, J.E., Folsom, R.C., Cone-Wesson, B., et al., 2000. Identification of neonatal hearing impairment: hearing status at 8 to 12 months corrected age using a visual reinforcement audiometry protocol. Ear Hear. 21, 471–487.

Wood, S.A., Sutton, G.J., Davis, A.C., 2015. Performance and characteristics of the Newborn Hearing Screening Programme in England: the first seven years. Int. J. Audiol. 54, 353–358.

Yoshinaga-Itano, C., Sedey, A.L., Coulter, D.K., Mehl, A.L., 1998. Language of early- and later-identified children with hearing loss. Pediatrics 102 (5), 1161–1171.

Part IV

The Treatments

Chapter 9

Hearing Aids

What do people expect from hearing aids (HAs)? HAs are not—as many new users expect—the auditory equivalent of contact lenses or glasses that instantly restore all aspects of hearing. That would only apply for their use to compensate for a pure conductive hearing loss. HAs in their basic form consist of a microphone, an amplifier with frequency specific gain and amplitude compression, and a miniature speaker. Currently, digital HAs, besides adding an analog-to-digital (A/D) converter after the microphone and a digital-to-analog converter before the speaker, allow sophisticated adjustments, tailored to the individual needs. Some of these are based on digital filters and algorithms for feedback cancellation, noise reduction, frequency compression, and wireless interaction between the HAs in case of binaural amplification (Chung, 2004a; Levitt, 2007). Currently, most HAs also have wireless interaction (e.g., Bluetooth) with many external devices including remote controls and smart phones (http://www.healthyhearing.com/help/hearing-aids/bluetooth; last accessed March 25, 2016).

HAs are mostly prescribed for sensorineural hearing loss (SNHL) and are designed to ameliorate effects of hearing sensitivity loss for speech perception. Bilateral HAs also aim to restore sound localization. It is thus imperative to first briefly survey some of the problems that hearing loss causes in audibility, sound localization and identification, and in communication (see chapter: Hearing Problems). The added burden of cognitive decline in older persons also needs considerable attention in outcome measures of HA use and benefit. We will see that most HAs cannot fully replace the loss of the nonlinear cochlear amplifier, cannot provide signal-to-noise ratios (SNRs) that allow speech understanding in background noise, and do not have a sufficient dynamic range to fully enjoy music. However, in 2015 HAs became available featuring large input dynamic ranges that can handle the higher overall and peak levels of music. There are a number of different ways that this has been accomplished but one device uses an 18-bit system that allows an input dynamic range of 108 dB (Kuk et al., 2015).

Hearing Loss. DOI: http://dx.doi.org/10.1016/B978-0-12-805398-0.00009-8

9.1 EFFECTS OF HEARING LOSS

9.1.1 Early Model Predictions on Speech Understanding

Here some of the pioneering work on the understanding of speech in noise conducted by Reinier Plomp and colleagues in the 1970s and 1980s will be presented. This discussion will apply to the hard of hearing and the elderly, the problems they face with speech understanding in noise, and the limited benefit they may experience from HAs in these conditions. This work may seem outdated, but the reader is reminded that the principles still apply today, and not only for HAs but also for cochlear implants (see chapter: Cochlear Implants). I start with a quote from Plomp (1978) arguing the need for his approach:

> *Our insights into why hearing-impaired people appear to be so seriously handicapped in everyday listening situations seem to be very scanty. This lack of knowledge particularly manifests itself in the uncritical way in which hearing aids are assumed to be of benefit. Since most conductive defects in the transmission chain up to the cochlea can nowadays be successfully rehabilitated by means of surgery, the great majority of the remaining inoperable cases are sensorineural hearing impairments. Although it is generally recognized that electronic amplification cannot compensate satisfactorily for these losses, it is remarkable how much hearing-aid prescribers expect from careful selection and fitting followed by good training. On the other hand, many hearing impaired appear to be rather disappointed about their hearing aids.*

This was amplified 30 years later by Killion (2008):

> *There are two difficulties that accompany hearing loss: loss of ability to hear quiet sounds (usually accompanied by a much smaller loss of ability to hear stronger sounds), and loss of ability to understand speech, especially speech in noise. The two losses often but not always go together: either loss can occur without the other. Someone who cannot even detect quiet speech may be able to understand almost as well as someone with normal hearing at an extremely noisy party, while another person who can still detect quiet speech may be unable to understand speech in the presence of noise at any presentation level.*

Plomp's (1978, 1986) approach to this problem provided a scheme for modeling the observed differences between normal hearing (NH) and hearing-impaired persons. This approach is as relevant today as it was in 1978 and may be applied to the understanding of speech perception with HAs. The effects of hearing loss are well illustrated by Fig. 9.1, where the speech-reception threshold (SRT) is plotted against the level of the interfering noise. The SRT is the average A-weighted (see chapter: Hearing Problems) speech level at which 50% of two-syllable words are repeated correctly by the listener. These days mostly dB SPL, instead of dB(A), is used, but that hardly makes any difference. The figure shows a reference curve for

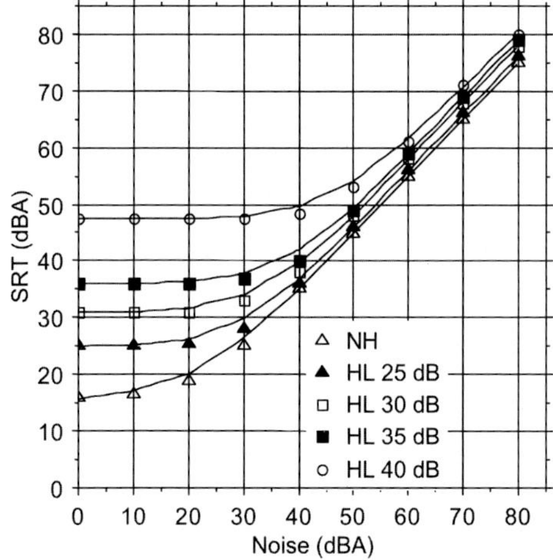

FIGURE 9.1 SRT for sentences as a function of noise level. The lowest curve (NH) applies to NH listeners, the other curves to four groups of listeners with increasing hearing losses. Note that for higher noise levels the SRT corresponds to a constant SNR, which increases with increasing hearing losses. *Data from Duquesnoy, A.J., 1983. The intelligibility of sentences in quiet and in noise in aged listeners. J. Acoust. Soc. Am. 74, 1136–1144.*

NH persons, and curves for four groups of people with SNHLs ranging from 25 to 40 dB. The hearing loss results in an increasing SRT in quiet (0 dB(A) masking). The linear rising part of the NH reference curve (for levels >30 dB(A)) shows that, over a large range of noise levels, the SRT corresponds to a constant SNR of typically −5 dB. This becomes about 0 dB for people with 40 dB hearing loss.

Plomp's model uses two parameters to characterize the SRT–noise level curves (Fig. 9.2). Parameter *A* (for attenuation) is equal to the hearing loss as determined by the pure-tone audiogram and is mainly responsible for the substantially higher speech levels required by the hearing impaired at low noise levels. At higher noise (and speech) levels, well above the elevated hearing threshold (i.e., in the linear rising parts of the curves), there remains a difference between the various curves. This is quantified by the parameter *D* (for distortion), indicating that hearing-impaired persons typically require a better SNR for achieving the 50% correct score. According to Plomp's framework the *D*-term, also called hearing loss for speech in noise, reflects the main problem in speech communication for the hearing impaired. As background noise is common in daily life, HAs are only of limited benefit in compensating for the underlying distortion caused by the hearing loss.

Plomp (1978) considered two (idealized) classes of hearing impairment: hearing loss of class *A* with attenuation in quiet and hearing loss of class *D*, comparable with a speech deficit in noise. Hearing loss of

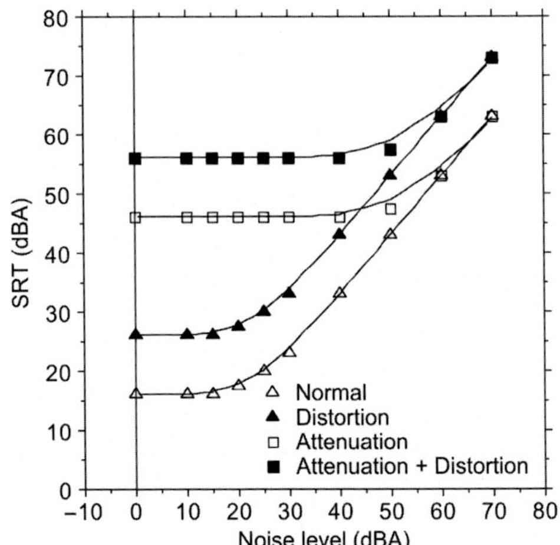

FIGURE 9.2 SRT for sentences in a typical, everyday listening situation as a function of noise level (dBA). The lower curve (Δ) holds for NH, the "Distortion" condition reflecting problems with speech in noise by ▲. The two upper curves represent the effects of 30 dB hearing loss for speech in silence (□) and in noise (■). An NH listener would need a 50 dB(A) SRT for noise levels up to 40 dB(A), slowly increasing to SRT = 60 dB(A) for 65 dB(A) noise. *Data from Plomp, R., 1978. Auditory handicap of hearing impairment and the limited benefit of hearing aids. J. Acoust. Soc. Am. 63, 533–549.*

class *A* (□) shifts the normal SRT curve (Fig. 9.2) by the amount of hearing loss in quiet (in this example by 30 dB) but approaches the normal curve at higher background noise levels. The listener with hearing loss would need a 50 dB(A) SRT for noise levels up to 40 dB(A), slowly increasing to SRT = 60 dB(A) for 65 dB(A) noise. We see that this represents a considerable SRT loss in quiet but nearly normal SRTs at a 60-dB(A) background noise level. The pure class *D* (▲) represents a parallel upward shift (10 dB in Fig. 9.2) of the normal SRT–noise level curve. So this represents a minor loss of SRT for normal speech levels (~65 dB(A)) in quiet, but a substantial handicap above a 60-dB background noise level, unless the speech is substantially amplified. The more realistic combination of both class *A* and *D* (■) shows a substantial loss of speech understanding both in quiet and in noise.

In line with these findings (Ching et al., 1998), it is clear that audibility (e.g., the result of amplification) cannot explain the less than optimal speech recognition of people with severe losses listening at high sensation levels. The data from Ching et al. (1998) suggested that audibility as quantified by the speech intelligibility index (SII) over-predicts speech performance at high sensation levels for listeners with severe hearing losses. The SII also underestimated speech scores at low sensation levels in many cases. The SII is computed as $SII(f) = (SNR(f) + 15)/30$, here f stands for the various

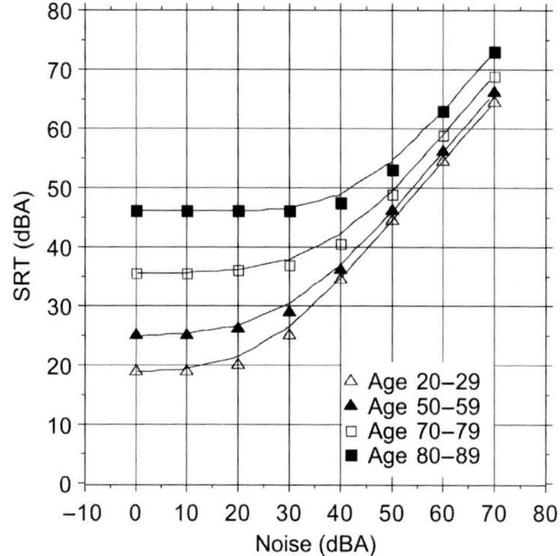

FIGURE 9.3 Median SRT for sentences as a function of noise level for male subjects in the age intervals 20–29, 50–59, 70–79, and 80–89. *Data from Plomp, R., Mimpen, A.M., 1979. Speech-reception threshold for sentences as a function of age and noise level. J. Acoust. Soc. Am. 66, 1333–1342.*

frequency bands. As the total range of SII ($0 \leq \text{SII} \leq 1$) corresponds to 30 dB, every 3 dB increase in the *D*-effect means a decrease of 10% in SII.

Perhaps the most important consequence of the decline in hearing sensitivity with aging is difficulty in understanding speech. The distortion factor *D* (Fig. 9.3) increases sharply with age and adds to the problems of loss in hearing sensitivity (cf. Fig. 9.1). In 140 male subjects (20 per decade between the ages 20 and 89) and 72 female subjects (20 per decade between 60 and 89, and 12 for the age interval 90–96), Plomp and Mimpen (1979) measured the monaural SRT for sentences in quiet and at four noise levels (22.5, 37.5, 52.5, and 67.5 dB(A) noise with long-term average speech spectra). The data were described in terms of the model shown in Fig. 9.2. $A + D$ effects increased progressively above age 50 reaching values of 20–40 dB for subjects between 80 and 90 years old (Fig. 9.3). *D*-effects also increased progressively above age 50 reaching values of 5–10 dB for subjects between 80 and 90. However subjects with the same hearing loss for speech in quiet may differ considerably in their ability to understand speech in noise. Thus, SRT in quiet is a poor predictor of SRT in noise, indicating that the SRT in noise should also be measured when a good picture of a person's hearing ability is required. The data confirm that the hearing handicap of many elderly subjects manifests itself primarily in a noisy environment. Noise levels in rooms used by the aged must be 5–10 dB lower than those for NH subjects for acceptable speech perception.

9.1.2 Age Effects on Aided Hearing in Noisy Environments

The most important reason for dissatisfaction by those with SNR loss in the 10–15 dB range is their resultant inability to understand speech in the presence of noise at restaurants, parties, etc. Some report that they take off their hearing aids in loud social situations where audibility is not a problem but their inability to separate speech from noise (typically interfering with speech) is.

Killion, 2008

Dubno et al. (2003) measured intelligibility for nonsense syllables in modulated noise as a function of modulation frequency for young and elderly (clinically) NH listeners. Speech recognition in interrupted noise was poorer for older than younger subjects (compare Festen and Plomp, 1990; Plomp et al., 1990; Fig. 9.3). Small age-related differences were observed in the decrease in score with interrupted noise relative to the score without interrupted noise. Not only elderly, but also middle-aged listeners often complain about difficulties with conversation in social settings, even when they have normal audiograms (Humes, 2007). Ruggles et al. (2012) investigated whether early aging influences an individual's ability to communicate in everyday settings. They found that age affects which auditory evoked potential component predicts communication performance in reverberant conditions. Whereas in younger adults, envelope cues (as reflected in the auditory steady-state response) predicted performance, in middle-aged listeners reliance relied heavily on the temporal fine structure (as measured in the frequency-following response), which is more disrupted by reverberant energy than temporal envelope structure is. Anderson et al. (2013) assumed that hearing loss results in an unbalanced neural representation of speech: The slowly varying envelope is enhanced, dominating representation in the auditory pathway and perceptual salience at the cost of the rapidly varying fine structure. Under that assumption they envisioned to ameliorate this by auditory–cognitive training to reduce the emphasis on the speech envelope in older adults (ages 55–79) with NH and with hearing loss. They found that the group with hearing loss experienced a reduction in the neural representation of the speech envelope presented in noise, approaching levels observed in NH older adults. No changes were noted in the NH group.

Kramer et al. (1997) combined pupillometry, which measures processing load as reflected by pupil dilation (Beatty, 1982), with an SRT task and could show that the SNR affected processing load as reflected by changes in the pupil dilation response. They observed that the pupil response was larger in the single-talker masker conditions than in the fluctuating noise conditions. They suggested that this reflects increased processing load evoked by semantic interference during the perception of speech, independent of intelligibility level. Koelewijn et al. (2012) assumed that the benefit of HAs is not primarily reflected in better speech performance, but that it is reflected in less effortful listening in the aided than in the unaided condition. They measured pupil dilation in 32 NH participants while listening to sentences masked by fluctuating noise or interfering speech at either 50% and 84% intelligibility. Koelewijn et al. (2012) concluded that better cognitive abilities not only relate to better speech perception, but also partly explain the ability to carry the higher processing load in complex listening conditions.

A similar approach was followed by Pichora-Fuller (2003) who stressed that: "Age-related problems in understanding spoken language are exacerbated by perceptual stressors such as noise and by cognitive stressors such as memory load." Signal-processing technologies in HAs, designed for older adults, have to include not only improved audibility but also reducing stress on the listener during this information processing. Consequently, Pichora-Fuller and Singh (2006) argued that: "Long-standing approaches to rehabilitative audiology should be revitalized to emphasize the important role that training and therapy play in promoting compensatory brain reorganization as older adults acclimatize to new technologies." Pichora-Fuller and Singh (2006) also noted that: "When the listening conditions are matched so that it is as difficult for younger adults to identify individual words as it is for older adults, apparent age-related declines in cognitive performance on measures of memory, attention, and discourse comprehension largely disappear." Singh et al. (2015) found: "Evidence to suggest that perceived social support is a significant predictor of satisfaction with hearing aids."

9.1.3 Effects of Hearing Aids on Sound Localization

In their study on the evolution of human hearing, Masterton et al. (1969) concluded that high-frequency sensitivity in mammals evolved principally as a consequence of the marked improvement it provided in sound localization (see chapter: Hearing Basics). Besides high-frequency hearing, localizing spatially discrete acoustical sources in a noisy world is the most effective means of suppressing noise by the spatial filtering provided by binaural hearing. Such a filter passes acoustical signals and noise from the general direction of the sound source, while attenuating most of the noise power from other directions (Lewis, 1987). High-frequency hearing loss is very detrimental for sound localization based on interaural loudness differences. In order to preserve the ability to localize sound based on interaural time (phase) differences, the sound levels at both ears should be comparable also in the case of hearing loss, hence the need for bilateral HAs. Van den Bogaert et al. (2011) investigated the effect of various commercial HAs on the ability to resolve front-back confusions and on sound localization in the frontal horizontal and vertical plane. Corroborating the role of audibility across a wide frequency range, they found that "hearing-impaired subjects reached the same performance with and without the different hearing aids, *if* in the unaided condition, a frequency-specific audibility correction was applied."

9.1.4 Hearing Aids at the Cocktail Party

In Chapter 4, Hearing Problems, we showed that hearing-impaired listeners perform more poorly in cocktail party conditions compared to NH people, and that this problem increases with age. Do HAs help under these conditions? Kidd et al. (2015) listed a variety of signal processing strategies implemented by HAs may assist the listener with SNHL. Among these are algorithms that implement environmental noise reduction, and thus attenuate unwanted sound sources. In addition there is directional amplification, which emphasizes a source originating from a specific azimuth relative to the head, directly related to improving SNR and enhancing source selection. However, Kidd et al. (2015) warned that these two forms of noise reduction are effective for certain types of unwanted sounds, however, they do not help the listener in choosing among competing talkers. For that purpose, cognitive factors such as attention become a dominant factor.

Marrone et al. (2008) had found that spatial release from masking (SRM) with bilateral HAs (mean ~4 dB) was negatively correlated with the amount of hearing loss. With a single HA, SRM was lower (mean ~2.5 dB) and related to the level of the stimulus in the unaided ear. In NH subjects the SRM was on average approximately 10 dB. Kidd et al. (2015) determined the benefit provided to listeners with SNHL by an acoustic beamforming microphone array, which provides directional amplification, in a speech-on-speech masking experiment. They compared this with bilateral amplification. They found that acoustic beamforming provided a large (mean ~9 dB) spatial release from speech-on-speech masking for SNHL listeners. This is about the same as for natural NH. Kidd et al. (2015) reported that for most SNHL listeners in the wider masker-signal separation condition, lower thresholds were obtained through the microphone array than through bilateral amplification. Kidd et al. (2015) suggested that consequently candidacy for highly spatially tuned amplification might depend on performance with conventional bilateral amplification. Especially, individuals with poor performance using natural cues are more likely to benefit from using a beamforming microphone array.

9.2 ACCLIMATIZATION AND PLASTICITY

Hearing loss can produce plastic changes in the adult central auditory system; this plasticity potentially also allows continuous adjustment to further changes in the perceived acoustic environment induced by HAs. The degree of these changes depends on the duration of use of the HA and is generally called acclimatization (see chapter: Brain Plasticity and Perceptual Learning). The generally accepted definition is: "Auditory acclimatization is a systematic change in auditory performance with time, linked to a change in the acoustic information available to the listener. It involves an improvement in performance that cannot be attributed purely to task, procedural or training effects" (Arlinger et al., 1996).

Arlinger et al. (1996) concluded from their review that acclimatization is not always observed for current (in 1996!) linear HAs when the dependent variable is a measure of speech identification ability. The mean reported improvement in benefit over time was 0–10% across a wide range of speech materials and presentation conditions. Acclimatization, when it happens, is not completed until after at least a number of months. Auditory rehabilitation of hearing-impaired adults may thus involve use-dependent plasticity. Philibert et al. (2002) compared intensity-related performance between two groups of subjects matched for age, gender, and absolute thresholds in both ears. One group comprised long-term binaural HA users and the other non-HA users. The effect of HA use was measured in two intensity-related tasks, an intensity discrimination threshold (IDT) task and a loudness-scaling task. Results indicated that significant differences existed in loudness perception between long-term HA users and non-HA users, the latter rating intensity as louder than the former which experience a down-regulated central gain (in agreement with Formby et al., 2003; chapter: Brain Plasticity and Perceptual Learning). Intensity discrimination performance showed only a tendency to lower IDTs in long-term HA compared to non-HA users, suggesting that the moderate changes that occurred in loudness scaling had no effect on these comparisons. This study suggested that significant perceptual modification occurred and thus that a possible functional plasticity resulted from HA use. In a follow-up study, Philibert et al. (2005) fitted eight listeners with symmetrical SNHL with binaural HAs for the first time. Perceptual performances were measured four times during auditory rehabilitation, again using an intensity discrimination task and a loudness-scaling task. Pure tones of two different frequencies were used, one well amplified by HAs and the other weakly amplified. Two intensity levels were tested, one rated "soft" by the listeners and the other "loud." Auditory brainstem responses (ABRs) to click stimulation were recorded without HA. There was no effect for ABR amplitude, or for waves I and III latency. However, wave V latency became significantly shorter over the HA fitting time course in right ears. The results were considered consistent with the auditory acclimatization effect because most modifications induced by HA fitting were found at high sound levels and at high frequency, i.e., for acoustic information that was newly available to the listener as a result of HA use. Since wave III is generated in the lower brainstem and wave V in the lateral lemniscus providing input to the inferior colliculus, the acclimatization effect is already visible in the upper brainstem. However, as the cochlear region that contributes to click-evoked wave V is different from that of wave III (Don and Eggermont, 1978), and this effect is level dependent (Eggermont and Don, 1980), non-acclimatization mechanisms such as effects of amplification need to be considered.

Two studies by Dawes et al. (2013, 2014) have also cast some doubt on the general presence of acclimatization. Dawes et al. (2013) tested SRM within the first week of fitting and after 12 weeks HA use for unilateral and bilateral adult HA users. A control group of experienced HA users completed testing over a similar time frame. HA users were tested with and without HAs, with SRM calculated as the 50% speech recognition threshold advantage when maskers and target are spatially separated at $\pm 90°$ azimuth to the listener compared to a colocated masker–target condition. Dawes et al. (2013) found (1) that on average there was no improvement over time in familiar aided listening conditions; (2) that greater improvement was associated with better cognitive ability and younger age, but not associated with HA use; and (3) that overall, bilateral aids facilitated better SRM performance than unilateral aids. The latter is expected based on increased spatial filtering (Section 9.1.3). Dawes et al. (2014) then investigated changes in central auditory processing following unilateral and bilateral HA fitting using late cortical auditory evoked potentials (CAEPs). The N_1 and P_2 components (cf. Fig. 8.1) were recorded to 500 and 3000 Hz tones presented at 65, 75, and 85 dB SPL to either the left or right ear. New unilateral and new bilateral HA users were tested at the time of first fitting and after 12 weeks HA use. A control group of long-term HA users was tested over the same time frame. No significant changes in the CAEP were observed for any group. Dawes et al. (2014)'s study does not appear to support an acclimatization effect observable in CAEPs following 12 weeks HA use, however, this use period may have been on the short side for acclimatization to materialize.

9.3 SATISFACTION AND QUALITY OF LIFE

Chien and Lin (2012) estimated that 14.2% of Americans 50 years or older with hearing loss from 1999 through 2006 did wear HAs. The prevalence of HA use was consistently low (<4%) in individuals with mild hearing loss across all age decades but generally increased with older age in individuals with moderate or greater hearing loss. Overall, the prevalence of HA use in individuals with hearing loss of at least 25 dB increased with every age decade, from 4.3% in individuals aged 50–59 years to 22.1% in individuals 80 years and older. There are an estimated 22.9 million older Americans with audiometric hearing loss who do not use HAs. Chien and Lin (2012) claimed that this was the first national estimate of HA prevalence in the US population based on audiometric data and a large, well-characterized representative sample. They considered a previous estimate of 10% from the Farmington Cohort (Gates et al., 1990) likely not representative of the US population.

The prevailing approach to treatment of age-related hearing loss is compensation of peripheral functional deficits by HAs and cochlear implants. This does not address that aging affects both the peripheral and central auditory systems. It is tempting to associate these with the *A* (attenuation) and *D* (distortion) factors respectively as introduced by Plomp (1978) (cf. Figs. 9.2 and 9.3). There is also growing evidence for an association between age-related hearing loss and cognitive decline (Hua et al., 2013; chapter: Hearing Problems). Thus, diagnostic evaluation should go beyond standard audiometric testing and include measures of central auditory function, including dichotic tasks and speech-in-noise testing (Parham et al., 2013).

Kaplan-Neeman et al. (2012) surveyed 177 hearing-impaired adults who were fitted with advanced digital HAs. Eighty-three percent used their HAs regularly, whereas 17% were nonusers. Of the users, 92% were satisfied to some degree with their HAs. The HA users gave as the main reason for non-use to be excessive amplification of background noise and therefore not providing a functional benefit. Boi et al. (2012) assessed the effects of HAs on mood, quality of life, and caregiver burden in the elderly. Fifteen patients older than 70 years and suffering from hearing loss and depressive mood were recruited. HA use clearly improved depressive symptoms, general health, and social interactivity.

Yamada et al. (2012) investigated whether self-reported hearing loss in older adults is associated with a decline in their ability to perform activities of daily living (ADL) or a decline in social participation. They enrolled 921 participants with a perfect baseline ADL score and valid follow-up scores and found that 105 self-reported hearing loss at baseline. Continuing with this hearing loss group, 44.8% reported a decline in their ADL score over the 3-year follow-up period. Yamada et al. (2012) found a statistically significant difference in ADL decline over the 3-year period for those with hearing loss at baseline compared to those without (odds ratio (OR) = 1.79, confidence interval (CI) = 1.12–2.87). Self-reported hearing loss at baseline did not have a statistically significant effect on decline in social participation (OR = 1.05, CI = 0.63–1.76) over the 3-year follow-up period. Refer to Chapter 7, Epidemiology and Genetics of Hearing Loss and Tinnitus for definition of OR and CI. Gopinath et al. (2012) also assessed the association between hearing impairment with activity limitations as assessed by the ADL scale. Out of a total of 1952 Blue Mountains Hearing Study participants aged at least 60 years, 164 reported ADL difficulty. Particularly, increased severity of hearing loss was significantly associated with impaired ADL. Pryce and Gooberman-Hill (2012) found that hearing loss affected whether people in residential care were able to access social opportunities, largely due to contextual issues that compounded communication difficulties, and environmental noise that restricted the residents' communication choices. The consensus in the "care people" reflected this as "there is a hell of a noise." This was particularly observed at every mealtime and during formal and informal group activities. Not surprisingly, HA use did not improve social engagement in these poor SNR conditions.

Two recent studies amplified these findings. Harrison-Bush et al. (2015) found that peripheral hearing, measured as the 0.5, 1, and 2 kHz pure-tone average in the better ear, explained a significant part of the variance in measures of speed of processing, executive function, and memory, as well as global cognitive status. Doherty and Desjardins (2015) assessed the effect of HA use on auditory working memory function in middle-aged and young–older adults with mild to moderate SNHL. Their participants were tested on two objective measures of auditory working memory in aided and unaided listening conditions. An age-matched control group without HAs followed the same experimental protocol. The aided scores on the auditory working memory tests were significantly improved while wearing HAs in all participants.

As we have seen in Chapter 4, Hearing Problems, hearing loss is associated with declining cognitive performance and incident dementia. Dawes et al. (2015) investigated whether use of HAs was associated with better cognitive performance, and if this relationship was the consequence of social isolation and/or depression. They carried out a structural equation modeling of associations between hearing loss, cognitive performance, social isolation, depression, and HA use in a subsample of the UK Biobank data set ($n = 164{,}770$) of UK adults aged 40–69 years who completed a hearing test. Age, sex, general health, and socioeconomic status were controlled for. They found that HA use was associated with better cognition, independently of social isolation and depression.

Summarizing, hearing impairment correlates with impaired daily activity, and the use of HAs improves depressive systems, general health, cognitive status, and social interaction. The latter does not apply to noisy residential care units where it would be most needed!

9.4 TYPES OF HEARING AIDS

There are many types of HAs, which vary in size, power, and circuitry. The different sizes and models are described in https://en.wikipedia.org/wiki/Hearing_aid, accessed October 1, 2015. The following short overview is based on the information presented in this link.

9.4.1 Behind-the-Ear Aids

Behind-the-ear (BTE) aids consist of a case, an earmold or dome, and a connection between them. The case contains the electronics, controls, battery, microphone(s), and often the loudspeaker. Generally, the case sits behind the pinna with the connection from the case coming down the front into the ear. The sound from the instrument can be routed acoustically or electrically to the ear. If the sound is routed electrically, the speaker (receiver) is located in the earmold or an open-fit dome, while acoustically coupled instruments use a plastic tube to deliver the sound from the case's loudspeaker to the earmold. BTE aids can be used for mild to profound hearing loss. As the electrical components are located outside the ear, the chance of moisture and earwax damaging the components is reduced, which can increase the durability of the instrument. BTE aids are also easily connected to assistive listening devices, such as FM systems, to directly integrate sound sources with the instrument. BTE aids are commonly worn by children who need a durable type of HA.

9.4.2 In-the-Ear Aids

In-the-ear (ITE) aids devices fit in the outer ear bowl (called the concha); they are sometimes visible when standing face to face with someone. ITE HAs are custom-made to fit each individual's ear. They can be used in mild to some severe hearing losses. Feedback, a squealing/whistling caused by sound (particularly high-frequency sound) leaking and being amplified again, may be a problem for severe hearing losses. Some modern circuits are able to provide feedback management or cancellation to assist with this. Venting may also cause feedback. A vent is a tube primarily placed to offer pressure equalization. However, different vent styles and sizes can be used to influence and prevent feedback. Traditionally, ITE aids have not been recommended for young children.

9.4.3 In-the-Canal Aids

In-the-canal (ITC) HAs fit inside the ear canal completely, leaving little to no trace of an installed HA visible. A comfortable fit is achieved because the shell of the aid is custom-made to the individual ear canal after taking a mold. Invisible HA types use venting and their deep placement in the ear canal to give a more natural experience of hearing. The ITC aid does not block the majority of the ear. This means that sound can be collected more naturally by the shape of the ear and can travel down into the ear canal as it would with unassisted hearing. All models allow the wearer to use a mobile phone as a remote control to alter memory and volume settings, instead of taking the ITC out to do this. ITC types are most suitable for users up to middle age, but are not suitable for more elderly people.

9.4.4 Open-Fit Aids

"Open-fit" or "over-the-ear" (OTE) HAs are small BTE type devices. This type is characterized by a minimal amount of effect on the ear canal resonances, as it traditionally leaves the ear canal as open as possible, often only being plugged up by a small speaker resting in the middle of the ear canal space. Traditionally, these HAs have a small plastic case behind the ear and a small clear tube running into the ear canal. Inside the ear canal, a small soft silicone dome or a molded, highly vented acrylic tip holds the tube in place. This design is intended to reduce the occlusion effect. Conversely, because of the increased possibility of feedback, and because an open fit allows low-frequency sounds to leak out of the ear canal, they are limited to high-frequency hearing losses. Thus, open-fit HAs are appropriate for individuals with good low-frequency hearing and mild to moderate high-frequency hearing loss, the individual will hear the low frequencies because the physical HA is not blocking the sound (Palmer, 2009).

9.4.5 Bone Conduction Hearing Aids

Bone conduction HAs are suited for people with conductive hearing and/or mixed hearing losses, and for people who have hearing loss in one ear (unilateral hearing loss). A bone conduction HA helps people with a conductive or mixed hearing loss by picking up the sound, amplifying it, and changing it to a vibration. The vibration is picked up by the cochlea as sound (cf. chapter: Implantable Hearing Aids). A bone conduction HA usually consists of a BTE HA that pick ups and amplifies sound across a bone conductor often incorporated into a headband or spectacles. A bone conduction HA is often a suitable solution for a temporary conductive hearing loss or for people awaiting surgery for a Bone-Anchored Hearing Aid (see chapter: Implantable Hearing Aids). See also www.hearnet.org.au/bone-conduction-hearing-aids and www.snikimplants.nl.

9.5 PROCESSING

Every electronic HA has at minimum a microphone, a loudspeaker (commonly called a receiver), a battery, and electronic circuitry. The electronic circuitry varies among devices, even if they are the same style. Some of the following is again based on https://en.wikipedia.org/wiki/Hearing_aid.

9.5.1 Digital Audio, Programmable Control

Both the audio circuit and the additional control circuits are fully digital. The hearing professional programs the HA with an external computer temporarily connected to the device and can adjust all processing characteristics on an individual basis. Fully digital HAs can be programmed with multiple programs that can be invoked by the wearer or that operate automatically and adaptively. These programs reduce acoustic feedback (whistling), reduce background noise, detect and automatically accommodate different listening environments, control additional components such as multiple microphones to improve spatial hearing, transpose frequencies (shift high frequencies that a wearer may not hear to lower frequency regions where hearing may be better), and implement many other features. Fully digital circuitry also allows control over wireless transmission capability for both the audio and the control circuitry. Control signals in a HA on one ear can be sent wirelessly to the control circuitry in the HA on the opposite ear, thus ensuring that the audio in both ears is either matched directly or that the audio contains intentional differences that mimic the differences in normal binaural hearing to preserve spatial hearing ability.

9.5.2 The Benefit of Bilateral Amplification

Palmer (2009) described the benefits of two HAs as: "The human brain is a far better signal detector than any hearing aid algorithm, and it is the input from both ears to the brainstem that allows the brain to detect the primary signal versus the noise." However, many individuals with SNHL will continue to have difficulty hearing in a noisy surround even when both ears are amplified due to frequency and temporal resolution problems that result in poor word recognition and are not solved through amplification. Other benefits include better sound localization (Section 9.1).

9.6 HIGH-FREQUENCY HEARING LOSS, LOUDNESS RECRUITMENT, AND REDUCED SNR

HAs improve sound audibility for people with hearing loss, but the ability to make use of the amplified signal, especially in the presence of competing noise, varies across people. We have seen that high-frequency amplification also aids in sound localization, but if the hearing loss in that range is too severe, frequency compression may be useful. The problem of loudness recruitment (see chapter: Types of Hearing Loss) may be ameliorated somewhat by amplitude compression, however this may also result in SNR reduction.

9.6.1 High-Frequency Amplification

Is high-frequency amplification always useful? Hogan and Turner (1998) found that in some cases for more severely impaired listeners, increasing the audibility of high-frequency speech information resulted in no further improvement in speech recognition, or even decreases in speech recognition. This occurred specifically when the hearing loss at a given frequency (>4 kHz) increased beyond 55 dB HL, and when additional audibility to that frequency region was diminished. However, Stelmachowicz et al. (2001) observed that whereas for understanding a male talker optimum performance was reached at a bandwidth of approximately 4–5 kHz, optimum performance for understanding females and children needed a bandwidth of 9 kHz. Stelmachowicz et al. (2001) and Pittman (2008) provided compelling data that children need audibility through 9 kHz to adequately hear the /s/ sound.

Horwitz et al. (2008) reported that adults have improved intelligibility with increased bandwidth past 4 kHz. However, most commercially available HAs do not provide usable gain past 4–5 kHz. Constraints to bandwidth come from the sampling rate used by the digital chip and the resulting battery drain (Palmer, 2009). Thus, the problem can potentially be solved with longer-life batteries.

9.6.2 Frequency Compression

There are two distinct frequency-lowering techniques to compensate in part for the loss of high-frequency hearing. These are linear frequency transposition and nonlinear frequency compression (NLFC), which maps a wide frequency range in the input signal onto a narrow frequency range in the output. Two programmable parameters, cutoff frequency and compression ratio, define how frequency compression is applied to the input signal. The energy that is above a particular cutoff frequency is compressed and shifted to the lower frequency region. Sounds below the cutoff frequency are not affected by NLFC processing, thereby preserving a natural sound quality (Zhang et al., 2014). McDermott (2011) analyzed the linear and nonlinear methods and found each effective for certain high-frequency acoustic signals, although both techniques distorted some signals. Bentler et al. (2014) confirmed this. Note that in digital HAs these filters cause time delays, in the order of a few milliseconds, which can be a problem for understanding if only one HA is used and the other ear is still functional.

Parsa et al. (2013) investigated the impact of an NLFC algorithm on perceived sound quality and found that the cutoff frequency parameter had more impact on sound quality ratings than the compression ratio, and that the hearing impaired adults were more tolerant to increased frequency compression than NH adults. There were no statistically significant differences in the sound quality ratings of speech-in-noise and music stimuli processed through various NLFC settings by hearing-impaired listeners. Wolfe et al. (2011) evaluated NLFC as a means to improve speech recognition for children with moderate to severe hearing loss following a 6-month acclimatization period. They found that in these children NLFC improved audibility for and recognition of high-frequency speech sounds. In many cases, they reported that perceptual improvements with NLFC increased with a longer period of acclimatization. Subsequently, Wolfe et al. (2015) tested 11 children with mild to moderate hearing loss with: (1) Phonak BTE without NLFC, (2) Phonak BTE with NLFC, and (3) Oticon BTE with wideband response extending to 8000 Hz. Wolfe et al. (2015) found that such children have good access to high-frequency phonemes presented at fixed levels (e.g., 50–60 dB(A)) with both wideband and NLFC technology. Similarly, sentence recognition in noise was similar with wideband and NLFC. However, a study by Zhang et al. (2014) in young children with bilateral hearing loss found that the mean speech intelligibility rating and the number of cortical responses present for /s/ were significantly higher when children were using NLFC processing than conventional processing in their HAs. For consonants, the score was higher with NLFC processing compared to conventional processing, but the difference did not reach the 5% significance level. Ellis and Munro (2015) studied 12 experienced HA users (aged 65–84 years) with moderate to severe high-frequency hearing loss. They noted that only auditory factors were significantly correlated with the degree of benefit obtained from NLFC. The strongest predictor of aided speech recognition, both with and without NLFC, was high-frequency hearing loss.

9.6.3 Amplitude Compression

In modern HAs, audibility for soft, moderate, and loud inputs without the need for constant volume adjustments is obtained through the use of amplitude compression. Currently, aspects of compression (e.g., compression threshold, compression ratio, attack time, and release time) can be controlled separately in over 20 frequency channels (Palmer, 2009). However, as the result of amplitude compression, HAs unfortunately decrease the SNR, largely by more amplification of the (lower level) background noise (Tremblay and Miller, 2014; Fig. 9.4).

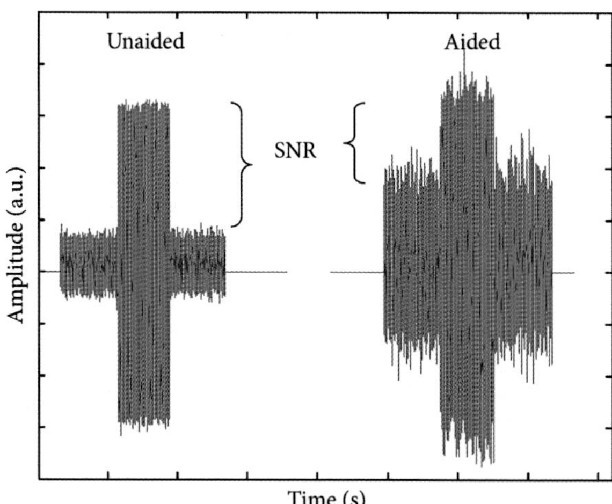

FIGURE 9.4 Time waveforms of ITC acoustic recordings for one individual. The unaided (left) and aided (right) conditions are shown together. Signal output as measured at the 1000 Hz centered one-third octave band was approximately equivalent at 73 and 74 dB SPL for the unaided and aided conditions. However, noise levels in the same one-third octave band were approximately 26 and 54 dB SPL, demonstrating the significant change in SNR. *From Tremblay, K.L., Miller, C.W., 2014. How neuroscience relates to hearing aid amplification. Int. J. Otolaryngol. 2014, 641–652.*

9.6.4 Binaural Aids and Directional Microphones

The use of modern microphone technology allows spatial filtering and potentially provides significant benefit for listening in noise (see also Section 9.1.4). Kim et al. (2014) found that users around 64 years of age accepted more noise with binaural amplification than with monaural amplification, regardless of the type of competing speech. The individuals' binaural advantages were significantly greater for longer experience of HAs, yet not related to their age or hearing thresholds. They concluded that binaural directional (DIR) microphone processing allowed HA users to accept a greater amount of background noise. However, Brimijoin et al. (2014) noted that the amount of benefit from DIR microphones is dependent on how close the signal of interest is to the front of the user. Their findings were that for larger off-axis target angles, listeners using DIR microphones took longer to orient to targets than they did when using omnidirectional (OMNI) microphones, although they were just as accurate. In contrast, for smaller off-axis target angles, listeners using DIR microphones were more quickly to orient to the targets compared to using OMNI microphones (Brimijoin et al., 2014).

9.6.5 Noise Reduction

The ability to understand speech in noise can be expressed in an SNR for understanding 50% of speech, i.e., the SRT (Section 9.1.1). The SRT of people with hearing loss may be as much as 30 dB higher than that of people with NH. This means that for a given background noise, the speech needs to be as much as 30 dB higher for people with hearing loss to achieve the same level of understanding as people with NH (Fig. 9.1; Chung, 2004a). There are two strategies for improving SNR by environmental noise reduction: DIR microphones and noise reduction algorithms. As we have seen, DIR microphones are more sensitive to sounds coming from the front than sounds coming from the back and the sides. DIR microphones take advantage of spatial separation between speech and noise. Noise reduction algorithms take advantage of the temporal separation and spectral differences between speech and noise (Chung, 2004a).

9.6.6 Combatting Wind Noise

> *Wind noise can be a nuisance to hearing aid users. With the advent of sophisticated feedback reduction algorithms, people with higher degrees of hearing loss are fit with larger vents than previously allowed, and more people with lesser degrees of hearing loss are fit with open hearing aids.*
>
> Chung, 2012c

Zakis (2011) investigated the variation of wind noise at HA microphones with wind speed, wind azimuth, and type of HA. Comparisons were made for BTE and in-the-canal devices, and between microphones within BTE devices. One ITC device and two BTE devices were placed on a Knowles Electronics Manikin for Acoustic Research. Zakis (2011) found that wind noise levels differed by up to 12 dB between microphones within the same BTE device, and across BTE devices by up to 6 or 8 dB for front or rear microphones, respectively. On average, wind noise levels were lowest with the ITC device and highest at the rear microphone of the smaller BTE device.

Chung (2012a) found that different noise reduction algorithms provided different amounts of wind noise reduction in different microphone modes, frequency regions, flow velocities, and head angles. Chung (2012b) observed that the adaptive DIR microphone is the most versatile microphone for use in wind. Following that, Chung (2012c) programmed BTE aids to have linear amplification and matching frequency responses between the DIR and OMNI modes. Comparison of wind noise showed that DIR generally produced higher noise levels than OMNI for all HAs. Chung (2013) concluded: "The open vent condition, however, yielded the lowest wind noise levels, which could not be entirely predicted by the frequency response changes of the hearing aids."

9.7 HEARING AIDS AND MUSIC PERCEPTION

Music amplified through hearing aids has some interesting characteristics but high fidelity is not typically one of them. This poses a serious problem for the investigator who wants to perform research on music with hearing impaired individuals who wear hearing aids. If the signal at the tympanic membrane is somewhat distorted then this has consequences for the assessment of music processing when examining both the peripheral and the central auditory system.

Chasin and Hockley, 2014

Hearing instrument design focuses on the amplification of speech. Many musicians and passive music listeners also require their hearing instruments to perform well when listening to the frequent, high amplitude peaks of live music. In most current digital hearing instruments with 16-bit A/D converters, the compressor output before the A/D conversion is limited to 95 dB (SPL) or less at the input. Whereas this covers the dynamic range of speech, this does not accommodate the amplitude peaks present in live music (Fig. 9.5) (Hockley et al., 2012). As Chasin (2012) remarked: "Amplified music tends to be of rather poor fidelity. The increased sound level relative to that of speech, and the much larger crest factor—the difference in dB between the instantaneous peak of a signal and its RMS value makes it difficult to transduce music without significant distortion."

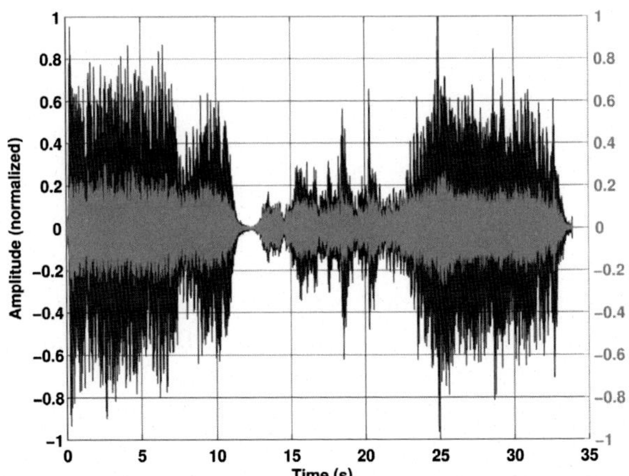

FIGURE 9.5 Recording of input (110 dB SPL) and output of a HA. The black waveform is the original input file. *Reprinted from Chasin, M., Hockley, N.S., 2014. Some characteristics of amplified music through hearing aids. Hear. Res. 306, 2–12, with permission from Elsevier.*

One potential solution is to use a HA microphone that is less sensitive to some of the lower frequency intense components of music, thereby providing the A/D converter with an input that is within its optimal operating region. The "missing" low-frequency information can still enter through an unoccluded earmold as unamplified sound and be part of the entire music listening experience (Schmidt, 2012). Madsen and Moore (2014) found that listening to music with HAs could be improved by an increase of the input and output dynamic range, extension of the low-frequency response, and improvement of feedback cancellation and automatic gain control systems. Mussoi and Bentler (2015) determined the impact of previous music training and hearing status on the effects of frequency compression in music perception. They found that although less frequency compression was in general preferred, there was more variability in the comparisons involving the default settings for a 50 dB hearing loss (i.e., start frequency 4000 Hz, compression ratio 2.5) and no compression, suggesting that mild amounts of compression may not be detrimental to perceived sound quality.

To make things worse, recorded music often undergoes compression limiting in the music industry, i.e., using the MP3 format. HA users may thus experience twofold effects of compression when listening to recorded music: namely, compression limiting during recording and HA wide dynamic range compression (WDRC). Croghan et al. (2014) found that for classical music, linear processing and slow WDRC were equally preferred, and the main effect of number of channels was not significant. For rock music, linear processing was preferred over slow WDRC, and three channels were preferred to 18 channels. This conclusion may not hold universally as it only went down to three channels and there is some evidence that a one-channel system for string heavy music would be even better than a three-channel one (Chasin, personal communication).

9.8 HEARING AIDS AND TINNITUS

HAs, after first being proposed for the alleviation of tinnitus more than 60 years ago (Saltzmann and Ersner, 1947), were only considered as a viable option for relief of tinnitus from the late 1970s on (e.g., Vernon et al., 1977; Del Bo and Ambrosetti, 2007) and still remain controversial. Amplification provides environmental masking in the frequency range of tinnitus, especially if the HAs have a good high-frequency response, and in addition it allows brain plasticity to play a role. It is important that the desired amplification in the more than 4 kHz range is high enough to balance the activity of the auditory nerve fibers across frequency. There is, however, some doubt that HAs really suppress or mask tinnitus. Moffat et al. (2009) compared effects of conventional HA amplification and high-bandwidth amplification, the idea being that the high bandwidth would provide more amplification in the tinnitus frequency range and thus would induce more plastic changes in the tinnitus spectrum. However, after up to 30 days post-fitting, the tinnitus percept was affected only weakly in the conventional amplification group where the low-frequency components of the tinnitus (<2 kHz) were attenuated and was not at all affected in the high-bandwidth group. This was explained as due to more robust changes in cortical reorganization in the high frequency, i.e., the hearing loss, range, that would not be malleable over the relatively short (30 days) treatment period.

A more positive outlook was provided by Schaette et al. (2010) who demonstrated that HAs (amplification up to 6 kHz) worn for 6 months had a significant effect—tinnitus loudness decrease, and decrease of annoyance as measured by the Tinnitus Questionnaire—on persons with tinnitus pitch less than 6 kHz but not for those with higher tinnitus pitch. Searchfield et al. (2010) observed improved scores on the Tinnitus Handicap Questionnaire when counseling was accompanied by wearing a HA compared to counseling only. This was also found by Parazzini et al. (2011) who compared the effectiveness of tinnitus retraining therapy (TRT) with sound generators or with open ear HAs in the rehabilitation of tinnitus. There was a highly significant improvement in both tinnitus treatments starting from the first 3 months and up to 1 year of therapy, with a progressive and statistically significant decrease in the disability every 3 months. TRT was equally effective with sound generator or open ear HAs.

Reporting a surprise finding, Peltier et al. (2012) observed that: "Over the last three years of hearing aid dispensing that among 74 subjects fitted with a linear octave frequency transposition (LOFT) hearing aid, 60 reported partial or complete tinnitus suppression during day and night, an effect still lasting after several months or years of daily use. … Tinnitus suppression started after a few days of LOFT hearing aid use and reached a maximum after a few weeks of daily use. … When the use of the LOFT hearing aid was stopped tinnitus reappeared within a day, and after re-using the LOFT aid it disappeared again within a day. For about one-third of the 38 subjects a classical amplification or a nonlinear frequency compression aid was also tried, and no such tinnitus suppression was observed. Besides improvements in audiometric sensitivity to high frequencies and in speech discrimination scores, LOFT can be considered as a remarkable opportunity to suppress tinnitus over a long time scale."

9.9 SUMMARY

Hearing loss results in less ability to understand quiet sounds and loss of speech understanding, especially in noise. HAs are designed to remediate this. Whereas they alleviate the loss of hearing low-level sounds and speech, they may aggravate the understanding of speech in noise by reducing the SNR. This results from the compression of the amplification of higher-level sounds compared to low-level sounds such as background noise. In addition, limited dynamic range in the amplification in addition impairs music appreciation. Modern HAs with larger dynamic ranges may alleviate this problem. Nevertheless, nearly all studies show an increased quality of life, including improved cognition. The expected improvement in social engagement was not obvious. HA processing has greatly improved with the arrival of digital HAs and the more frequent binaural HA fitting. HAs may also alleviate tinnitus, albeit that this is not a general finding.

REFERENCES

Anderson, S., Parbery-Clark, A., White-Schwoch, T., Kraus, N., 2013. Aging affects neural precision of speech encoding. J. Neurosci. 32, 14156–14164.

Arlinger, S., Gatehouse, S., Bentler, R.A., Byrne, D., Cox, R.M., Dirks, D.D., et al., 1996. Report of the Eriksholm Workshop on auditory deprivation and acclimatization. Ear Hear. 17 (3 Suppl), 87S–98S.

Beatty, J., 1982. Task-evoked pupillary responses, processing load, and the structure of processing resources. Psychol. Bull. 91, 276–292.

Bentler, R., Walker, E., McCreery, R., Arenas, R.M., Roush, P., 2014. Nonlinear frequency compression in hearing aids: impact on speech and language development nonlinear frequency compression in hearing aids: impact on speech and language development. Ear Hear. 35, e143–e152.

Boi, R., Racca, L., Cavallero, A., Carpaneto, V., Racca, M., Dall' Acqua, F., et al., 2012. Hearing loss and depressive symptoms in elderly patients. Geriatr. Gerontol. Int. 12, 440–445.

Brimijoin, W.O., Whitmer, W.M., McShefferty, D., Akeroyd, M.A., 2014. The effect of hearing aid microphone mode on performance in an auditory orienting task. Ear Hear. 35, e204–e212.

Chasin, M., 2012. Music and hearing aids—an introduction. Trends Amplif. 16, 136–139.

Chasin, M., Hockley, N.S., 2014. Some characteristics of amplified music through hearing aids. Hear. Res. 308, 2–12.

Chien, W., Lin, F.R., 2012. Prevalence of hearing aid use among older adults in the United States. Arch. Intern. Med. 172, 292–293.

Ching, T.Y.C., Dillon, H., Byrne, D., 1998. Speech recognition of hearing-impaired listeners: predictions from audibility and the limited role of high-frequency amplification. J. Acoust. Soc. Am. 103, 1128–1140.

Chung, K., 2004a. Challenges and recent developments in hearing aids. Part I. Speech understanding in noise, microphone technologies and noise reduction algorithms. Trends Amplif. 8, 83–124.

Chung, K., 2012a. Wind noise in hearing aids: I. Effect of wide dynamic range compression and modulation-based noise reduction. Int. J. Audiol. 51, 16–28.

Chung, K., 2012b. Wind noise in hearing aids: II. Effect of microphone directivity. Int. J. Audiol. 51, 29–42.

Chung, K., 2012c. Comparisons of spectral characteristics of wind noise between omnidirectional and directional microphones. J. Acoust. Soc. Am. 131, 4508–4517.

Chung, K., 2013. Effects of venting on wind noise levels measured at the eardrum. Ear Hear. 34, 470–481.

Croghan, N.B., Arehart, K.H., Kates, J.M., 2014. Music preferences with hearing aids: effects of signal properties, compression settings, and listener characteristics. Ear Hear. 35, e170–e184.

Dawes, P., Munro, K.J., Kalluri, S., Edwards, B., 2013. Unilateral and bilateral hearing aids, spatial release from masking and auditory acclimatization. J. Acoust. Soc. Am. 134, 596–606.

Dawes, P., Munro, K.J., Kalluri, S., Edwards, B., 2014. Auditory acclimatization and hearing aids: late auditory evoked potentials and speech recognition following unilateral and bilateral amplification. J. Acoust. Soc. Am. 135, 3560–3569.

Dawes, P., Emsley, R., Cruickshanks, K.J., Moore, D.R., Fortnum, H., Edmondson-Jones, M., et al., 2015. Hearing loss and cognition: the role of hearing aids, social isolation and depression. PLoS One 10 (3), e0119616.

Del Bo, L., Ambrosetti, U., 2007. Hearing aids for the treatment of tinnitus. Prog. Brain Res. 166, 341–345.

Doherty, K.A., Desjardins, J.L., 2015. The benefit of amplification on auditory working memory function in middle-aged and young-older hearing impaired adults. Front. Psychol. 6, 721.

Don, M., Eggermont, J.J., 1978. Analysis of the click-evoked brainstem potentials in man using high-pass noise masking. J. Acoust. Soc. Am. 63, 1084–1092.

Dubno, J.R., Horwitz, A.R., Ahlstrom, J.B., 2003. Recovery from prior stimulation: masking of speech by interrupted noise for younger and older adults with normal hearing. J. Acoust. Soc. Am. 113, 2084–2094.

Duquesnoy, A.J., 1983. The intelligibility of sentences in quiet and in noise in aged listeners. J. Acoust. Soc. Am. 74, 1136–1144.

Eggermont, J.J., Don, M., 1980. Analysis of click-evoked brainstem potentials in humans using high-pass noise masking. II. Effect of click intensity. J. Acoust. Soc. Am. 68, 1671–1675.

Ellis, R.J., Munro, K.J., 2015. Predictors of aided speech recognition, with and without frequency compression, in older adults. Int. J. Audiol. 54, 467–475.

Festen, J.M., Plomp, R., 1990. Effects of fluctuating noise and interfering speech on the speech-reception threshold for impaired and normal hearing. J. Acoust. Soc. Am. 88, 1725–1736.

Formby, C., Sherlock, L.P., Gold, S.L., 2003. Adaptive plasticity of loudness induced by chronic attenuation and enhancement of the acoustic background. J. Acoust. Soc. Am. 114, 55–58.

Gates, G.A., Cooper Jr., J.C., Kannel, W.B., Miller, N.J., 1990. Hearing in the elderly: the Framingham Cohort, 1983–1985. Part 1. Basic audiometric test results. Ear Hear. 11, 247–256.

Gopinath, B., Schneider, J., McMahon, C.M., Teber, E., Leeder, S.R., Mitchell, P., 2012. Severity of age-related hearing loss is associated with impaired activities of daily living. Age Ageing 41, 195–200.

Harrison Bush, A.L., Lister, J.J., Lin, F.R., Betz, J., Edwards, J.D., 2015. Peripheral hearing and cognition: evidence from the Staying Keen in Later Life (SKILL) study. Ear Hear. 36, 395–407.

Hockley, N.S., Bahlmann, F., Fulton, B., 2012. Analog-to-digital conversion to accommodate the dynamics of live music in hearing instruments. Trends Amplif. 16, 146–158.

Hogan, C.A., Turner, C.W., 1998. High-frequency audibility: benefits for hearing-impaired listeners. J. Acoust. Soc. Am. 104, 332–341.

Horwitz, A., Ahlstrom, J., Dubno, J., 2008. Factors affecting the benefits of high-frequency amplification. J. Speech Lang. Hear. Res. 51, 798–813.

Hua, H., Karlsson, J., Widén, S., Möller, C., Lyxell, B., 2013. Quality of life, effort and disturbance perceived in noise: a comparison between employees with aided hearing impairment and normal hearing. Int. J. Audiol. 52, 642–649.

Humes, L.E., 2007. The contributions of audibility and cognitive factors to the benefit provided by amplified speech to older adults. J. Am. Acad. Audiol. 18, 590–603.

Kaplan-Neeman, R., Muchnik, C., Hildesheimer, M., Henkin, Y., 2012. Hearing aid satisfaction and use in the advanced digital era. Laryngoscope 122, 2029–2036.

Kidd Jr., G., Mason, C.R., Best, V., Swaminathan, J., 2015. Benefits of acoustic beamforming for solving the cocktail party problem. Trends Hear. 19, 1–15.

Killion, M.C., 2008. Hearing Loss and Hearing Aids: A Perspective. Chapter 3.27 in Sensory Systems. Elsevier, pp. 475–483.

Kim, J.-H., Lee, J.H., Lee, H.-K., 2014. Advantages of binaural amplification to acceptable noise level of directional hearing aid users. Clin. Exp. Otorhinolaryngol. 7, 94–101.

Koelewijn, T., Zekveld, A.A., Festen, J.M., Kramer, S.E., 2012. Pupil dilation uncovers extra listening effort in the presence of a single-talker masker. Ear Hear. 33, 291–300.

Kramer, S.E., Kapteyn, T.S., Festen, J.M., Kuik, D.J., 1997. Assessing aspects of auditory handicap by means of pupil dilatation. Audiology 36, 155–164.

Kuk, F., Schmidt, E., Jessen, A.H., Sonne, M., et al., 2015. The professional practice: three ways to stay competitive in a changing market. Hear. Rev. 22 (11), 32.

Levitt, H., 2007. A historical perspective on digital hearing aids: how digital technology has changed modern hearing aids. Trends Amplif. 11, 7–24.

Lewis, E.R., 1987. Speculations about noise and the evolution of vertebrate hearing. Hear. Res. 25, 83–90.

Madsen, S.M.K., Moore, B.C.J., 2014. Music and hearing aids. Trends Hear. 18, pii: 2331216514558271.

Marrone, N., Mason, C.R., Kidd Jr., G., 2008. Evaluating the benefit of hearing aids in solving the cocktail party problem. Trends Amplif. 12, 300–315.

Masterton, B., Heffner, H., Ravizza, R., 1969. The evolution of human hearing. J. Acoust. Soc. Am. 45, 966–985.

McDermott, H.J., 2011. A technical comparison of digital frequency-lowering algorithms available in two current hearing aids. PLoS One 6, e22358.

Moffat, G., Adjout, K., Gallego, S., Thai-Van, H., Collet, L., Noreña, A.J., 2009. Effects of hearing aid fitting on the perceptual characteristics of tinnitus. Hear. Res. 254, 82–91.

Mussoi, B.S.S., Bentler, R.A., 2015. Impact of frequency compression on music perception. Int. J. Audiol. 54, 627–633.

Palmer, C.V., 2009. A contemporary review of hearing aids. Laryngoscope 119, 2195–2204.

Parazzini, M., Del, Bo, L., Jastreboff, M., Tognola, G., Ravazzani, P., 2011. Open ear hearing aids in tinnitus therapy: an efficacy comparison with sound generators. Int. J. Audiol. 50, 548–553.

Parham, K., Lin, F.R., Coelho, D.H., Sataloff, R.T., Gates, G.A., 2013. Comprehensive management of presbycusis: central and peripheral. Otolaryngol. Head Neck Surg. 148, 537–539.

Parsa, V., Scollie, S., Glista, D., Seelisch, A., 2013. Nonlinear frequency compression: effects on sound quality ratings of speech and music. Trends Amplif. 17, 54–68.

Peltier, E., Peltier, C., Tahar, S., Alliot-Lugaz, E., Cazals, Y., 2012. Long-term tinnitus suppression with linear octave frequency transposition hearing aids. PLoS One 7, e51915.

Philibert, B., Collet, L., Vesson, J.F., Veuillet, E., 2002. Intensity-related performances are modified by long-term hearing aid use: a functional plasticity? Hear. Res. 165, 142–151.

Philibert, B., Collet, L., Vesson, J.F., Veuillet, E., 2005. The auditory acclimatization effect in sensorineural hearing-impaired listeners: evidence for functional plasticity. Hear. Res. 205, 131–142.

Pichora-Fuller, M.K., 2003. Cognitive aging and auditory information processing. Int. J. Audiol. 42 (Suppl 2), 26–32.

Pichora-Fuller, M.K., Singh, G., 2006. Effects of age on auditory and cognitive processing: implications for hearing aid fitting and audiologic rehabilitation. Trends Amplif. 10, 29–59.

Pittman, A., 2008. Short-term word-learning rate in children with normal hearing and children with hearing loss in limited and extended high-frequency bandwidths. J. Speech Lang. Hear. Res. 51, 785–797.

Plomp, R., 1978. Auditory handicap of hearing impairment and the limited benefit of hearing aids. J. Acoust. Soc. Am. 63, 533–549.

Plomp, R., 1986. A signal-to-noise ratio model for the speech-reception threshold of the hearing impaired. J. Speech Hear. Res. 29, 146–154.

Plomp, R., Mimpen, A.M., 1979. Speech-reception threshold for sentences as a function of age and noise level. J. Acoust. Soc. Am. 66, 1333–1342.

Plomp, R., Festen, J.M., Bronkhorst, A.W., 1990. Noise as a problem for the hearing impaired. Environ. Int. 16, 393–398.

Pryce, H., Gooberman-Hill, R., 2012. 'There's a hell of a noise': living with a hearing loss in residential care. Age Ageing 41, 40–46.

Ruggles, D., Bharadwaj, H., Shinn-Cunningham, B.G., 2012. Why middle-aged listeners have trouble hearing in everyday settings. Curr. Biol. 22, 1417–1422.

Saltzmann, M., Ersner, M.S., 1947. A hearing aid for relief of tinnitus aurium. Laryngoscope 57, 358–366.

Schaette, R., König, O., Hornig, D., Gross, M., Kempter, R., 2010. Acoustic stimulation treatments against tinnitus could be most effective when tinnitus pitch is within the stimulated frequency range. Hear. Res. 269, 95–101.

Schmidt, M., 2012. Musicians and hearing aid design—is your hearing instrument being overworked? Trends Amplif. 16, 140–145.

Searchfield, G.D., Kaur, M., Martin, W.H., 2010. Hearing aids as an adjunct to counseling: tinnitus patients who choose amplification do better than those that don't. Int. J. Audiol. 49, 574–579.

Singh, G., Lau, S.-T., PIchora-Fuller, M.K., 2015. Social support predicts hearing aid satisfaction. Ear Hear. 36, 664–676.

Stelmachowicz, P.G., Pittman, A.L., Hoover, B.M., Lewis, D.E., 2001. Effect of stimulus bandwidth on the perception of /s/ in normal- and hearing-impaired children and adults. J. Acoust. Soc. Am. 110, 2183–2190.

Tremblay, K.L., Miller, C.W., 2014. How neuroscience relates to hearing aid amplification. Int. J. Otolaryngol. 2014, 641–652.

Van den Bogaert, T., Carette, E., Wouters, J., 2011. Sound source localization using hearing aids with microphones placed behind-the-ear, in-the-canal, and in-the-pinna. Int. J. Audiol. 50, 164–176.

Vernon, J., Schleuning, A., Well, I., Hughes, F., 1977. A tinnitus clinic. Ear Nose Throat J. 56, 58–71.

Wolfe, J., John, A., Schafer, E., Nyffeler, M., Boretzki, M., Caraway, T., et al., 2011. Long-term effects of non-linear frequency compression for children with moderate hearing loss. Int. J. Audiol. 50, 396–404.

Wolfe, J., John, A., Schafer, E., Hudson, M., Boretzki, M., Scollie, S., et al., 2015. Evaluation of wideband frequency responses and non-linear frequency compression for children with mild to moderate high-frequency hearing loss. Int. J. Audiol. 54, 170–181.

Yamada, M., Nishiwaki, Y., Michikawa, T., Takebayashi, T., 2012. Self-reported hearing loss in older adults is associated with future decline in instrumental activities of daily living but not in social participation. J. Am. Geriatr. Soc. 60, 1304–1309.

Zakis, J.A., 2011. Wind noise at microphones within and across hearing aids at wind speeds below and above microphone saturation. J. Acoust. Soc. Am. 129, 3897–3907.

Zhang, V.W., Ching, T.Y., Van Buynder, P., Hou, S., Flynn, C., Burns, L., et al., 2014. Aided cortical response, speech intelligibility, consonant perception and functional performance of young children using conventional amplification or nonlinear frequency compression. Int. J. Pediatr. Otorhinolaryngol. 78, 1692–1700.

Chapter 10

Implantable Hearing Aids

Implantable hearing aids (IHAs) can roughly be divided in bone-anchored hearing aids (BAHAs) and middle ear implants, both are mostly indicated for conductive hearing loss. However, middle ear implants are also a good solution for patients with sensorineural hearing loss who cannot tolerate ear molds because of external otitis. At the end of the 20th century, Chasin (1997) noted that "this is an extremely dynamic field and there is rarely a month that passes without the granting of a new patent for some aspect of an IHA." Here we first review mechanism for bone conduction as a basis for understanding the action of BAHAs. Then, I present an evaluation of the various available BAHAs and middle ear implants, based largely on a series of published findings in small or larger numbers of patients. Throughout this chapter I will use quotes extensively to highlight individual conclusions and will when possible summarize these findings. An authoritative overview of all IHAs can be found in www.snikimplants.nl.

10.1 BONE CONDUCTION MECHANISMS

One of the most important, and likely the most important, finding in bone conduction physiology research was when von Békésy (1960) in 1932 reported the cancellation of the perception of a bone conducted (BC) tone by an air conducted (AC) tone. ... The perception of a 400 Hz tone, 57 dB above threshold, delivered to both ears by a BC transducer, could be cancelled by careful adjustment of the amplitude and phase of the AC stimuli produced by binaural earphones. ... The accomplishment of canceling a BC tone by an AC tone of the same frequency lead von Békésy to conclude that, although the transmission to the inner ear is different, the final processes for AC and BC stimulation are the same. Stenfelt (2007)

Stenfelt and Håkansson (2002) aiming to further test the findings of von Békésy, carried out loudness matching at each frequency and at 30–80 dB hearing level (HL). They fixed the sound pressure from the earphones and the subject adjusted the output level of the bone transducer for equal loudness. They found a non-unity relation between the different loudness functions for air-conducted (AC) and bone-conducted (BC) sound with slopes between

Hearing Loss. DOI: http://dx.doi.org/10.1016/B978-0-12-805398-0.00010-4

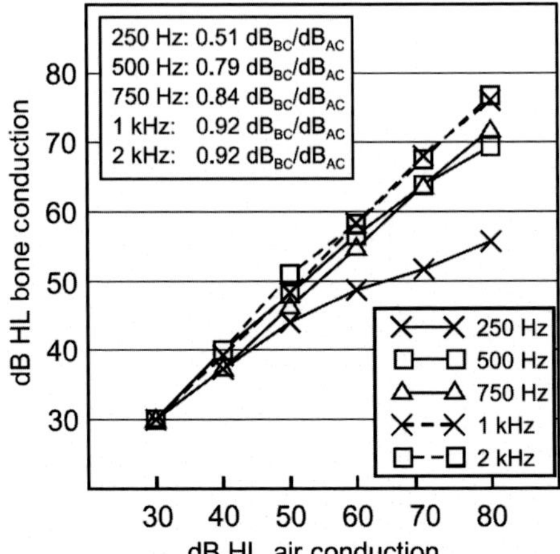

FIGURE 10.1 The mean results of the loudness balance test for the sensorineural hearing-impaired group. 250 Hz: crosses connected with solid lines; 500 Hz: squares connected with solid lines; 750 Hz: triangles connected with solid lines; 1 kHz: crosses connected with dotted lines; and 2 kHz: squares connected with dotted lines. *Reprinted from Stenfelt, S., Håkansson, B., 2002. Air versus bone conduction: an equal loudness investigation. Hear. Res. 167, 1–12, with permission from Elsevier.*

0.51 and 0.92. This resulted in a 6–10 dB difference in the AC and BC loudness functions for the normal hearing group at 250–750 Hz. At 1–4 kHz the difference was only 4–5 dB over the same dynamic range (Fig. 10.1). Similar results were obtained for the sensorineural hearing-impaired group.

Stenfelt and Zeitooni (2013) further investigated this discrepancy between air and bone conduction at low frequencies by using another loudness estimation method (adaptive categorical loudness scaling) in 20 normal hearing subjects. When the stimulation was by bone conduction, the loudness functions were steeper and the ratios between the slopes of the AC and BC loudness functions were 0.88 for the low-frequency sound and 0.92 for the high-frequency sound. These results were almost identical to those of Stenfelt and Håkansson (2002) using the equal loudness estimation procedure. Thus, the findings could not be attributed to the loudness estimation procedure. Stenfelt and Håkansson (2002) described the AC and BC paths as schematically illustrated in Fig. 10.2. The transmission of sound for the AC path is by way of the outer ear canal to the middle ear via the ossicles, resulting in a motion of the stapes that gives a motion of the cochlear fluid. The AC sound pressure at the different parts of the ear induces some BC sound via pathways indicated in Fig. 10.2. Due to the large impedance difference between air and bone this sound transfer is negligible for a normal hearing ear. The BC path has, as indicated in Fig. 10.2, a rather complex form. The transducer is applied either directly to the bone or to the skin covering the bone. The vibrations at the temporal bone radiate sounds into the outer ear canal by way of relative jaw movements and sound radiation from the cartilage and soft

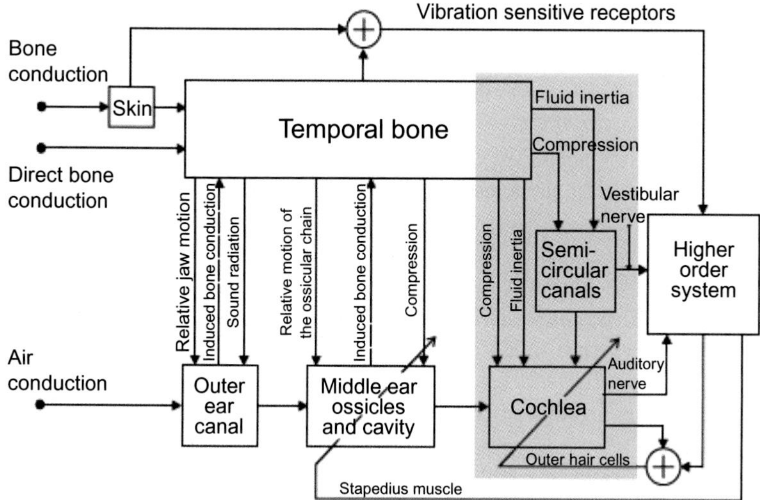

FIGURE 10.2 A schematic illustrating possible AC and BC paths. The AC path is straightforward with a sound outside the ear that is transmitted through the outer ear canal, via the middle ear ossicles to the cochlea. The BC sound transmission path is rather complex: inertial effects and compressional effects transmit a vibration of the temporal bone to the outer ear canal, middle ear ossicles and cavity, and to the semicircular canals. All are transmitted to the cochlea and added to the direct stimulation of the cochlea by BC sound. The vibration of the temporal bone and skin influences vibrotactile receptors that transmit stimulation to the higher order system that can influence other parts of the system and/or add to the total perceived sound. Further, there are some transmission parameters, such as the stapedius muscle and the outer hair cells that are affected by the stimulation of the higher order system. The most important mechanisms, encompassing fluid inertia and compression, are highlighted (pink). *Reprinted from Stenfelt, S., Håkansson, B., 2002. Air versus bone conduction: an equal loudness investigation. Hear. Res. 167, 1–12, with permission from Elsevier.*

tissues and may induce sounds in the middle ear cavity by compression of the cavity. More importantly, they also cause the relative motion of the middle ear ossicles and the compression of the cochlea together with the fluid inertia (Bárány, 1938; Kirikae, 1959; Tonndorf, 1966; von Békésy, 1960).

For understanding the action of bone conduction devices, the dominant contributions to perception are the inertial component of the cochlear fluid and the compression of these fluids (highlighted in Fig. 10.2). Important here is the mobility of the cochlear windows. Compression of the cochlear shell occurs for wavelengths smaller than the size of the audiovestibular system. Compression of cochlear fluids requires one mobile cochlear window, whereas the inertial response requires two mobile windows. Because cochlear fluids are incompressible, the inertial component in bone conduction is dominant up to approximately 4 kHz (Stenfelt, 2015). In addition, based on a model study, Stenfelt (2016) concluded that: "Inner ear compression and middle ear inertia were within 10 dB for almost the entire frequency range of 0.1 to 10 kHz. Ear canal sound pressure gave some contribution at the low and high frequencies, but was around 15 dB below the total contribution at the mid frequencies. Intracranial sound pressure gave responses similar to the others at low frequencies, but decreased with frequency to a level of 55 dB below the total contribution at 10 kHz."

As we have seen, the middle ear ossicles contribute to BC hearing primarily by the inertial forces acting on them when the skull is vibrating. This middle ear inertia is effective in the mid frequencies (1–3 kHz), i.e., the resonance frequencies of the middle ear. Stenfelt and Goode (2005) described that for a fixed stapes footplate, as in otosclerosis, the ossicular inertia is removed and a loss in BC sensitivity of approximately 20 dB around 2 kHz is normally seen, referred to as the Carhart notch. This depends on the mobility of the oval window, which is typically absent for stapes fixation (Stenfelt, 2015). No major alteration of the BC thresholds at the high and low frequencies is expected from otosclerosis with fixation of the stapes. Other lesions of the middle ear usually affect the BC thresholds only 10 dB or less. A conductive impairment in the middle ear affects the BC thresholds similarly whether the stimulation position is on the mastoid or on the forehead. Stenfelt and Goode (2005) noted that when a bone conductor is applied to the head: "the skull develops vibrations in all three planes and rotational motion. This complex motion of the skull is reflected in the motion of the cochlea; it also moves in all three dimensions in space without any dominating direction. The transcranial attenuation of BC vibration energy is on the average −5 to 10 dB. This indicates that during BC testing, masking of the non-test ear is always required if the hearing status of a single ear is tested. The skull has several resonances and anti resonances in the frequency range of hearing; the first appears around 0.8 to 0.9 kHz. The resonances are highly damped and do not affect hearing by BC, whereas the anti resonances can cause up to 20 dB attenuation for narrow bandwidths. The sensitivity of BC sound depends on the position of the transducer; the mastoid site is approximately 10 dB more sensitive than the forehead."

10.2 BONE-ANCHORED HEARING AIDS

A BAHA is an auditory prosthetic based on bone conduction, which can be surgically implanted. It is an option for patients without external ear canals and when conventional hearing aids (CHAs) with a mold in the ear cannot be used. BAHAs are especially useful in case of chronic otitis media (www.snikimplants.nl). The BAHA uses the skull as a pathway for sound to travel to the inner ear. For people with conductive hearing loss, the BAHA bypasses the external auditory canal and middle ear, stimulating the functioning cochlea. For people with unilateral complete deafness, often referred to as single-sided deafness (SSD), the BAHA is placed on the deaf side and uses the skull to conduct the sound from the deaf side to the side with the functioning cochlea. Individuals under the age of 2 (five in the United States) typically wear the BAHA device on a Softband. This can be worn

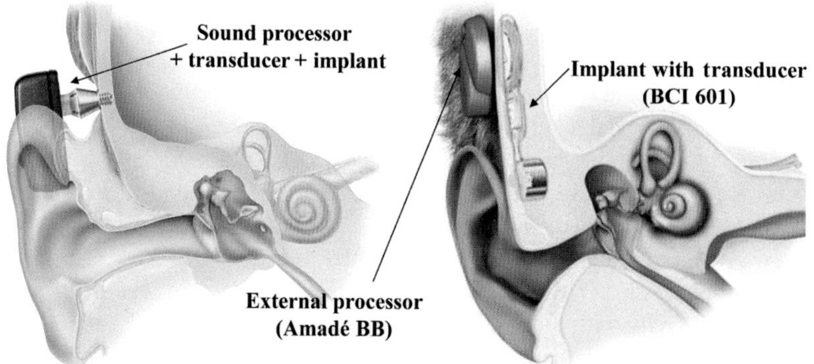

FIGURE 10.3 Schematic representation of the BAHA and the Bonebridge system. While the sound processor and transducer of the BAHA are percutaneously connected to the implant, the Bonebridge consists of an external audio processor (Amadé BB) and a transcutaneously stimulated transducer implant (BCI 601). *Reprinted from Huber, A.M., Sim, J.H., Xie, Y.Z., Chatzimichalis, M., Ullrich, O., Röösli, C., 2013. The Bonebridge: preclinical evaluation of a new transcutaneously activated bone anchored hearing device. Hear. Res. 301, 93–99, with permission from Elsevier.*

from the age of 1 month as babies tend to tolerate this arrangement very well. When the child's skull bone is sufficiently thick, a titanium percutaneous abutment can be surgically embedded into the skull with a small abutment exposed outside the skin. The BAHA sound processor sits on this abutment and transmits sound vibrations to the external abutment of the titanium implant (Fig. 10.3, left). The implant vibrates the skull and inner ear, which stimulate the nerve fibers of the inner ear, allowing hearing. Two companies manufacture BAHAs today—Cochlear and Oticon (https://en.wikipedia.org/wiki/Hearing_aid; accessed November 1, 2015).

Bone conduction implants (BCIs) come in two forms: percutaneous (BAHA) and transcutaneous. Chasin (1997) described these as "while the advantage of the transcutaneous approach such as that used with the Audiant™ is one of cosmetics (i.e., no abutment protruding through the skin), the improved transduction (by up to 20 dB) of a 'hard wired' percutaneous implant such as the BAHA™ appears to be a much more significant factor in the successful fitting of these patients." Other early comparisons of the two types (BAHA vs a transcutaneous temporal bone stimulator) resulted in the BAHA as the better choice (Snik et al., 1998).

10.2.1 General Performance

10.2.1.1 Single-Sided Deafness

BCIs are often used as a "Contralateral Routing of Signals" device, which improve directional hearing in single-sided deafness (SSD). Agterberg et al.

(2011) examined sound localization in azimuth in patients with acquired severe unilateral conductive hearing loss. All patients were fitted with a BAHA to restore bilateral hearing. The patients were tested in the unaided (monaural) and aided (binaural) hearing condition. They found that the BAHA significantly improved sound localization in 8/12 of the unilateral hearing loss patients. van Wieringen et al. (2011) collected data of six patients with SSD, seven with a mild to severe hearing loss at the BAHA side and (near-) normal hearing at the other side. They found that SSD patients listened mainly with their normal ear, although the BAHA lifted the head shadow effect. van Wieringen et al. (2011) also observed that these patients even regained limited binaural sensitivity with the device. The six patients with severe bilateral hearing loss in this study listened predominantly with their BAHA and only regained limited directional hearing. Battista et al. (2013) also evaluated the sound localization capabilities of patients with unilateral, profound sensorineural hearing loss who had been implanted with either a bone-anchored hearing device (BAHA BP100) or a TransEar 380-HF bone conduction hearing device. They found that: "Neither the BP100 nor the TransEar device improved sound localization accuracy in patients with unilateral, profound sensorineural hearing loss compared with performance in the unaided condition."

May et al. (2014) found that restoration of aural sensitivity in the deaf hemifield with an integrated bone conduction hearing aid enhances speech intelligibility under complex listening conditions after 3 months of unstructured use. Desmet et al. (2012) evaluated 196 patients with SSD, 93% of these patients suffered from an acquired hearing loss, who were enrolled for a trial period of 2 weeks. The most important reason mentioned by 66% of all the patients who declined the BAHA was lack of improvement of speech understanding in noise. Pai et al. (2012) evaluated the efficacy of BAHA for SSD in unilateral profound hearing loss with normal or mild high-frequency hearing loss in the hearing ear (pure-tone average (PTA) ≤25 dB HL measured at 0.5, 1, 2, and 3 kHz). After a 6-month trial, these adult patients benefitted most in speech understanding in challenging listening situations.

Laske et al. (2015) implanted nine adults with SSD for more than 1 year and normal hearing on the contralateral side with a Bonebridge (cf. Fig. 10.3, right). They found that: "Speech discrimination scores showed a mean signal-to-noise ratio improvement of 1.7 dB SPL for the aided condition compared with the unaided condition in the setting where the sound signal is presented on the side of the implanted ear and the noise source was in the front." It is interesting that the benefit was only 1.7 dB when one would expect an improvement of about 5 dB (the head shadow for broadband signals, Snik, personal communication). The potential underlying variability is described in the work of Lin et al. (2006), who noted that patients with a moderate SNHL in the functioning ear perceived greater increments in benefit, especially in background noise, and demonstrated greater improvements in speech understanding with BAHA amplification.

Summarizing, in unilateral hearing loss sound localization accuracy and speech understanding in noise can be improved by a BAHA, in SSD only when the speaker is localized on the deaf side.

10.2.1.2 Bilateral Hearing Loss

Janssen et al. (2012) systematically reviewed the literature and found some evidence that bilateral BAHAs provided additional benefit compared to unilateral BAHA. Desmet et al. (2014) investigated the subjective benefit from a BCI sound processor in 14 patients, who were fitted with a Cochlear BAHA Compact, 23 with a BAHA Divino, or 7 with a BAHA Intenso. The used a survey with a median follow-up time of 50 months. At that time, 86% of the patients still used their sound processor and reported that: "Speech understanding in noisy situations is rated rather low, and 58% of all patients reported that their BCI benefit was less than expected."

Riss et al. (2014) tested 24 successive patients equipped with the Bonebridge. They measured the overall average functional hearing gain of all patients ($N = 23$) was 28.8 ± 16.1 dB (mean ± SD). Monosyllabic word scores at 65 dB SPL in quiet increased statistically significantly from 4.6 ± 7.4% to 53.7 ± 23.0%. Evaluation of preoperative bone conduction thresholds revealed three patients with thresholds higher than 45 dB HL in the high frequencies starting at 2 kHz. These three patients had a very limited benefit of their BCIs. Riss et al. (2014) concluded that: "The Bonebridge bone-conduction implant provides satisfactory results concerning functional gain and speech perception if preoperative bone conduction lies within 45 dB HL." Mertens et al. (2014) found that "the maximum output of the Bonebridge ranges from 55 to 71 dB HL, depending on frequency. Accepting a minimum dynamic range of 35 dB with the Bonebridge, fitting of the Bonebridge in a linear program is advocated in patients with a sensorineural hearing loss component of up to 30 dB HL."

Hol et al. (2013) compared the new transcutaneous bone conduction hearing aid, the Sophono Alpha 1, with the percutaneous BAHA system (Fig. 10.4). They found: "The BAHA-based outcome was slightly better compared with Sophono-based results in sound field thresholds, speech recognition threshold, and speech comprehension at 65 dB." They also remarked that "the Sophono offers appealing clinical benefits of transcutaneous bone conduction hearing; however, the audiologic challenges of transcutaneous application remain, as the Sophono does not exceed percutaneous application regarding audiologic output." See also www.snikimplants.nl.

10.2.2 Application in Children

Doshi et al. (2012) reviewed the use of BAHAs in children, they concluded: "The latest generation of percutaneous implants have been designed to

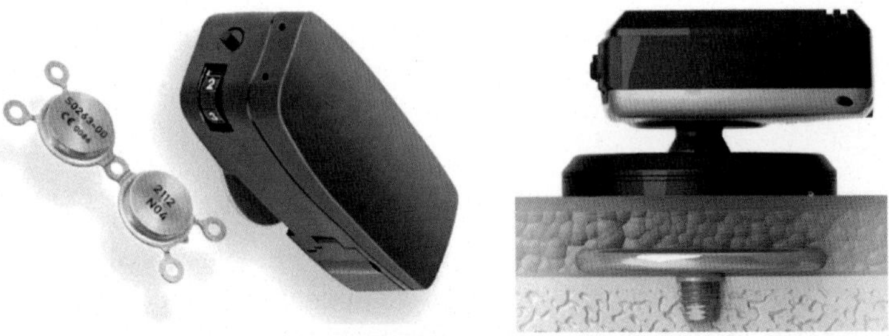

FIGURE 10.4 Sophono Alpha (left) and BAHA Attract (right). The Sophono Alpha bone conduction hearing device lies completely under skin and has low risk of skin issues. The BAHA Attract is a percutaneous device that leaves the skin intact; it uses a magnetic connection to attract the sound processor to the implant. *Images reproduced from the Internet.*

reduce skin complications, promote better osseo-integration and earlier loading of the sound processor. Alternative devices without a skin-penetrating abutment are now available and have shown promising results in the paediatric population." However, Kraai et al. (2011) noted that in pediatric patients "Complications of bone-anchored hearing aid implants are common in our experience and obesity and socioeconomic factors appear to contribute to a higher risk for complications." Dun et al. (2013) investigated "whether children with bilateral conductive hearing loss benefit from their second device (i.e., the bilateral bone conduction devices BAHA Divino or BAHA Compact). ... Children demonstrated an improvement in speech recognition when speech was presented from the front and noise was presented from the right-hand side as compared with both speech and noise being presented from the front. The minimum audible angle decreased from 57° in the best monaural condition to 13° in the bilateral condition. ... The audiological outcomes demonstrate the advantage of bilateral BCD fitting in children with bilateral conductive hearing loss."

Powell et al. (2015) compared the Sophono Alpha with the BAHA Attract (Fig. 10.4) in 10 children and 2 adults (six recipients for each device) and reported: "The unaided four-frequency average air conduction for affected ears was ~61 dB HL for the BAHA group and ~58 dB HL for the Sophono group; these improved to mean aided thresholds of 31 dB HL and 30 dB HL, respectively. There was no statistical difference between the speech discrimination scores for the two devices in quiet at 55 dB SPL or in noise. ... Both systems provide audiologic benefit compared with the unaided situation. ... All BAHA and Sophono users reported improvement in quality of life and would recommend their device to others in a similar situation." Baker et al. (2015) conducted a retrospective chart review for the first 11 Sophono implanted children and for the first 6 patients implanted with the BAHA Attract at their clinic. They found that: "Average improvement for the BAHA Attract in pure-tone average and speech reception threshold was 41 dB hearing level (dB HL) and 56 dB HL, respectively. The

Sophono average improvement in PTA and SRT was 38 dB HL and 39 dB HL, respectively." Baker et al. (2015) concluded that: "Significant improvements in both pure-tone averages and speech reception threshold for both devices were achieved. In direct comparison of the two separate devices using the chi-square test, the PTA and SRT data between the two devices do not show a statistically significant difference."

Denoyelle et al. (2015) studied the gain and cutaneous tolerance of the Sophono Alpha1 implant that was used for unilateral hearing rehabilitation in 15 children aged from 61 to 129 months with ear atresia. They also demonstrated close similarity in outcomes with the referral closed skin device, BAHA on a test band. Preimplantation, the patients had a pure conductive deafness with a mean air conduction pure-tone average (ACPTA) of 69 ± 9 dB and a mean speech reception threshold (SRT) of 72 ± 9 dB. At 6 months after implantation, the mean aided ACPTA was 34 ± 5 dB, the mean aided SRT 38 ± 5 dB, and the mean aided SRT in noise was significantly improved (−7.8 dB). The Sophono Alpha1 performance was similar to BAHA on a test band. At 12 months after implantation, the mean aided ACPTA was significantly (2.9 dB) higher but the mean SRT (+0.7 dB) was not significantly different. At M12, all children used the implant 5–12 h daily without cutaneous complications. Both children and parents reported being satisfied or very satisfied. The score for 7/10 questions in silence or noisy environment was statistically improved when wearing the device. Denoyelle et al. (2015) concluded that the aided threshold for the Sophono Alpha1 is between that obtained with BAHA on a softband and the percutaneous BAHA Attract.

10.3 IMPLANTABLE ACTIVE MIDDLE EAR DEVICES

There are fully implantable and semi-implantable active middle ear devices. The basic components include a microphone, audio processor, battery, receiver, and vibration transducer which attach to the ossicular chain or to the round window (Fig. 10.5). Active middle ear implants (AMEIs) are surgically implanted into the middle ear and do not obstruct the external ear canal (Butler et al., 2013). In semi-implantable devices, the microphone, processing electronics, and power source are kept externally. Sound signals and power are transmitted transcutaneously to the transducer via a telecommunication coil or magnetic induction (Jenkins and Uhler, 2014).

10.3.1 First Results

Suzuki et al. (1983) introduced the "Rion" IHA as the world's first such hearing device. "Most patients reported high satisfaction from reduced feedback and more natural, clear sound compared with their conventional hearing aids. In 2005, the RION Device was discontinued, because the company was unable to maintain a sufficient profit in Japan's socialized medical insurance

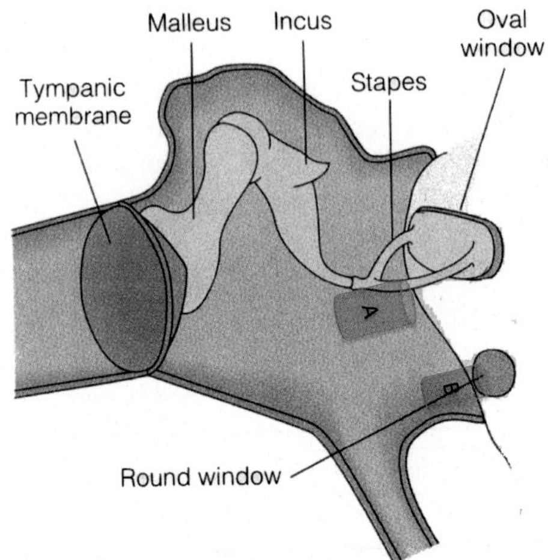

FIGURE 10.5 Schematic for VSB. Here a floating mass driver attaches to the stapes (A) or against the round window (B). The transducers are driven via an electric wire connector with the external processor or driven by a processor and coil in the ear canal.

system" (Carlson et al., 2014). Also in the early 1980s, Hough et al. (1986) began a series of experiments that led to the development of a Direct Drive Hearing System, recently reintroduced as the Maxum Hearing Implant. Hough et al. (1987) phrased it as: "This device stimulates, by an electromagnetic field, an independent electromagnetic sensitive prosthesis attached to the ossicular chain. This direct energy transfer to the ossicular chain provides a high degree of sound amplification and fidelity, thus providing benefit for those with various degrees of sensorineural hearing impairment." First clinical experiences, however, were mixed (Roush and Rauch, 1990; van der Hulst et al., 1993).

10.3.2 General Performance

The hearing loss indication for application of an AMEI is overlapping with that for CHAs (Fig. 10.6). Boeheim et al. (2010) found that "open ear hearing solutions provide benefit to patients with sloping high-frequency SNHL while leaving the ear canal unoccluded and with little to no chance of acoustic feedback. Although aided hearing thresholds and speech recognition in quiet and noise were successfully improved with both open-ear hearing solutions, performance with the AMEI (Vibrant Soundbridge) was significantly better than with the open-fit HA."

10.3.2.1 The Vibrant Soundbridge

The Vibrant Soundbridge (VSB) is a semi-implantable device. Carlson et al. (2014) described it as "the VSB consists of 2 primary components: an

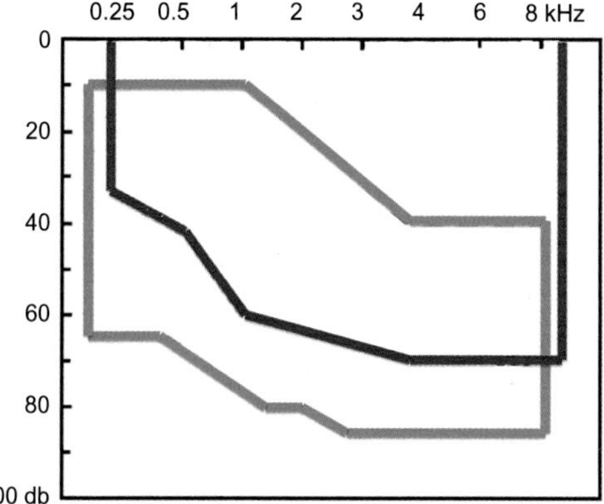

FIGURE 10.6 Overlapping indication fields for hearing aids (red lines) and two AMEIs (VSB and Bone-bridge; green lines). *Data from Boeheim, K., Pok, S.-M., Schloegel, M., Filzmoser, P., 2010. Active middle ear implant compared with open-fit hearing aid in sloping high-frequency sensorineural hearing loss. Otol. Neurotol. 31, 424–429.*

external audio processor and a surgically implanted vibrating ossicular prosthesis" (Fig. 10.5). The external audio processor contains a microphone, signal processor, telemetry coil, and a replaceable battery, all housed within a single unit. The external unit is designed to sit against the postauricular skin, in the hairline, through magnetic attraction with the implanted receiver. The floating mass transducer (FMT) shown here is attached to the stapes head, but newer versions also include a round-window driver.

Two of the first tests of the VSB (Snik and Cremers, 1999, 2001) with both an analog and a digital sound processor concluded that the device is suitable for patients with hearing loss of up to 70 dB HL. They also reported that "On average, results obtained with the Vibrant Soundbridge were not as good as those obtained with the conventional device. Nevertheless, their patients, all with chronic otitis externa, were satisfied with the Vibrant Soundbridge because they could use it all day without pain or itching." Schmuziger et al. (2006) noted that: "Satisfaction with the VSB was not superior to conventional hearing aids in subjective and in audiometric terms. Because of its impact on residual hearing and the requirement of implantation middle ear surgery, implantation of the VSB should be limited to patients with relevant side effects of hearing aids, e.g., severe chronic otitis externa."

Two early multicenter studies in France (Fraysse et al., 2001; Sterkers et al., 2003) concluded that: "The VBS surgical implantation procedure does not affect the residual hearing level in the implanted ear, nor does it present any unacceptable risk. Measurable benefit from the VBS in comparison with conventional amplification was demonstrated with regard to the provision of superior usable amplification and greater ease in communication in daily listening environments for the majority of patients." Cremers et al. (2010) in another multicenter study reported that AMEIs such as the VSB had been implanted in more than 60 children and adolescents worldwide by the end of 2008. They concluded that: "Taken together, the VSB offers another viable treatment for children and adolescents with compromised hearing." Garin et al. (2010) in a multicenter and retrospective study of 15 patients with symmetrical sensorineural hearing loss who were implanted sequentially in both ears, evaluated the benefit of bilateral VSB middle ear implantation as compared with unilateral implantation in quiet and noisy environments. They concluded that qualitative and quantitative assessments demonstrated improvement in speech intelligibility, especially in background noise, but also for low voice intensity in quiet. Sziklai and Szilvássy (2011) found "no significant difference in speech recognition scores obtained with the Vibrant Soundbridge and the open-fit hearing aid. However, the Vibrant Soundbridge may be superior to open-fit hearing aids in improving hearing at high frequencies (4–8 kHz)."

Later versions of the VSB also included direct stimulation of the round window. Marino et al. (2013) determined the efficacy of this round window application of the VSB in patients with mixed or conductive hearing loss. Sixteen of the 18 subjects were successful VSB users. The VSB caused no significant deterioration in bone conduction preoperatively versus postoperatively. Speech recognition in quiet was not significantly different to performance wearing hearing aids. Marino et al. (2013) found that speech-in-noise performance was substantially improved with use of the VSB.

Verhaegen et al. (2012) assessed the long-term benefit of the VSB in patients with severe mixed hearing loss and to compare it with other hearing devices. Six patients with severe mixed hearing loss and a mean sensorineural hearing loss component between 40 and 70 dB received a VSB with the FMT coupled to the round window or to the oval window via a residual stapes. They found a "large variance in functional gain between the patients suggesting high variability in the effectivity of the FMT coupling. The speech recognition results for the experimental group were not systematically better than in either control group. ... On the average, speech recognition results were not better or worse than those found in patients with similar hearing loss fitted with bone-anchored hearing devices."

10.3.2.2 MET, Carina and Esteem

Design of a partially implantable Middle Ear Transducer (MET) with a favorable linear input–output curve started in the 1970s (Carlson et al., 2014). "The semi-implantable MET uses an external unit called the button external audio processor, containing a microphone, battery, signal processor, and transmitter. The implanted portion consisted of a receiver, electronics package, and electromagnetic driver. ... The fully implantable Carina uses the same electromagnetic transduction system, but includes an implantable battery, sound processor, and receiving coil for device charging and programming. A separate microphone that is connected to the implanted sound processor is placed in a postauricular subcutaneous pocket" (Carlson et al., 2014). However, "patients fitted with a VSB or an Otologics MET middle ear implant do not demonstrate better speech recognition scores than patients fitted with today's conventional hearing aids. Results might even been worse. Still, the VSB and Otologics MET are a good option in patients with moderate (VSB) to severe (Otologics MET) sensorineural hearing loss and external otitis" (Verhaegen et al., 2008).

Another fully implantable device is the Esteem II system, which is composed of a hermetically sealed titanium dual-sound processor with an integrated nonrechargeable lithium battery, along with two separate piezoelectric transducers. The sound processor is implanted in a 2-mm bony well and secured similar to how a cochlear implant receiver–stimulator package is secured. The piezoelectric receiver acting as a microphone and the piezoelectric stimulator are both located along the ossicular chain, and the driver is cemented to the head of the stapes. The nonrechargeable battery has an estimated 5- to 9-year life span and requires a small surgical procedure for replacement after expiration. An external device controls volume and programming (Carlson et al., 2014). Barbara et al. (2011) found that: "The Esteem® device proved to offer beneficial results in subjects suffering from high frequency, severe bilateral sensorineural hearing loss and may be considered as an alternative procedure to conventional hearing aids or electroacoustic stimulation systems."

10.3.2.3 The Maxum Hearing Implant

Hough et al. (1986) began a series of experiments that led to the development of the SOUNDTEC Direct Drive Hearing. Upgrades from the earlier prototypes included a stronger and lighter magnet driver, a hermetically sound titanium laser welded canister, and a wire clip for attachment at the incudostapedial joint, designed to provide optimal coaxial alignment with the in-the-canal electromagnetic coil. In 2009 Ototronix purchased the SOUNDTEC technology, and after several upgrades, including a miniaturized integrated digital processor and coil and a self-crimping Nitinol wire, the device was rereleased as the Maxum Hearing Implant (Carlson et al., 2014; Pelosi et al., 2014).

10.3.3 Safety Issues

Klein et al. (2012) examined the safety and effectiveness of fully implantable middle ear devices in the treatment of hearing loss. They found that: "The majority of studies were quasi-experimental, pre-post comparisons of aided and unaided conditions. Complication rates with the Esteem® were higher than with the Carina®, and most commonly included taste disturbance. However, device failure was common with the Carina®, predominately related to charging difficulties. For both devices, clinically significant improvements in functional gain, speech reception, and speech recognition over the unaided condition were found [but not compared to behind the ear aids]. In studies comparing the Esteem® or Carina® to hearing aids, findings were mixed. ... Despite limited evidence, these devices seem to offer a relatively safe and effective treatment option, particularly for patients who are medically unable to wear conventional hearing aids." Verhaert et al. (2013) performed a systematic review of literature to determine the clinical outcome and safety of the range of hearing implants in adults with mixed hearing loss. They found that hearing implants and their different coupling strategies in the treatment of mixed hearing loss were "beneficial in terms of speech in quiet, patient reported outcome measures, and safety regarding residual hearing. Overall, the level of evidence and the quality of the included studies were judged to be moderate to low."

Carlson et al. (2014) concluded that: "The initial motivation for the development of AMEIs was to provide an alternative to conventional hearing aids for patients with moderate to severe hearing loss. Although most commercially approved devices show a satisfactory risk profile, several series have reported a high rate of implant extrusion, temporary and permanent facial paralysis, and device malfunction ... The few devices that use a fully implantable design frequently require partial ossicular resection to minimize reverse feedback. After device failure, these patients may be left with a considerable conductive hearing loss, which compounds preexisting hearing loss. There are also significant concerns regarding long-term device reliability and the need for revision surgery."

10.3.4 Middle Ear Implants Versus Conventional Hearing Aids

In a systematic review, Tysome et al. (2010) investigated whether AMEIs improved hearing as much as CHAs. From inspecting 17 papers they found that "the quality of studies was moderate to poor with short follow-up. The evidence supports the use of AMEI because, overall, they do not decrease residual hearing, result in a functional gain in hearing comparable to CHA, and may improve perception of speech in noise and sound quality." Butler et al. (2013) undertook a systematic review to advise on the effectiveness of the AMEI in patients with sensorineural hearing loss, compared with external hearing aids. The studies employed a variety of AMEI devices, including the Envoy Esteem, Otologics MET, Soundtec Direct Drive Hearing System (Maxum), and VSB. They found that: "Generally, the active middle-ear implant appears to be as effective as the external hearing aid in improving hearing outcomes in patients with sensorineural hearing loss."

Carlson et al. (2014) summarized the anticipated versus realized benefits of IHAs as: "To justify the greater cost and increased risk associated with AMEIs, there must be a substantial benefit to the user compared with conventional air conduction hearing aids. The theoretic advantages of an implantable hearing aid include improved functional gain, reduced risk of feedback, enhanced sound quality, greater speech understanding in noise, improved concealment, lack of occlusion, and greater freedom to lead an active lifestyle. Although, clearly, substantial improvements have been made over the past several decades, many of these theoretic advantages have not yet been realized."

10.4 SUMMARY

IHAs can be divided in percutaneous and transcutaneous BAHAs and middle ear implants. They are mostly applied for conductive hearing loss and are also a good solution for patients with sensorineural hearing loss who cannot tolerate ear molds because of external otitis. I started this chapter with a review on the mechanism of bone conduction and describing the nonlinear relation between AC and BC sound loudness for certain frequencies. This functions as a background for the BAHAs and their increasing application in older children. The use in bilateral hearing loss and single-sided hearing loss is described as well as the general performance of various brands. Middle ear implants also come in two types: fully- and semi-implantable. In the latter, the microphone, processing electronics, and power source are kept externally. The available implants (VSB, MET, Carina and Esteem, Maxum) are described in relation to application criteria and their performance is compared with CHAs. The semi-implantable middle ear implants appear to have less performance problems compared to the fully implantable ones.

REFERENCES

Agterberg, M.J., Snik, A.F., Hol, M.K., van Esch, T.E., Cremers, C.W., Van Wanrooij, M.M., et al., 2011. Improved horizontal directional hearing in bone conduction device users with acquired unilateral conductive hearing loss. J. Assoc. Res. Otolaryngol. 12, 1–11.

Baker, S., Centric, A., Chennupati, S.K., 2015. Innovation in abutment-free bone-anchored hearing devices in children: updated results and experience. Int. J. Pediatr. Otorhinolaryngol. 79, 1667–1672.

Bárány, E., 1938. A contribution to the physiology of bone conduction. Acta Otolaryngol. Suppl. 26, 1–223.

Barbara, M., Biagini, M., Monini, S., 2011. The totally implantable middle ear device 'Esteem' for rehabilitation of severe sensorineural hearing loss. Acta Otolaryngol. 131, 399–404.

Battista, R.A., Mullins, K., Wiet, R.M., Sabin, A., Kim, J., Rauch, V., 2013. Sound localization in unilateral deafness with the Baha or TransEar device. JAMA Otolaryngol. Head Neck Surg. 139, 64–70.

Boeheim, K., Pok, S.-M., Schloegel, M., Filzmoser, P., 2010. Active middle ear implant compared with open-fit hearing aid in sloping high-frequency sensorineural hearing loss. Otol. Neurotol. 31, 424–429.

Butler, C.L., Thavaneswaran, P., Lee, I.H., 2013. Efficacy of the active middle-ear implant in patients with sensorineural hearing loss. J. Laryngol. Otol. 127 (Suppl. 2), S8–S16.

Carlson, M.L., Pelosi, S., Haynes, D.S., 2014. Historical development of active middle ear implants. Otolaryngol. Clin. North Am. 47, 893–914.

Chasin, M., 1997. Current trends in implantable hearing aids. Trends Amplif. 2, 84–107.

Cremers, C.W., O'Connor, A.F., Helms, J., Roberson, J., Clarós, P., Frenzel, H., et al., 2010. International consensus on Vibrant Soundbridge® implantation in children and adolescents. Int. J. Pediatr. Otorhinolaryngol. 74, 1267–1269.

Denoyelle, F., Coudert, C., Thierry, B., Parodi, M., Mazzaschi, O., Vicaut, E., et al., 2015. Hearing rehabilitation with the closed skin bone-anchored implant Sophono Alpha1: results of a prospective study in 15 children with ear atresia. Int. J. Pediatr. Otorhinolaryngol. 79, 382–387.

Desmet, J.B., Wouters, K., De Bodt, M., Van de Heyning, P., 2012. Comparison of 2 implantable bone conduction devices in patients with single-sided deafness using a daily alternating method. Otol. Neurotol. 33, 1018–1026.

Desmet, J., Wouters, K., De Bodt, M., Van de Heyning, P., 2014. Long-term subjective benefit with a bone conduction implant sound processor in 44 patients with single-sided deafness. Otol. Neurotol. 35, 1017–1025.

Doshi, J., Sheehan, P., McDermott, A.L., 2012. Bone anchored hearing aids in children: an update. Int. J. Pediatr. Otorhinolaryngol. 76, 618–622.

Dun, C.A., Agterberg, M.J., Cremers, C.W., Hol, M.K., Snik, A.F., 2013. Bilateral bone conduction devices: improved hearing ability in children with bilateral conductive hearing loss. Ear Hear. 34, 806–808.

Fraysse, B., Lavieille, J.P., Schmerber, S., Enée, V., Truy, E., Vincent, C., et al., 2001. A multicenter study of the Vibrant Soundbridge middle ear implant: early clinical results and experience. Otol. Neurotol. 22, 952–961.

Garin, P., Schmerber, S., Magnan, J., Truy, E., Uziel, A., Triglia, J.M., et al., 2010. Bilateral vibrant soundbridge implantation: audiologic and subjective benefits in quiet and noisy environments. Acta Otolaryngol. 130, 1370–1378.

Hol, M.K., Nelissen, R.C., Agterberg, M.J., Cremers, C.W., Snik, A.F., 2013. Comparison between a new implantable transcutaneous bone conductor and percutaneous bone-conduction hearing implant. Otol. Neurotol. 34, 1071–1075.

Hough, J., Vernon, J., Johnson, B., et al., 1986. Experiences with implantable hearing devices and a presentation of a new device. Ann. Otol. Rhinol. Laryngol. 95, 60–65.

Hough, J., Vernon, J., Himelick, T., Meikel, M., Richard, G., Dormer, K., 1987. A middle ear implantable hearing device for controlled amplification of sound in the human: a preliminary report. Laryngoscope 97, 141–151.

Huber, A.M., Sim, J.H., Xie, Y.Z., Chatzimichalis, M., Ullrich, O., Röösli, C., 2013. The Bonebridge: preclinical evaluation of a new transcutaneously activated bone anchored hearing device. Hear. Res. 301, 93–99.

Janssen, R.M., Hong, P., Chadha, N.K., 2012. Bilateral bone-anchored hearing aids for bilateral permanent conductive hearing loss: a systematic review. Otolaryngol. Head Neck Surg. 147, 412–422.

Jenkins, H.A., Uhler, K., 2014. Otologics active middle ear implants. Otolaryngol. Clin. North Am. 47, 967–978.

Kirikae, I., 1959. An experimental study on the fundamental mechanism of bone conduction. Acta Otolaryngol. Suppl. 145, 1–111.

Klein, K., Nardelli, A., Stafinski, T., 2012. A systematic review of the safety and effectiveness of fully implantable middle ear hearing devices: the Carina and Esteem systems. Otol. Neurotol. 33, 916–921.

Kraai, T., Brown, C., Neeff, M., Fisher, K., 2011. Complications of bone-anchored hearing aids in pediatric patients. Int. J. Pediatr. Otorhinolaryngol. 75, 749–753.

Laske, R.D., Röösli, C., Pfiffner, F., Veraguth, D., Huber, A.M., 2015. Functional results and subjective benefit of a transcutaneous bone conduction device in patients with single-sided deafness. Otol. Neurotol. 36, 1151–1156.

Lin, L.M., Bowditch, S., Anderson, M.J., May, B., Cox, K.M., Niparlo, J.K., 2006. Amplification in the rehabilitation of unilateral deafness: speech in noise and directional hearing effects with bone-anchored hearing and contralateral routing of signal amplification. Otol. Neurotol. 27, 172–182.

Marino, R., Linton, N., Eikelboom, R.H., Statham, E., Rajan, G.P., 2013. A comparative study of hearing aids and round window application of the vibrant sound bridge (VSB) for patients with mixed or conductive hearing loss. Int. J. Audiol. 52, 209–218.

May, B.J., Bowditch, S., Liu, Y., Eisen, M., Niparko, J.K., 2014. Mitigation of informational masking in individuals with single-sided deafness by integrated bone conduction hearing aids. Ear Hear. 35, 41–48.

Mertens, G., Desmet, J., Snik, A.F., Van de Heyning, P., 2014. An experimental objective method to determine maximum output and dynamic range of an active bone conduction implant: the Bonebridge. Otol. Neurotol. 35, 1126–1130.

Pai, I., Kelleher, C., Nunn, T., Pathak, N., Jindal, M., O'Connor, A.F., et al., 2012. Outcome of bone-anchored hearing aids for single-sided deafness: a prospective study. Acta Otolaryngol. 132, 751–755.

Pelosi, S., Carlson, M.L., Glasscock III, M.E., 2014. The Ototronix MAXUM system. In: Mankeka, G. (Ed.), Implantable Hearing Devices Other Than Cochlear Implants. Springer, India, p. 97.

Powell, H.R., Rolfe, A.M., Birman, C.S., 2015. A comparative study of audiologic outcomes for two transcutaneous bone-anchored hearing devices. Otol. Neurotol. 36, 1525–1531.

Riss, D., Arnolder, C., Baumgartner, W.-D., Blineder, M., Flak, S., Bacher, A., et al., 2014. Indication criteria and outcomes with the Bonebridge transcutaneous bone-conduction implant. Laryngoscope 124, 2802–2806.

Roush, J., Rauch, S.D., 1990. Clinical application of an implantable bone conduction hearing device. Laryngoscope 100, 281–285.

Schmuziger, N., Schimmann, F., Wengen, D., Patscheke, J., Probst, R., 2006. Long-term assessment after implantation of the Vibrant Soundbridge device. Otol. Neurotol. 27, 183–188.

Snik, A.F., Cremers, C.W., 1999. First audiometric results with the Vibrant Soundbridge, a semi-implantable hearing device for sensorineural hearing loss. Audiology 38, 335–338.

Snik, A.F., Cremers, C.W., 2001. Vibrant semi-implantable hearing device with digital sound processing: effective gain and speech perception. Arch. Otolaryngol. Head Neck Surg. 127, 1433–1437.

Snik, A.F., Dreschler, W.A., Tange, R.A., Cremers, C.W., 1998. Short- and long-term results with implantable transcutaneous and percutaneous bone-conduction devices. Arch. Otolaryngol. Head Neck Surg. 124, 265–268.

Stenfelt, S., 2007. Simultaneous cancellation of air and bone conduction tones at two frequencies: extension of the famous experiment by von Békésy. Hear. Res. 225, 105–116.

Stenfelt, S., 2015. Inner ear contribution to bone conduction hearing in the human. Hear. Res. 329, 41–51.

Stenfelt, S., 2016. Model predictions for bone conduction perception in the human. Hear. Res. 340, 135–143.

Stenfelt, S., Goode, R., 2005. Bone conducted sound: physiological and clinical aspects. Otol. Neurotol. 26, 1245–1261.

Stenfelt, S., Håkansson, B., 2002. Air versus bone conduction: an equal loudness investigation. Hear. Res. 167, 1–12.

Stenfelt, S., Zeitooni, M., 2013. Binaural hearing ability with mastoid applied bilateral bone conduction stimulation in normal hearing subjects. J. Acoust. Soc. Am. 134, 481–493.

Sterkers, O., Boucarra, D., Labassi, S., Bebear, J.P., Dubreuil, C., Frachet, B., et al., 2003. A middle ear implant, the Symphonix Vibrant Soundbridge: retrospective study of the first 125 patients implanted in France. Otol. Neurotol. 24, 427–436.

Suzuki, J., Kodera, K., Yanagihara, N., 1983. Evaluation of middle-ear implant: a six-month observation in cats. Acta Otolaryngol. 95, 646–650.

Sziklai, I., Szilvássy, J., 2011. Functional gain and speech understanding obtained by Vibrant Soundbridge or by open-fit hearing aid. Acta Otolaryngol. 131, 428–433.

Tonndorf, J., 1966. Bone conduction. Studies in experimental animals. Acta Otolaryngol. Suppl. 213.

Tysome, J.R., Moorthy, R., Lee, A., Jiang, D., O'Connor, A.F., 2010. Systematic review of middle ear implants: do they improve hearing as much as conventional hearing aids? Otol. Neurotol. 31, 1369–1375.

van der Hulst, R.J., Dreschler, W.A., Tange, R.A., 1993. First clinical experiences with an implantable bone conduction hearing aid at the University of Amsterdam. Eur. Arch. Otorhinolaryngol. 250, 69–72.

van Wieringen, A., De Voecht, K., Bosman, A.J., Wouters, J., 2011. Functional benefit of the bone-anchored hearing aid with different auditory profiles: objective and subjective measures. Clin. Otolaryngol. 36, 114–120.

Verhaegen, V.J., Mylanus, E.A., Cremers, C.W., Snik, A.F., 2008. Audiological application criteria for implantable hearing aid devices: a clinical experience at the Nijmegen ORL clinic. Laryngoscope 118, 1645–1649.

Verhaegen, V.J., Mulder, J.J., Cremers, C.W., Snik, A.F., 2012. Application of active middle ear implants in patients with severe mixed hearing loss. Otol. Neurotol. 33, 297–301.

Verhaert, N., Desloovere, C., Wouters, J., 2013. Acoustic hearing implants for mixed hearing loss: a systematic review. Otol. Neurotol. 34, 1201–1209.

von Békésy, G., 1960. Experiments in Hearing. McGraw-Hill, New York.

Chapter 11

Cochlear Implants

Cochlear implants are among the great success stories of modern medicine. Thirty years ago these devices provided little more than a sensation of sound and sound cadences—they were useful as an aid to lip-reading. In the 1980s, however, systems with multiple channels of processing and multiple sites of stimulation in the cochlea were developed and these systems supported significantly higher levels of speech reception than their single-channel and single-site predecessors. In the late 1980s and continuing to the present, new and better processing strategies, in conjunction with multielectrode implants, have produced further large improvements.

Wilson and Dorman, 2008

11.1 BASICS OF COCHLEAR IMPLANTS

Cochlear implants (CIs) consist of essentially three parts: an external sound processor, a subcutaneous receiver, and an intracochlear electrode array.

11.1.1 The Electrode Array

Most CI systems use electrode arrays that extend 1–1.5 turns (3–4 octaves in frequency) from the basal entry point into the scala tympani, but one manufacturer (MED-EL GmbH) uses an electrode array that is considerably longer, but not necessarily extends further into the cochlea. Typically, a CI electrode consists of a linear array of 12–22 metal electrode contacts, depending on the device. An implant channel comprises one active electrode along the cochlear array and one or more return electrodes, which may be inside or outside the cochlea. The active electrode typically delivers a train of biphasic pulses. Most of the current flows along the fluid-filled scala tympani, but some also flows into the less conductive osseous spiral lamina adjacent to it. Within this bone lie the spiral ganglion neurons, whose axons form the auditory nerve. A high enough current triggers a volley of action potentials in the auditory nerve (Bierer, 2010).

Fig. 11.1 shows micrographs of four of the currently available electrode arrays and their extent of insertion in the cochlea. The MED-EL Flex31

Hearing Loss. DOI: http://dx.doi.org/10.1016/B978-0-12-805398-0.00011-6

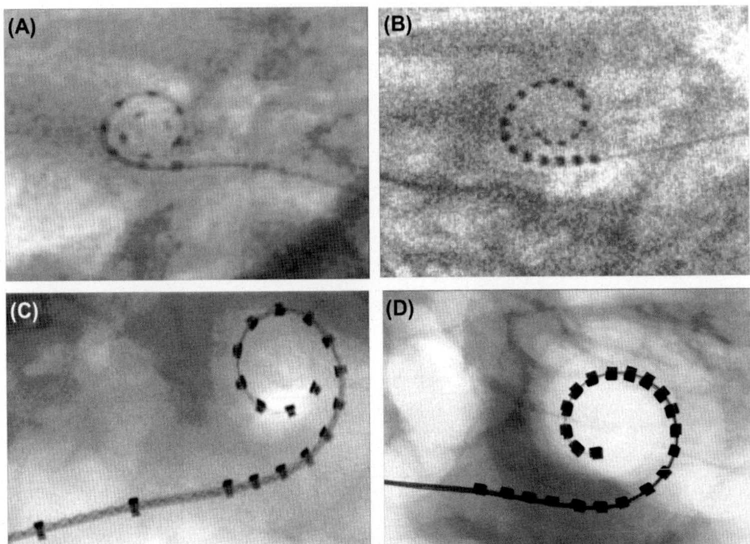

FIGURE 11.1 X-ray views of several currently available CI electrode arrays, showing extent of typical insertions. (A) MED-EL Flex31, (B) Neurelec Standard Array, (C) Advanced Bionics Helix, and (D) Cochlear Contour Advance. *Reprinted from Boyd, P.J., 2011. Potential benefits from deeply inserted cochlear implant electrodes. Ear Hear. 32, 411–427.*

(Fig. 11.1A) and Neurelec Standard Array (Fig. 11.1B) arrays are flexible straight designs that curve during insertion, while the Advanced Bionics Helix (Fig. 11.1C) and Cochlear Contour Advance (Fig. 11.1D) arrays have precurved designs. The Advanced Bionics HiFocus Helix array and the Cochlear Contour Advance array have their active electrodes spaced over the shortest distance (13 and 15 mm, respectively). The MED-EL Flex31 array has the widest "electrode spread" at 26.4 mm (Boyd, 2011).

11.1.2 The Sound Processor

The sound processor splits the sound in up to 120 channels (note, this is not the same as electrodes) by band-pass filtering. In the original analog devices the output of the various channels was compressed and send to a transmitter coil. On the receiver end, the output of these channels was distributed to individual electrodes (or pairs) in a frequency specific manner; high frequencies to electrodes ending in the basal turn and low frequencies to more apical ones. In the current digital devices the same is done, but now with an extra step of sound envelope detection between the analog filtered signals and the compression to result in a signal that modulates a train of bipolar pulses either in amplitude or width (Fig. 11.2).

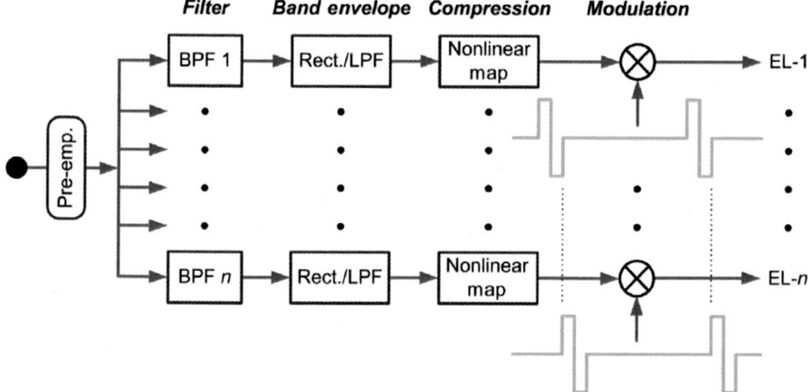

FIGURE 11.2 Block diagram of the continuous interleaved sampling (CIS) processing strategy for CIs. The input is at the left-most part of the diagram. Following the input, a pre-emphasis filter (Pre-emp.) is used to attenuate strong components in the input at frequencies below 1.2 kHz. This filter is followed by multiple channels of processing. Each channel includes stages of band-pass filtering (BPF), envelope detection, compression, and modulation. The envelope detectors generally use a full-wave or half-wave rectifier (Rect.) followed by a low-pass filter (LPF). Carrier waveforms for two of the modulators are shown immediately below the two corresponding multiplier blocks (circles with an " × " mark within them). The outputs of the multipliers are directed to intracochlear electrodes (EL-1 to EL-*n*), via a transcutaneous link. *Reprinted from Wilson, B.S., Dorman, M.F., 2008. Cochlear implants: a remarkable past and a brilliant future. Hear. Res. 242, 3–21, with permission from Elsevier.*

Several steps can be isolated from this general processing scheme: (1) spectral shape representation by band-pass filtering. (2) Coding of single tone frequency and complex sounds by taking the band envelope. (3) Amplitude compression to map the more than 100 dB range of normal hearing to the less that 15 dB in electric stimulation provided by CIs. These three steps result in nonlinear mapping of bipolar pulse rains, which are sent to the electrodes.

11.1.3 Spectral Sound Shape Representation

Frequency information in the acoustic signal is encoded in the spatial pattern of stimulation along the implant array. Each electrode carries information in a frequency range that, in normal hearing, would be transduced by hair cells and nerve fibers at a portion of the basilar membrane in the vicinity of the electrode. More apical electrodes are activated by lower frequencies and basal electrodes by higher frequencies (Bierer, 2010). This tonotopic ordering is set in the CI by the clinical mapping procedure, which assigns each analyzed frequency band to each channel. The particular subset of electrodes that form a channel affects how the electrical current delivered by that channel activates the auditory nerve fibers (ANFs) (Bierer, 2010). Most implant listeners appear to perform as well with six to eight channels as with all channels on speech perception measured in quiet or background noise (Fishman et al., 1997; Friesen et al., 2001).

11.1.4 Coding of Single Frequencies and Complex Sounds

Because of the mapping, a single frequency will ideally activate a single channel. This place coding of sound in a CI is crude because of the small number of electrodes and the electrical cross talk between them. Low frequencies may also be coded by the periodicity in the signal applied to a single electrode. Coding of the periodicity of complex sounds in implants depends almost entirely on a temporal code and does not extend above about 300 Hz (Moore, 2003).

11.1.5 Amplitude Compression

A nonlinear (typically logarithmic) mapping function is used in each channel to compress the wide dynamic range of sounds in the environment (e.g., 90 or 100 dB), into the narrow dynamic range of electrically evoked hearing, which for short-duration pulses usually is less than 15 dB. The mapping still preserves a high number of discriminable loudness levels across the dynamic range of the input. The output of this compression stage modulates the train of biphasic pulses for each channel (Wilson, 2015).

11.1.6 Measurement of the Electrically Evoked Compound Action Potential

Telemetry capabilities became commercially available in 1998 (e.g., Shallop et al., 1999) for the measurement of the electrically evoked compound action potential (eCAP) from the auditory nerve in CI recipients. Several names are used for this: Neural Response Telemetry (Cochlear), Auditory Response Telemetry (Med-El), or Neural Response Imaging (Advanced Bionics). The eCAP is recorded via the intracochlear electrodes of the implant. Because the eCAP is a short-latency evoked potential, it overlaps with the stimulus artifact. All newer CI systems are equipped with two-way telemetry capabilities that allow for quick and easy measurement of electrode impedance and the eCAP. Just as the acoustically evoked CAP (see chapter: Hearing Basics), the eCAP is a synchronous response from electrically stimulated ANFs and is essentially the electrical version of the acoustically evoked CAP. The eCAP is recorded as a negative peak (N_1) at about 0.2–0.4 ms following stimulus onset, followed by a much smaller positive peak or plateau (P_2) occurring at about 0.6–0.8 ms. The amplitude of the eCAP can be as large as 1 mV (Gordon, personal communication), which is much larger in magnitude than the electrically evoked auditory brainstem response (eABR)

(Shallop et al., 1999) and the CAP (up to 30 μV) recorded by transtympanic electrocochleography in normal ears (see chapter: Hearing Basics; Appendix).

11.2 A LITTLE HISTORY

> *For their work on the development of the modern cochlear implant, which bestows hearing to individuals with profound deafness, Ingeborg Hochmair, Graeme Clark, and Blake Wilson are the 2013 recipients of the Lasker–DeBakey Clinical Medical Research Award. Ingeborg Hochmair, Graeme Clark, and Blake Wilson each lead a team of professionals dedicated to aiding the many individuals with hearing impairment. The Lasker Foundation now recognizes their scientific contributions to the development of cochlear implants, and in so doing acknowledges the effort of many individuals, both past and present, in the field of cochlear implantation, widely recognized as one of the most successful achievements in modern medicine.*
>
> Roland and Tobey, 2013

It all started in France when Djourno and colleagues, using electric stimulation, reported sound sensations in two totally deaf patients (Djourno and Eyries, 1957; Djourno et al., 1957a,b). They used an "induction coil with one end placed on the stump of the auditory nerve or adjacent brainstem and the other end within the temporalis muscle. … The patient was able to sense the presence of environmental sounds but could not understand speech or discriminate among speakers or many sounds" (Wilson and Dorman, 2008; Eshraghi et al., 2012).

After the French attempts, a patient in the United States was implanted at the Ear Research Institute (Los Angeles, CA). By 1969, a multiple-electrode system had been developed, and three more subjects were implanted. The first House-Urban single-electrode induction coil implant was placed in October, 1972 (House et al., 1979) and was also the first FDA approved one. Using a psychoacoustic approach to evaluate the effects of cochlear implantation, Simmons et al. (1965) at Stanford University studied auditory perceptions produced by electrically exciting the auditory nerve in deaf patients through (six) permanently implanted electrodes in the modiolus portion of the auditory nerve. They reported that: "at pulse repetition rates of about 1 pulse/sec the sound produced was usually described as either a "ping" or "ding." As the rate was increased to 3 to 4 pulse/sec, pulses were described as individually heard "clicks", which seemed to merge into a "buzz" at rates above about 10 per second. At times, 10-pulse/sec rates were described as "telephone ringing," but this description was usually volunteered only as the stimulus rate increased beyond about 20 per second, depending upon stimulus intensity and electrode selection. As stimulus rates were increased beyond 30 pulse/sec, the predominant descriptive term changed to 'car horn', 'telephone ring muffled by pillow', or 'bee buzz'. A transition in the 'nature of the sound' (but not necessarily its 'pitch') from 'a buzzing sound' to a 'steady sound' took place between 50 and 80 pulse/sec. Rates between 100 and 300 pulse/sec were typically described as 'steady high-pitched ringings', 'whistles,' or 'buzzes'."

In the normal hearing auditory system, periodicities in sound envelope give rise to three dominant percepts: rhythm, roughness, and periodicity pitch. Rhythm, also called fluctuation strength, is perceived for repetition rates below 20 Hz, the sensation of roughness occurs for rates of 20–300 Hz and is strongest around 70 Hz. Periodicity pitch starting above 30 Hz looses much of its perceptual strength above 3 kHz (Zwicker and Fastl, 1990). Thus, there is an overlap of the perception of roughness and pitch, likely due to the coexistence of both temporal (in the ANF firing times) and spectral (place in the cochlea) representations of pitch (Eggermont, 2015a). As in CIs there is limited place pitch, pitch (e.g., ringing) and roughness (e.g., buzzing) dominate the repetition rate range between 30 and 300 Hz. Pulse rates in current CIs are typically beyond these rates, and periodicity is conveyed by the modulation of these trains (Section 11.3.2).

The next steps in the development of the CI were made by several groups, such as Michelson, Schindler, and Merzenich at the University of California, San Francisco (Michelson et al., 1973); Chouard and MacLeod (1973) in France; and Clark et al. (1973) in Australia. Somewhat later they were joined by Burian's group, including the Hochmairs, in Austria (Burian et al., 1979). The work of the San Francisco group would ultimately lead to the establishment of Advanced Bionics, the Australian group to Cochlear, the Austrian group to MED-EL, and the French group to Neurelec (these days Oticon). These four are at the time of this writing the dominating CI manufacturers. Recently a 26 electrode CI (Nurotron) was developed by Fan-Gang Zeng and colleagues (Zeng et al., 2015) for, but not limited to, distribution in China.

The performance of CIs requires a good speech processing strategy. This work started with Wilson et al. (1988) who compared several speech processing strategies for multichannel auditory prostheses in studies of two patients implanted with the UCSF/Storz electrode array. They compared the compressed analog processor of the UCSF/Storz prosthesis with several interleaved pulses via processors with pulse–amplitude modulation. For these patients test scores were better with the interleaved pulses processors. Improvements gave rise to the CIS technique from which all subsequent processing strategies are derived (Fig. 11.2).

11.3 SOUND PROCESSING STRATEGIES

11.3.1 The Long Way to Speech Understanding With a Cochlear Implant

Before the introduction of modern digital processing strategies, speech understanding with a CI was rare as the following quotes may illustrate:

In the case of total deafness, the multichannel cochlear implant prosthesis constitutes a substantial improvement. It allows known words to be recognized without lip reading, with a great percentage of good responses (Pialoux et al., 1979).

These audiological findings indicate that the improvements obtained with multiple-channel electrical stimulation [with F0, F2 multiple-channel speech-processor], as shown by the phoneme and feature scores, carry over to the perception of words, sentences, and connected discourse. They also show that multiple-channel electrical stimulation combined with lip reading produced very significant improvements in patients' abilities to communicate when compared to results with lip reading alone (Clark et al., 1983).

Clinical experience is now considerable with the single-electrode cochlear implant. ... Auditory thresholds with the implant are such that patients can hear conversational level speech and a wide variety of sounds in the environment. Although they cannot understand speech, they can discriminate at an above-chance level on closed-set discrimination tasks (Eisenberg et al., 1983).

Among the various single-channel coding strategies investigated, stimulation schemes using a somewhat modified analog speech signal rank highest in speech understanding (Hochmair and Hochmair-Desoyer, 1983).

Using the four channel cochlear implant system with a vocoder-based processor developed at UCSF ... eight of the nine patients obtaining some open-set auditory only speech understanding (Schindler et al., 1986).

Effective processing strategies for CIs were developed in the late 1980s and early 1990s. Among these were the CIS (Wilson et al., 1991), *n*-of-*m* (Wilson et al., 1988), and spectral peak (SPEAK) (Skinner et al., 1991) strategies. Speech reception performance increased significantly with these strategies.

11.3.2 Description of Common Processor Strategies

A current review (Wouters et al., 2015) nicely condenses the various strategies and describes the general principle as follows based on the CIS strategy.

11.3.2.1 Continuous Interleaved Sampling

CIS (Fig. 11.2) is based on a running spectral analysis of the preprocessed digital input sound signal performed by a bank of band-pass filters (blue blocks in Fig. 11.3) or a fast Fourier transform. The filter bank has an overall bandwidth from approximately 100 to 8000 Hz, and the number of filters usually equals the number of stimulation channels at the electrode array−neuron interface. The filters have partially overlapping frequency responses and bandwidths that are proportional to frequency. After the filter bank, the magnitude of the envelope in each channel is determined (blue blocks in Fig. 11.3), e.g., by using rectification or using a Hilbert transformation followed by low-pass filtering. Typical cutoff frequencies are between 125 and 300 Hz. Only one pulse is delivered at any time, and all channels

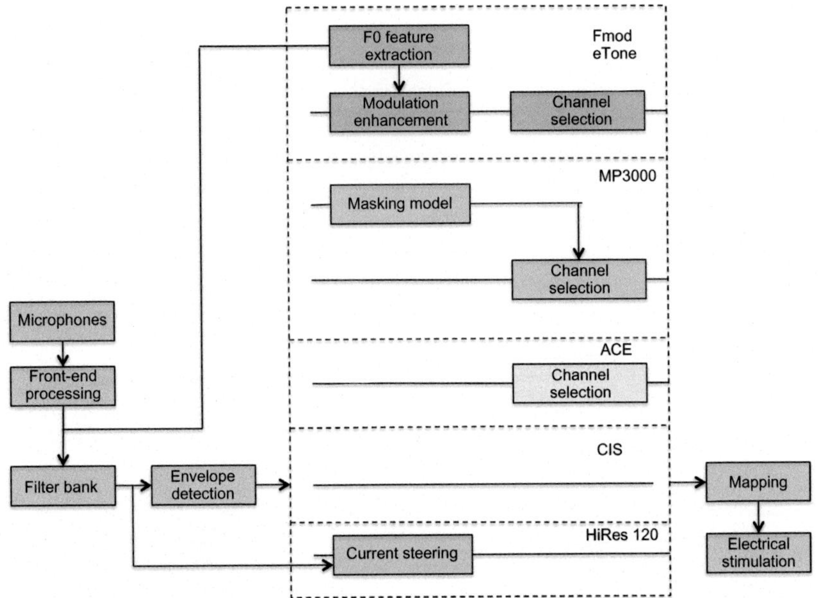

FIGURE 11.3 A block diagram of most monaural strategies. Common elements are shown in blue boxes, while strategy-specific elements are shown with individual colors in the dotted box. Details in the text. *Based on Wouters, J., McDermott, H.J., Francart, T., 2015. Sound coding in cochlear implants. IEEE Signal Proc. Mag. 3, 67–80.*

are activated in a temporally nonoverlapping sequence. A fixed stimulation carrier rate is used typically 500–2000 pulses/s (Wouters et al., 2015).

11.3.2.2 SPEAK and ACE

The channel selection (n-of-m) or "spectral peak picking" scheme used in the Advanced Combination Encoder (ACE) and SPEAK strategies (yellow block labeled ACE in Fig. 11.3) was designed in part to reduce the density of stimulation, while still representing the most important aspects of the acoustic environment. The n-of-m strategy uses a channel-selection scheme, in which the envelope signals for the different channels are "scanned" prior to each frame of stimulation across the intracochlear electrodes, to identify the signals with the n-highest amplitudes (n is fixed) from among m processing channels (and associated electrodes). Stimulus pulses are delivered only to the electrodes that correspond to the channels with those highest amplitudes. The deletion of low-amplitude channels (and associated stimuli) for each frame of stimulation may reduce the overall level of masking or interference across electrode and excitation regions in the cochlea (Wilson and Dorman, 2008). The SPEAK strategy uses much lower rates of stimulation than CIS. In ACE (Vandali et al., 2000; Milczynski et al., 2009) only a specified number of channels with maximal magnitudes are selected in every stimulation cycle. The ACE strategy combines the advantages of both SPEAK and CIS strategies by using high rates of stimulation (600–1800 pulses/s, as in CIS) with dynamic electrode selection and a large number of available electrodes (as in SPEAK; Kiefer et al., 2001). Nonetheless, there is a high degree of variability around the rates used clinically and there is often user preference for lower rates (e.g., Vandali et al., 2000). In a modeling study, Verschuur (2009) suggested that consonant recognition within the ACE strategy with the Nucleus 24 device is rather robust to interference with stationary noise.

11.3.2.3 HiRes 120: Current Steering

Several studies have shown that speech perception in quiet does not improve above about seven electrode contacts (Fishman et al., 1997; Fu et al., 1998; Friesen et al., 2001; Baskent, 2006), although a greater number is generally considered beneficial for listening in noise (Frijns et al., 2003; Nie et al., 2006). This asymptote of performance with increasing electrode contact number is thought to result from channel interactions and limited spatial selectivity of stimulation, which act to limit the degree of spectral discrimination (Fu and Nogaki, 2005). The production of intermediate pitches by stimulation of adjacent electrodes has been reported in several early reports (e.g., Tong et al., 1979; Tong and Clark, 1985; Townshend et al., 1987). Simultaneous stimulation of two electrodes became available with the Clarion device and provided a way to "steer" current by delivering different levels of current from adjacent electrodes. Shifting the relative current delivered to a pair of electrodes, current steering (green block in Fig. 11.3), has been shown to produce pitch percepts that are between those obtained using discrete electrode sites (Büchner et al., 2012). Current steering requires two independent current channels because two electrodes have to be stimulated simultaneously but with different current levels and is only available on the Advanced Bionics devices using the HiRes 120 strategy.

Firszt et al. (2007) reported that users of the Advanced Bionics CII and HiRes 90K implants were able to discriminate between 8 and 451 pitch percepts over the entire electrode array, with an average of 63. The production of intermediate pitches has also been demonstrated using sequential (interleaved) stimulation (McDermott and McKay, 1994; Kwon and van den Honert, 2006). Frijns et al. (2009) using modeling and psychoacoustics, compared dual- with single-electrode stimulation with respect to the site of stimulation in the cochlea, the spread of excitation, and channel interaction and found no differences in terms of channel interaction and spread of excitation.

11.3.3 Newer Coding Strategies

It should be noted that some of these new processing schemes have so far not been independently tested on a large number of implantees, so the stated optimistic conclusions must be tentative.

11.3.3.1 Multichannel Envelope Modulation

The multichannel envelope modulation (MEM) strategy was designed to provide F0 periodicity information synchronously across all activated electrodes (pink blocks in Fig. 11.3). MEM utilizes the envelope of the broadband signal, which contains F0 periodicity information to modulate the level of the filter bank channel signals derived from the ACE strategy (Vandali et al., 2005). They showed improvement in pitch ranking and no degradation in speech recognition (in quiet and in noise) as compared to ACE. Similar results for speech recognition were shown in a more recent study (Wong et al., 2008). In their MEM approach envelope fluctuations were extracted and additional control of modulation depth was provided. Modulation again took place in phase across channels.

11.3.3.2 MP3000

Nogueira et al. (2005) described a signal processing technique for CIs based on a psychoacoustic-masking model. This strategy is based on the same principles as the MP3 compression strategy used in music recording. They included spread of excitation and subsequent masking in the ACE coding strategy (orange block in Fig. 11.3). This coding strategy was later named MP3000. An advantage is that "with MP3000, it is not likely that an adjacent spectral component will be selected because the level of the second strongest component relative to the calculated masked threshold will be very small. ... Thus, MP3000 avoids repetitive stimulation of groups of neurons. It selects components that are dispersed more widely across the spectrum" (Nogueira et al., 2005). A comparative study in 221 subjects from 37 European Implant centers showed no significant difference for the speech scores and for coding preference between the SPEAK/ACE and MP3000 strategies (Buechner et al., 2011).

11.3.3.3 F0mod

Laneau et al. (2006) proposed a coding strategy called F0mod that used an enhanced coding of the envelope pitch cue (pink blocks in Fig. 11.3). The strategy incorporated a F0 estimator into the processing framework of the ACE scheme (yellow block). The F0mod strategy uses a sinusoidal modulator, synchronous across all channels, based on this estimate of F0. Francart et al. (2015) reported that "Immediately after switch-on of the F0mod strategy, speech recognition in quiet and noise were similar for ACE and F0mod, for four out of seven listeners. The remaining three listeners were subjected to a short training protocol with F0mod, after which their performance was reassessed, and a significant improvement was found."

11.3.3.4 Enhanced Envelope Encoded Tone (eTone)

Vandali and van Hoesel (2011) developed a sound-coding strategy named "eTone" to improve coding of fundamental frequency in the temporal envelopes of the electrical stimulus signals. The eTone strategy is based on the same principles as F0mod (pink blocks in Fig. 11.3) but includes a different F0 estimator based on harmonic sieves. Modulations are again synchronous across channels and an exponential decay modulation shape is used (Wouters et al., 2015).

From this overview one cannot escape the impression that the continued search for improved processor strategies suggests that none of those currently in use are optimal. Consequently, there is still something missing in the mimicking of normal processing by the peripheral auditory system. This missing aspect may be the spontaneous activity of the ANFs.

11.3.4 Mimicking Spontaneous Activity in the Auditory Nerve

A major difference between the deaf and hearing ears is the absence of spontaneous activity in the deafened cochlea ANFs (Kiang et al., 1970; Liberman and Dodds, 1984). There may be a functional advantage of spontaneous activity (Rubinstein et al., 1999). Previously, I wrote: "Spontaneous neural activity is very often considered as neural noise that sets limits on sensory performance. This neural noise idea has been the basis for the optimal processor model in psychoacoustics that typically worked on activity in auditory nerve fibers (Green, 1964; Siebert, 1965) and tried to extract the stimulus-induced activity from the spontaneous noise in these fibers. The concept of internal noise—albeit not limited to spontaneous firing—is fundamental in signal detection theory (Green and Swets, 1966). The alternative, already mentioned in Rodieck et al. (1962), investigating spontaneous firings in the cochlear nucleus, is that spontaneous firing is a carrier of information. The difference in these two hypotheses becomes visible when the effects of stimuli are taken into account. If spontaneous activity acts as noise one expects any stimulus induced activity to be additive to the spontaneous one. In contrast, when the spontaneous activity acts as carrier of information, one expects stimuli to modulate the spontaneous firings, i.e., a multiplicative action" (Eggermont, 2015b).

Litvak et al. (2003) recorded responses of ANFs to 10-min-long electric pulse trains (5000 pps)—called a desynchronizing pulse train (DPT)—from acutely deafened, anesthetized cats. Stimuli were delivered via an intracochlear electrode. Responses to pulse trains showed pronounced adaptation during the first 1–2 min, followed by either a sustained response or absence of spiking. Just like spontaneous activity in the normal ANFs, the adapted firing rates showed a broad range of values. Nearly half (46%) of ANFs responded to the DPT at rates less than 5 spikes/s, whereas 12% of the ANFs responded at higher rates than any spontaneously active ANF in a normal cochlea. Interspike interval histograms of sustained responses for some fibers had Poisson-like (exponential) shapes, resembling spontaneous activity in the normal ANFs, while others fired nearly periodically or showed burst firing. Simultaneous recordings from pairs of fibers showed no evidence of correlated activity, suggesting a desynchronizing action of the DPT. Litvak et al. (2003) concluded that responses to an ongoing DPT resemble spontaneous activity in a normal ear for a substantial fraction of the ANFs.

Applying this DPT to CIs, Hong et al. (2003) showed that all the tested cochlear implantees ($n = 28$; Clarion CII) demonstrated an increase in dynamic range in response to an appropriate level of unmodulated high-rate pulse train, mimicking spontaneous activity. The largest increase in dynamic range for each subject had a mean value of 6.7 dB.

11.4 TEMPORAL PROCESSING WITH A COCHLEAR IMPLANT

11.4.1 Refractoriness of Auditory Nerve Activity to Cochlear Implant Stimulation

Several studies have evaluated temporal response properties of the human auditory nerve by examining eCAP amplitudes in response to individual pulses in a train (Hay-McCutcheon et al., 2005; Rubinstein et al., 1999; Wilson et al., 1997; Hughes et al., 2012). Comparable studies in animals include those by Litvak et al. (2003), Zhang et al. (2007), and Miller et al. (2008). With these measurements, the relative eCAP amplitude reflects the total number of fibers responding to each pulse. Pulse rate has a dramatic effect on the eCAP time pattern (Fig. 11.4).

For 100–200 pps trains (Wilson et al., 1997), eCAP amplitudes are typically similar across individual pulses, suggesting that the same population of fibers is depolarized and then fully recovers following each pulse in the train. This is in-line with the earlier findings of Prijs (1980) in guinea pigs. For approximately 400–1500 pps an alternating response pattern appears (Fig. 11.4; Wilson et al., 1997), which reflects the variance in absolute and relative refractory-recovery times across individual ANFs (Rubinstein et al., 1999). The overall eCAP amplitudes are reduced and the alternating pattern typically diminishes as the stimulation rate is increased to 2000–5000 pps (Fig. 11.4). This is due to combined effects of refractory recovery, adaptation, and increased temporal jitter and represents stochastic independence among individual ANFs (Wilson et al., 1997; Rubinstein et al., 1999; Litvak et al., 2003; Miller et al., 2008). In this state auditory neurons may code

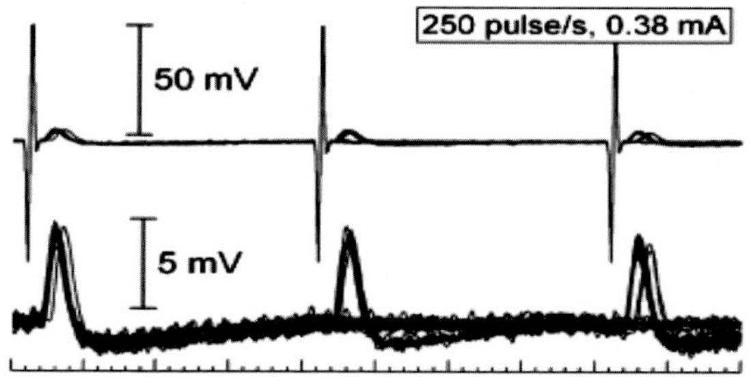

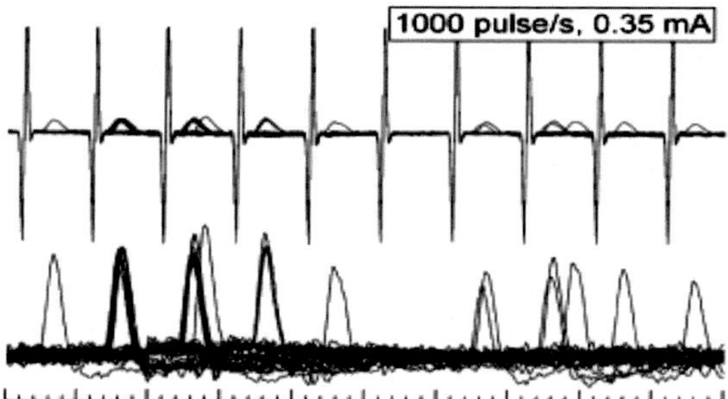

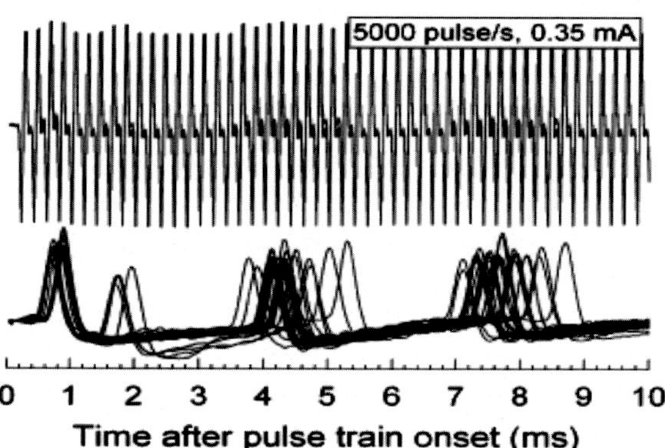

FIGURE 11.4 Examples of waveforms recorded in response to pulse trains presented at three rates. In each of the three cases shown, "raw" waveforms (with large electrical stimulus artifacts) are plotted above the processed waveforms. Each of the three sets of "raw" traces and each of the three sets of processed waveforms are plotted to the same scales (indicated by the two calibration bars in the upper plots). Stimulus parameters (pulse rate, stimulus level) are shown at the upper right of each graph. *From Zhang, F., Miller, C.A., Robinson, B.K., Abbas, P.J., Hu, N., 2007. Changes across time in spike rate and spike amplitude of auditory nerve fibers stimulated by electric pulse trains. J. Assoc. Res. Otolaryngol. 8, 356–372, with permission of Springer.*

information for faster rates of stimulation as a result of desynchronization of the whole-nerve response; consequently a subpopulation of fibers can respond at any given point in time (Hughes et al., 2012).

Hughes et al. (2012) calculated the absolute refractory period and the refractory-recovery time constant, τ, from masker-probe interval (MPI) functions. Data were available for 57 electrodes. The absolute refractory period was taken as the MPI for the peak of the function and ranged from 0.1 to 0.9 ms (mean: 0.5 ms). The relative refractory period was calculated by exponential curve fit to the eCap amplitude as a function of MPI. Overall, τ values ranged from ~0.4 to ~1.9 ms.

11.4.2 Adaptation to CI Stimulation

In the Litvak et al. (2003) study, ultra-high pulse rates (5000 pps) show either complete or partial adaptation in the evoked firing rate or eCAP amplitude. One could expect that the adaptation depends on the pulse rate, and that the level of adaptation affects temporal integration (Eggermont, 2015a; Ewert et al., 2007). Clay and Brown (2007) observed significant levels of adaptation in the eCAPs for all 21 CI subjects at stimulation rates of 80 and 300 Hz. Little or no adaptation was observed over the 5-min recording period when the 15-Hz rate was used. Hay-McCutcheon et al. (2005) determined the impact of auditory nerve adaptation on behavioral measures of temporal integration in CI recipients. Neural adaptation was measured from amplitude changes of the eCAP in response to 1000-pps biphasic pulse trains of varying durations. Results showed no correlation between the degree of neural adaptation and psychophysical measures of temporal integration. Chatterjee (1999) in a behavioral forward masking experiment found that recovery from a 300-ms-long pulse train presented at 1000 pps was fastest in the poorer CI performers. The shape of the recovery function was most strongly influenced by masker duration, suggesting that temporal integration plays a prominent role in recovery from forward masking. The recovery functions were reasonably well described by a sum of two exponentially decaying processes with rapid time constants of 2–5 ms and short-term time constants ranging from 50 to 200 ms.

Recording from ANFs in acutely deafened cats, Zhang et al. (2007) found that pulse integration can be seen (Fig. 11.5) in the spikes evoked by the 1000 pulse/s train, i.e., there is a greater increase in response probability to the second or third pulse relative to that of the first pulse. Responses to the 5000 pulse/s trains were clustered at intervals of about 4 ms, suggesting a strong influence of refractoriness, particularly shortly after pulse-train onset. For the pulse rates greater than 250 pulse/s, adaptation obtained from deafened and electrically excited ANFs (Fig. 11.5) is, in many cases, characterized by both "rapid" and "short-term" components. Zhang et al. (2007) found that these were similar to those reported by Westerman and Smith (1984) in ANFs for acoustic stimulation. Averaging data across response

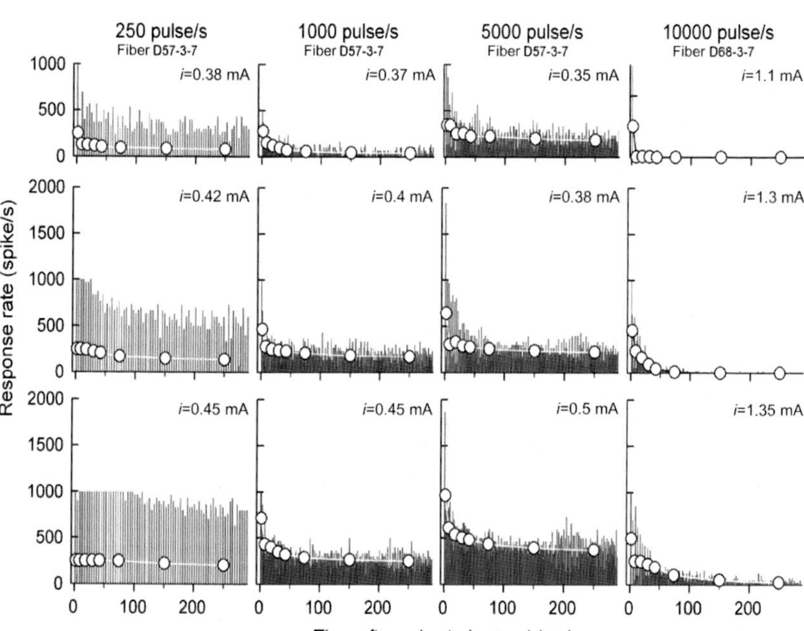

FIGURE 11.5 Examples of spike-rate adaptation observed in the poststimulus time histograms (PSTHs) from the responses of two ANFs at the four stimulus rates used in this study. Histograms for stimulus rates of 250, 1000, and 5000 pulse/s were provided by one fiber, whereas a second fiber provided the 10,000 pulse/s rate data. Each column contains PSTHs obtained at three stimulus levels. Histograms based on 1 ms bins are plotted with vertical bars whereas those based on progressively wider bins (defined in the text) are plotted using open circles. *From Zhang, F., Miller, C.A., Robinson, B.K., Abbas, P.J., Hu, N., 2007. Changes across time in spike rate and spike amplitude of auditory nerve fibers stimulated by electric pulse trains. J. Assoc. Res. Otolaryngol. 8, 356–372, with permission of Springer.*

rates and stimulus rates, the rapid time constant for eCAPs was 8 ms, whereas the mean (across level) rapid-time constant from ANFs in the Westerman and Smith (1984) data was 3 ms. Median short-term time constant for eCAPs was 80 ms, whereas the mean acoustic value was 55 ms. One has to recall that in ANFs, recovery time constants as measured in (acoustial) forward masking are always longer than perstimulatory adaptation time constants (Eggermont, 1985). The fact that electrically activated ANFs produce roughly comparable time constants compared to acoustic simulation is surprising, since adaptation in the hair cell–nerve fiber synapse is not involved.

11.4.3 Amplitude Modulation Detection

Temporal modulation detection ability matures over many years after birth and may be particularly sensitive to acoustic experience during this period (Eggermont, 2015a). Profound hearing loss during early childhood might

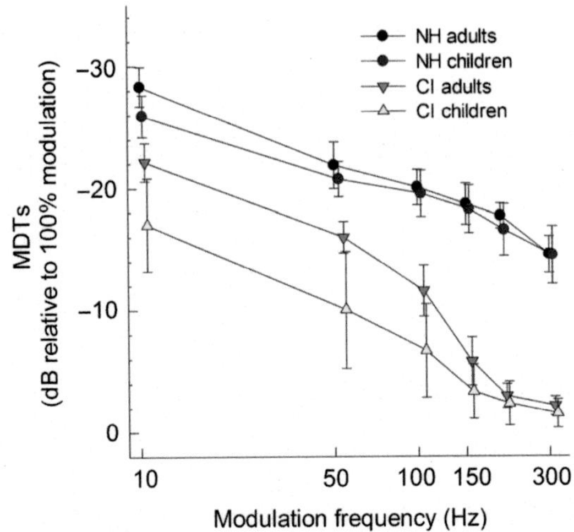

FIGURE 11.6 Comparison of the average temporal modulation transfer functions measured in four subject groups. Error bars indicate 95% confidence intervals across the subjects for modulation detection thresholds (MDTs) at each modulation frequency. *From Park, M-H., Won, J.H., Horn, D.L., Rubinstein, J.T., 2015. Acoustic temporal modulation detection in normal-hearing and cochlear implanted listeners: effects of hearing mechanism and development. J. Assoc. Res. Otolaryngol. 16, 389–399, with permission of Springer.*

result in greater perceptual deficits than a similar loss beginning in adulthood. Park et al. (2015) measured temporal MDTs in profoundly deaf children and adults fitted with CIs. They considered two independent variables affecting temporal modulation detection performance in children with CIs; altered encoding of modulation information resulting from CI stimulation, and atypical development of central processing of sound. Altered encoding was investigated by testing normal hearing versus CI and the atypical development was studied by testing two different age groups. Comparison of MDTs among the four subject groups (Fig. 11.6) showed that temporal resolution was mainly constrained by hearing mechanisms. Normal hearing listeners could detect smaller amplitude modulations at high modulation frequencies than CI users. A comparison of the MDT levels at the lowest modulation frequency showed that modulation sensitivity was significantly poorer in children with CIs, relative to the other three groups. This suggests that developmental aspects play a role as well.

11.4.4 Spectral-Ripple Detection

Spectral-ripple stimuli consist of rippled spectra noise signals in which the frequency positions of the spectral peaks and valleys alternate (Fig. 11.7).

Spectral-ripple discrimination has been used in behavioral experiments including the ability to discriminate a phase reversal of the rippled shape,

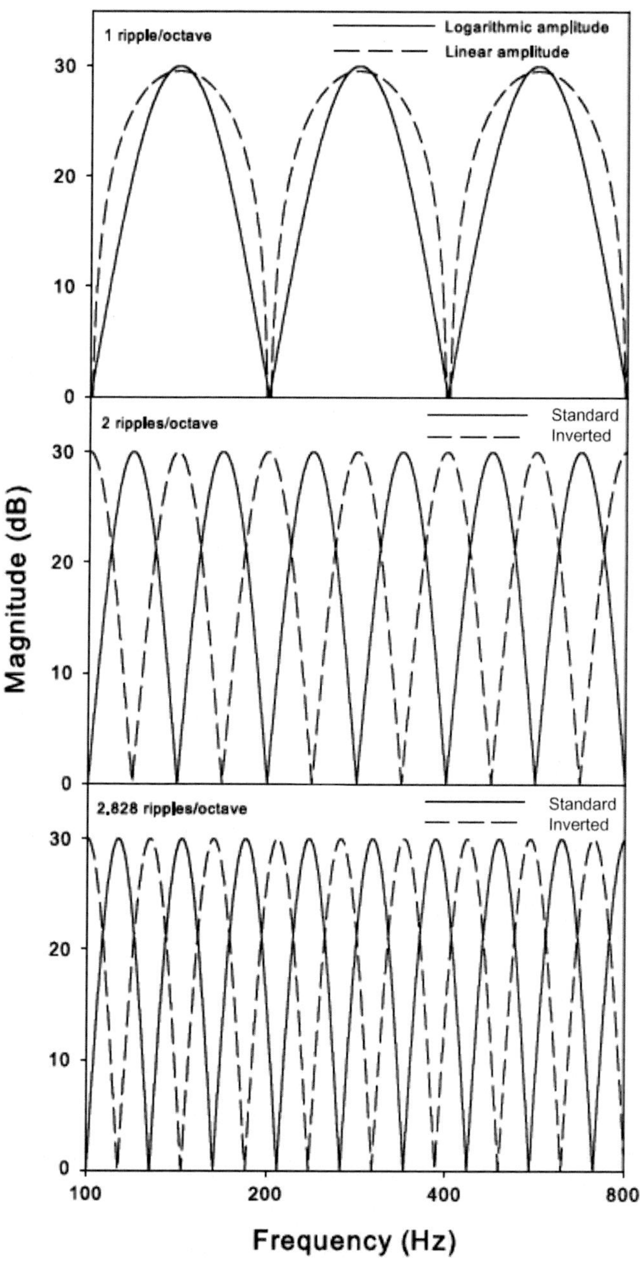

FIGURE 11.7 The figure shows the spectra of spectral-ripple stimulus from 100 to 800 Hz. Linear and logarithmic amplitude ripples with 1 ripple/octave are shown in the first panel. Standard and inverted ripples with 2 and 2.828 ripples/octave (with logarithmic amplitude) are shown in the second and third panels, respectively. *From Won, J.H., Drennan, W.R., Rubinstein, J.T., 2007. Spectral-ripple resolution correlates with speech reception in noise in cochlear implant users. J. Assoc. Res. Otolaryngol. 8, 384–392, with permission of Springer.*

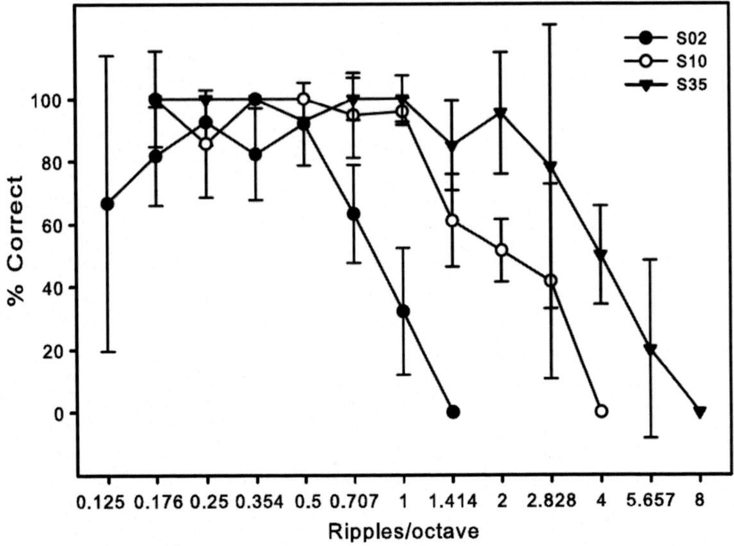

FIGURE 11.8 Psychometric function of the performance for three cochlear-implant subjects. One could call this the ripple modulation transfer function. Error bars represent ± SD. *From Won, J.H., Drennan, W.R., Rubinstein, J.T., 2007. Spectral-ripple resolution correlates with speech reception in noise in cochlear implant users. J. Assoc. Res. Otolaryngol. 8, 384–392, with permission of Springer.*

and the ability to differentiate between a spectral-ripple stimulus and white noise (Fig. 11.8). Henry et al. (2005) found that spectral peak resolution, based on detectability of ripple-phase reversal, varied widely among listeners, from 0.13 to 7.55 ripples/octave. The average spectral peak resolution was 4.84 ripples/octave in normal hearing listeners (2.03–7.55 ripples/octave), 1.77 ripples/octave in hearing-impaired listeners (0.33–4.97 ripples/octave), and 0.62 ripples/octave in CI listeners (0.13–1.66 ripples/octave). Spectral-ripple discrimination ability correlates with vowel and consonant recognition in quiet (Henry et al., 2005; Litvak et al., 2007), with word recognition in noise (Won et al., 2007), and music perception abilities (Won et al., 2010) for CI users.

Won et al. (2011) showed that the cortical P_1–N_1–P_2 response (cf. Fig. 8.1) could be elicited by the changes within the spectral-ripple stimulus. They found that MDT as well as the P_1–N_1–P_2 response correlated positively with the number of vocoder-processing channels, i.e., better discrimination ability at higher ripple densities was observed as more channels were added. Won et al. (2011) also showed that the amplitudes of the P_1–N_1–P_2 "change" responses were significantly correlated with d' values from the single-interval behavioral procedure. Recently, Drennan et al. (2016) used spectral-ripple discrimination to assess progression of adult CI users over the first year of implantation. They found that: "Speech understanding in quiet improved between 1 and 12 months postactivation (mean 8% improvement). Speech in noise performance showed no statistically significant improvement. Mean spectral-ripple discrimination thresholds and temporal-modulation detection thresholds for modulation frequencies of 100 Hz and above also showed no significant improvement. Spectral-ripple discrimination thresholds were significantly correlated with speech understanding."

11.5 EFFECTS OF AGE ON IMPLANTATION

To appreciate what early-onset deafness does to the auditory cortex we have to briefly recall the maturational properties of ABRs in the normal hearing child (see chapter: Early Diagnosis and Prevention of Hearing Loss). The ABR can be reliably recorded in premature infants from the 28th to 29th weeks of conceptional age (CA). At 30–35 weeks CA, the ABR wave I, the negative wave following wave II, wave V, and the vertex contralateral mastoid-recorded waves II and V were the most consistently present with detection rates of 87–100%. At ages 35 weeks CA and older waves I, III, and V at both sides were clearly present (Ponton et al., 1992).

11.5.1 Effects of Early Cochlear Implantation: Electrophysiological Measures

Does the immature auditory system lose the ability to function normally and mature when it does not receive auditory stimulation? To answer that question, Gordon et al. (2006) recorded neural responses evoked by stimulation with CIs in 75 prelingually deafened children and 11 adults. eABR latencies significantly decreased with duration of CI use and were not significantly affected by the age at implant activation. Significant decreases in early latency waves and between waves occurred within the first 1–2 months of implant use, whereas longer-term changes (6–12 months) were found for waves eV and eIII–eV, which reflect activity from sources between cochlear nucleus and inferior colliculus. Comparisons with acoustically evoked ABRs in normal hearing children revealed shorter interwave eABR latencies, potentially reflecting increased neural synchrony, but similar rates of change in the longer latency eV and eIII–eV interpeak latency with time in sound. It appears that normal-like development of the upper auditory brainstem is promoted by CI use in children of a wide range of ages.

Lammers et al. (2015a) compared auditory brainstem activation in prelingually deaf and late-implanted adult CI users to postlingually deaf CI users. eABRs were recorded by monopolar stimulation. Wave eV was significantly delayed in the prelingually deaf CI users, and the eIII–eV interwave interval was also significantly longer in the prelingually deaf group. This may be caused by slower neural conduction in the auditory brainstem.

The electrically evoked middle latency response (eMLR) reflects primary auditory cortical activity. Gordon et al. (2005) recorded this response repeatedly in 50 children over the first year of CI use and in 31 children with 5.4 ± 2.9 years of implant experience. The eMLR was detected in only 35% of awake children at initial device stimulation (Fig. 11.9).

However, the detectability of the eMLR increased over the first year of implant use, becoming 100% detectable in children after at least 1 year. eMLRs shortly after implantation were often found in older children despite longer periods of auditory deprivation. Within 6 months of implant use, most children had detectable eMLRs. At early stages of device use, eMLR amplitudes were lower in children implanted below the age of 5 years compared to children implanted at older ages. Latencies after 6 months of implant use were prolonged in the younger group and decreased with implant use. These eMLR changes with chronic CI use reflect activity-dependent neural plasticity. In

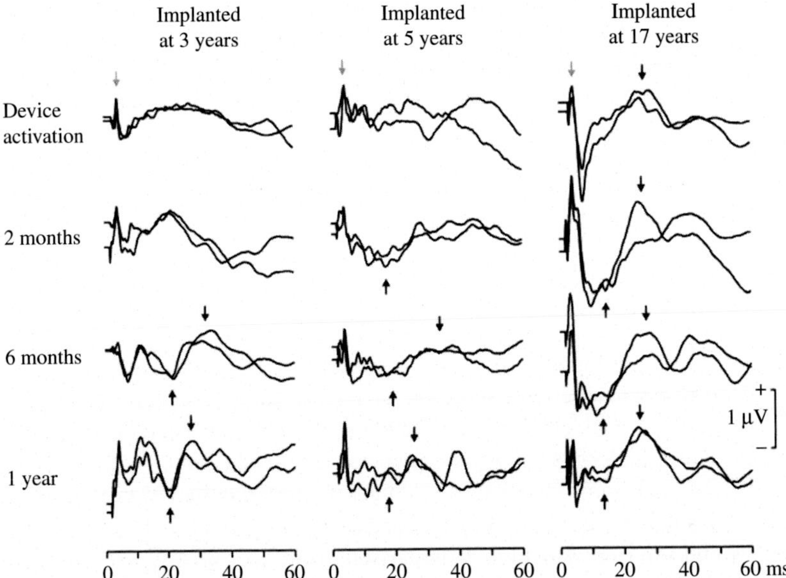

FIGURE 11.9 Repeated measures of the eMLR waveforms evoked by a basal electrode from three children in the longitudinal group. Replicated waveforms are shown at each time for each child. The duration of implant use at time of recording is indicated. The eABR wave eV, indicated by a light gray downward arrow at the first time point, is present in all waveforms. The black upward arrows mark eMLR wave eNa and the black downward arrows mark eMLR wave ePa wherever present. In the youngest child, eMLR peaks are not detectable at initial device activation nor after 2 months of implant use (despite the clear presence of wave eV from the eABR) but can be identified at 6 months and 1 year of implant use. EMLR peaks cannot be detected with acute stimulation in the 5-year old but are seen at subsequent tests. In the oldest postlingually deaf child, despite a much longer period of auditory deprivation, eMLRs are detectable at all test times. *Reprinted from Gordon, K.A., Papsin, B.C., Harrison, R.V., 2005. Effects of cochlear implant use on the electrically evoked middle latency response in children. Hear. Res. 204, 78–89, with permission from Elsevier.*

postlingually-deaf implanted patients, eMLRs were already detected at the time of CI activation. These findings suggest that the pattern and development of electrically evoked activity in the auditory thalamocortical pathways depend on the duration of auditory deprivation occurring in early childhood.

11.5.2 Auditory Deprivation Effects on Auditory Cortex

Whereas delayed occurrence of eMLRs reflect immaturities in primary cortex on Heschl's gyrus, longer latency AEPs measure activity from the auditory cortical areas downstream from Heschl's gyrus (Ponton et al., 2000). These long-latency cortical AEPs (CAEPs) change dramatically in morphology with age (Fig. 11.10, left-hand column). The CAEP morphology at 5–6

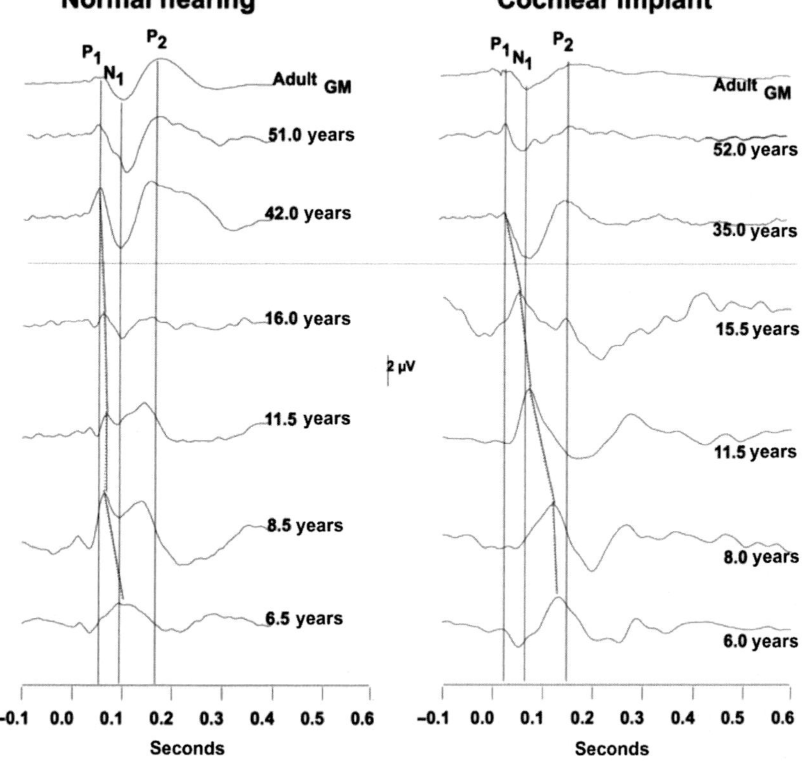

FIGURE 11.10 CAEP waveforms in normal hearing (left) and persons with CIs (right). The waveforms in late-implanted adults, with previous normal hearing are very similar to those in normal hearing adults. Note the absence of N_1 in the late-implanted children. Lines connect the P_1 peaks. GM, grand mean. *From Eggermont, J.J., 2014. Noise and the Brain. Experience Dependent Developmental and Adult Plasticity, Academic Press, London; based on data from Ponton, C.W., Don, M., Eggermont, J.J., Waring, M.D., Masuda, A., 1996b. Maturation of human cortical auditory function: differences between normal-hearing children and children with cochlear implants. Ear Hear. 17, 430–437.*

years of age is very different from that in a young adult and CAEPs consist mostly of a broad positive peak comprised of a fusion of the P_1 and the very early maturing P_2 (Ponton et al., 2000). This early fusion of P_1 and P_2 is the consequence of the late detection of the N_1, which occurs only reliably around 8–9 years of age. N_1 is assumed to originate in the planum temporale, that is, in the parabelt area of auditory cortex. In normal hearing children, N_1 manifests itself initially as a small dip between P_1 and P_2 that continues to increase in size and in adulthood dominates the CAEP.

What happens to the normal cortical maturation when the cortex does not receive specific auditory afferent inputs? What are the consequences for maturation of subsequent stimulation using a CI? Fig. 11.10 shows a comparison of individual recordings at the centrally located electrode Cz, age matched for normal hearing (left column) and CI (right column) subjects. Note that all the CI adults in this study became deaf after adolescence and have very similar CAEPs as the normal hearing controls. Thus, deafness occurring after complete maturation of auditory cortex does not seem to affect the morphology of CAEPs. The grand means differ only slightly in amplitude, and the latencies are comparable albeit shorter in the CI group, likely the result of increased neural synchrony provided by the CI compared with acoustic stimulation. A particular picture emerged in children who became deaf early in life (before age 3) and were implanted at age 6 or later. For one such patient, one notices the decrease in P_1 latency with age that indicates some maturational aspects, and potentially the emergence of N_1. But the most dramatic finding is that in these CI children, even in late adolescence, there is no sign of the N_1, the most common and largest component of the CAEP.

Lammers et al. (2015b) recorded CAEP waveforms in response to electric stimuli in prelingually deaf adults, who received their CI after the age of 21 years. Stimulus rate was comparable to those in the study of Ponton et al. (1996b). Waveform morphology and peak latencies were compared to the CAEP responses obtained in postlingually deaf adults, who became deaf after the age of 16. Lammers et al. (2015b) reported: "Unexpectedly, typical CAEP waveforms with adult-like P_1-N_1-P_2 morphology could be recorded in the prelingually deaf adult CI users. ... Also, latencies of the N_1 peak were significantly shorter and amplitudes were significantly larger in the prelingual group than in the postlingual group." In light of the Ponton et al. (1996b) findings as well as those of Sharma et al. (2005) in the next paragraph, this unexpected result, particularly the observation of an N_1 in the prelingually deaf could be understandable if these had residual low-frequency hearing and had used hearing aids for some time.

Sharma et al. (2005) examined the longitudinal development of the CAEP in 21 children who were fitted with unilateral CIs and in two children, who were fitted with bilateral CIs either before age 3.5 years or after age 7 years. They noted that early-implanted children showed rapid development in CAEP waveform morphology and P_1 latency. Late-implanted children showed aberrant waveform morphology and significantly slower reduction in P_1 latency postimplantation. Sharma et al. (2005) found that: "The relative change in P_1 latency between hookup and one month was the same (35%) for early and late-implanted children (see Fig. 11.11). Thus an early, input-dependent maturation process in P_1 latency is comparable between early- and late-implanted children. However, after approximately 4 weeks of stimulation, the maturation process becomes arrested in late-implanted children, but continues in early-implanted children." Sandmann et al. (2015) observed that postlingually deafened "CI users revealed a remarkable improvement in auditory discrimination ability, which was most pronounced over the first eight weeks of CI experience. At the same time, CI users developed N_1 auditory event-related potentials with significantly enhanced amplitude and decreased latency, both in the auditory cortex contralateral and ipsilateral to the CI."

Population studies in congenitally deaf children have shown that these children benefit most when cochlear implantation takes place within the first

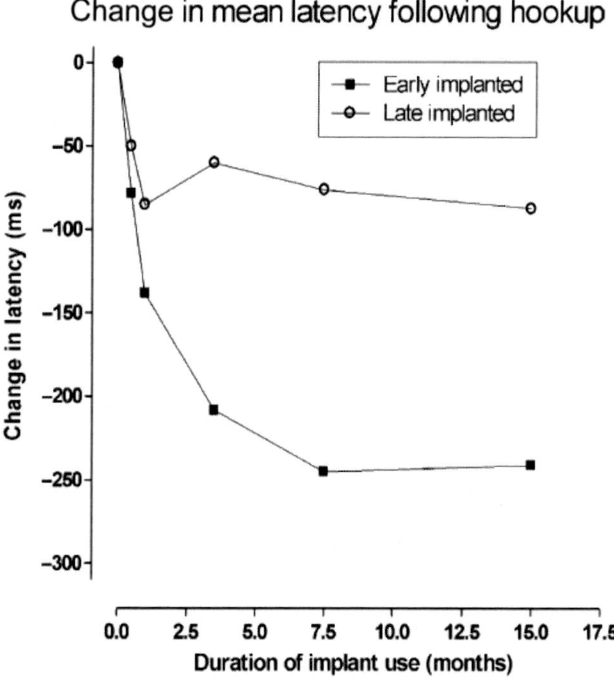

FIGURE 11.11 Changes in mean P_1 latency following stimulation for the early- and late-implanted children are shown following normalization of starting latencies. *Reprinted from Sharma, A., Dorman, M.F., Kral, A., 2005. The influence of a sensitive period on central auditory development in children with unilateral and bilateral cochlear implants. Hear. Res. 203, 134–143, with permission from Elsevier.*

3.5 years of life, when the central auditory pathways show maximal plasticity. The latency of the P_1 component as recorded at Cz of the cortical auditory evoked potential decreases rapidly and reaches the normal age range in children who receive an implant before 3.5 years of age (Kral and Sharma, 2012). By contrast, children who receive implants after the age of 7 show abnormal cortical responses, even after many years of CI use (Ponton et al., 1996a). These age cutoffs, determined by electrophysiological studies, correspond closely to studies of positron emission tomography (PET) measurements, which showed decreased spontaneous glucose metabolism (owing to years of auditory deprivation) in the auditory cortices of children implanted before 4 years of age. By contrast, children implanted after 6.5–7.5 years of deafness show normal metabolism in higher-order auditory cortices, suggesting that these areas have been taken over by other modalities because they were unavailable for auditory processing, leading to functional respecialization of these areas (Lee et al., 2001). A more recent study (Lee et al., 2007) showed by assessing PET activity at 3 years after implantation that: "Whereas age at implantation was positively correlated with increased activity in the right superior temporal gyrus, speech scores were selectively associated with enhanced metabolic activity in the left prefrontal cortex and decreased metabolic activity in right Heschl's gyrus and in the posterior superior temporal sulcus." There is also a close correspondence between the age cutoffs described in the electrophysiological studies and the speech and language performance of congenitally deaf, implanted children. The eventual end of the sensitive period (estimated at ~6.5–7.0 years of age in humans but likely shorter for early deafness as the N1 does not develop for deprivation periods longer than 3 years under the age of 6) has consequences for the reorganization of cortical areas and pathways (Ponton et al., 1996a).

11.5.3 Effects of Early Implantation on Speech and Language

Children who receive implants before 3–4 years of age show significantly higher speech perception scores and better language skills compared with children implanted after 6–7 years of age (Svirsky et al., 2004). Recent studies investigated whether children implanted before 1 year of age demonstrate even greater benefit in behavioral auditory performance, with outcomes depending on the experimental paradigm (Niparko et al., 2010). They found that age at implantation and residual hearing are associated with rate increases in the acquisition of spoken language in children with CIs. In multivariable analyses, greater residual hearing prior to cochlear implantation, higher ratings of parent–child interactions, and higher socioeconomic status were associated with greater rates of improvement in comprehension and expression. These findings underscore the need to develop objective tools that can monitor the benefit of amplification in supporting spoken language acquisition and guide timely intervention with cochlear implantation. Boons et al. (2012b) investigated receptive and expressive language development of 288 children who received CIs by age 5 in a retrospective multicenter study. They found that "simple linear regressions with age at first fitting and independent samples *t*-tests demonstrated that children implanted before the age of two performed significantly better on all tests than children who were implanted at an older age." Van Dijkhuizen et al. (2011) collected speech samples from 25 CI candidates with prelingual deafness and presented those to two normal hearing listeners, who judged the intelligibility. Subsequently, from this group, nine participants with above-average preimplant intelligibility were selected for implantation. Speech perception data with a CI collected at 1-year postimplantation and compared to preimplant speech intelligibility. This suggested that preimplant speech intelligibility is a promising tool for guiding CI candidacy decisions. A follow-up study (Van Dijkhuizen et al., 2016) studied 92 prelingually deafened adults to be enrolled in a CI program. They took part in a vowel intelligibility test, in which recordings of their pronunciation of the vowel stimuli were presented to two normal hearing listeners. One-year postimplantation speech perception scores obtained correlated significantly ($r = 0.79$) with the preimplant intelligibility of the patient's speech. Van Dijkhuizen et al. (2016) concluded that for prelingually deaf adults, preimplant intelligibility of the patient's speech (to normal hearing listeners) is a valid predictor of postimplant speech perception.

11.5.4 Cochlear Implantation in the Elderly

Cochlear implantation in the elderly population results in improved quality of life (Di Nardo et al., 2014; Vermeire et al., 2005). However, there are reports of poor hearing outcomes in the elderly (Lenarz et al., 2012; Roberts et al., 2013), while others have found good results (Haensel et al., 2005; Holden et al., 2013). A recent study by Hiel et al. (2016) clearly shows that age is not a limiting factor for good functional CI results. They compared hearing outcomes in quiet conditions between CI users implanted over and under the age of 70 years. They found that in quiet, older CI users perform as well as middle-aged and younger subjects. However, the duration of severe to profound hearing loss negatively impacted CI outcomes in all cases. Hiel et al. (2016) concluded that age should not be a decisive criterion for or against cochlear implantation.

11.6 COCHLEAR IMPLANTS AND MUSIC PERCEPTION

Auditory discrimination accuracy in adult CI users (Nucleus Freedom) and normal hearing controls was quantified by behavioral tasks and mismatch negativity (MMN) recordings (Sandmann et al., 2010). They obtained discrimination profiles for original and vocoded clarinet sounds varying along frequency, intensity, and duration and four levels of deviation magnitudes. Behavioral results as well as MMN recordings showed reduced auditory discrimination accuracy in CI users. MMN amplitude was inversely related to the duration of profound deafness. Petersen et al. (2015) recorded the MMN to changes in musical features in adolescent CI users (Nucleus Freedom) and in normal hearing age-matched controls. MMN recordings and behavioral testing were carried out before and after a 2-week music training program for the CI users and in two sessions equally spaced in time for normal hearing controls. Significant MMNs were obtained in CI users for deviations in timbre, intensity, and rhythm. In contrast, only one of the two pitch deviants elicited an MMN in CI users. This pitch discrimination deficit was also found by behavioral measures. Overall, MMN amplitudes were significantly smaller in CI users than in normal hearing controls, suggesting poorer music discrimination ability. Despite compliance from the CI participants, Petersen et al. (2015) found no effect of the music training and assumed that it resulted from the brevity of the program. It would be interesting to see if current steering or the F0mod strategy to increase pitch perception would make any difference. Having residual low-frequency hearing that is spared by short electrode implantation appears to preserve the ability accurate perception on three pitch-based music listening tasks (Driscoll et al., 2016). In this study the electroacoustic group was significantly better than standard implant groups and not significantly different from normal hearing listeners.

11.7 ONE-SIDED OR BILATERAL IMPLANTATION?

Binaural performance with CIs was significantly better than monaural performance for all subjects, for all stimulus types, and for different sound sources. Only small differences in performance with different stimuli were observed (Verschuur et al., 2005). They recruited 20 subjects from the UK multicenter trial of bilateral cochlear implantation with Nucleus 24K/M device. Sound localization was assessed for each of five stimuli (speech, tones, noise, transients, and reverberant speech) in an anechoic room with an 11-loudspeaker array under four test conditions: right CI, left CI, binaural CI, and dual microphone. They found that the mean localization error with bilateral implants was 24° compared with 67° (near chance performance) for monaural implant and dual microphone conditions. Normal controls average 2° to 3° in similar conditions (Verschuur et al., 2005). These findings were partly corroborated by Summerfield et al. (2006) and Firszt et al. (2008).

Gordon et al. (2007, 2012) investigated whether the deaf and immature human auditory system is able to integrate input delivered from bilateral CIs. Using eABR measures that included the binaural-difference response, they showed that a period of unilateral deprivation before bilateral CI use resulted in asymmetric response latencies but that amplitudes of the binaural-difference response were not significantly affected. Many children receiving unilateral CIs are developing mature-like brainstem and thalamo-cortical responses to sound with long-term use despite these sources of variability; however, there remain considerable abnormalities in cortical function (Gordon et al., 2013). The most apparent, determined by implanting the other ear and measuring responses to acute stimulation, is a loss of normal cortical response from the deprived ear. Recent data reveal that this can be avoided in children by early implantation of both ears simultaneously or with limited delay. Gordon et al. (2013) concluded that auditory development requires input early in development and from both ears. The same group (Jiwani et al., 2013) recorded CAEPs in 79 children who received one CI within on average approximately 2 years of bilateral deafness and had up to approximately 16 years of time-in-sound experience, and in 58 normal hearing peers. Differences in responses from CI users compared to normal hearing controls decreased over time, but were not eliminated even after 10 years of time-in-sound. The typical $P_1-N_1-P_2-N_2$ complex of a normally mature response began to emerge by 10 years of time-in-sound experience, but the amplitudes of peaks P_2 and N_2 became abnormally large. Jiwani et al. (2013) reported that normal-like cortical responses were found in children after long-term unilateral CI use, but recent data show that there are considerable hemispheric differences from normal which underlie these normal-like response peaks (Jiwani et al., 2016). The main conclusion is that an "aural preference" for the first implanted ear develops (Gordon et al., 2013, 2015) and disrupts normal hemispheric dominance patterns (Jiwani et al., 2016). The deprived ear shows effects of long-term deprivation (Jiwani et al., 2016) with abnormal responses similar to those found in children implanted at older ages/longer bilateral deprivation (Gordon et al., 2010).

Boons et al. (2012a) studied 25 children with 1 CI matched with 25 children with 2 CIs who underwent cochlear implantation before 5 years of age. They found that "the use of bilateral cochlear implants is associated with better spoken language learning. The interval between the first and second implantation correlates negatively with language scores. On expressive language development, we find an advantage for simultaneous compared with sequential implantation." Sparreboom et al. (2014) studied 30 prelingually deaf children who received their first implant at a mean age of 1.8 years and their second implant at a mean age of 5.3 years, using CAEPs. They observed that responses evoked by the second implant still lag behind that of the longer used first implant. Especially for the children with longer interimplant delays they found no evidence that CAEPs will become similar on both implant sides over time.

Two systematic reviews qualify these findings. Van Schoonhoven et al. (2013) conducted a systematic review on the clinical effectiveness of bilateral cochlear implantation compared with unilateral cochlear implantation or bimodal stimulation as reported in studies published between 2006 and 2011. They concluded that: "All studies showed a significant bilateral benefit in localization over unilateral cochlear implantation. Bilateral cochlear implants were beneficial for speech perception in noise under certain conditions and several self-reported measures. Most speech perception in quiet outcomes did not show a bilateral benefit." Blamey et al. (2015) analyzed basic speech audiometry in a retrospective multicenter study and noted that on average, a second CI is likely to provide slightly better postoperative speech outcome than an additional hearing aid for people with very low preoperative performance.

11.8 COCHLEAR IMPLANTATION AND TINNITUS

Tinnitus is major issue in cochlear implantation as there are probably benefits (reduction of tinnitus related to deafness) as well as problems (tinnitus emerging with implant use).

11.8.1 Tinnitus in the CI Population

People considered for cochlear implantation have a high prevalence of tinnitus (Baguley and Atlas, 2007) as illustrated in Fig. 11.12A. Here, for each study, the number of patients with tinnitus are plotted against the number implanted. The regression line, and the very high correlation coefficient, indicates that the sample size, ranging from seven to several hundred, in the reviewed studies did not affect this prevalence. Consequently, we might say that the prevalence of tinnitus in all these groups is about 79%. This is much

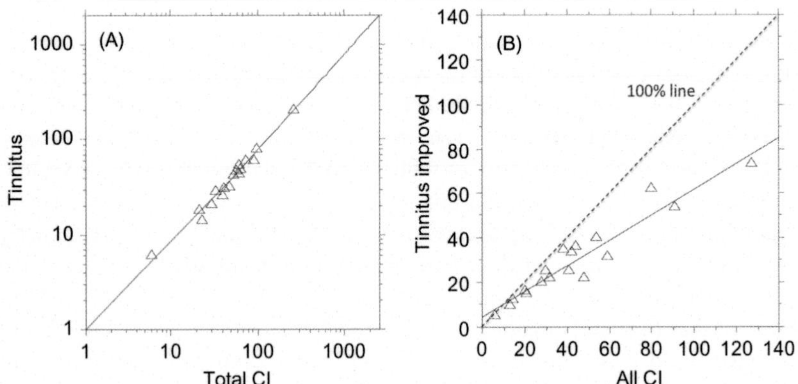

FIGURE 11.12 (A) Number of patients with tinnitus presenting for CI as a function of sample size of the various studies. The regression line, tinnitus $= -1.02 + 0.787 \times$ (total CI), $r^2 = 0.99$, suggests a universal prevalence of 79% in this population. (B) Number of patients with improved tinnitus after CI as a function of sample size in the various studies. The regression line, tinnitus improved $= 4.82 + 0.573 \times$ (all CI), $r^2 = 0.91$, suggests a universal prevalence of approximately 57% in this population. *Data from Baguley, D.M., Atlas, M.D., 2007. Cochlear implants and tinnitus. Prog. Brain Res. 166, 347–355.*

higher than the prevalence in the entire population, which is 10–15% (Eggermont, 2012). CIs are successful in improving tinnitus as long as the CI is switched on. On average 58% of all cochlear implantees show improvement, which tends to be higher in studies with fewer patients (Fig. 11.12B), potentially due to a more relaxed patient selection in the large group studies. For many patients, utilizing a CI in one ear can also markedly reduce the perceived intensity of tinnitus in the contralateral ear (Baguley and Atlas, 2007). I first will review some of these earlier studies and then proceed with those not covered in the Baguley and Atlas (2007) survey.

The earliest studies suggesting that electrical stimulation of the cochlea could alleviate tinnitus came from House (1976) and Portmann et al. (1979). They found that stimulating the wall of the cochlea (the promontory), or even better the round window (Portmann et al., 1983), with negative-current pulses induced sensations of sound and enhanced tinnitus, whereas positive pulses reduced tinnitus. This polarity sensitivity will determine the performance of CIs, which typically use bipolar pulses, on potentially suppressing tinnitus.

One of the earlier studies on tinnitus suppression using a CI was by McKerrow et al. (1991) from the San Francisco group that devised the UCSF/Storz CI. They found that tinnitus was suppressed in five of six patients with the device on and with noise input, and reduction of perceived tinnitus loudness occurred in four of six with the device on but without acoustic input. They were likely the first to note that unilateral CI stimulation often resulted in bilateral tinnitus suppression. A year later a larger study using the Nucleus 22 device in 33 postlingually deafened patients (Souliere et al., 1992) reported that preoperatively tinnitus was present in 85% of patients. In their study, 15 patients (54%) with preoperative tinnitus demonstrated a loudness decrease of 30% or more; 43% demonstrated an annoyance decrease of 30% or more. Contralateral tinnitus suppression was reported by 42% of patients. Ito and Sakakihara (1994) subsequently reported that 87% of the CI candidates experienced tinnitus. Following cochlear implantation (no device mentioned), tinnitus was reduced in 77% of the patients. In 60% of the patients whose tinnitus was suppressed in the implanted ear, tinnitus was also suppressed in the contralateral ear. So all these early findings were very similar and fit with the general results presented in Fig. 11.12.

In more recent studies, Ruckenstein et al. (2001) found a significant reduction for 35 of 38 patients (92%) in tinnitus intensity using CIs. More recently (Di Nardo et al., 2007), a comparison between pre- and postimplantation Tinnitus Handicap Inventory (THI) scores showed a decreased score in 13 of 20 patients with preimplantation tinnitus, an unchanged score in six and increased score in one. Pan et al. (2009) found in 153 implanted patients with tinnitus that 61% had total suppression and 39% reported a partial reduction. Of 91 people without tinnitus receiving a CI, only 12% reported tinnitus postoperatively. Olze et al. (2012) studied 55 postlingually deafened adults who were unilaterally implanted with a multichannel CI for at least 6 months. Twenty patients were aged 70–84 years, and 35 patients were 19–67 years old. Forty-eight (87%) patients reported having chronic tinnitus before CI. Tinnitus annoyance and perceived stress were reduced in elderly patients to the same extent as in younger patients in the case of high initial severity level. Blasco and Redleaf (2014) reviewed subjective changes of tinnitus in 27 patients and found 96% were improved. No patient reported worsening of tinnitus after cochlear implantation. Kim et al. (2013) reviewed medical records for 35 patients who underwent CI. Of them, 22 had tinnitus prior to CI (62.9 %) and the tinnitus group was older than the non-tinnitus group. Tinnitus was completely eliminated in 10 patients (45.5%) although the most severe cases had the greatest benefit.

In children, implanted between 3 and 15 years (Chadha et al., 2009), tinnitus occurred most commonly in the implanted ear, when the implants were not in use (e.g., in bed at night). Tinnitus was most frequent in children aged 6–8 years (8/17, 47%) and in bilateral implantees with an interprocedure delay of at least 2 years (6/10, 60%). Tinnitus was least reported in those implanted bilaterally simultaneously (1/6, 17%) and in those 5 years old or younger (3/11, 27%). They found no relationship with etiology.

11.8.2 Tinnitus in Single-Sided Deafness

Alleviating tinnitus in the deaf ear in single-sided deafness (SSD) has provided a new application of CIs, largely as a result of the pioneering studies by the Antwerp group. The first study (Van de Heyning et al., 2008) reported on 21 subjects who complained of severe intractable tinnitus that was unresponsive to treatment received a CI. CI stimulation resulted in a significant reduction in tinnitus loudness 1 and 2 years after implantation. With the device deactivated, tinnitus loudness was still reduced over 24 months. In a more recent study, the same group (Kleine Punte et al., 2013) implanted seven patients who met the criteria of severe tinnitus due to SSD. All the tinnitus outcome measures remained unchanged with 1, 2, 3, or 4 activated electrode pairs. With complete CI activation, the tinnitus decreased significantly comparable with earlier reports. The tinnitus loudness was reduced at 6 months postimplantation. Mertens et al. (2016) conducted a long-term evaluation of CIs in SSD and asymmetric hearing loss patients who wear their CI 7 days a week. They reported that tinnitus reduced significantly up to 3 months after the first-fitting and then remained stable up to the test interval at on average 8 (3–10) years after implantation.

11.9 MODELING STUDIES

Modeling CI activation basically starts with myelinated nerve fibers (Frijns and Ten Kate, 1994; Frijns et al., 1994). Their research suggested that a multiple nonlinear node model of a myelinated nerve fiber is superior to single-node models and that the choice of model is particularly relevant for electrical prosthesis design. Subsequently, Frijns et al. (1995) developed a rotationally symmetric volume conductor model of the electrically implanted cochlea, incorporating two submodels, one computing the potential distribution due to stimulating electrodes in the complex geometry of the cochlea, and the other simulating the neural reaction to this potential field. There were two major differences with previous models: (1) it represented a more realistic geometry of the modiolus albeit in guinea pigs and (2) it included an active model of an auditory nerve fiber with mammalian nodal kinetics. The model could use realistic temporal stimulus waveforms, such as biphasic pulses. In an update, Frijns et al. (1996) further studied this model with respect to the spatial selectivity of realistic electrical stimulation by CIs. They concluded that the application of charge-balanced asymmetric biphasic current pulses is probably a way to realize the relatively localized excitation pattern produced by monophasic current pulses while complying with biological safety considerations.

Rattay et al. (2001) then introduced an active compartment model of a human cochlear neuron. They noted that in contrast to all investigated animals most of the cell bodies as well as the pre- and post-somatic regions of the afferent human cochlear neurons are unmyelinated. Incorporating this, their model showed that the long peripheral processes in human cochlear neurons cause first excitation in the periphery and, consequently, neurons with lost dendrite need higher stimuli. Other notable models with "human" nerve fiber and volume conductor properties are from the South African group (Hanekom, 2001; Smit et al., 2010, Hanekom and Hanekom, 2016). For an up to date review on single-node and cable models see O'Brien and Rubinstein (2016).

Briaire and Frijns (2005) recalled the more recent developments in their two-part computational model of the cochlea (Frijns et al., 2000; Briaire and Frijns, 2000a). The 3D volume conduction part provides insight into the distribution of the current through the cochlea (Briaire and Frijns, 2000b) and how this distribution can be influenced by, for example, electrode orientation. Frijns et al. (2001) extended the original guinea pig volume conduction model so that it could use realistic electrode geometries and match the human cochlear anatomy. Briaire and Frijns (2005) then extended the three-dimensional human cochlea model further by incorporating back measuring, i.e., "telemetry" capabilities. This would allow investigation of the contribution of single-fiber action potentials to the measured eCAP. Briaire and Frijns (2005) noted that the resulting peak eCAP latencies were too short and did not match the measured human eCAP using neural response telemetry (Frijns et al., 2002). Briaire and Frijns (2006) used this model to show that a large P_0 peak in the eCAP occurs before the N_1P_1 complex when the dendrites are not degenerated. They suggested that the absence of this peak

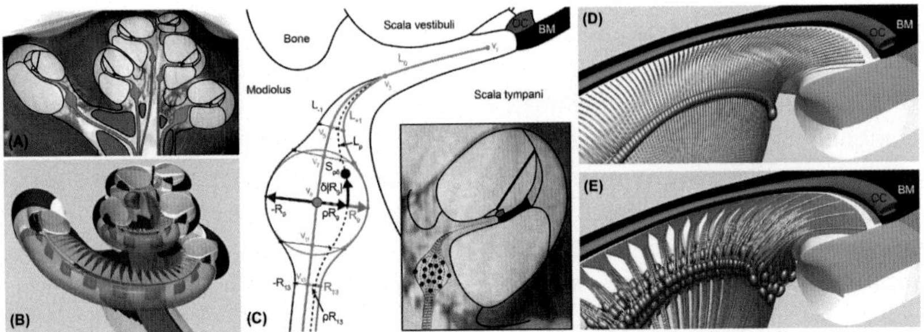

FIGURE 11.13 Visual representations of one of the model's cochlear geometries. (A) It shows a mid-modiolar cross-section from μCT imaging data of a human temporal bone. The lines overlaid on the μCT reconstruction represent the boundaries of the modeled cochlear geometry and the modeled neurons. (B) It is a ray-traced image of the three-dimensional model cochlea, cut open to reveal the neurons and a laterally inserted electrode array. (C–E) These illustrate the implementation of spatially distributed cell bodies. (C) It shows a mid-modiolar cross-section of one of the model geometries at 360° from the round window (black lines). Colored curves indicate neural trajectories; purple circles in the insert indicate modeled cell body locations. (D) It is a ray-traced image of a section of the geometry corresponding to (C) containing the nerve fibers with spiral ganglion cells aligned along a spiraling curve, as they were modeled in previous studies. (E) It shows the same section with the updated nerve fibers with spatially distributed cell bodies. *Reprinted from Kalkman, R.K., Briaire, J.J., Frijns, J.H.M., 2015. Current focussing in cochlear implants: An analysis of neural recruitment in a computational model. Hear. Res. 322, 89–98, with permission from Elsevier.*

might be used as an indicator for neural degeneration (more detail in the Appendix).

The Briaire and Frijns (2006) model was extended by Kalkman et al. (2014), who introduced a realistic computational model for the implanted human cochlea (Fig. 11.13A and B). They incorporated realistic nerve fiber trajectories in the cochlea and evaluated its application for the pitch-place map that underlies the mapping procedures in CI fitting. Kalkman et al. (2015) extended this further to study current focusing strategies. They modified the model to include variability in the location of the spiral ganglion cell bodies, which was essential for adequate modeling of spread of excitation at higher stimulus levels (Fig. 11.13C–E). A recent overview on stimulation strategies and electrode design as studied by computational modeling is provided by Kalkman et al. (2016).

11.10 SUMMARY

CIs are a great success story of modern medicine by being able to restore speech understanding and communication ability in the totally deaf. We describe some of its approximately 50-year history, the considerable advances made in the last 25 years in electrode design, sound processing strategies, and neural response telemetry allowing the measurement of eCAPs. The decrease in implantation age together with the increase in early binaural implantation are examples on how to overcome and/or prevent the effects of hearing deprivation on the brain. Quality of life issues led to increasing implantation in the elderly, but also to implanting the deaf ear in SSD to successfully alleviate tinnitus in that ear. We conclude with a review of some modeling studies that aim at designs for better electrodes and current delivery, assessing the effects of certain processing strategies, and providing information about the diagnostic value of eCAPs to assess the status of the spiral ganglion.

REFERENCES

Baguley, D.M., Atlas, M.D., 2007. Cochlear implants and tinnitus. Prog. Brain Res. 166, 347–355.

Baskent, D., 2006. Speech recognition in normal hearing and sensorineural hearing loss as a function of the number of spectral channels. J. Acoust. Soc. Am. 120, 2908–2925.

Bierer, J.A., 2010. Probing the electrode-neuron interface with focused cochlear implant stimulation. Trends Amplif. 14, 84–95.

Blamey, P.J., Maat, B., Başkent, D., Mawman, D., Burke, E., Dillier, N., et al., 2015. A retrospective multicenter study comparing speech perception outcomes for bilateral implantation and bimodal rehabilitation. Ear Hear. 36, 408–416.

Blasco, M.A., Redleaf, M.I., 2014. Cochlear implantation in unilateral sudden deafness improves tinnitus and speech comprehension: meta-analysis and systematic review. Otol. Neurotol. 35, 1426–1432.

Boons, T., Brokx, J.P.L., Frijns, J.H.M., Peeraer, L., Philips, B., Vermeulen, A., et al., 2012a. Effect of pediatric bilateral cochlear implantation on language development. Arch. Pediatr. Adolesc. Med. 166, 28–34.

Boons, T., Brokx, J.P.L., Dhooge, I., Frijns, J.H.M., Peeraer, L., Vermeulen, A., et al., 2012b. Predictors of spoken language development following pediatric cochlear implantation. Ear Hear. 33, 627–639.

Boyd, P.J., 2011. Potential benefits from deeply inserted cochlear implant electrodes. Ear Hear. 32, 411–427.

Briaire, J.J., Frijns, J.H.M., 2000a. 3D mesh generation to solve the electrical volume conduction problem in the implanted inner ear. Simulat. Pract. Theory 8, 57–73.

Briaire, J.J., Frijns, J.H.M., 2000b. Field patterns in a 3d tapered spiral model of the electrically stimulated cochlea. Hear. Res. 148, 18–30.

Briaire, J.J., Frijns, J.H.M., 2005. Unraveling the electrically evoked compound action potential. Hear. Res. 205 (1–2), 143–156.

Briaire, J.J., Frijns, J.H.M., 2006. The consequences of neural degeneration regarding optimal cochlear implant position in scala tympani: a model approach. Hear. Res. 214, 17–27.

Büchner, A., Lenarz, T., Boermans, P.P., Frijns, J.H., Mancini, P., Filipo, R., et al., 2012. Benefits of the HiRes 120 coding strategy combined with the Harmony processor in an adult European multicentre study. Acta Otolaryngol. 132, 179–187.

Buechner, A., Beynon, A., Szyfter, W., Niemczyk, K., Hoppe, U., Hey, M., et al., 2011. Clinical evaluation of cochlear implant sound coding taking into account conjectural masking functions, MP3000™. Cochlear Implants Int. 12, 194–204.

Burian, K., Hochmair, E., Hochmair-Desoyer, I., Lessel, M.R., 1979. Clinical observations in electric stimulation of the ear. Acta Otorhinolaryngol. 223, 139–166.

Chadha, N.K., Gordon, K.A., James, A.L., Papsin, B.C., 2009. Tinnitus is prevalent in children with cochlear implants. Int. J. Pediatr. Otorhinolaryngol. 73, 671–675.

Chatterjee, M., 1999. Temporal mechanisms underlying recovery from forward masking in multielectrode-implant listeners. J. Acoust. Soc. Am. 105, 1853–1863.

Chouard, C.H., MacLeod, P., 1973. La réhabilitation des surdités totales: essai de l'implantation cochléaire d'électrodes multiples. La Nouv. Presse Médicale 44, 2958.

Clark, G.M., Kranz, H.G., Minas, H., 1973. Behavioral thresholds in the cat to frequency modulated sound and electrical stimulation of the auditory nerve. Exp. Neurol. 41, 190–200.

Clark, G.M., Tong, Y.C., Dowell, R.C., 1983. Clinical results with a multichannel pseudobipolar system. Ann. N. Y. Acad. Sci. 405, 370–376.

Clay, K.M., Brown, C.J., 2007. Adaptation of the electrically evoked compound action potential (ECAP) recorded from nucleus CI24 cochlear implant users. Ear Hear. 28, 850–861.

Di Nardo, W., Cantore, I., Cianfrone, F., Melillo, P., Scorpecci, A., Paludetti, G., 2007. Tinnitus modifications after cochlear implantation. Eur. Arch. Otorhinolaryngol. 264, 1145–1149.

Di Nardo, W., Anzivino, R., Giannantonio, S., Schiaia, L., Paludetti, G., 2014. The effects of cochlear implantation on quality of life in the elderly. Eur. Arch. Otorhinolaryngol. 271, 65–73.

Djourno, A., Eyries, C., 1957. Auditory prosthesis by means of a distant electrical stimulation of the sensory nerve with the use of an indwelt coiling. Presse Med. 65, 1417.

Djourno, A., Eyries, C., Vallancien, P., 1957a. Preliminary attempts of electrical excitation of the auditory nerve in man, by permanently inserted micro-apparatus. Bull. Acad. Natl. Med. 141, 481–483.

Djourno, A., Eyries, C., Vallancien, P., 1957b. Electric excitation of the cochlear nerve in man by induction at a distance with the aid of micro-coil included in the fixture. C. R. Seances Soc. Biol. Fil. 151, 423–425.

Drennan, W.R., Won, J.H., Timme, A.O., Rubinstein, J.T., 2016. Nonlinguistic outcome measures in adult cochlear implant users over the first year of implantation. Ear Hear. 37, 354–364.

Driscoll, V.D., Welhaven, A.E., Gfeller, K., Oleson, J., Olszewski, C.P., 2016. Music perception of adolescents using electroacoustic hearing. Otol. Neurotol. 37, e141–e147.

Eggermont, J.J., 1985. Peripheral auditory adaptation and fatigue: a model oriented review. Hear. Res. 18, 57–71.

Eggermont, J.J., 2012. The Neuroscience of Tinnitus. Oxford University Press, Oxford, UK.

Eggermont, J.J., 2014. Noise and the Brain. Experience Dependent Developmental and Adult Plasticity. Academic Press, London.

Eggermont, J.J., 2015a. Auditory Temporal Processing and its Disorders. Oxford University Press, Oxford, UK.

Eggermont, J.J., 2015b. Animal models of spontaneous activity in the healthy and impaired auditory system. Front. Neural Circuits 9, 19.

Eisenberg, L.S., Berliner, K.I., House, W.F., Edgerton, B.J., 1983. Status of the adults' and children's cochlear implant programs at the House Ear Institute. Ann. N. Y. Acad. Sci. 405, 323–331.

Eshraghi, A.A., Nazarian, R., Telischi, F.F., Rajguru, S.M., Truy, E., Gupta, C., 2012. The cochlear implant: historical aspects and future prospects. Anat. Rec. (Hoboken) 295, 1967–1980.

Ewert, S.D., Hau, O., Dau, T., 2007. Forward masking: temporal integration or adaptation? In: Kollmeier, B., Klump, G.V., Hohmann, U., et al.,Hearing—From Sensory Processing to Perception. Springer-Verlag, Berlin, Heidelberg, pp. 165–172.

Firszt, J.B., Koch, D.B., Downing, M., Litvak, L., 2007. Current steering creates additional pitch percepts in adult cochlear implant recipients. Otol. Neurotol. 28, 629–636.

Firszt, J.B., Reeder, R.M., Skinner, M.W., 2008. Restoring hearing symmetry with two cochlear implants or one cochlear implant and a contralateral hearing aid. J. Rehabil. Res. Dev. 45, 749–767.

Fishman, K.E., Shannon, R.V., Slattery, W.H., 1997. Speech recognition as a function of the number of electrodes used in the SPEAK cochlear implant speech processor. J. Speech Lang. Hear. Res. 40, 1201–1215.

Francart, T., Osses, A., Wouters, J., 2015. Speech perception with F0mod, a cochlear implant pitch coding strategy. Int. J. Audiol. 20, 1–9.

Friesen, L.M., Shannon, R.V., Baskent, D., Wang, X., 2001. Speech recognition in noise as a function of the number of spectral channels: comparison of acoustic hearing and cochlear implants. J. Acoust. Soc. Am. 110, 1150–1163.

Frijns, J.H., Ten Kate, J.K., 1994. A model of myelinated nerve fibres for electrical prosthesis design. Med. Biol. Eng. Comput. 32, 391–398.

Frijns, J.H., Mooij, J., Ten Kate, J.H., 1994. A quantitative approach to modeling mammalian myelinated nerve fibers for electrical prosthesis design. IEEE Trans. Biomed. Eng. 41, 556–566.

Frijns, J.H., De Snoo, S.L., Schoonhoven, R., 1995. Potential distributions and neural excitation patterns in a rotationally symmetric model of the electrically stimulated cochlea. Hear. Res. 87, 170–186.

Frijns, J.H., De Snoo, S.L., Ten Kate, J.H., 1996. Spatial selectivity in a rotationally symmetric model of the electrically stimulated cochlea. Hear. Res. 95, 33–48.

Frijns, J.H., Briaire, J.J., Schoonhoven, R., 2000. Integrated use of volume conduction and neural models to simulate the response to cochlear implants. Simulat. Pract. Theory 8, 75–97.

Frijns, J.H., Briaire, J.J., Grote, J.J., 2001. The importance of human cochlear anatomy for the results with modiolus hugging multi-channel cochlear implants. Otol. Neurotol. 22, 340–349.

Frijns, J.H., Briaire, J.J., de Laat, J.A., Grote, J.J., 2002. Initial evaluation of the Clarion CII cochlear implant: speech perception and neural response imaging. Ear Hear. 23, 184–197.

Frijns, J.H., Klop, W.M., Bonnet, R.M., Briaire, J.J., 2003. Optimizing the number of electrodes with high-rate stimulation of the clarion CII cochlear implant. Acta Otolaryngol. 123, 138–142.

Frijns, J.H., Kalkman, R.K., Vanpoucke, F.J., Snel Bongers, J., Briaire, J.J., 2009. Simultaneous and non-simultaneous dual electrode stimulation in cochlear implants: evidence for two neural response modalities. Acta Otolaryngol. 129, 433–439.

Fu, Q.J., Nogaki, G., 2005. Noise susceptibility of cochlear implant users: the role of spectral resolution and smearing. J. Assoc. Res. Otolaryngol. 6, 19–27.

Fu, Q.J., Zeng, F.G., Shannon, R.V., Soli, S.D., 1998. Importance of tonal envelope cues in Chinese speech recognition. J. Acoust. Soc. Am. 104, 505–510.

Gordon, K., Henkin, Y., Kral, A., 2015. Asymmetric hearing during development: the aural preference syndrome and treatment options. Pediatrics 136, 141–153.

Gordon, K.A., Papsin, B.C., Harrison, R.V., 2005. Effects of cochlear implant use on the electrically evoked middle latency response in children. Hear. Res. 204, 78–89.

Gordon, K.A., Papsin, B.C., Harrison, R.V., 2006. An evoked potential study of the developmental time course of the auditory nerve and brainstem in children using cochlear implants. Audiol. Neurootol. 11, 7–23.

Gordon, K.A., Valero, J., Papsin, B.C., 2007. Binaural processing in children using bilateral cochlear implants. Neuroreport 18, 613–617.

Gordon, K.A., Wong, D.D., Papsin, B.C., 2010. Cortical function in children receiving bilateral cochlear implants simultaneously or after a period of interimplant delay. Otol. Neurotol. 31, 1293–1299.

Gordon, K.A., Salloum, C., Toor, G.S., van Hoesel, R., Papsin, B.C., 2012. Binaural interactions develop in the auditory brainstem of children who are deaf: effects of place and level of bilateral electrical stimulation. J. Neurosci. 32, 4212–4223.

Gordon, K.A., Wong, D.D., Papsin, B.C., 2013. Bilateral input protects the cortex from unilaterally-driven reorganization in children who are deaf. Brain 136, 1609–1625.

Green, D.M., 1964. Consistency of auditory detection judgments. Psychol. Rev. 71, 392–407.

Green, D.M., Swets, J.A., 1966. Signal Detection Theory and Psychophysics. Wiley, New York, NY.

Haensel, J., Ilgner, J., Chen, Y.S., Thuermer, C., Westhofen, M., 2005. Speech perception in elderly patients following cochlear implantation. Acta Otolaryngol. 125, 1272–1276.

Hanekom, T., 2001. Three-dimensional spiraling finite element model of the electrically stimulated cochlea. Ear Hear. 22, 300–315.

Hanekom, T., Hanekom, J.J., 2016. Three-dimensional models of cochlear implants: a review of their development and how they could support management and maintenance of cochlear implant performance. Network 27, 67–106.

Hay-McCutcheon, M.J., Brown, C.J., Abbas, P.J., 2005. An analysis of the impact of auditory-nerve adaptation on behavioral measures of temporal integration in cochlear implant recipients. J. Acoust. Soc. Am. 118, 2444–2457.

Henry, B.A., Turner, C.W., Behrens, A., 2005. Spectral peak resolution and speech recognition in quiet: normal hearing, hearing impaired, and cochlear implant listeners. J. Acoust. Soc. Am. 118, 1111–1121.

Hiel, A.-L., Gerard, J.-M., Decat, M., Deggouij, N., 2016. Is age a limiting factor for adaptation to cochlear implant? Eur. Arch. Otorhinolaryngol. 273 (9), 2495–2502.

Hochmair, E.S., Hochmair-Desoyer, I.J., 1983. Percepts elicited by different speech-coding strategies. Ann. N. Y. Acad. Sci. 405, 268–279.

Holden, L., Finley, C., Firszt, J., Holden, T.A., Brenner, C., Potts, L.G., et al., 2013. Factors affecting open set word recognition in adults with cochlear implants. Ear Hear. 34, 342–360.

Hong, R.S., Rubinstein, J.T., Wehner, D., Horn, D., 2003. Dynamic range enhancement for cochlear implants. Otol. Neurotol. 24, 590–595.

House, W.F., 1976. Cochlear implants. Ann. Otol. Rhinol. Laryngol 85 (Suppl. 27), 1–93.

House, W.F., Berliner, K.I., Eisenberg, L.S., 1979. Present status and future directions of the Ear Research Institute cochlear implant program. Acta Otolaryngol. 87, 176–184.

Hughes, M.L., Castioni, E.E., Goehring, J.L., Baudhuin, J.L., 2012. Temporal response properties of the auditory nerve: data from human cochlear-implant recipients. Hear. Res. 285, 46–57.

Ito, J., Sakakihara, J., 1994. Suppression of tinnitus by cochlear implantation. Am. J. Otolaryngol. 15, 145–148.

Jiwani, S., Papsin, B.C., Gordon, K.A., 2013. Central auditory development after long-term cochlear implant use. Clin. Neurophysiol. 124, 1868–1880.

Jiwani, S., Papsin, B.C., Gordon, K.A., 2016. Early unilateral cochlear implantation promotes mature cortical asymmetries in adolescents who are deaf. Hum. Brain Mapp. 37, 135–152.

Kalkman, R.K., Briaire, J.J., Dekker, D.M.T., Frijns, J.H.M., 2014. Place pitch versus electrode location in a realistic computational model of the implanted human cochlea. Hear. Res. 315, 10–24.

Kalkman, R.K., Briaire, J.J., Frijns, J.H.M., 2015. Current focussing in cochlear implants: an analysis of neural recruitment in a computational model. Hear. Res. 322, 89–98.

Kalkman, R.K., Briaire, J.J., Frijns, J.H.M., 2016. Stimulation strategies and electrode design in computational models of the electrically stimulated cochlea: an overview of existing literature. Network 27, 107–134.

Kiang, N.Y., Moxon, E.C., Levine, R.A., 1970. Auditory-nerve activity in cats with normal and abnormal cochleas. In: Sensorineural hearing loss. Ciba Found. Symp.241–273.

Kiefer, J., Hohl, S., Stürzebecher, E., Pfennigdorff, T., Gstöettner, W., 2001. Comparison of speech recognition with different speech coding strategies (SPEAK, CIS, and ACE) and their relationship to telemetric measures of compound action potentials in the Nucleus Cl 24M cochlear implant system. Audiology 40, 32–42.

Kim, D.K., Bae, S.C., Park, K.H., Jun, B.C., Lee, D.H., Yeo, S.W., et al., 2013. Tinnitus in patients with profound hearing loss and the effect of cochlear implantation. Eur. Arch. Otorhinolaryngol. 270, 1803–1808.

Kleine Punte, A., De Ridder, D., Van de Heyning, P., 2013. On the necessity of full length electrical cochlear stimulation to suppress severe tinnitus in single-sided deafness. Hear. Res. 295, 24–29.

Kral, A., Sharma, A., 2012. Developmental neuroplasticity after cochlear implantation. Trends Neurosci. 35, 111–122.

Kwon, B.J., van den Honert, C., 2006. Dual-electrode pitch discrimination with sequential interleaved stimulation by cochlear implant users. J. Acoust. Soc. Am. 120, EL1–EL6.

Lammers, M.J.W., van Eyl, R.H.M., van Zanten, G.A., Versnel, H., Grolman, W., 2015a. Delayed auditory brainstem responses in prelingually deaf and late-implanted cochlear implant users. J. Assoc. Res. Otolaryngol. 16, 669–678.

Lammers, M.J.W., Versnel, H., van Zanten, G.A., Grolman, W., 2015b. Altered cortical activity in prelingually deafened cochlear implant users following long periods of auditory deprivation. J. Assoc. Res. Otolaryngol. 16, 159–170.

Laneau, J., Wouters, J., Moonen, M., 2006. Improved music perception with explicit pitch coding in cochlear implants. Audiol. Neurootol. 11, 38–52.

Lee, D.S., Lee, J.S., Oh, S.-H., et al., 2001. Cross-modal plasticity and cochlear implants. Nature 409, 149–150.

Lee, H.-J., Giraud, A.-L., Kang, E., Oh, S.-H., Kang, H., Chong-Sun Kim, C.-S., et al., 2007. Cortical activity at rest predicts cochlear implantation outcome. Cereb. Cortex 17, 909–917.

Lenarz, M., Sönmez, H., Joseph, G., Büchner, A., Lenarz, T., 2012. Cochlear implant performance in geriatric patients. Laryngoscope 122, 1361–1365.

Liberman, M.C., Dodds, L.W., 1984. Single-neuron labeling and chronic cochlear pathology. II. Stereocilia damage and alterations of spontaneous discharge rates. Hear. Res. 16, 43–53.

Litvak, L.M., Smith, Z.M., Delgutte, B., Eddington, D.K., 2003. Desynchronization of electrically evoked auditory-nerve activity by high-frequency pulse trains of long duration. J. Acoust. Soc. Am. 114, 2066–2078.

Litvak, L.M., Spahr, A.J., Saoji, A.A., Fridman, G.Y., 2007. Relationship between perception of spectral ripple and speech recognition in cochlear implant and vocoder listeners. J. Acoust. Soc. Am. 122, 982–991.

McDermott, H.J., McKay, C.M., 1994. Pitch ranking with nonsimultaneous dual-electrode electrical stimulation of the cochlea. J. Acoust. Soc. Am. 96, 155–162.

McKerrow, W.S., Schreiner, C.E., Snyder, R.L., Merzenich, M.M., Toner, J.G., 1991. Tinnitus suppression by cochlear implants. Ann. Otol. Rhinol. Laryngol. 100, 552–558.

Mertens, G., De Bodt, M., Van de Heyning, P., 2016. Cochlear implantation as a long-term treatment for ipsilateral incapacitating tinnitus in subjects with unilateral hearing loss up to 10 years. Hear. Res. 331, 1–6.

Michelson, R.P., Merzenich, M.M., Pettit, C.R., Schindler, R.A., 1973. A cochlear prosthesis: further clinical observations; preliminary results of physiological studies. Laryngoscope 83, 1116–1122.

Miller, C.A., Hu, N., Zhang, F., Robinson, B.K., Abbas, P.J., 2008. Changes across time in the temporal responses of auditory nerve fibers stimulated by electric pulse trains. J. Assoc. Res. Otolaryngol. 9, 122–137.

Milczynski, M., Wouters, J., van Wieringen, A., 2009. Improved fundamental frequency coding in cochlear implant signal processing. J. Acoust. Soc. Am. 125, 2260–2271.

Moore, B.C., 2003. Coding of sounds in the auditory system and its relevance to signal processing and coding in cochlear implants. Otol. Neurotol. 24, 243–254.

Nie, K., Barco, A., Zeng, F.G., 2006. Spectral and temporal cues in cochlear implant speech perception. Ear Hear. 27, 208–217.

Niparko, J.K., Tobey, E.A., Thal, D.J., Eisenberg, L.S., Wang, N.Y., et al., 2010. Spoken language development in children following cochlear implantation. JAMA 303, 1498–1506.

Nogueira, W., Büchner, A., Lenarz, T., Edler, B., 2005. A psychoacoustic "nofm"-type speech coding strategy for cochlear implants. EURASIP J. Appl. Signal Process 18, 3044–3059.

O'Brien, G.E., Rubinstein, J.T., 2016. The development of biophysical models of the electrically stimulated auditory nerve: single-node and cable models. Network 27, 135–156.

Olze, H., Gräbel, S., Förster, U., Zirke, N., Huhnd, L.E., Haupt, H., et al., 2012. Elderly patients benefit from cochlear implantation regarding auditory rehabilitation, quality of life, tinnitus, and stress. Laryngoscope 122, 196–203.

Pan, T., Tyler, R.S., Ji, H., Coelho, C., Gehringer, A.K., Gogel, S.A., 2009. Changes in the tinnitus handicap questionnaire after cochlear implantation. Am. J. Audiol. 18, 144–151.

Park, M.-H., Won, J.H., Horn, D.L., Rubinstein, J.T., 2015. Acoustic temporal modulation detection in normal-hearing and cochlear implanted listeners: effects of hearing mechanism and development. J. Assoc. Res. Otolaryngol. 16, 389–399.

Petersen, B., Weed, E., Sandmann, P., Brattico, E., Hansen, M., Sørensen, S.D., et al., 2015. Brain responses to musical feature changes in adolescent cochlear implant users. Front. Hum. Neurosci. 9, 7.

Pialoux, P., Chouard, C.H., Meyer, B., Fugain, C., 1979. Indications and results of the multichannel cochlear implant. Acta Otolaryngol. 87, 185–189.

Ponton, C.W., Eggermont, J.J., Coupland, S.G., Winkelaar, R., 1992. Frequency-specific maturation of the eighth nerve and brain-stem auditory pathway: evidence from derived auditory brain-stem responses (ABRs), J. Acoust. Soc. Am., 91. pp. 1576–1586.

Ponton, C.W., Don, M., Eggermont, J.J., Waring, M.D., Kwong, B., Masuda, A., 1996a. Plasticity of the auditory system in children after long periods of complete deafness. Neuroreport 8, 61–65.

Ponton, C.W., Don, M., Eggermont, J.J., Waring, M.D., Masuda, A., 1996b. Maturation of human cortical auditory function: differences between normal-hearing children and children with cochlear implants. Ear Hear. 17, 430–437.

Ponton, C.W., Eggermont, J.J., Kwong, B., Don, M., 2000. Maturation of human central auditory system activity: evidence from multi-channel evoked potentials. Clin. Neurophysiol. 111, 220–236.

Portmann, M., Cazals, Y., Negrevergne, M., Aran, J.M., 1979. Temporary tinnitus suppression in man through electrical stimulation of the cochlea. Acta Otolaryngol. 87, 294–299.

Portmann, M., Nègrevergne, M., Aran, J.M., Cazals, Y., 1983. Electrical stimulation of the ear: clinical applications. Ann. Otol. Rhinol. Laryngol. 92, 621–622.

Prijs, V.F., 1980. On peripheral auditory adaptation. II. Comparison of electrically and acoustically evoked action potentials in the guinea pig. Acustica 45, 1–13.

Rattay, F., Lutter, P., Felix, H., 2001. A model of the electrically excited human cochlear neuron I. Contribution of neural substructures to the generation and propagation of spikes. Hear. Res. 153, 43–63.

Roberts, D.S., Lin, H.W., Herrmann, B.S., Lee, D.J., 2013. Differential cochlear implant outcomes in older adults. Laryngoscope 123, 1952–1956.

Rodieck, R.W., Kiang, N.Y., Gerstein, G.L., 1962. Some quantitative methods for the study of spontaneous activity of single neurons. Biophys. J. 2, 351–368.

Roland, P.S., Tobey, E., 2013. A tribute to a remarkably sound solution. Cell 154, 1175–1177.

Rubinstein, J.T., Wilson, B.S., Finley, C.C., Abbas, P.J., 1999. Pseudo-spontaneous activity: stochastic independence of auditory nerve fibers with electrical stimulation. Hear. Res. 127, 108–118.

Ruckenstein, M.J., Hedgepeth, C., Rafter, K.O., Montes, M.L., Bigelow, D.C., 2001. Tinnitus suppression in patients with cochlear implants. Otol. Neurotol. 22, 200–204.

Sandmann, P., Kegel, A., Eichele, T., Dillier, N., Lai, W., Bendixen, A., et al., 2010. Neurophysiological evidence of impaired musical sound perception in cochlear-implant users. Clin. Neurophysiol. 121, 2070–2082.

Sandmann, P., Plotz, K., Hauthal, N., de Vos, M., Schönfeld, R., Debener, S., 2015. Rapid bilateral improvement in auditory cortex activity in postlingually deafened adults following cochlear implantation. Clin. Neurophysiol. 126, 594–607.

Schindler, R.A., Kessler, D.K., Rebscher, S.J., Yanda, J.L., Jackler, R.K., 1986. The UCSF/Storz multichannel cochlear implant: patient results. Laryngoscope 96, 597–603.

Shallop, J.K., Facer, G.W., Peterson, A., 1999. Neural response telemetry with the nucleus CI24M cochlear implant. Laryngoscope 109, 1755–1759.

Sharma, A., Dorman, M.F., Kral, A., 2005. The influence of a sensitive period on central auditory development in children with unilateral and bilateral cochlear implants. Hear. Res. 203, 134–143.

Siebert, W.M., 1965. Some implications of the stochastic behavior of primary auditory neurons. Kybernetik 2, 206–215.

Simmons, F.B., Epley, J.M., Lummis, R.C., Guttman, N., Frishkopf, L.S., Harmon, L.D., et al., 1965. Auditory nerve: electrical stimulation in man. Science 148, 104–106.

Skinner, M., Holden, L., Holden, T., Dowell, R., Seligman, P., et al., 1991. Performance of postlinguistically deaf adults with the wearable speech processor (WSP III) and mini speech processor (MSP) of the Nucleus multi-electrode cochlear implant. Ear Hear. 12, 3–22.

Smit, J.E., Hanekom, T., can Wieringen, A., Wouters, J., Hanekom, J.J., 2010. Threshold predictions of different pulse shapes using a human auditory nerve fibre model containing persistent sodium and slow potassium currents. Hear. Res. 269, 12–22.

Souliere Jr., C.R., Kileny, P.R., Zwolan, T.A., Kemink, J.L., 1992. Tinnitus suppression following cochlear implantation. A multifactorial investigation. Arch. Otolaryngol. Head Neck Surg. 118, 1291–1297.

Sparreboom, M., Beynon, A.J., Snik, A.F., Mylanus, E.A., 2014. Auditory cortical maturation in children with sequential bilateral cochlear implants. Otol. Neurotol. 35, 35–42.

Summerfield, A.Q., Barton, G.R., Toner, J., McAnallen, C., Proops, D., Harries, C., et al., 2006. Self-reported benefits from successive bilateral cochlear implantation in post-lingually deafened adults: randomised controlled trial. Int. J. Audiol. 45 (Suppl. 1), S99–S107.

Svirsky, M.A., Teoh, S.-W., Neuburger, H., 2004. Development of language and speech perception in congenitally, profound deaf children as a function of age at cochlear implantation. Audiol. Neurootol. 9, 224–233.

Tong, Y.C., Clark, G.M., 1985. Absolute identification of electric pulse rates and electrode positions by cochlear implant patients. J. Acoust. Soc. Am. 77, 1881–1888.

Tong, Y.C., Black, R.C., Clark, G.M., Forster, I.C., Millar, J.B., O'Loughlin, B.J., et al., 1979. A preliminary report on a multiple-channel cochlear implant operation. J. Laryngol. Otol. 93, 679–695.

Townshend, B., Cotter, N., Van Compernolle, D., White, R.L., 1987. Pitch perception by cochlear implant subjects. J. Acoust. Soc. Am. 82, 106–115.

Vandali, A.E., van Hoesel, R.J., 2011. Development of a temporal fundamental frequency coding strategy for cochlear implants. J. Acoust. Soc. Am. 129, 4023–4036.

Vandali, A.E., Whitford, L.A., Plant, K.L., Clark, G.M., 2000. Speech perception as a function of electrical stimulation rate: using the Nucleus 24 cochlear implant system. Ear Hear. 21, 608–624.

Vandali, A.E., Sucher, C., Tsang, D.J., McKay, C.M., Chew, J.W., McDermott, H.J., 2005. Pitch ranking ability of cochlear implant recipients: a comparison of sound-processing strategies. J. Acoust. Soc. Am. 117, 3126–3138.

Van de Heyning, P., Vermeire, K., Diebl, M., Nopp, P., Anderson, I., De Ridder, D., 2008. Incapacitating unilateral tinnitus in single-sided deafness treated by cochlear implantation. Ann. Otol. Rhinol. Laryngol. 117, 645–652.

Van Dijkhuizen, J.N., Beers, M., Boermans, P.-P., Briaire, J.J., Frijns, J.H.M., 2011. Speech intelligibility as a predictor of cochlear implant outcome in prelingually deafened adults. Ear Hear. 32, 445–458.

Van Dijkhuizen, J.N., Boermans, P.-P., Briaire, J.J., Frijns, J.H.M., 2016. Intelligibility of the patient's speech predicts the likelihood of cochlear implant success in prelingually deaf adults. Ear Hear. 37, e302–e310.

Van Schoonhoven, J., Sparreboom, M., van Zanten, B.G.A., Scholten, R.J.P.M., Mylanus, E.A.M., Dreschler, W.A., et al., 2013. The effectiveness of bilateral cochlear implants for severe-to-profound deafness in adults: a systematic review. Otol. Neurotol. 34, 190–198.

Vermeire, K., Brokx, J.P., Wuyts, F.L., Cochet, E., Hofkens, A., Van de Heyning, P., 2005. Quality-of-life benefit from cochlear implantation in the elderly. Otol. Neurotol. 26, 188–195.

Verschuur, C., 2009. Modeling the effect of channel number and interaction on consonant recognition in a cochlear implant peak-picking strategy. J. Acoust. Soc. Am. 125, 1723–1736.

Verschuur, C.A., Lutman, M.E., Ramsden, R., Greenham, P., O'Driscoll, M., 2005. Auditory localization abilities in bilateral cochlear implant recipients. Otol. Neurotol. 26, 965–971.

Westerman, L.A., Smith, R.L., 1984. Rapid and short-term adaptation in auditory nerve responses. Hear. Res. 15, 249–260.

Wilson, B.S., 2015. Getting a decent (but sparse) signal to the brain for users of cochlear implants. Hear. Res. 322, 24–38.

Wilson, B.S., Dorman, M.F., 2008. Cochlear implants: a remarkable past and a brilliant future. Hear. Res. 242, 3–21.

Wilson, B.S., Finley, C.C., Farmer Jr., J.C., Lawson, D.T., Weber, B.A., Wolford, R.D., et al., 1988. Comparative studies of speech processing strategies for cochlear implants. Laryngoscope 98, 1069–1077.

Wilson, B.S., Finley, C.C., Lawson, D.T., Wolford, R.D., Eddington, D.K., Rabinowitz, W.M., 1991. Better speech recognition with cochlear implants. Nature 352, 236–238.

Wilson, B.S., Finley, C.C., Lawson, D.T., Zerbi, M., 1997. Temporal representations with cochlear implants. Am. J. Otol. 18 (Suppl), S30–S34.

Won, J.H., Drennan, W.R., Rubinstein, J.T., 2007. Spectral-ripple resolution correlates with speech reception in noise in cochlear implant users. J. Assoc. Res. Otolaryngol. 8, 384–392.

Won, J.H., Drennan, W.R., Kang, R.S., Rubinstein, J.T., 2010. Psychoacoustic abilities associated with music perception in cochlear implant users. Ear Hear. 31, 796–805.

Won, J.H., Clinard, C.G., Kwon, S., Dasika, V.K., Nie, K., Drennan, W.R., et al., 2011. Relationship between behavioral and physiological spectral-ripple discrimination. J. Assoc. Res. Otolaryngol. 12, 375–393.

Wong, L.L., Vandali, A.E., Ciocca, V., Luk, B., Ip, V.W., Murray, B., et al., 2008. New cochlear implant coding strategy for tonal language speakers. Int. J. Audiol. 47, 337–347.

Wouters, J., McDermott, H.J., Francart, T., 2015. Sound coding in cochlear implants. IEEE Signal Proc. Mag. 3, 67–80.

Zeng, F.-G., Rebscher, S.J., Fu, Q.-J., Chen, H., Sun, X., Yin, L., et al., 2015. Development and evaluation of the Nurotron 26-electrode cochlear implant system. Hear. Res. 322, 188–199.

Zhang, F., Miller, C.A., Robinson, B.K., Abbas, P.J., Hu, N., 2007. Changes across time in spike rate and spike amplitude of auditory nerve fibers stimulated by electric pulse trains. J. Assoc. Res. Otolaryngol. 8, 356–372.

Zwicker, H., Fastl, H., 1990. Psychoacoustics. Facts and Models. Springer Verlag, Berlin.

Part V

The Future

Chapter 12

Auditory Brainstem and Midbrain Implants

Persons who lack an auditory nerve (AN) cannot benefit from cochlear implants (CIs), but a prosthesis utilizing an electrode array implanted on the surface or in the cochlear nucleus (CN) can restore some hearing. Worldwide, more than 1200 persons (Lim and Lenarz, 2015) have received these auditory brainstem implants (ABIs), most commonly after removal of the tumors that occur with bilateral vestibular schwannomas (Type 2 neurofibromatosis, NF2). Most ABI patients who lose their AN due to NF2 tumors have limited performance with the ABI presumably because of damage to the brainstem region either due to the tumor or during tumor removal surgery. If this is the case, then it may be necessary to bypass the damaged brainstem region to provide these patients with good speech recognition. The auditory midbrain implant (AMI) is a new central auditory prosthesis designed for stimulation of the human inferior colliculus that bypasses the brainstem.

12.1 AUDITORY BRAINSTEM IMPLANTS

The ABI typically consists of an electrode array placed on the surface of the CN, though there has also been a version with penetrating electrode arrays that was explored in a clinical trial.

12.1.1 Surface Electrodes

House and Hitselberger (2001),c reported the first long-time results of an auditory brainstem implantation performed in 1979. They had placed a single electrode on the surface of the CN via a translabyrinthine craniotomy at the time of the vestibular schwannoma removal. Their patient had benefit from this direct stimulation: "She reported being able to hear her dog bark, a knock on the door, an airplane overhead, and the sound of the sink garbage disposal. She had considerable relief of tinnitus. She noted increasing discrimination, and the unit was of considerable help in lipreading." Shannon and Otto (1990) explored the psychoacoustics of CN stimulation using an

Hearing Loss. DOI: http://dx.doi.org/10.1016/B978-0-12-805398-0.00012-8

ABI, consisting of two electrodes, each a 0.75×2.5 mm rectangular pad of platinum foil fixed to a Dacron mesh pad and 4.25 mm apart. They found that: "The usable range of electrical amplitudes above threshold was comparable with that of cochlear implants, typically 10–15 dB. Little temporal integration occurred over a range of stimulus durations from 2–1000 ms and patients' ability to detect amplitude modulation as a function of modulation frequency was similar to that of cochlear implant patients and normal listeners." Brackmann et al. (1993) and Shannon et al. (1993) reported on results with a three-channel device. The ABI was positioned in the lateral recess of the fourth ventricle, adjacent to the cochlear nuclei, and consisted of three platinum plates mounted on a Dacron mesh backing. The electrodes remained stable for over 10 years. These ABIs had comparable psychophysical and speech performance to single-channel CIs. ABI patients had significant enhancement of speech understanding when the sound from the ABI was combined with lip reading. A detailed history of ABI application can be found in the study of Sennaroglu and Ziyal (2012).

Modern ABIs use the same electrode stimulator and processing strategies as a standard CI (see chapter: Cochlear Implants), except that the electrode array is placed on the CN. Each electrode potentially activates a variety of neuron types, excitatory as well as inhibitory, possibly with different characteristic frequencies (see chapter: Hearing Basics). The first company-based multichannel electrode array was developed by Cochlear Ltd. and tested in Europe and the United States. At the same time Laszig et al. (1991) devised an array consisting of 20 platinum electrodes that were assembled inside a silastic carrier 7 mm in diameter. This allowed them to use the Nucleus mini-22 device CI stimulator. Each electrode was a tiny round plate with a diameter of 0.6–0.7 mm. Laszig et al. (1995) found that in the nine implanted NF2 patients the speech perception results and patient satisfaction were encouraging, and the data showed even limited open speech recognition.

The House group (Otto et al., 1998) then developed an eight-electrode multichannel ABI that was evaluated in 20 patients who had at least 3 months' experience with the device. Perceptual performance indicated benefit from the device for communication purposes, including sound-only sentence recognition scores in three patients ranging from 49% to 58% and ability to converse on the telephone. In a later evaluation of this ABI type (Otto et al., 2002) reported that the multichannel ABI was effective and safe and gave useful auditory sensations in most patients with NF2. They found that: "The ABI improved patients' ability to communicate compared with the lipreading-only condition, it allowed the detection and recognition of many environmental sounds, and in some cases it provided significant ability to understand speech by using just the sound from the ABI (with no lipreading cues). Its performance in most patients has continued to improve for up to 8 years after implantation." Another follow-up of the same device

(Kuchta et al., 2004) concluded that at least three spectral channels, programmed in the appropriate individual tonotopic order, were required for satisfactory speech recognition in most patients with ABI. However, patients with ABIs did not receive more frequency information with more than five stimulated surface electrodes. Kuchta et al. (2004) assumed that this was largely due to limited access to the tonotopic axis of the CN.

Current ABIs comprise an external part similar to a CI device and an implanted portion designed for surface plate electrodes. There are differences in the number of electrode channels and shapes of the pads between the different companies (Fig. 12.1). Each pad has a mesh to help secure the pad to the brainstem surface and in which tissue can grow over time.

Existing ABI devices use the same processing strategies as CIs. The electrodes of the matrix array are assumed to stimulate different tonotopic layers of the CN, and the variations of electrical level on each electrode encode envelope variations of the signal in a narrow frequency band associated with that electrode. However, the simple place and amplitude codes that were originally developed for stimulating the AN in CI devices may not adequately convey the important information in the speech signal with an ABI. The CN is located higher in the auditory processing hierarchy than the AN, and its direct stimulation may bypass many important processes required for transmitting speech information (Colletti et al., 2009). Furthermore, the electrodes are placed on the surface of the CN and may not sufficiently access its tonotopic organization that exist within deeper regions.

Lenarz et al. (2001) reported that between May 1996 and April 2000, 14 patients with NF2 underwent implantation with a multichannel ABI. Three

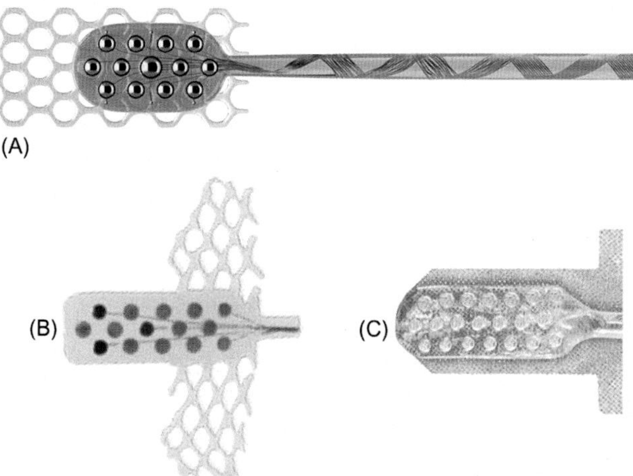

FIGURE 12.1 ABI electrodes from Med-El (A), Neurelec (B), and Cochlear (C). *Images downloaded from the Internet.*

different types of multichannel ABIs, composed of eight Nucleus M22, one Nucleus M24 (Cochlear Ltd., Sydney, Australia), and five Clarion-1.2 (magnet-free with 16 contact electrodes; Advanced Bionics Corp., Sylmar, CA), were implanted in their patients. Lenarz et al. (2001) found that "lip-reading enhancement improved within the first 6 months and then entered a plateau phase. In the auditory alone mode, Lenarz et al. (2001) showed that more than half of the patients showed their first positive result in the vowel test 3 months after device activation, but it took about 6 months until half of the patients revealed a result above the chance level in the consonant and Freiburger numbers tests. Open set speech recognition in the auditory alone mode (in the speech-tracking test) was not common and happened relatively late (within 1 year or later)." For the same patients, Lenarz et al. (2002) "concluded that the auditory brainstem implant is an effective support for receiving and, to some degree, differentiating environmental sounds, and that as an adjuvant to lip-reading, it enhances speech perception, especially in quiet surroundings."

In a multicenter study, Nevison et al. (2002) described findings in 27 subjects who received a Nucleus 20- or 21-channel ABI. All subjects involved in the trial had NF2 bilateral tumors removed. The majority of users had environmental sound awareness and used stress and rhythm cues in speech that assisted with lip reading. Only two subjects from this series did receive sufficient benefit from the ABI in conversation without lip reading.

More promising results came from Colletti and Shannon (2005) who compared ABI performance in 10 NF2 patients and 10 nontumor patients (e.g., those with cochlear nerve aplasia/avulsion or cochlear ossification). They concluded: "The difference in modulation detection between the two groups suggests a difference in the survival of specific cells in the cochlear nucleus that support modulation. The pattern of results indicates a separate pathway of auditory processing that is specialized for modulated sounds, and that pathway is critical for speech understanding. *In NF2 patients, the tumor and surgery may selectively damage this pathway*, resulting in poor speech recognition with prosthetic stimulation." In a follow-up paper on ABIs implanted in 112 patients (83 adults and 29 children) with tumor (T) and nontumor disorders (NT). Of the 112 patients, 15 have previously had a CI elsewhere with no sound detection (Colletti et al., 2009). They found that "at the most recent follow-up, NT adults scored from 10 to 100% in open-set speech perception tests (average, 59%), and T patients scored from 5 to 31% (average, 10%). The differences between these results are statistically significant. The best performance was observed in patients who lost their nerve VIII from head trauma or severe ossification." However, in a later review of ABI use, Colletti et al. (2012) stated: "It is now clear that ABIs can produce excellent speech recognition in some patients with NF2, allowing even conversational telephone use. Although the factors leading to this improved performance are not completely clear, these new results show that excellent hearing is possible for NF2 patients with the ABI."

12.1.2 A Note on Electrode Placement

McCreery et al. (1998) showed that central regions of the posteroventral CN in the cat are suitable for penetrating auditory brainstem implantation. McCreery et al. (2007) found that the rostrolateral and rostromedial region of the ventral cochlear nucleus (VCN) when stimulated electrically and chronically elicited significantly lower degrees of frequency specificity in receiving ICC neurons compared with the caudolateral and caudomedial VCN regions. This suggested that the electrodes had to access the tonotopic organization of the VCN to achieve good performance. Shivdasani et al. (2008) used "multielectrode recordings [in the rat] to assess the frequency specificity of activation in the central nucleus of the inferior colliculus (ICC) produced by electrical stimulation of localized regions within the ventral cochlear nucleus. In 26% of the 193 ICC sites, they found a high correlation between acoustic tuning and electrical tuning obtained through VCN stimulation. A high degree of frequency specificity was found in 58% of the 118 lowest threshold VCN–ICC pairs. Frequency specific stimulation was obtained for medial, central, and posterolateral VCN regions rather than more anterolateral regions." Shivdasani et al. (2008) concluded that "if a surface array is placed on the posterolateral surface of the VCN in a dorsoventral direction using the present commercial ABI, some frequency-specific stimulation may be achievable but a penetrating array in the PVCN would result in an increased likelihood of frequency-specific ICC activation."

12.1.3 Penetrating Electrodes

Because many patients equipped with surface electrodes on the CN did not achieve a sufficient and/or systematic range of pitches, penetrating electrodes for ABI (PABI) were developed to better access the tonotopy of the CN. The PABI uses 8 or 10 penetrating microelectrodes in conjunction with a separate array of 10 or 12 surface electrodes. Typically with PABIs, the threshold for auditory sensation is low, because the penetrating electrodes are not separated from the CN by the glia-pia membrane (McCreery, 2008). Despite better tonotopic access has been reported with PABI, there is still a relatively poor coverage of the low frequencies with these penetrating electrodes (Moore and Shannon, 2009; Vincent, 2012).

Otto et al. (2008) reported that in a prospective clinical trial, 10 individuals, all with NF2, received a PABI after vestibular schwannoma removal via a translabyrinthine approach. Approximately 8 weeks after implantation, PABI devices were activated and tested. Mean follow-up time was 33.8 months and showed that less than 25% of penetrating electrodes resulted in auditory sensations, whereas more than 60% of surface electrodes were effective. Even after more than 3 years of experience, patients using penetrating electrodes did not achieve improved speech recognition compared with those using surface electrode ABIs. They concluded that: "The PABI met the goals of lower threshold, increased pitch range, and high selectivity, but these properties did not result in improved speech recognition." One key challenge for the PABI was in placing the electrodes into the appropriate regions of the CN. This was more or less the last clinical paper published on PABIs, suggesting that this technique is no longer used.

12.1.4 Performance With Auditory Brainstem Implants

Moore and Shannon (2009) compiled data from the Colletti and Shannon (2005) study that clearly showed the ABI performance compared to CIs and the effects of etiology (Fig. 12.2). Only some (~20%) non-NF2 patients could perform as well as users with CIs. Recent studies amplify these findings of Moore and Shannon (2009). Behr et al. (2014) emphasized that brainstem trauma is a primary factor in the variability of outcomes in NF2 patients. They noted that high levels of speech recognition, including high levels of open-set speech recognition, are possible with the ABI even in patients with NF2 and large tumors. Corroborating this, Matthies et al. (2014) conducted a prospective study on ABI operations performed with the

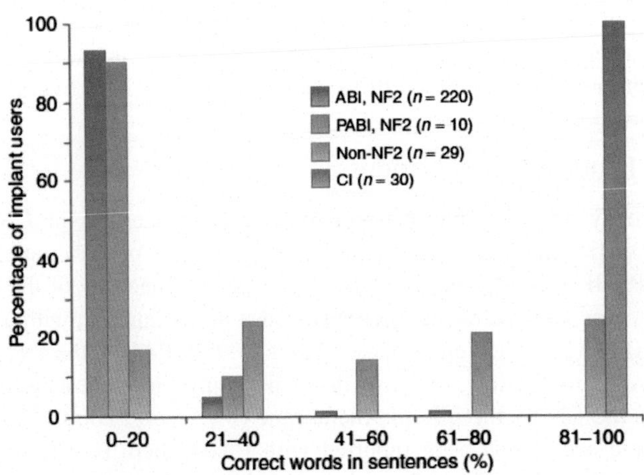

FIGURE 12.2 Word recognition performance of implant users. Data are the percentage of users of the various implant types achieving different levels of word-in-sentence performance on standardized tests in two laboratories. Most ABI users have the NF2 variety of Schwann cell pathology. Data for ABI, PABI, and CI are from the House Ear Institute and those for non-NF2 are from the University of Verona (R.V.S., unpublished data). *Reprinted by permission from Macmillan Publishers Ltd.: Moore, D.R., Shannon, R.V., 2009. Beyond cochlear implants: awakening the deafened brain. Nat. Neurosci. 12, 686–691.*

aid of multimodality neuromonitoring between 2005 and 2009 in 18 patients with NF2. They found "that open-set speech recognition in pure auditory mode is feasible in patients with ABIs. Large tumor volumes do not prevent good outcome. Positive preconditions are short ipsilateral and short bilateral deafness periods and high number of auditory electrodes." A study by McSorley et al. (2014) "demonstrated that ABI users make use of their device for at least 12 hours per day on average and obtain considerable subjective benefit in speech discrimination when using the device with familiar speakers in a quiet environment. Benefit is less significant in noise and with an unfamiliar speaker. Lip reading enhances subjective benefit ..., and lip reading training may have a role to play in maximizing subjective benefit."

To investigate other contributions to the variability and limitations in hearing performance in NF2 ABI patients, Azadpour and McKay (2014) compared six CI users and four ABI users of Nucleus implants with the ACE processing strategy (see chapter: Cochlear Implants). They reported that: "The difference in the performance of ABI users could be related to the location of their electrode array on the CN, anatomy, and physiology of their CN or the damage to their auditory brainstem due to tumor or surgery." McKay et al. (2015) then evaluated whether speech understanding in ABI users, who have a tumor pathology, could be improved by the selection of a subset of electrodes that were appropriately pitch ranked and distinguishable. They hypothesized that disordered pitch or spectral percepts and channel interactions may contribute significantly to the poor outcomes in most ABI users. McKay et al. (2015) reported that "the results of the pitch ranking and multidimensional scaling procedures confirmed that the ABI users did not have a well-ordered set of percepts related to electrode position, thus supporting the proposal that difficulty in processing of spectral information may contribute to poor speech understanding. However, none of the subjects benefited from a map that reduced the stimulation electrode set to a smaller number of electrodes that were well ordered in place pitch. Thus, although poor spectral processing may contribute to poor understanding in ABI users, it is not likely to be the sole contributor to poor outcomes."

McKay et al. (2013) stressed the loss of processing power when bypassing the AN and its parallel activation of three areas in the VCN, and even more so when also bypassing the auditory brainstem and directly stimulating the auditory midbrain. They noted that for electrical stimulation of the AN, central processing of intensity information over time can be usefully described using the same temporal integration (TI) model developed for normal acoustic hearing (McKay and McDermott, 1998). Because there are different types of neurons in these structures that play varying roles in the parallel processing of different stimulus features, and that have complex excitatory and inhibitory connections related to both afferent and efferent inputs, the neural coding of stimulus intensity information in the CN and IC is more complex than in the AN. Therefore, when considering appropriate ways to electrically stimulate these structures to convey intensity information (including its temporal modulation), both the types of neurons activated and the appropriate temporal or spatial patterns of activation will be important.

With this in mind we proceed to some promising findings obtained with AMIs.

12.2 AUDITORY MIDBRAIN IMPLANTS

The fact that the ABI is able to produce high levels of speech perception in nontumor patients (e.g., those with inaccessible cochleae or posttraumatic damage to the cochlear nerve) suggests that limitations in ABI performance in NF2 patients may be associated with CN damage caused by the tumors or the tumor removal process. Thus, stimulation of the auditory midbrain proximal to the damaged CN may be a better alternative for hearing restoration in NF2 patients. The AMI is a new central auditory prosthesis designed for penetrating stimulation of the human inferior colliculus (Lenarz et al., 2006a,b; Lim and Lenarz, 2015).

12.2.1 First Results

In collaboration with Cochlear Ltd., Lenarz et al. (2006a) developed a human prototype AMI (Fig. 12.3A), which was designed for electrical stimulation along the well-defined tonotopic gradient of the ICC. Considering that better speech perception and hearing performance has been correlated with a greater number of discriminable frequency channels of information available,

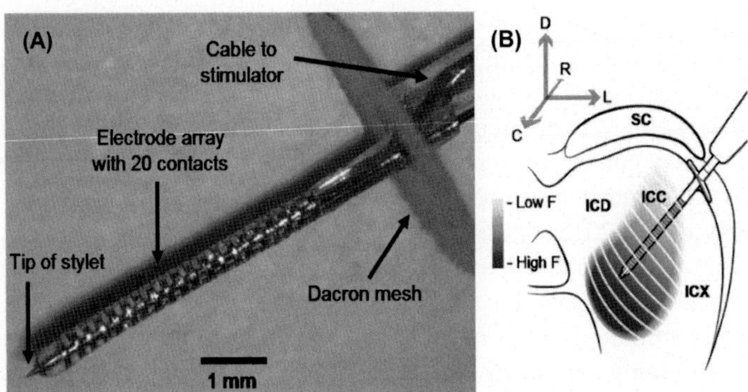

FIGURE 12.3 (A) Single-shank AMI array developed by Cochlear Ltd. (Australia) with 20 ring sites (200 μm spacing, 200 μm thickness, 400 μm diameter) along a silicone carrier. Dacron mesh prevents over-insertion of the array into the IC and tethers it to the brain. The cable connects to an implanted stimulator. (B) Drawing of the AMI array positioned along the tonotopic gradient of ICC. *From Lenarz, M., Lim, H.H., Patrick, J.F., Anderson, D.J., Lenarz, T., 2006. Electrophysiological validation of a human prototype auditory midbrain implant in a guinea pig model. J. Assoc. Res. Otolaryngol. 7, 383–398, with permission of Springer.*

the ability of the AMI to effectively activate discrete frequency regions within the ICC may enable better hearing performance than achieved by the ABI. Lenarz et al. (2006a), in an animal model, investigated if their AMI array (Fig. 12.3B) "could achieve low threshold, frequency-specific activation within the ICC, and whether the levels for ICC activation via AMI stimulation were within safe limits for human application." They electrically stimulated different frequency regions within the ICC via the AMI array and recorded the evoked neural activity in the primary auditory cortex in ketamine-anesthetized guinea pigs. Lenarz et al. (2006a) found that AMI stimulation achieved lower thresholds and more localized, frequency-specific activation of primary auditory cortex than CI stimulation. Furthermore, AMI stimulation achieves cortical activation with current levels that were within safe limits for central nervous system stimulation. A follow-up chronic implant study in cats further confirmed the safety of the AMI in terms of surgical implantation and long-term stimulation within the auditory midbrain (Lenarz et al., 2007).

The year 2007 was stellar for the AMI. Colletti et al. (2007) reported the first success of midbrain stimulation in a human volunteer. They used an ABI array (Med-El ABI) and placed it on the dorsal surface of the inferior colliculus. Electrical stimulation produced auditory sensations on all 12 electrodes. Auditory threshold levels indicated the stability of the electrode array over time. Middle latency responses showed activation in the contralateral auditory cortex but none in ipsilateral cortex. Lim et al. (2007) implanted three patients with the AMI, and it has proven to be safe with minimal movement over time. They found that "thus far, all three patients obtain enhancements in lip reading capabilities and environmental awareness and some improvements in speech perception comparable with that of NF2 ABI patients."

In a review by Lim et al. (2008a), the authors explained that the electrode array ended up in different locations across patients (i.e., ICC, dorsal cortex of the IC (ICX), lateral lemniscus), although the intended target was the ICC. They found that "the patient implanted within the intended target, the ICC, exhibited the greatest improvements in hearing performance. However, this patient has not yet achieved open-set speech perception to the performance level typically observed for cochlear implant patients." Lim et al. (2008b) stated that although each of these three patients was implanted into a different region, they generally exhibited similar threshold versus phase duration, threshold versus pulse rate, and pitch versus pulse rate curves. However, Lim et al. (2008b) "observed large differences across patients in loudness adaptation to continuous pulse stimulation over long time scales. One patient (implanted in dorsal cortex of IC) even experienced complete loudness decay and elevation of thresholds with daily stimulation." Lim et al. (2009) reviewed the initial surgical, psychophysical, and speech results from the first three implanted patients. They summarized these as: "All patients obtain improvements in hearing capabilities on a daily basis. However, performance varies dramatically across patients depending on the implant location within the midbrain with the best performer still not able to achieve open set speech perception without lip-reading cues. Stimulation of the auditory midbrain provides a wide range of level, spectral, and temporal cues, all of which are important for speech understanding, but they do not appear to sufficiently fuse together to enable open set speech perception with the currently used stimulation strategies." More recently, Lim and Lenarz (2015) summarized the results across five patients who were implanted between 2006 and 2008 with the single-shank AMI. Only one of these five patients was implanted with the array appropriately aligned along the tonotopic gradient of the ICC. However, this one patient has continued to improve in hearing performance within the upper range of performance of ABI patients and achieves some open-set speech perception, emphasizing the critical importance of placement of sites along the tonotopy of the ICC and in learning over time.

Using a midbrain implant, Lim et al. (2013) further "identified an ordering of pitch percepts for electrical stimulation of sites across the human inferior colliculus (IC) that was consistent with the IC tonotopy shown in animals. ... Interestingly, this pitch ordering was not initially observed for stimulation across the IC, possibly due to central changes caused by prior hearing loss. Daily implant stimulation for about 4 months altered the pitch percepts from being predominantly low to exhibiting the expected ordering across the stimulated IC," again emphasizing the immense capability of the brain to learn or adapt over time.

12.2.2 Toward a Better Auditory Midbrain Implant Design

Calixto et al. (2012) expected that stimulation of a reasonable number of frequency regions of the ICC (Lim and Anderson, 2006; Lim et al., 2007) and the ability of ICC neurons to generally synchronize to the temporal envelope of the signal (Rees and Langner, 2005; Rode et al., 2013) would enable sufficient speech understanding with the AMI using CI-based stimulation strategies (see chapter: Cochlear Implants). AMI stimulation can achieve frequency-specific activation (Lenarz et al., 2006b; Lim and Anderson, 2006). However, as shown in the study of Lim et al. (2008a,b) and Calixto et al. (2012) "Stimulation of a single site within a given frequency lamina elicits strong refractory effects for short delays between electrical pulses as well as greater suppressive effects for longer delays than typically observed for acoustic stimulation that is likely limiting temporal coding abilities."

The single-shank AMI (Fig. 12.3) only stimulates one site in any given ICC lamina and does not exhibit enhanced activity (i.e., louder percepts or lower thresholds) for repeated pulses on the same site with intervals less than 2–5 ms, as occurs for CI pulse or acoustic click stimulation. This enhanced activation, related to short-term TI, is important for tracking the rapid temporal fluctuations of a speech signal. Therefore, Calixto et al. (2012) investigated the effects of coactivation of different regions within an ICC lamina on primary auditory cortex activity in ketamine-anesthetized guinea pigs (Fig. 12.4A). This coactivation revealed an enhancement mechanism for integrating converging inputs from an ICC lamina on a fast scale (<6 ms window) that is compromised when stimulating just a single ICC location. Coactivation of two ICC regions also

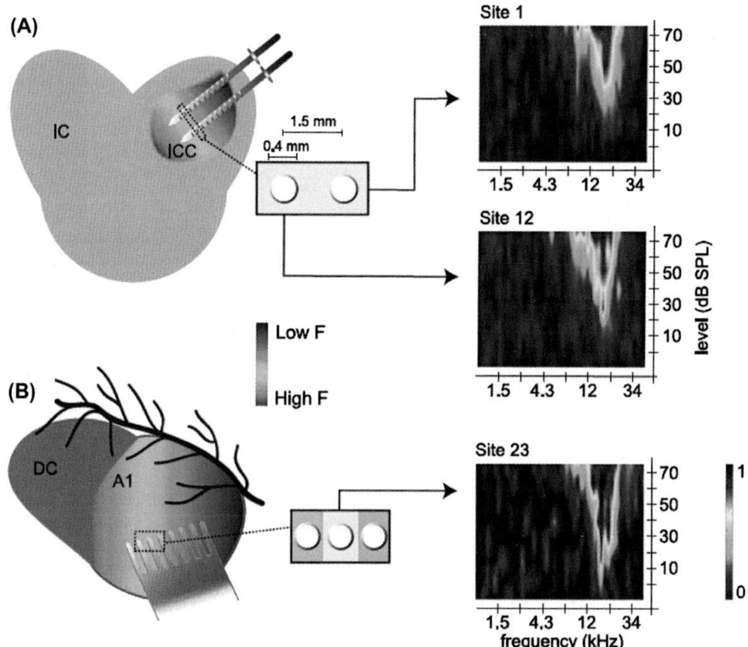

FIGURE 12.4 Schematic of array placements. (A) Drawing of a coronal midbrain cut showing the left and right IC. On the right half is a schematic of two single-shank AMI arrays that are both aligned along the frequency gradient of the ICC with examples of frequency response maps (FRMs) for two selected sites (one from each shank). Both FRMs have a best frequency of approximately 15 kHz. (B) Drawing of the primary auditory cortex (A1) and our eight-shank recording array. The array was positioned perpendicular to the A1 surface with each shank aligned along a different best-frequency column. With current source density plots, the depth of the shanks was adjusted to have at least one site on each shank within the main input layer (III/IV) of A1. Only one site in the main input layer with the closest best frequency (~15 kHz) to the selected AMI sites was chosen for final analysis. The colored scale in the FRMs corresponds to normalized spike count from 0 to 1. For further details on FRM and current source density analysis, see Placement of Electrode Arrays. DC, dorsocaudal auditory cortex; F, frequency. *From Calixto, R., Lenarz, M., Neuheiser, A., Scheper, V., Lenarz, T., Lim, H.H., 2012. Coactivation of different neurons within an isofrequency lamina of the inferior colliculus elicits enhanced auditory cortical activation. J. Neurophysiol. 108, 1199–1210, with permission.*

reduces the strong and long-term (>100 ms) suppressive effects induced by repeated stimulation of just a single location.

On the basis of these findings, Calixto et al. (2012, 2013) investigated new types of two-shank and multi-shank AMI arrays to stimulate not only within different frequency regions but also within different iso-frequency locations within the ICC. They reported "a significant improvement in the ability to overcome strong suppressive effects and to rapidly modulate A1 activity by varying the [inter-pulse interval] IPI between two stimulated sites within a similar lamina."

In order to understand the different requirements for AMI stimulation compared to that for CIs, McKay et al. (2013) conducted experiments to measure the effect of IPIs on detection thresholds and loudness; temporal modulation transfer functions (TMTFs); effect of duration on detection thresholds; and forward masking decay. The CI data were consistent with a phenomenological model that based detection or loudness decisions on the output of a sliding TI window (Oxenham and Moore, 1994), based on the input to which was the hypothetical AN response to each stimulus pulse. They found that: "AMI data were consistent with a neural response that decreased more steeply compared to CI stimulation as the pulse rate increased or interpulse interval decreased." McKay et al. (2013) found this "consistent with the midbrain neurons having a different refractory behavior from that of the AN neurons, with very little or no activity (or contribution) evoked by a pulse that follows another within 2 ms. ... [Thus,] the AMI model required an integration window that was significantly wider (i.e., decreased temporal resolution) than that for CI data, the latter being well fit using the same integration window shape as derived from normal-hearing data."

The poorer modulation detection of AMI subjects compared to CI subjects as reflected in the TMTF (Fig. 12.5) is "consistent with a significantly longer TI time window than for CI users based on a higher weighting for the longer time constant and/or an increase in the actual time constants ... Poor temporal resolution and increased forward masking are both likely to impair speech perception. Envelope modulation in speech signals conveys information about the manner of articulation and voicing in consonants, vowel duration information and voice pitch information. Slow decay of forward masking is likely to produce masking of one phoneme that follows another in the speech stream, and interfere with the segmentation of words and syllables" (McKay et al., 2013).

Overall, these psychophysical findings are consistent with the neurophysiological findings by Calixto et al. (2012) and Straka et al. (2013) in which stimulation of a single site in a given ICC lamina causes strong refractory or suppressive effects and multi-site stimulation across a lamina can greatly reduce these negative effects. As a result of these animal and human studies, a new two-shank AMI array and improved surgical approach were developed to stimulate at least two sites in each ICC lamina with varying IPI pulses

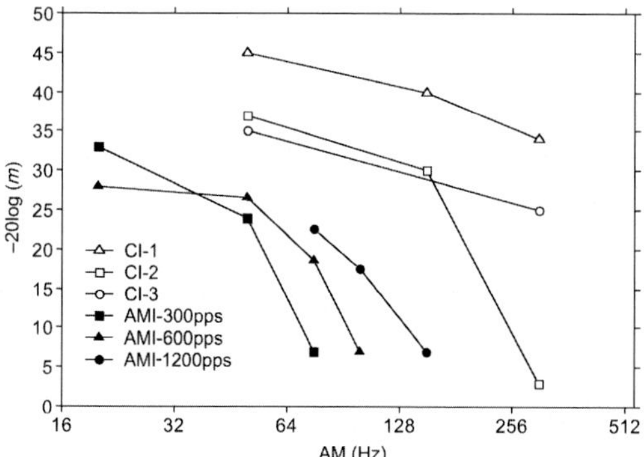

FIGURE 12.5 AM detection ability is lower for the AMI patient than for three typical CI users. The modulation depth (*m*; higher ordinate value means less depth and better detection) for 70% correct was identified for each modulation frequency (Hz) and carrier rate (pps: pulses per second). Red circle: maximum depth used. *Based on data from McKay, C.M., Lim, H.H., Lenarz, T., 2013. Temporal processing in the auditory system: insights from cochlear and auditory midbrain implantees. J. Assoc. Res. Otolaryngol. 14, 103–124.*

(Lim and Lenarz, 2015). This second-generation AMI device will be evaluated in an upcoming clinical trial funded by the National Institutes of Health.

12.3 SUMMARY

Persons who lack an AN cannot benefit from CIs, but a prosthesis utilizing an electrode array implanted on the surface or in the CN, an ABI can restore some hearing. Existing ABI devices use the same processing strategies as CIs. Both surface and penetrating electrodes are currently used. Some speech understanding without lip reading is obtained with ABTs, but on average at most 30% of words in sentences, and only in nontumor patients. The AMI is a new central auditory prosthesis designed for stimulation of the human inferior colliculus and bypassing the brainstem, which may be damaged by the tumor or the surgery to remove it. From the initial studies, showing pitch perception after long-term use but no speech understanding, it became clear that tonotopic stimulation of the inferior colliculus was insufficient. Electrophysiological studies in animals suggested the need for simultaneously stimulating different parts of the ICC, whereas psychoacoustic studies into the temporal processing of AMI stimulation indicated the different refractory behavior of the ICC compared to AN fibers. Solving these problems is ongoing.

REFERENCES

Azadpour, M., McKay, C.M., 2014. Processing of speech temporal and spectral information by users of auditory brainstem implants and cochlear implants. Ear Hear. 35, e192–e203.

Behr, R., Colletti, V., Matthies, C., Morita, A., Nakatomi, H., Dominique, L., et al., 2014. New outcomes with auditory brainstem implants in NF2 patients. Otol. Neurotol. 35, 1844–1851.

Brackmann, D.E., Hitselberger, W.E., Nelson, R.A., Moore, J., Waring, M.D., Portillo, F., et al., 1993. Auditory brainstem implant: I. Issues in surgical implantation. Otolaryngol. Head Neck Surg. 108, 624–633.

Calixto, R., Lenarz, M., Neuheiser, A., Scheper, V., Lenarz, T., Lim, H.H., 2012. Coactivation of different neurons within an isofrequency lamina of the inferior colliculus elicits enhanced auditory cortical activation. J. Neurophysiol. 108, 1199–1210.

Calixto, R., Salamat, B., Rode, T., Hartmann, T., Volckaerts, B., Ruther, P., et al., 2013. Investigation of a new electrode array technology for a central auditory prosthesis. PLoS One 8, e82148.

Colletti, L., Shannon, R., Colletti, V., 2012. Auditory brainstem implants for neurofibromatosis type 2. Curr. Opin. Otolaryngol. Head Neck Surg. 20, 353–357.

Colletti, V., Shannon, R.V., 2005. Open set speech perception with auditory brainstem implant? Laryngoscope 115, 1974–1978.

Colletti, V., Shannon, R., Carner, M., Sacchetto, L., Turazzi, S., Masotto, B., et al., 2007. The first successful case of hearing produced by electrical stimulation of the human midbrain. Otol. Neurotol. 28, 39–43.

Colletti, V., Shannon, R., Carner, M., Veronese, S., Colletti, L., 2009. Outcomes in nontumor adults fitted with the auditory brainstem implant: 10 years' experience. Otol. Neurotol. 30, 614–618.

House, W.F., Hitselberger, W.E., 2001. Twenty-year report of the first auditory brain stem nucleus implant. Ann. Otol. Rhinol. Laryngol. 110, 103–104.

Kuchta, J., Otto, S.R., Shannon, R.V., Hitselberger, W.E., Brackmann, D.E., 2004. The multichannel auditory brainstem implant: how many electrodes make sense? J. Neurosurg. 100, 16–23.

Laszig, R., Kuzma, J., Seifert, V., Lehnhardt, E., 1991. The Hannover auditory brainstem implant: a multiple-electrode prosthesis. Eur. Arch. Otorhinolaryngol. 248, 420–421.

Laszig, R., Sollmann, W.P., Marangos, N., Charachon, R., Ramsden, R., 1995. Nucleus 20-channel and 21-channel auditory brain stem implants: first European experiences. Ann. Otol. Rhinol. Laryngol. 166 (Suppl), 28–30.

Lenarz, M., Lim, H.H., Patrick, J.F., Anderson, D.J., Lenarz, T., 2006a. Electrophysiological validation of a human prototype auditory midbrain implant in a guinea pig model. J. Assoc. Res. Otolaryngol. 7, 383–398.

Lenarz, M., Lim, H.H., Lenarz, T., Reich, U., Marquardt, N., Klingberg, M.N., et al., 2007. Auditory midbrain implant: histomorphologic effects of long-term implantation and electric stimulation of a new deep brain stimulation array. Otol. Neurotol. 28 (8), 1045–1052.

Lenarz, T., Moshrefi, M., Matthies, C., Frohne, C., Lesinski-Schiedat, A., Illg, A., et al., 2001. Auditory brainstem implant: part I. Auditory performance and its evolution over time. Otol. Neurotol. 22, 823–833.

Lenarz, M., Matthies, C., Lesinski-Schiedat, A., Frohne, C., Rost, U., Illg, A., et al., 2002. Auditory brainstem implant part II: subjective assessment of functional outcome. Otol. Neurotol. 23, 694–697.

Lenarz, T., Lim, H.H., Reuter, G., Patrick, J.F., Lenarz, M., 2006b. The auditory midbrain implant: a new auditory prosthesis for neural deafness—concept and device description. Otol. Neurotol. 27, 838–843.

Lim, H.H., Anderson, D.J., 2006. Auditory cortical responses to electrical stimulation of the inferior colliculus: implications for an auditory midbrain implant. J. Neurophysiol. 96, 975–988.

Lim, H.H., Lenarz, T., 2015. Auditory midbrain implant: research and development towards a second clinical trial. Hear. Res. 322, 212–223.

Lim, H.H., Lenarz, T., Joseph, G., Battmer, R.D., Samii, A., Samii, M., et al., 2007. Electrical stimulation of the midbrain for hearing restoration: insight into the functional organization of the human central auditory system. J. Neurosci. 27, 13541–13551.

Lim, H.H., Lenarz, T., Anderson, D.J., Lenarz, M., 2008a. The auditory midbrain implant: effects of electrode location. Hear. Res. 242, 74–85.

Lim, H.H., Lenarz, T., Joseph, G., Battmer, R.D., Patrick, J.F., Lenarz, M., 2008b. Effects of phase duration and pulse rate on loudness and pitch percepts in the first auditory midbrain implant patients: comparison to cochlear implant and auditory brainstem implant results. Neuroscience 154, 370–380.

Lim, H.H., Lenarz, M., Lenarz, T., 2009. Auditory midbrain implant: a review. Trends Amplif. 13, 149–180.

Lim, H.H., Lenarz, M., Joseph, G., Lenarz, T., 2013. Frequency representation within the human brain: stability versus plasticity. Sci. Rep. 3, 1474.

Matthies, C., Brill, S., Varallyay, C., Solymosi, L., Gelbrich, G., Roosen, K., et al., 2014. Auditory brainstem implants in neurofibromatosis Type 2: is open speech perception feasible? J. Neurosurg. 120, 546–558.

McCreery, D.B., 2008. Cochlear nucleus auditory prostheses. Hear. Res. 242, 64–73.

McCreery, D.B., Shannon, R.V., Moore, J.K., Chatterjee, M., 1998. Accessing the tonotopic organization of the ventral cochlear nucleus by intranuclear microstimulation. IEEE Trans. Rehabil. Eng. 6, 391–399.

McCreery, D.B., Lossinsky, A., Pikov, V., 2007. Performance of multisite silicon microprobes implanted chronically in the ventral cochlear nucleus of the cat. IEEE Trans. Biomed. Eng. 54, 1042–1051.

McKay, C.M., McDermott, H.J., 1998. Loudness perception with pulsatile electrical stimulation: the effect of interpulse intervals. J. Acoust. Soc. Am. 104, 1061–1074.

McKay, C.M., Lim, H.H., Lenarz, T., 2013. Temporal processing in the auditory system: insights from cochlear and auditory midbrain implantees. J. Assoc. Res. Otolaryngol. 14, 103–124.

McKay, C.M., Azadpour, M., Jayewardene-Aston, D., O'Driscoll, M., El-Deredy, W., 2015. Electrode selection and speech understanding in patients with auditory brainstem implants. Ear Hear. 36, 454–463.

McSorley, A., Freeman, S.R., Ramsden, R.T., Motion, J., King, A.T., Rutherford, S.A., et al., 2014. Subjective outcomes of auditory brainstem implantation. Otol. Neurotol. 36, 873–878.

Moore, D.R., Shannon, R.V., 2009. Beyond cochlear implants: awakening the deafened brain. Nat. Neurosci. 12, 686–691.

Nevison, B., Laszig, R., Sollmann, W.P., Lenarz, T., Sterkers, O., Ramsden, R., et al., 2002. Results from a European clinical investigation of the Nucleus multichannel auditory brainstem implant. Ear Hear. 23, 170–183.

Otto, S.R., Shannon, R.V., Brackmann, D.E., Hitselberger, W.E., Staller, S., Menapace, C., 1998. The multichannel auditory brain stem implant: performance in twenty patients. Otolaryngol. Head Neck Surg. 118, 291–303.

Otto, S.R., Brackmann, D.E., Hitselberger, W.E., Shannon, R.V., Kuchta, J., 2002. Multichannel auditory brainstem implant: update on performance in 61 patients. J. Neurosurg. 96, 1063–1071.

Otto, S.R., Shannon, R.V., Wilkinson, E.P., Hitselberger, W.E., McCreery, D.B., Moore, J.K., et al., 2008. Audiologic outcomes with the penetrating electrode auditory brainstem implant. Otol. Neurotol. 29, 1147–1154.

Oxenham, A.J., Moore, B.C., 1994. Modeling the additivity of nonsimultaneous masking. Hear. Res. 80, 105–118.

Rees, A., Langner, G., 2005. Temporal coding in the auditory midbrain. In: Winer, J.A., Schreiner, C.E. (Eds.), The Inferior Colliculus. Springer Science & Business Media, New York, pp. 346–376.

Rode, T., Hartmann, T., Hubka, P., Scheper, V., Lenarz, M., Lenarz, T., et al., 2013. Neural representation in the auditory midbrain of the envelope of vocalizations based on a peripheral ear model. Front. Neural Circuits 7, 166.

Sennaroglu, L., Ziyal, I., 2012. Auditory brainstem implantation. Auris Nasus Larynx 39, 439–450.

Shannon, R.V., Otto, S.R., 1990. Psychophysical measures from electrical stimulation of the human cochlear nucleus. Hear. Res. 47, 159–168.

Shannon, R.V., Fayad, J., Moore, J., Lo, W.W., Otto, S., Nelson, R.A., et al., 1993. Auditory brainstem implant: II. Postsurgical issues and performance. Otolaryngol. Head Neck Surg. 108, 634–642.

Shivdasani, M.N., Mauger, S.J., Rathbone, G.D., Paolini, A.G., 2008. Inferior colliculus responses to multichannel microstimulation of the ventral cochlear nucleus: implications for auditory brain stem implants. J. Neurophysiol. 99, 1–13.

Straka, M.M., Schendel, D., Lim, H.H., 2013. Neural integration and enhancement from the inferior colliculus up to different layers of auditory cortex. J. Neurophysiol. 110 (4), 1009–1020.

Vincent, C., 2012. Auditory brainstem implants: how do they work? Anat. Rec. 295, 1981–1986.

Chapter 13

Repairing and Building New Ears

First we briefly review the potential of gene therapy for hereditary hearing loss and for generating new hair cells in deaf ears.

13.1 GENE THERAPY FOR HEREDITARY HEARING LOSS

About 15% of hereditary deafness is inherited as autosomal dominant nonsyndromic hearing loss (see chapter: Epidemiology and Genetics of Hearing Loss and Tinnitus). Examples include *GJB3* (DFNA2), *GJB2* (DFNA3), *GJB6* (DFNA3), and possibly also DFNA4, DFNA9, and DFNA20/26. Since inheritance is autosomal dominant, silencing of the mutated allele could potentially preserve hearing (Hildebrand et al., 2008). A recent proof-of-principle study validated this prediction—a small interfering RNA (siRNA) suppressed expression of the *R75W* allele of human *GJB2* in a mouse model (Maeda et al., 2005). By using a construct containing $GJB2_{R75W}$ that interferes with the functioning of the wild-type gap junction protein, Maeda et al. (2005) could recapitulate human deafness (DFNA3) in a mouse model. In subsequent experiments (Maeda et al., 2005), the same construct and specific anti-$GJB2_{R75W}$ siRNAs specifically reduced expression of the $GJB2_{R75W}$ allele and prevented occurrence of the hearing loss phenotype. Hildebrand et al. (2008) stated that: "Based on the results achieved with *GJB2*, it is highly likely that alleles of other genes that cause autosomal dominant nonsyndromic hearing loss can be targeted by RNAi therapy delivered at different developmental time-points through different surgical approaches."

Preventing cell death may also be an option for treatment. Several mutations in apoptosis genes cause monogenic hearing impairment (Op de Beeck et al., 2011). These genes include *TJP2*, *DFNA5*, and *MSRB3*. *TJP2* encodes the tight junction protein ZO-2. Tight junction proteins (TJPs) belong to a family of membrane-associated guanylate kinase homologs that are involved in the organization of epithelial and endothelial intercellular junctions. TJPs bind to the cytoplasmic C terminals of junctional transmembrane proteins and link them to the actin cytoskeleton.

Nonsyndromic hearing impairment is associated with a mutation in DFNA5, which encodes the hearing-impairment protein 5. *MSRB3* encodes

Hearing Loss. DOI: http://dx.doi.org/10.1016/B978-0-12-805398-0.00013-X

methionine sulfoxide reductase B3 and catalyzes the reduction of free and protein-bound methionine sulfoxide to methionine, which is essential for hearing. This implies that apoptosis not only contributes to the pathology of acquired forms of hearing impairment, but also to genetic hearing impairment. These genes may constitute a new target in the prevention of hearing loss (Op de Beeck et al., 2011).

13.2 REGENERATING HAIR CELLS

Cochlear hair cells are susceptible to damage from a variety of sources (see chapters: Causes of Acquired Hearing Loss; Epidemiology and Genetics of Hearing Loss and Tinnitus). The consequence of this damage in humans often is permanent hearing loss. The discovery that hair cells can regenerate in birds and other non-mammalian vertebrates has led to various attempts of restoring hearing after such damage. After reviewing the early findings that lead to these studies, we will describe the various ways in which inner ear function in humans may eventually be restored.

According to Fukui and Raphael (2013), once the ear shows hair cell loss, protection is no longer an option and efforts need to be dedicated to hair cell regeneration. This may be accomplished by manipulating cell proliferation control (Chen and Segil, 1999; Lowenheim et al., 1999) or by influencing expression of genes that specify hair cell differentiation. The latter studies mostly involved regulating expression of *Atoh1* for which it was shown that the duration of expression is critical for hair cell survival and for the type of hair cell that is generated (Pan et al., 2012). The hair cells that are generated by induced *Atoh1* expression result from transdifferentiation of nonsensory cells in the organ of Corti. It has been demonstrated that the possibility for this transdifferentiation is gradually reduced as the cochlea matures (Kelly et al., 2012; Liu et al., 2012).

13.3 BIRDS CAN DO IT

> *When Cotanche presented striking scanning electron microscope images at the meeting of the Association for Research in Otolaryngology in February of 1986 and 1987 (Cotanche et al., 1986), many others began to join in the effort and some expanded the interpretation of older data (Cruz et al., 1985, 1987). Cotanche's (1986, 1987a,b) SEM images brought greater credibility to the idea that hair cell regeneration was worthy of study, because they provided vividly clear and incontrovertible evidence that rapid and remarkably complete self repair had occurred in chicken auditory epithelia within days after they had been damaged by loud sound.*
>
> Burns and Corwin, 2013

13.3.1 Structural Recovery After Noise Trauma in Birds

The avian acoustic trauma and hair cell regeneration literature (Dooling et al., 2008; Ryals and Rubel, 1988; Smolders, 1999) yields some important observations. In Saunders' laboratory, Cotanche et al. (1986) "noted that the structures of the basilar papilla underwent a remarkable postexposure recovery, even to the extent that cochlear nerve fibers reconnected to the new hair cells and successfully sent afferent information into the auditory CNS" (Fig. 13.1). Saunders (2010) noted that: "As remarkable as this recovery was, it did not restore the basilar papilla to its preexposure condition. The

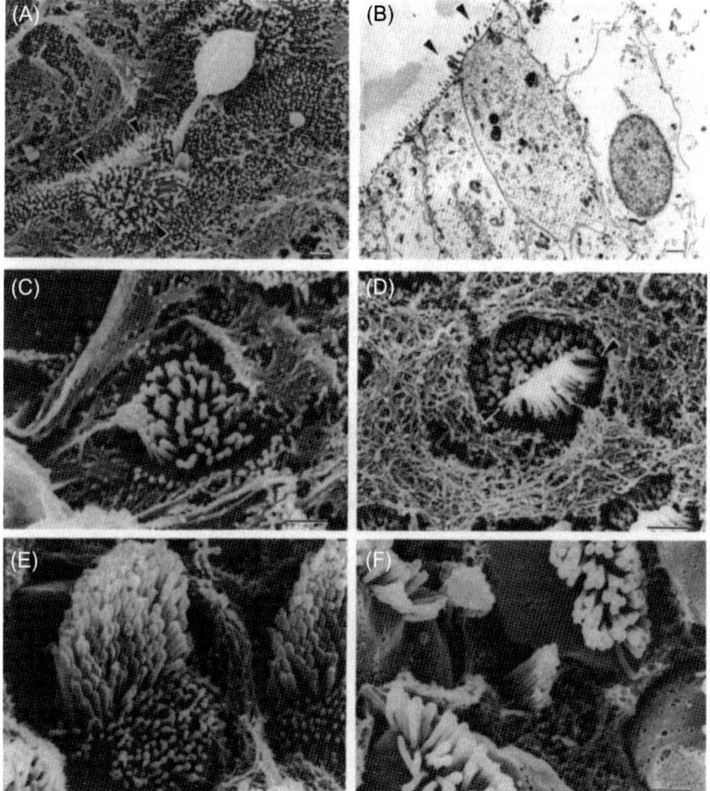

FIGURE 13.1 Regenerating stereociliary bundles in the recovering cochlea. (A) Newly forming bundle at 48 h of recovery (arrowheads); (B) TEM of newly forming hair cell at 48 h of recovery (arrowheads), stereocilia; (C) regenerating bundle after 4 days of recovery; (D) new stereociliary bundle at 6 days of recovery (arrowheads), tallest row of stereocilia; (E) stereociliary bundle after 10 days of recovery; and (F) newly regenerating hair cell in the strip region of damage at 10 days of recovery. Bars: 1 μm. *Reprinted from Cotanche, D.A., 1987. Regeneration of hair cell stereociliary bundles in the chick cochlea following severe acoustic trauma. Hear. Res. 30, 181–196, with permission from Elsevier.*

process of hair cell regeneration replaced 90% of the hair cells destroyed by the exposure, and the 10% reduction in hair cell numbers in the patch lesion appeared to persist. Moreover, the upper layer of the tectorial membrane never recovered."

The rate of functional recovery and the emergence of regenerated hair cells appear to be related. New hair cells are first seen 3–4 days after exposure, the evoked response thresholds also reduced significantly within 4 days (Saunders, 2010). The new hair cells were not immediately functionally mature. It took an additional day or so before they were working normally, and perhaps even longer before afferent fibers reconnected to the hair cells (Wang and Raphael, 1996). Thus, functional recovery may not simply be concluded from the visual appearance of new hair cells and not even from an apparent innervation by the auditory nerve fibers (Saunders, 2010). Furthermore, the degree to which hearing thresholds recover is frequency dependent; lower frequencies recover to preregeneration levels while higher frequency hearing recovers less completely (Ryals et al., 2013). Hair cell regeneration in birds is also not just a delayed developmental response, as 6-year-old quails (which are 3 years beyond their average lifespan) can regenerate hair cells as readily as newborn chicks (Ryals and Rubel, 1988). If this regeneration could be induced in the mammalian cochlea, it would offer a potential therapeutic treatment for sensorineural hearing loss in humans (Cotanche and Kaiser, 2010).

13.3.2 Functional Recovery After Noise Trauma in Birds

The ultimate goal for hair cell regeneration as a therapeutic approach is not only to recover hearing sensitivity, but also to allow speech understanding. Behavioral studies in birds with simple stimuli suggest that intensity discrimination remains unchanged after hair cell regeneration, while frequency and temporal resolution remain compromised. Budgerigars trained to recognize species-specific contact calls respond to the same "old" contact calls after regeneration as though they were "new" calls being heard for the first time. However, the birds could be retrained to recognize contact calls after hair cell regeneration. So even though auditory memory for the previous calls is impaired, relearning through the regenerated auditory channel is not (Ryals et al., 2013).

13.4 TRIALS IN MAMMALS

13.4.1 The Problem

In non-mammalian vertebrates, the supporting cell response to hair cell loss includes two events; one or more rounds of cellular proliferation to generate new cells in the epithelium followed by the differentiation of some of those cells

> *into regenerated hair cells. ... The loss of regenerative ability in mammalian inner ears presumably reflects a fundamental change in the ability of supporting cells to mount a regenerative response.*
>
> Matsui and Ryals, 2005

Achieving mammalian hair cell regeneration for the purpose of restoring hearing and balance can be explored in several ways. These include: (1) transplantation of inner ear stem cells, (2) cell cycle reentry, and (3) transdifferentiation of supporting cells into hair cells. This is schematically illustrated in Fig. 13.2.

13.4.2 Transplantation of Inner Ear Stem Cells

Restoration of function may be done by replacement of damaged tissues or organs, either by transplantation of tissue or implantation of artificial materials. Functional restoration must include generation of an adequate number of cells to reverse the defect, differentiation of cells to the correct phenotype, formation of appropriate three-dimensional tissue structures, production of cells/tissues that are mechanically and structurally compliant with the native tissue, and integration of the transplanted tissue with the native tissue without immunological rejection (Vats et al., 2002).

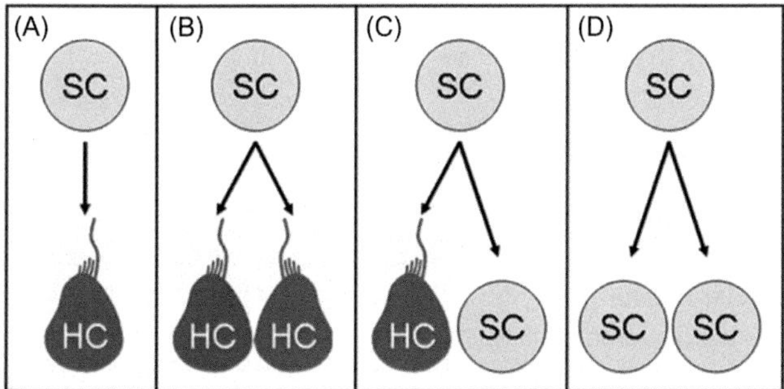

FIGURE 13.2 Methods of hair cell replacement. The production of hair cells may occur by several methods. (A) Supporting cells (SCs) may rapidly produce hair cells (HCs) by direct transdifferentiation; direct, phenotypic conversion to a HC occurs without the requirement for mitosis. When HC replacement depends on mitosis, there are several possible mechanisms. (B) Symmetric division of one SC produces two HCs, rapidly replacing HCs but eventually leading to a depletion of SCs. (C) Asymmetric SC division produces one HC and one SC, replacing lost HCs more slowly but replenishing the SC pool. (D) Symmetric SC division may produce two SCs as a method of maintaining the SC population. This symmetric division could occur in tandem with or following symmetric SC divisions, resulting in two HCs. One final alternative, not depicted, is that SCs produce HC precursors distinct from a fully differentiated HC and thus introduce a middle stage to all of the methods depicted above. *Reprinted from Brignull, H.R., Raibke, D.W., Stone, J.S., 2009. Feathers and fins: non-mammalian models for hair cell regeneration. Brain Res. 1277, 12–23, with permission from Elsevier.*

"Transplantation of inner ear stem cells has great promise provided that they can be expanded for directional differentiation in vivo, which requires identification and characterization of inner ear stem cell genes and an understanding of the cascades required for differentiation" (Edge and Chen, 2008). Pluripotent stem cells (derived from fibroblasts) are able to differentiate advanced features of hair cells in culture, including hair bundles and mechano-transduction currents (Oshima et al., 2010). Recently, (Ronaghi et al., 2014) found that stem cells from human embryos were capable of forming hair cell–like cells in culture.

13.4.3 Cell Cycle Reentry

Cell cycle reentry by endogenous hair cells or supporting cells for regeneration has distinct advantages, as the cells are produced in situ and are likely to attract neurons to form a synapse, but control of cell cycle remains a challenge. The mechanisms of avian hair cell regeneration have recently been reviewed in detail (Stone and Cotanche, 2007), and only a brief summary will be given here. In non-mammalian ears, "loss of hair cells triggers hair cell replacement by the supporting cells that surround each hair cell. Initially, supporting cells appear to transdifferentiate directly into hair cells in the absence of any cell division (Adler and Raphael, 1996; Roberson et al., 2004) starting before the ejection of dying hair cells from the damaged sensory epithelium (Cafaro et al., 2007). After 3–4 days, some supporting cells begin to re-enter the cell cycle and give rise to both hair cells and supporting cells after division" (Groves, 2010).

13.4.4 Transdifferentiation of Supporting Cells into Hair Cells

Transdifferentiation of supporting cells into hair cells occurs in non-mammalian vertebrates and may be possible in mammalian cochlea. In mammals, relatively normal hair cells and supporting cells can be regenerated during early development but this cannot currently be done in the adult. The activation of a group of stem cell genes can reprogram fully mature postmitotic cells to become induced pluripotent stem cells (Hanna et al., 2008), and a subset of stem/progenitor genes might be recruited to reprogram mature hair cells or supporting cells to an embryonic or immature phenotype that can reenter the cell cycle while remaining differentiated (Edge and Chen, 2008). Some supporting cells can directly differentiate into hair cells through a process called transdifferentiation (Forge et al., 1993, 1998; Adler and Raphael, 1996; Roberson et al., 2004). Thus the ability to reenter the cell cycle and/or to switch fates from a supporting cell to a hair cell may have been lost in mammals (White et al., 2006). Therefore, identifying the specific molecular factors that regulate cell cycle and cell fate in the inner ear may suggest potential targets for regenerative therapies (Kelley, 2006).

The most promising advance in the application of gene therapy to restore auditory function has been the discovery that the *Atoh1* (*Math1*) gene induces hair cell differentiation (Bermingham et al., 1999; Kawamoto et al., 2003, 2009; Woods et al., 2004). Atonal homolog 1 (Atoh1) promotes hair cell regeneration. When the *Atoh1* gene that encodes Atoh1 is deleted, hair cells in the organ of Corti do not form. Izumikawa et al. (2005) had shown that "transfer of adenoviral (AV) vectors expressing Atoh1 resulted in the formation of "hair cell–like" cells in the guinea pig organ of Corti 5 weeks post-inoculation. In 2 months, the surface of the auditory epithelium contained numerous cells with mature-looking stereocilia bundles: cross sections of the organ of Corti revealed inner hair cells (IHCs) with normal morphology. New outer hair cells, however, remained poorly differentiated suggesting that additional factors are required to specify outer hair cell development. Functional testing at 2 months postinoculation revealed significantly improved auditory brainstem responses consistent with Atoh1-induced regeneration of "hair cell–like" cells (Izumikawa et al., 2005). This study also showed that in addition to being effective, the in vivo delivery of AV was safe" (Hildebrand et al., 2008).

Other aspects play a role as well particularly the interaction between the Notch receptor on the surface of cells and Atoh1. Notch is activated by molecules on adjacent cells. The most pertinent role of Notch is that it inhibits hair cell formation. The transdifferentiation cycle proceeds like this: In undamaged epithelia, there is high Notch receptor activity in supporting cells inhibiting Atoh1 expression. Following hair cell damage, Notch receptor activity in nearby supporting cells is reduced and Atoh1 levels increase allowing supporting cells to transdifferentiate into hair cells. Maturation of the hair cells again increases Notch activity again and consequently reduces Atoh1 transcription to normal levels (Lewis et al., 2016).

A recent promising attempt was reported by Wise et al. (2015) who deafened guinea pigs by noise exposure (130 dB, 11–13 kHz, 2 h). After 2 weeks, the left cochleae were injected with an adenoviral vector containing the *Atoh1* gene. Control animals were injected with a control adenoviral vector. Three weeks after injection cochleae were assessed for hair cell density, maturity, and hair cell synaptogenesis with auditory neurons. Hearing thresholds were assessed throughout. Wise et al. (2015) found significantly more myosin VIIa-positive hair cells in cochleae that received *Atoh1* gene therapy compared to contralateral cochleae and compared to cochleae that received control gene therapy. However, the number of hair cells in *Atoh1*-treated animals was far below normal. Expression of Atoh1 had a significant preservation effect on the cytoarchitecture of the sensory epithelium compared to controls. Expression of the synaptic protein CtBP2 was present in some transfected cells from *Atoh1*-injected guinea pigs but at a reduced density compared to normal cochleae. There was evidence of auditory neuron preservation near transfected hair cells in *Atoh1*-injected cochleae, but there were no improvements in hearing thresholds.

The findings can be summarized as (based on Richardson and Atkinson, 2015):

1. The *Atohl* gene directs hair cell and supporting cell development in the cochlea, through hair cell differentiation, hair cell maturation, and stereocilia formation.
2. In neonatal animals, there is spontaneous regeneration of hair cells after damage to the sensory epithelium, but this becomes very limited as the cochlea matures.
3. *Atohl* gene therapy in the mature cochlea forces supporting cells to transdifferentiate into immature hair cells that express multiple hair cell markers with functional mechano-transducer channels.

13.5 OUTLOOK

Fujioka et al. (2015) listed the following outstanding questions: "Is the dual regeneration of functional inner and outer hair cells possible? Can supporting cells be reprogrammed to generate new hair cells rather than glial scars? Can a proliferative response be induced in the adult cochlea without leading to apoptosis? Will this cell division replace supporting cells that have transdifferentiated to hair cells? Are there options for treatment long after damage has occurred? Can the findings in mice be extended to primates? And are there treatment options for presbycusis?"

This suggests that application to these procedures to humans is likely still many years, if not decades away.

REFERENCES

Adler, H.J., Raphael, Y., 1996. New hair cells arise from supporting cell conversion in the acoustically damaged chick inner ear. Neurosci. Lett. 205, 17–20.

Bermingham, N.A., Hassan, B.A., Price, S.D., Vollrath, M.A., Ben-Arie, N., Eatock, R.,A., et al., 1999. Math1: an essential gene for the generation of inner ear hair cells. Science 284, 1837–1841.

Burns, J.C., Corwin, J.T., 2013. A historical to present-day account of efforts to answer the question: "what puts the brakes on mammalian hair cell regeneration?". Hear. Res. 297, 52–67.

Cafaro, J., Lee, G.S., Stone, J.S., 2007. Atoh1 expression defines activated progenitors as well as differentiating hair cells during avian hair cell regeneration. Dev. Dyn. 236, 156–170.

Chen, P., Segil, N., 1999. P27(Kip1) links cell proliferation to morphogenesis in the developing organ of Corti. Development 126, 1581–1590.

Cotanche, D.A., 1987a. Regeneration of hair cell stereociliary bundles in the chick cochlea following severe acoustic trauma. Hear. Res. 30, 181–196.

Cotanche, D.A., 1987b. Regeneration of the tectorial membrane in the chick cochlea following severe acoustic trauma. Hear. Res. 30, 197–206.

Cotanche, D.A., Kaiser, C.L., 2010. Hair cell fate decisions in cochlear development and regeneration. Hear. Res. 266, 18–25.

Cotanche, D.A., Saunders, J.C., Tilney, L.G., 1986. Hair cell recovery after severe acoustic trauma to the chick basilar papilla. Abstr. Assoc. Res. Otolaryngol. 9, 14.

Cruz, R.M., Lambert, P.R., Rubel, E.W., 1985. Temporal patterns of gentamicin induced hair cell loss in the chick basilar papilla. Am. Acad. Otolaryngol. Res. Forum Abstr.

Cruz, R.M., Lambert, P.M., Rubel, E.W., 1987. Light microscopic evidence of hair cell regeneration after gentamicin toxicity in chick cochlea. Arch. Otolaryngol. Head Neck Surg. 113, 1058–1062.

Dooling, R.J., Dent, M.L., Lauer, A.M., Ryals, B.M., 2008. Functional recovery after hair cell regeneration in birds. In: Salvi, R.J., Popper, A.N., Fay, R.R. (Eds.), Hair Cell Regeneration, Repair and Protection. Springer, New York, pp. 117–140.

Edge, A.S., Chen, Z.Y., 2008. Hair cell regeneration. Curr. Opin. Neurobiol. 18, 377–382.

Forge, A., Li, L., Corwin, J.T., Nevill, G., 1993. Ultrastructural evidence for hair cell regeneration in the mammalian inner ear. Science 259, 1616–1619.

Forge, A., Li, L., Nevill, G., 1998. Hair cell recovery in the vestibular sensory epithelia of mature guinea pigs. J. Comp. Neurol. 397, 69–88.

Fujioka, M., Okano, H., Edge, A.S.B., 2015. Manipulating cell fate in the cochlea: a feasible therapy for hearing loss. Trends Neurosci. 38, 139–144.

Fukui, H., Raphael, Y., 2013. Gene therapy for the inner ear. Hear. Res. 297, 99–105.

Groves, A.K., 2010. The challenge of hair cell regeneration. Exp. Biol. Med. (Maywood) 235, 434–446.

Hanna, J., Markoulaki, S., Schorderet, P., Carey, B.W., Beard, C., Wernig, M., et al., 2008. Direct reprogramming of terminally differentiated mature B lymphocytes to pluripotency. Cell 133, 250–264.

Hildebrand, M.S., Newton, S.S., Gubbels, S.P., Sheffield, A.M., Kochhar, A., de Silva, M.G., et al., 2008. Advances in molecular and cellular therapies for hearing loss. Mol. Ther. 16, 224–236.

Izumikawa, M., Minoda, R., Kawamoto, K., Abrashkin, K.A., Swiderski, D.L., Dolan, D.F., et al., 2005. Auditory hair cell replacement and hearing improvement by *Atoh1* gene therapy in deaf mammals. Nat. Med. 11, 271–276.

Kawamoto, K., Ishimoto, S., Minoda, R., Brough, D.E., Raphael, Y., 2003. Math1 gene transfer generates new cochlear hair cells in mature guinea pigs in vivo. J. Neurosci. 23, 4395–4400.

Kawamoto, K., Izumikawa, M., Beyer, L.A., Atkin, G.M., Raphael, Y., 2009. Spontaneous hair cell regeneration in the mouse utricle following gentamicin ototoxicity. Hear. Res. 247, 17–26.

Kelley, M.W., 2006. Hair cell development: commitment through differentiation. Brain Res. 109, 172–185.

Kelly, M.C., Chang, Q., Pan, A., Lin, X., Chen, P., 2012. Atoh1 directs the formation of sensory mosaics and induces cell proliferation in the postnatal mammalian cochlea in vivo. J. Neurosci. 32, 6699–6710.

Lewis, R.M., Rubel, E.W., Stone, J.S., 2016. Regeneration of auditory hair cells: a potential treatment for hearing loss on the horizon. Acoust. Today 12 (2), 40–48.

Liu, Z., Walters, B.J., Owen, T., Brimble, M.A., Steigelman, K.A., Zhang, L., et al., 2012. Regulation of p27Kip1 by Sox2 maintains quiescence of inner pillar cells in the murine auditory sensory epithelium. J. Neurosci. 32, 10530–10540.

Lowenheim, H., Furness, D.N., Kil, J., Zinn, C., Gultig, K., Fero, M.L., et al., 1999. Gene disruption of p27 (Kip1) allows cell proliferation in the postnatal and adult organ of Corti. Proc. Natl. Acad. Sci. U.S.A. 30, 4084–4088.

Maeda, Y., Fukushima, K., Nishizaki, K., Smith, R.J., 2005. In vitro and in vivo suppression of GJB2 expression by RNA interference. Hum. Mol. Genet. 14, 1641–1650.

Matsui, J.I., Ryals, B.M., 2005. Hair cell regeneration: an exciting phenomenon...but will restoring hearing and balance be possible? J. Rehabil. Res. Dev. 42 (4 Suppl. 2), 187–198.

Op de Beeck, K., Schacht, L., Van Camp, G., 2011. Apoptosis in acquired and genetic hearing impairment: the programmed death of the hair cell. Hear. Res. 281, 18–27.

Oshima, K., Shin, K., Diensthuber, M., Peng, A.W., Ricci, A.J., Heller, S., 2010. Mechanosensitive hair cell-like cells from embryonic and induced pluripotent stem cells. Cell 141, 704–716.

Pan, N., Kopecky, B., Jahan, I., Fritzsch, B., 2012. Understanding the evolution and development of neurosensory transcription factors of the ear to enhance therapeutic translation. Cell Tissue Res 349, 415–432.

Richardson, R., Atkinson, P.J., 2015. Atoh1 gene therapy in the cochlea for hair cell regeneration. Expert. Opin. Biol. Ther. 15 (3), 417–430.

Roberson, D.W., Alosi, J.A., Cotanche, D.A., 2004. Direct transdifferentiation gives rise to the earliest new hair cells in regenerating avian auditory epithelium. J. Neurosci. Res. 78, 461–471.

Ronaghi, M., Nasr, M., Ealy, M., Durruthy-Durruthy, R., Waldhaus, J., Diaz, G.H., et al., 2014. Inner ear hair cell like cells from human embryonic stem cells. Stem Cells Dev. 23, 1275–1284.

Ryals, B.M., Rubel, E.W., 1988. Hair cell regeneration after acoustic trauma in adult Coturnix quail. Science 240, 1774–1776.

Ryals, B.M., Dent, M.L., Dooling, R.J., 2013. Return of function after hair cell regeneration. Hear. Res. 297, 113–120.

Saunders, J.C., 2010. The role of hair cell regeneration in an avian model of inner ear injury and repair from acoustic trauma. ILAR J. 51, 326–337.

Smolders, J.W.T., 1999. Functional recovery in the avian ear after hair cell regeneration. Audiol. Neurootol. 4, 286–302.

Stone, J.S., Cotanche, D.A., 2007. Hair cell regeneration in the avian auditory epithelium. Int. J. Dev. Neurosci. 51, 633–647.

Vats, A., Tolley, N.S., Polak, J.M., Buttery, L.D., 2002. Stem cells: sources and applications. Clin. Otolaryngol. 27, 227–232.

Wang, Y., Raphael, Y., 1996. Re-innervation patterns for chick auditory sensory epithelium after acoustic overstimulation. Hear. Res. 97, 11–18.

White, P.M., Doetzlhofer, A., Lee, Y.S., Groves, A.K., Segil, N., 2006. Mammalian cochlear supporting cells can divide and trans-differentiate into hair cells. Nature 441, 984–987.

Wise, A.K., Flynn, B.O., Atkinson, P.J., Fallon, J.B., Nicholson, M., Richardson, R.T., 2015. Regeneration of cochlear hair cells with Atoh1 gene therapy after noise-induced hearing loss. J. Regen. Med. 4, 1.

Woods, C., Montcouquiol, M., Kelley, M.W., 2004. Math1 regulates development of the sensory epithelium in the mammalian cochlea. Nat. Neurosci. 7, 1310–1318.

Appendix A

Electrocochleography From the Promontory and via a Cochlear Implant

A.1 INTRODUCTION

Electrocochleography (ECochG) is the recording of stimulus-related potentials generated in the human cochlea and the auditory nerve. These potentials are the cochlear microphonics (CM), the summating potentials (SPs), and the compound action potential (CAP). There are three important applications of ECochG: (1) correlation of physiological and psychoacoustic properties, (2) investigation of certain diseases, and (3) objective diagnosis of individual cases of deafness (Ruben, 1967). Currently, auditory brainstem response (ABR) recording has taken over the role of ECochG for items 1 and 3. ECochG remains as an excellent tool for diagnosing and studying among others Ménière's disease and various forms of auditory neuropathy (see chapter: Types of Hearing Loss). In addition the use of neural response telemetry in cochlear implants resulting in electrically evoked CAPs (eCAPs) is used for testing electrode impedance and neural survival (see chapter: Cochlear Implants). We will compare these two recording methods in Section A.5 but first present a brief introduction to ECochG.

A.2 METHODS

A.2.1 Stimuli

Sound stimuli capable of evoking CAPs from the auditory nerve have to be rapidly changing in time since a well-defined triggering of the auditory nerve fiber activity is required. ECochG with clicks provides a general impression of cochlear functioning rather quickly. In cases of a flat hearing loss, one obtains an exact threshold determination and considerable information about

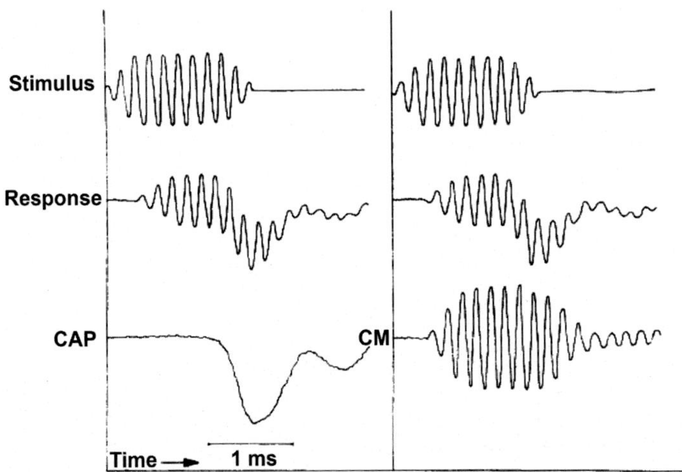

FIGURE A.1 Alternating phase presentation. Odd and even stimuli have opposite phase so in the corresponding responses the CM are in opposite phase. Summation leads to cancellation of the CM. The CAP is canceled simultaneously in the other channel by reversing the phase of the even responses before summation. *From Eggermont, J.J., Odenthal, D.W., Schmidt, P.R., Spoor, A., 1974. Electrocochleography. Basic principles and clinical application. Acta Otolaryngol. Suppl. 316, 1–84.*

the suprathreshold behavior of the cochlea. Eggermont et al. (1974), however, used short trapezoidal tone bursts, which results in good threshold frequency specificity. In addition, the longer duration and the flat plateau of the tone burst considerably facilitates the quantification of the CM and SP. These pure tone bursts had two periods of the sine wave during the rise time and fall time and at least six periods during the plateau. This makes the duration of the tone burst dependent on the stimulus frequency. However, the shortest rise time used was 0.33 ms and the plateau duration was never made less than 4 ms to assure reliable CM and SP measurements.

When tone bursts are used, canceling of the CM becomes important. Two methods can be employed: rolling (random) phase stimulation or presentation of the tone bursts alternately in phase and counter-phase. Both methods have advantages. In rolling phase stimulation, the even harmonics present in the CM at high intensity levels are also canceled, which is not the case using alternating phase stimuli; however, the CM information is completely lost. In alternating-phase presentation (Fig. A.1), the CM can also be obtained free of CAP by additionally commutating the responses recorded and averaging them separately.

A.2.2 Recording Sites

Stimulus-related potentials from the cochlea and eighth nerve might be recorded from various sites around the cochlea. The most remote places are

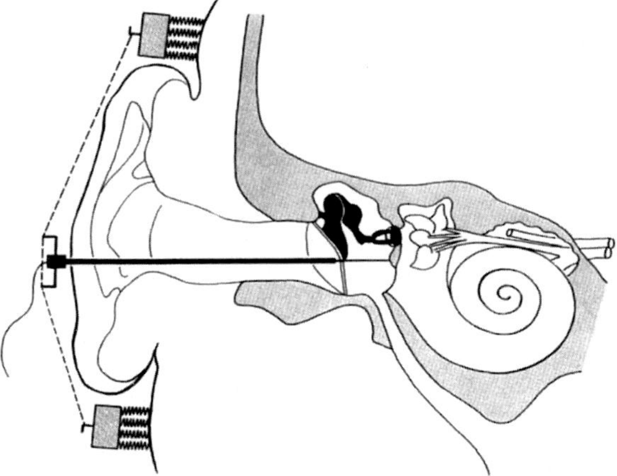

FIGURE A.2 Trans-tympanic electrode placement. The electrode is pierced through the eardrum and placed on the promontory; the reference electrode is a hypodermic needle placed in the earlobe. *From Eggermont, J.J., Odenthal, D.W., Schmidt, P.R., Spoor, A., 1974. Electrocochleography. Basic principles and clinical application. Acta Otolaryngol. Suppl. 316, 1–84.*

on the surface of the skull (e.g., the mastoid) and the earlobe, which are not especially suited for ECochG but favorable for ABR. Large responses are obtained when transtympanic recordings are performed (Fig. A.2). The first recording of auditory nerve potentials in humans from the external ear canal was reported by Yoshie et al. (1967); they used a 0.3 mm hypodermic needle, coated except at the tip, which was placed in the posterior meatal wall within a distance of about 5 mm from the rim of the tympanic membrane. The reference electrode was placed on the earlobe. At 90 dB SL, clicks produced CAPs of about 3.5 μV in amplitude in normal ears, and about 500 responses were used for averaging.

Eggermont et al. (1974) used a stainless steel electrode 0.2 mm in diameter and sharpened at the tip, pierced through the eardrum (which requires some type of anesthesia) and placed on the promontory between the oval and round windows. The CAP amplitude at 90 dB SL click stimulation ranged up to 30 μV, which is favorable for relatively quick measurements and also facilitates recording in patients with very large hearing losses.

A.3 RECEPTOR POTENTIALS

A.3.1 Cochlear Microphonics

The first quantitative data on the CM were reported by Yoshie and Yamura (1969), who recorded from the promontory as well as from the ear canal. One significant finding concerned the slope of the input–output curve: For the promontory recordings, a slope equal to unity was found, whereas it was distinctly lower for the external meatus recording. Elberling and Salomon (1973), also recording from the external ear canal, reported a slope of 0.7 in agreement with Yoshie and Yamura (1969). This finding is interpreted by both groups of authors as indicating that the external meatus is more favorable for recording contributions of more apically situated hair cells than is the promontory recording. The 0.1 and 0.3 μV isopotential curves recorded from the meatus were in general 20 dB less sensitive than in the corresponding promontory recordings, and in normal hearing subjects CM could be obtained from 65 to 120 dB SPL.

A.3.2 Summating Potentials

In Fig. A.3A, the amplitude of the SP in response to 2000 Hz tone bursts is plotted as a function of stimulus intensity for 25 normal ears (Eggermont, 1976a). A drawn line connects the calculated means. The large standard

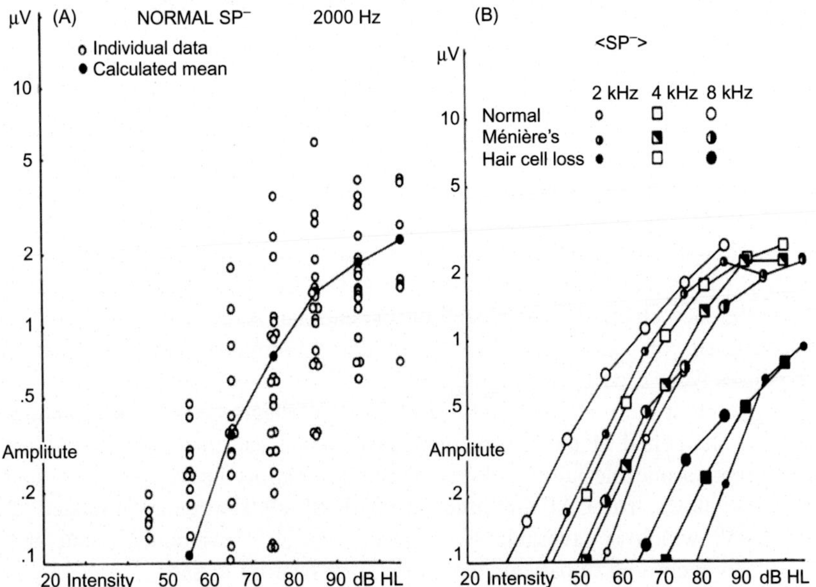

FIGURE A.3 (A) Amplitude of the SP in normal ears. For 20 normal ears, the SP amplitude is plotted as a function of stimulus intensity for 2000 Hz tone bursts. The large dispersion in the amplitude values is obvious; the mean value tends to saturate at about 2.5 μV for intensities above 85 dB HL. The detection threshold is around 55 dB HL. (B) Mean amplitude–intensity curves for the SP in normal and pathological ears at three tone burst frequencies. The amplitudes of the SP are significantly smaller for the ears with presumed hair cell loss, while potentials from ears of Ménière's patients show the same trend as those recorded in normal ears. *From Eggermont, J.J., 1976a. Electrocochleography. In: Keidel, W.D., Neff, W.D. (Eds.), Handbook of Sensory Physiology, Springer-Verlag, New York, pp. 626–705.*

TABLE A.1 SP Amplitudes at 85 dB HL

	2000 Hz	4000 Hz	8000 Hz
Normal hearing	1.5 ± 1.4 μV	1.8 ± 1.8 μV	3.0 ± 2.6 μV
Ménière's	3.0 ± 2.5 μV	1.7 ± 1.6 μV	1.4 ± 1.3 μV
Hair cell loss	0.2 ± 0.2 μV	0.25 ± 0.20 μV	0.45 ± 0.35 μV

deviation is obvious. Application of this type of analysis to a group of 25 Ménière ears, and a smaller group of 15 ears with hearing loss due to noise trauma or ototoxic drugs (hair cell loss group), results in mean input–output curves shown in Fig. A.3B. On average there is no difference between the normal hearing group and the Meniere group, but the SP output is distinctly lower for the hair cell loss group. Mean values and standard deviations were calculated for an intensity of 85 dB HL at three tone burst frequencies. The data for three groups, 70 Ménière's ears, 25 normal ears, and 22 hair cell loss ears, are presented in Table A.1.

Table A.1 shows that the mean values are nearly the same as the standard deviations, which points to an exponential distribution for the amplitude values of the SP irrespective of the state of the hearing organ. The same is found for the distribution of the CAP amplitude values at 85 dB HL.

From the data shown in Table A.1, it may be concluded that although the normal ears and the Ménière's ears cannot be differentiated on the basis of the magnitude of the SP the hair cell loss ears easily are. The combination of SP and CAP information, i.e., as a ratio, especially when tone bursts of different frequencies are used, offers far more evidence for a differential diagnosis (Schmidt et al., 1974).

A.4 THE COMPOUND ACTION POTENTIAL

The amount of synchronization of the auditory nerve fiber firings is very important for the shape of the CAP. For low-frequency signals (500 and 1000 Hz), the CAP is found at a fixed phase in the CM. When the tone frequency increases, adaptation gains progressively in importance, i.e., CAPs on subsequent cycles of a low-frequency tone burst become increasingly smaller until a final value is reached. This final value depends on the tone frequency: increasing frequency results in decreasing final values. For 1500 or 2000 Hz, the CAP still appears during a fixed phase of the sine wave, but, due to adaptation, only the CAPs on the first cycles are present. For higher frequencies only one CAP remains. This is called the "on-effect" (Fig. A.4). Measurements of the CAP from the human promontory or the external ear canal are in most cases based on this on-effect. The SP is visible as a short latency initial negative deflection, especially for 8000 Hz and click stimuli.

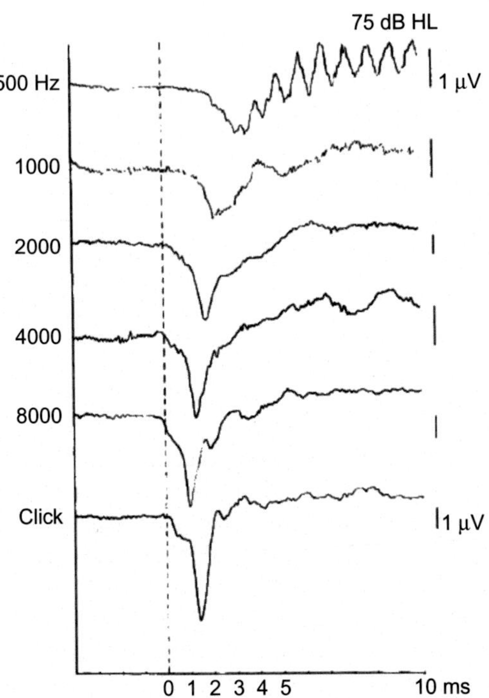

FIGURE A.4 The genesis of the "on-effect." In response to short tone bursts, the shape of the compound AP depends on the stimulus frequency. For low frequencies (below 2000 Hz), a phase lock appears and an AP is recorded on each cycle of the sine wave. Alternation of the phase of the sine wave results in a doubling of the number of APs. Due to adaptation and refractoriness, the AP amplitude diminishes with the successive number of the period. Above 2000 Hz, only the response to the start of the stimulus is recorded. This is the so-called "on effect." *From Eggermont, J.J., 1976a. Electrocochleography. In: Keidel, W.D., Neff, W.D. (Eds.), Handbook of Sensory Physiology, Springer-Verlag, New York, pp. 626–705.*

When the ear is stimulated with tone bursts presented alternately in phase, certain factors must be taken into account. With tone bursts having a tone frequency high enough to prevent phase locking between the nerve fiber discharges and the sine wave, alternation of the phase has little or no influence on the CAP waveform. This holds for frequencies of about 3000 Hz and above. For low stimulus frequencies (500 and 1000 Hz), alternation of the stimulus phase in the tone burst to abolish the CM leads to a half-period shift of the preferred discharge phase. This in turn causes a doubling of the number of AP's with respect to the sine wave period. These closely spaced CAPs overlap partially, thus forming a broad average response in which only the CAP tips indicate the phase lock. The CAP waveforms obtained for tone bursts of 1000 and 2000 Hz are shown in Fig. A.5A and B. It must be kept in mind that the tone burst duration and the rise-and-fall time depend on the

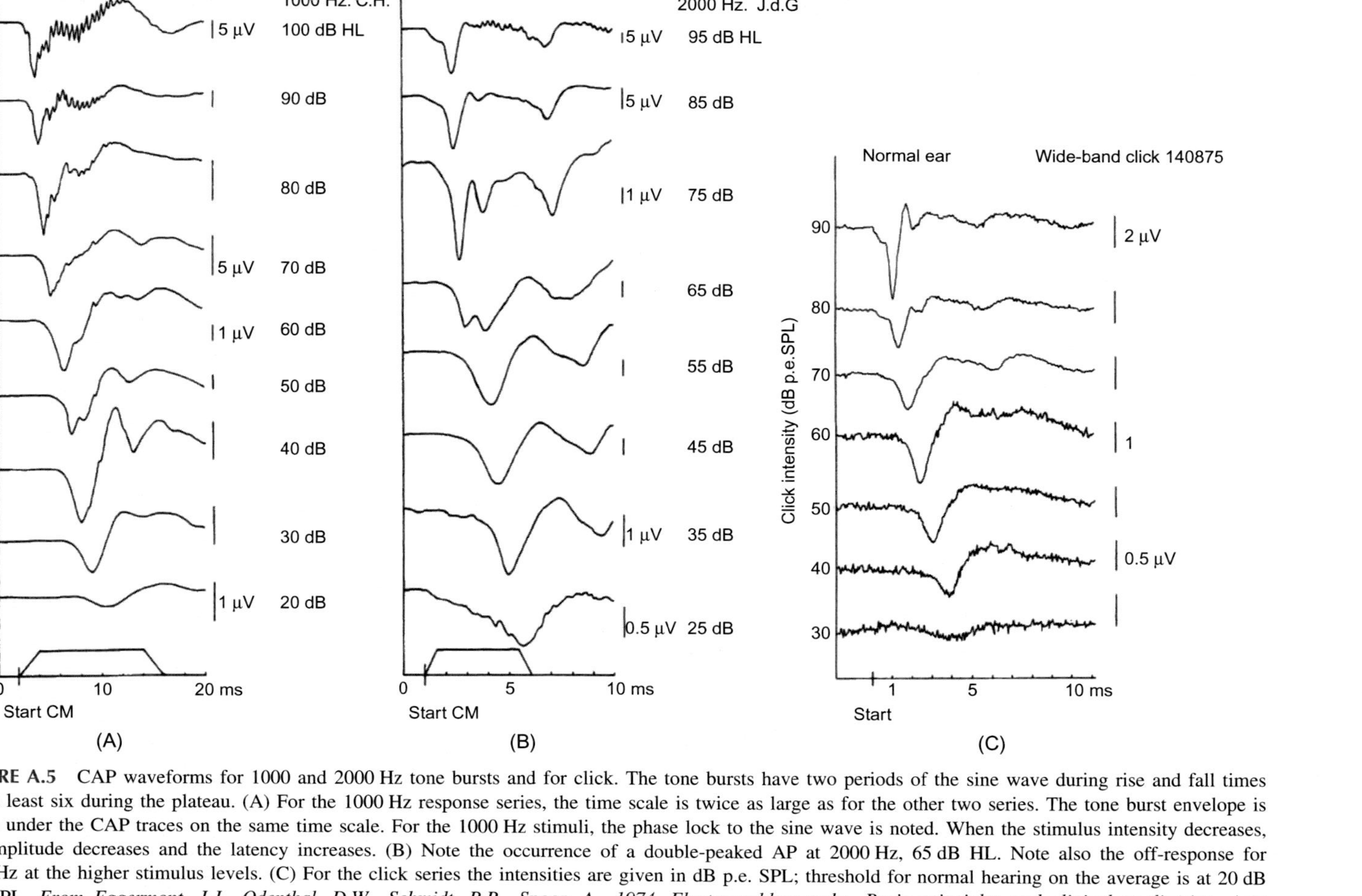

FIGURE A.5 CAP waveforms for 1000 and 2000 Hz tone bursts and for click. The tone bursts have two periods of the sine wave during rise and fall times and at least six during the plateau. (A) For the 1000 Hz response series, the time scale is twice as large as for the other two series. The tone burst envelope is drawn under the CAP traces on the same time scale. For the 1000 Hz stimuli, the phase lock to the sine wave is noted. When the stimulus intensity decreases, the amplitude decreases and the latency increases. (B) Note the occurrence of a double-peaked AP at 2000 Hz, 65 dB HL. Note also the off-response for 2000 Hz at the higher stimulus levels. (C) For the click series the intensities are given in dB p.e. SPL; threshold for normal hearing on the average is at 20 dB p.e. SPL. *From Eggermont, J.J., Odenthal, D.W., Schmidt, P.R., Spoor, A., 1974. Electrocochleography. Basic principles and clinical application. Acta Otolaryngol. Suppl. 316, 1–84.*

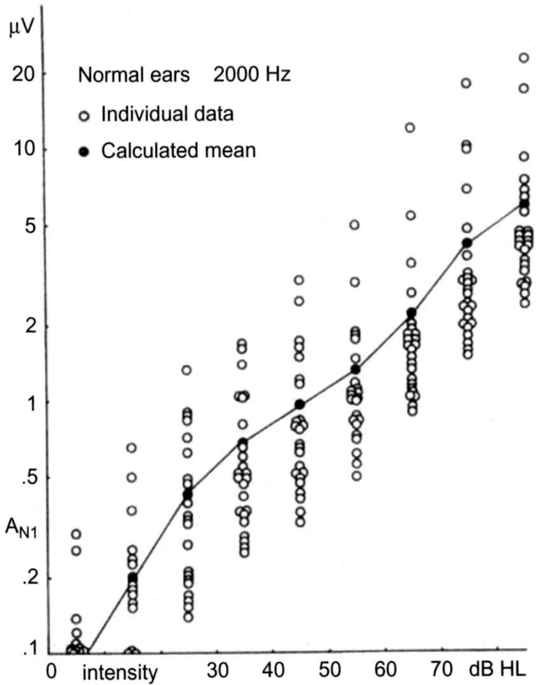

FIGURE A.6 CAP amplitudes as a function of intensity for 20 normal ears at 2000 Hz. The dispersion observed in the amplitudes at a given stimulus intensity is less than for the SP (cf. Fig. A.3), and the mean curve also shows the relatively flat region around 40 dB HL. *From Eggermont, J.J., 1976a. Electrocochleography. In: Keidel, W.D., Neff, W.D. (Eds), Handbook of Sensory Physiology, Springer-Verlag, New York, pp. 626–705.*

stimulus frequency, and it should be noted that the analysis interval is 20 ms for 1000 Hz and 10 ms for 2000 Hz series. A series of CAP responses to clicks alternating in phase are shown in Fig. A.5C.

The large variation in CAP amplitude for normal ears is illustrated in Fig. A.6 for stimulation with 2 kHz tone burst as a function of stimulus level. The arithmetic mean is indicated. As for the SP data (cf. Table A.1), the distribution of amplitude values at each level is nearly exponential.

A.5 COMPARING THE CAP AND THE eCAP

A.5.1 The Composition of the CAP Recorded From the Promontory

Back in 1939, Gasser and Grundfest (1939) used convolution to predict the waveform of the CAP for the saphenous nerve of the cat upon electrical stimulation from the distribution of nerve fiber diameters (resulting in a latency distribution) and a hypothetical individual fiber response. Twenty years later, Goldstein and Kiang (1958) pointed out that under the assumption that unit responses add with equal weight to the recording electrode at the round window the CAP waveform could be expressed as:

$$\mathrm{AP}(t) = N \int_0^t s(\tau)a(t-\tau)\mathrm{d}\tau$$

N being the number of nerve fibers, $s(\tau)$ the latency distribution function, and $a(t)$ the single-unit response.

A unit response (Fig. A.7), recorded from a nerve end, will be normally diphasic in shape as postulated for the auditory nerve by Teas et al. (1962), de Boer (1975), and Elberling (1976) and demonstrated by Kiang et al.

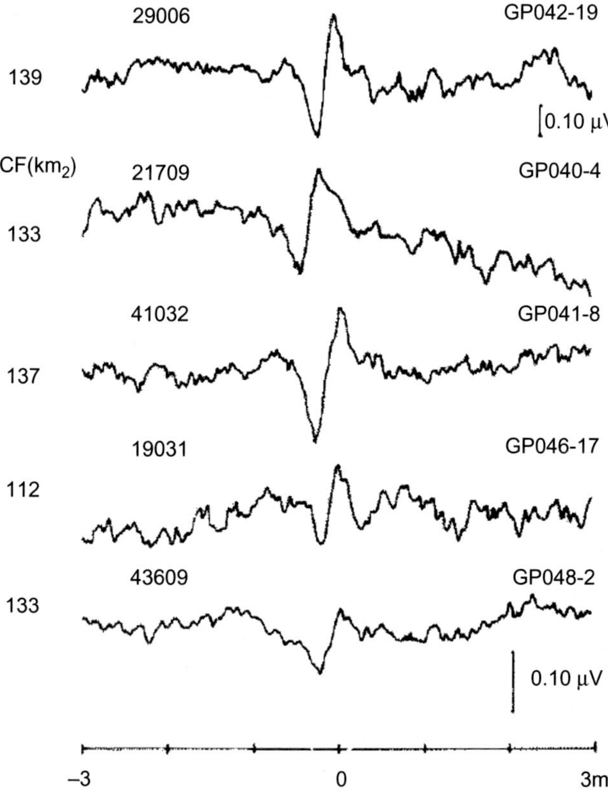

FIGURE A.7 Five examples of $a(t)$'s for fibers with almost identical CF in different animals. Computations were based on spontaneous activity. The number of counts is indicated on the left. Above each trace. In GP040 and GP041 triggering occurred at the leading edge of the spike in the auditory nerve, in the other GPs at the back edge. *Reprinted from Prijs, V.F., 1986. Single-unit response at the round window of the guinea pig. Hear. Res. 21, 127–133, with permission from Elsevier.*

(1976) and Prijs (1986). The convolution is allowed under the conditions of statistical independency of the individual nerve fiber contributions, $a(t)$.

For acoustic clicks as stimulus, the latency distribution function, $s(\tau)$, may be considered the impulse response function of the cochlea. For individual nerve fibers such impulse response functions may be obtained from compound PST histograms in response to condensation and rarefaction clicks (Pfeiffer and Kim, 1972) or from the cross-correlation between the nerve fiber response and a white noise stimulus evoking them (de Boer, 1975; Eggermont et al., 1983). In a practical situation, either in modeling or analyzing, the number of contributing units has to be restricted. Forming groups of nearly equivalent units may do this. For that purpose, it might be useful to divide the human cochlear partition into small regions about 3 mm long (corresponding to about half-an-octave in frequency) and study the narrow-band CAPs (NAPs) evoked on these small segments. Since the human cochlea is innervated by about 31,000 afferent nerve fibers (Rasmussen, 1940), such a segment is assumed to comprise about 3000 individual nerve fibers. The thresholds of the fibers in one segment are supposed to be nearly the same.

To obtain these segments, a click stimulus combined with steep (96 dB/octave) high-pass filtered noise maskers is used with various high-pass cutoff frequencies (Teas et al., 1962). Subtracting click-evoked CAP responses obtained in the presence of noise with cutoff frequencies being 1/2 octave apart results in NAPs, which can be assigned to particular narrow-band segments each characterized by a central frequency. This technique has first been used in human ECochG by Elberling (1976) for the analysis of click-evoked APs and by Eggermont et al. (1976b) for tone-burst evoked CAPs to elucidate the frequency-specific character of that type of stimulation. An example of such a separation of the CAP into NAPs for the human cochlea upon click stimulation with 90 dB peak equivalent (p.e.) SPL is shown in Fig. A.8. The NAPs are essentially diphasic in shape and their latencies range from 1.4 to 5.8 ms. The CAP latency is 1.4 ms and is therefore mainly dominated by the most basal contributions, due to the diphasic waveforms the contributions from segments with lower characteristic frequencies (CFs) tend to cancel each other and are therefore not seen in the CAP. See Chapter 5, Types of Hearing Loss for an application of this procedure for derived ABRs that are used in the diagnosis of vestibular schwannoma and Ménière's disease.

Each NAP may be considered as the convolution of a unit response and a latency distribution function for that narrow frequency band. For the ith narrow band one may write:

$$\mathrm{NAP}_i(t) = N_i \int_0^t S_i(\tau)a(t-\tau)\mathrm{d}\tau$$

The $S_i(\tau)$ for a click stimulus is the impulse response of maximally 3000 fibers and this may be more related to single-fiber impulse responses than

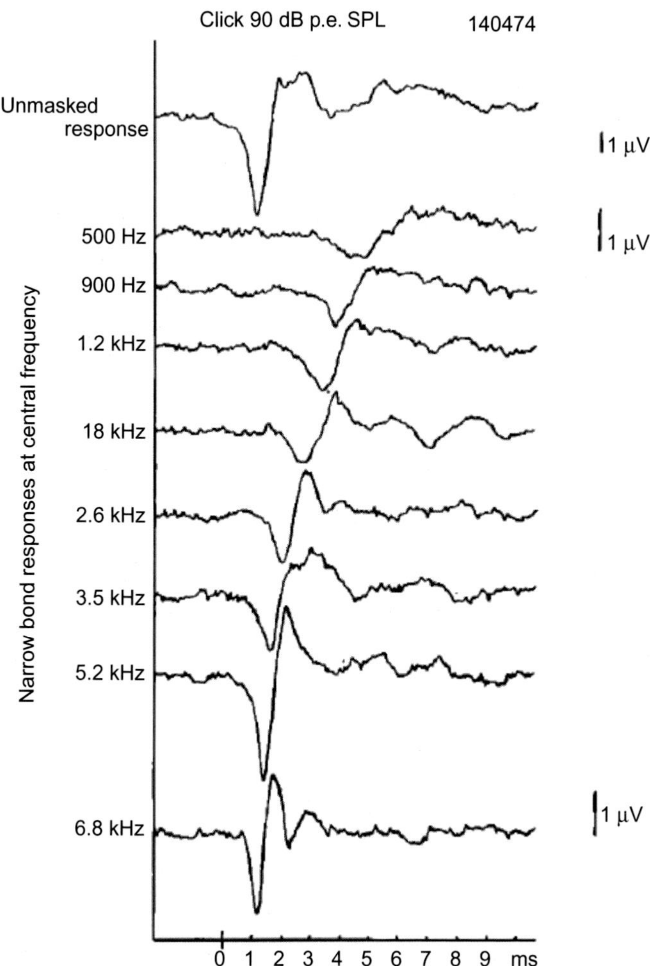

FIGURE A.8 Narrow-band analysis of a click-evoked, whole-nerve action potential. Using the high-pass noise technique to derive narrow-band AP responses, an analysis can be made of the composition of a whole-nerve AP. The particular responses, representing synchronous firings of fibers from a 1/2 octave wide region (i.e., about 3 mm wide) on the cochlear partition, add together to form the unmasked response. The central frequencies of the different narrow bands are indicated. Latency increases with distance from the stapes, and narrow band amplitude decreases. It is noted that the N1 component in the click response is entirely due to contributions from fibers with high CFs. *Reprinted with permission from Eggermont, J.J., 1976. Analysis of compound action potential responses to tone bursts in the human and guinea pig cochlea. J. Acoust. Soc. Am. 60, 1132–1139.*

the $s(\tau)$ of the whole cochlea. Because the $S_i(\tau)$ is narrowest for high CFs, the NAP at these high CFs may approximate the unit response. Comparing the human NAP derived for CF = 6.8 kHz (Fig. A.8) with the unit response from high CFs in guinea pigs (Fig. A.7) shows a close correspondence.

A.5.2 The eCAPs as Recorded by Cochlear Implants

On the assumption that a nerve fiber is infinitely long and lies in an infinite homogeneous medium, volume conductor theory predicts that a fiber's discharge causes in the surrounding medium a triphasic potential waveform with integral equal to zero (Fig. A.9A; Heringa et al., 1989). If such a waveform is recorded close to the site of spike initiation, which is approximately the case for an $a(t)$ recording at the round window, then, the waveform will be diphasic (Gydikov and Trayanova, 1986) and this is for most $a(t)$'s in the cat (Kiang et al., 1976) and in the guinea pig (Prijs, 1986; Versnel et al., 1992).

From their modeling studies, Briaire and Frijns (2005) noted that the calculated eCAPs (Fig. A.9B) based on the theoretical unit response (Fig. A.9A) did not match the measured human eCAP (Fig. A.9C) obtained using neural response telemetry (Frijns et al., 2002). Briaire and Frijns (2005) found the potential solution to the discrepancy from a study by Miller et al. (2004) that indicated that two action potentials are present, and that the initial positive peak in the eCAP originates from antidromic action potentials originating from a relatively central site on the nerve fiber, likely close the ganglion cell body, of action potential initiation. Thus, the dendrite may be responsible for the generation of the P_0 peak. Note that in acoustic stimulation the site of initial spike excitation is likely the proximal dendrite (Hossain et al., 2005) and hence resulting in a diphasic unit response and CAP.

The study by Miller et al. (2004) indicated that the state of neural degeneration of the fibers has a big influence on the presence of the P_0 peak in the unit response, as also implied by Rattay et al. (2001). Briaire and Frijns (2006) used this to show that a large P_0 peak in the eCAP occurs before the N_1P_1 complex (Fig. A.9B) when the fibers are not degenerated. They suggested that the absence of this peak might be used as an indicator for degeneration of the proximal dendrite. Westen et al. (2011) evaluated the use of the unit response as a unitary response in a convolution integral to predict the eCAP and found evidence for changes in the unit response with stimulus level. This suggested that the unit responses for different electrodes are not independent, likely caused by strong synchronization across fibers at high stimulus levels. Therefore, the eCAP cannot be predicted from the unit responses, and consequently, the inverse problem assessing the patency of the auditory nerve fibers on basis of the eCAP is not unambiguous.

Recently, Strahl et al. (2016) used a deconvolution model to estimate the nerve firing probability based on a biphasic unit response and the eCAP,

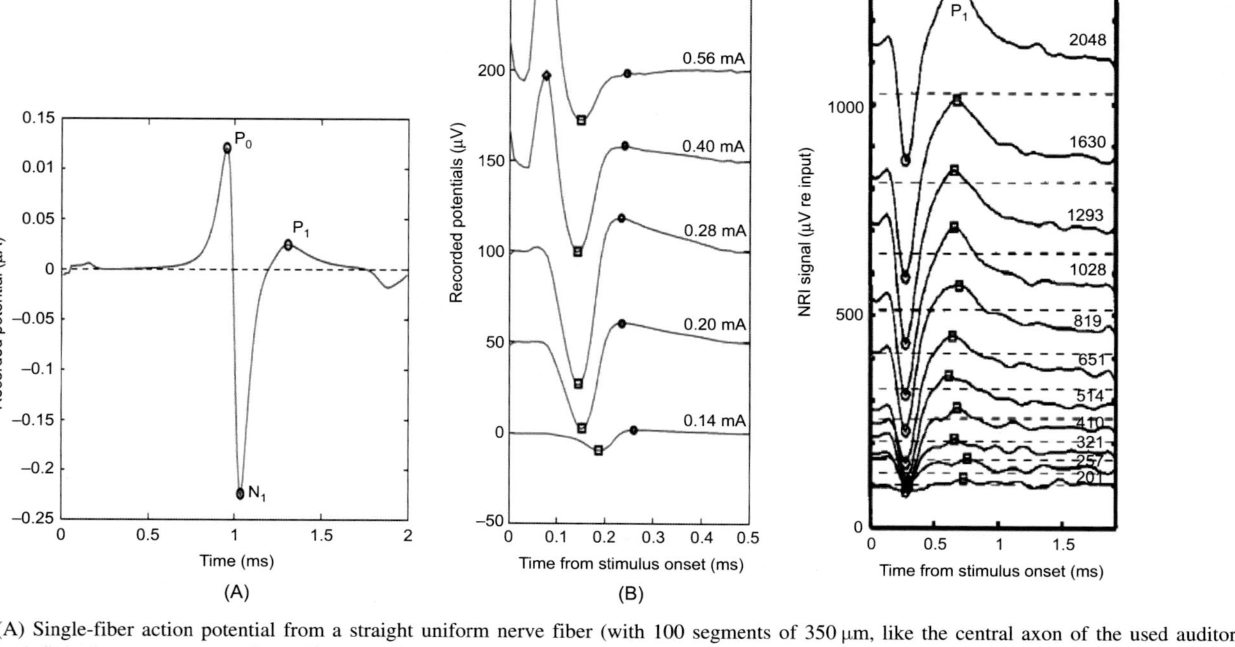

FIGURE A.9 (A) Single-fiber action potential from a straight uniform nerve fiber (with 100 segments of 350 μm, like the central axon of the used auditory nerve fiber) in an infinite homogeneous medium. The recording electrode is placed 17.5 mm from the tip of the nerve fiber and 0.6 mm from the fiber axis. (B) The calculated eCAP responses for the myelinated cell body condition, as induced by anodic-first biphasic current pulses for five current strengths. The N_1 and P_1 peaks are indicated by squares and circles, respectively. The P_0 visible at higher current strengths are indicated by a diamond. (C) Neural response imaging (NRI) recordings of patient F from the middle of the electrode array (alternating polarity paradigm, stimulating electrode 7, recording electrode 5). Traces for varying stimulus intensity recorded at the time of fitting. The N1 (circles) and P1 (squares) peaks have been determined automatically. To enhance the visibility the individual traces have been level shifted proportionally to the stimulus strength as indicated by the dashed lines. The numbers to the right indicate the stimulus current in μA. *Reprinted (A) and (B) from Briaire, J.J., Frijns, J.H.M., 2005. Unraveling the electrically evoked compound action potential. Hear. Res. 205, 143–156, with permission from Elsevier. (C) From Frijns, J.H., Briaire, J.J., de Laat, J.A., Grote, J.J., 2002. Initial evaluation of the Clarion CII cochlear implant: speech perception and neural response imaging. Ear Hear. 23, 184–197.*

both in guinea pigs and human implantees. They found that the estimated nerve firing probability was bimodal and could be parameterized by two Gaussian distributions with an average latency difference of 0.4 ms. The ratio of the scaling factors of the late and early component increased with neural degeneration in the guinea pig. The two-component firing probability was attributed to either latency differences in the population of nerve fibers resulting from late firing due to excitation of the proximal dendrite, compared to direct, central to the cell body, activation of the auditory nerve fibers. They suggested that the deconvolution of the eCAP could be used to reveal these two separate firing components in the auditory nerve, which may elucidate degeneration of the proximal dendrite.

REFERENCES

Briaire, J.J., Frijns, J.H.M., 2005. Unraveling the electrically evoked compound action potential. Hear. Res. 205 (1–2), 143–156.

Briaire, J.J., Frijns, J.H.M., 2006. The consequences of neural degeneration regarding optimal cochlear implant position in scala tympani: a model approach. Hear. Res. 214, 17–27.

de Boer, E., 1975. Synthetic whole-nerve action potentials for the cat. J. Acoust. Soc. Am. 53, 1030–1045.

Eggermont, J.J., 1976a. Electrocochleography. In: Keidel, W.D., Neff, W.D. (Eds.), Handbook of Sensory Physiology. Springer-Verlag, New York, pp. 626–705.

Eggermont, J.J., 1976b. Analysis of compound action potential responses to tone bursts in the human and guinea pig cochlea. J. Acoust. Soc. Am. 60, 1132–1139.

Eggermont, J.J., Odenthal, D.W., Schmidt, P.R., Spoor, A., 1974. Electrocochleography. Basic principles and clinical application. Acta Otolaryngol. Suppl. 316, 1–84.

Eggermont, J.J., Johannesma, P.I.M., Aertsen, A.M.H.J., 1983. Reverse correlation methods in auditory research. Quart. Rev. Biophys. 16, 341–414.

Elberling, C., 1976. Modeling action potentials. Rev. Laryngol. Otol. Rhinol. (Bord) 97, 527–537.

Elberling, C., Salomon, G., 1973. Cochlear microphonics recorded from the ear canal in man. Acta Otolaryngol. 75, 489–495.

Frijns, J.H., Briaire, J.J., de Laat, J.A., Grote, J.J., 2002. Initial evaluation of the Clarion CII cochlear implant: speech perception and neural response imaging. Ear Hear. 23, 184–197.

Gasser, H.S., Grundfest, H., 1939. Axon diameters in relation to the spike dimensions and the conduction velocity in mammalian A fibers. Am. J. Physiol. 127, 393–414.

Goldstein Jr., M.H., Kiang, N.Y.S., 1958. Synchrony of neural activity in electric responses evoked by transient acoustic stimuli. J. Acoust. Soc. Am. 30, 107–114.

Gydikov, A.A., Trayanova, N.A., 1986. Extracellular potentials of single active muscle fibres: effects of finite fibre length. Biol. Cybern. 53, 363–372.

Heringa, A., Stegeman, D.F., de Weerd, J.P.C., 1989. Calculated potential and electric field distributions around an active nerve fiber. J. Appl. Phys. 66, 2724–2731.

Hossain, W.A., Antic, S.D., Yang, Y., Rasband, M.N., Morest, D.K., 2005. Where is the spike generator of the cochlear nerve? Voltage-gated sodium channels in the mouse cochlea. J. Neurosci. 25, 6857–6868.

Kiang, N.Y.S., Moxon, E.C., Kahn, A.R., 1976. The relationship of gross potentials recorded from the cochlea to single unit activity in the auditory nerve. In: Ruben, R.J., Elberling,

C., Salomon, G. (Eds.), Electrocochleography. University Park Press, Baltimore, MD, pp. 95–115.

Miller, C.A., Abbas, P.J., Hay-McCutcheon, M.J., Robinson, B.K., Nourski, K.V., Jeng, F.C., 2004. Intracochlear and extracochlear eCAPs suggest antidromic action potentials. Hear. Res. 198, 75–86.

Pfeiffer, R.S., Kim, D.O., 1972. Response patterns of single cochlear nerve fibers to click stimuli: descriptions for cat. J. Acoust. Soc. Am. 52, 1669–1677.

Prijs, V.F., 1986. Single-unit response at the round window of the guinea pig. Hear. Res. 21, 127–133.

Rasmussen, A.T., 1940. Studies on the eight cranial nerve of man. Laryngoscope 50, 67–83.

Rattay, F., Lutter, P., Felix, H., 2001. A model of the electrically excited human cochlear neuron I. Contribution of neural substructures to the generation and propagation of spikes. Hear. Res. 153, 43–63.

Ruben, R.J., 1967. Cochlear potentials as a diagnostic test in deafness. Sensorineural Hearing Processes and Disorders. Little-Brown, Boston, MA.

Schmidt, P.H., Eggermont, J.J., Odenthal, D.W., 1974. Study of Meniere's disease by electrocochleography. Acta Otolaryngol. Suppl. 316, 75–84.

Strahl, S.B., Ramekers, D., Nagelkerke, M.M.B., Schwarz, K.E., Spitzer, P., Klis, S.F.L., et al., 2016. Assessing the firing properties of the electrically stimulated auditory nerve using a convolution model. In: van Dijk, P., et al., (Eds.), Physiology, Psychoacoustics and Cognition in Normal and Impaired Hearing, Advances in Experimental Medicine and Biology, 894. pp. 143–153.

Teas, D.C., Eldredge, D.H., Davis, H., 1962. Cochlear responses to acoustic transients: an interpretation of whole nerve action potentials. J. Acoust. Soc. Am. 34, 1433–1459.

Versnel, H., Prijs, V.F., Schoonhoven, R., 1992. Round-window recorded potential of single-fibre discharge (unit response) in normal and noise-damaged cochleas. Hear. Res. 59, 157–170.

Westen, A.A., Dekker, D.M.T., Briaire, J.J., Frijns, J.H.M., 2011. Stimulus level effects on neural excitation and eCAP amplitude. Hear. Res. 280, 166–176.

Yoshie, N., Yamura, K., 1969. Cochlear mocrophonic responses to pure tones in man recorded by a nonsurgical method. Acta Otolaryngol. Suppl. 202, 37–69.

Yoshie, N., Ohashi, T., Suzuki, T., 1967. Non surgical recording of auditory nerve action potentials in man. Laryngoscope 77, 76–85.

Index

Note: Page numbers followed by "*f*" and "*t*" refer to figures and tables, respectively.

B

C

D

E

I

N

T

U

V

W

X

Printed in the United States
By Bookmasters